Klaus Hermann

Crystallography and Surface Structure

Klaus Hermann

Crystallography and Surface Structure

An Introduction for Surface Scientists and Nanoscientists

Second, Revised and Expanded Edition

WILEY-VCH
Verlag GmbH & Co. KGaA

Author

Prof. Dr. Klaus Hermann
Fritz-Haber-Institut der MPG
Theory Department
Faradayweg 4-6
14195 Berlin
Germany

Cover
The book cover includes a scanning electron microscopy picture of a stepped Cu_2O surface (large circular area), kindly provided by Marc Willinger, Fritz-Haber Institute Berlin.
Medium circle: Interference structure of graphene on iridium, theory.
Small circle: Simulation of chiral molecules at a stepped surface.

■ All books published by **Wiley-VCH** are carefully produced. Nevertheless, authors, editors, and publisher do not warrant the information contained in these books, including this book, to be free of errors. Readers are advised to keep in mind that statements, data, illustrations, procedural details or other items may inadvertently be inaccurate.

Library of Congress Card No.: applied for

British Library Cataloguing-in-Publication Data
A catalogue record for this book is available from the British Library.

Bibliographic information published by the Deutsche Nationalbibliothek
The Deutsche Nationalbibliothek lists this publication in the Deutsche Nationalbibliografie; detailed bibliographic data are available on the Internet at <http://dnb.d-nb.de>.

© 2017 Wiley-VCH Verlag GmbH & Co. KGaA, Boschstr. 12, 69469 Weinheim, Germany

All rights reserved (including those of translation into other languages). No part of this book may be reproduced in any form – by photoprinting, microfilm, or any other means – nor transmitted or translated into a machine language without written permission from the publishers. Registered names, trademarks, etc. used in this book, even when not specifically marked as such, are not to be considered unprotected by law.

Print ISBN: 978-3-527-33970-9
ePDF ISBN: 978-3-527-69712-0
ePub ISBN: 978-3-527-69714-4
Mobi ISBN: 978-3-527-69715-1
oBook ISBN: 978-3-527-69713-7

Cover Design Grafik-Design Schulz
Typesetting SPi Global, Chennai, India
Printing and Binding Markono Print Media Pte Ltd, Singapore

Printed on acid-free paper

Contents

Preface to the Second Edition *IX*
Preface to the First Edition *XI*

1 **Introduction** *1*

2 **Bulk Crystals: Three-Dimensional Lattices** *7*
2.1 Basic Definition *7*
2.2 Representation of Bulk Crystals *11*
2.2.1 Alternative Descriptions Conserving the Lattice Representation *12*
2.2.2 Alternative Descriptions Affecting the Lattice Representation *14*
2.2.2.1 Cubic, Hexagonal, and Trigonal Lattices *16*
2.2.2.2 Superlattices and Repeated Slabs *25*
2.2.2.3 Linear Transformations of Lattice Vectors *29*
2.2.3 Centered Lattices *31*
2.3 Periodicity Cells of Lattices *35*
2.4 Lattice Symmetry *38*
2.5 Reciprocal Lattice *49*
2.6 Neighbor Shells *52*
2.7 Nanoparticles and Crystallites *63*
2.8 Incommensurate Crystals and Quasicrystals *71*
2.8.1 Modulated Structures *71*
2.8.2 Incommensurate Composite Crystals *73*
2.8.3 Quasicrystals *76*
2.9 Exercises *82*

3 **Crystal Layers: Two-Dimensional Lattices** *91*
3.1 Basic Definition, Miller Indices *91*
3.2 Netplane-Adapted Lattice Vectors *96*
3.3 Symmetrically Appropriate Lattice Vectors: Minkowski Reduction *98*
3.4 Miller Indices for Cubic and Trigonal Lattices *100*
3.5 Alternative Definition of Miller Indices and Miller–Bravais Indices *106*

3.6	Symmetry Properties of Netplanes	109
3.6.1	Centered Netplanes	110
3.6.2	Inversion	111
3.6.3	Rotation	114
3.6.4	Mirror Operation	119
3.6.5	Glide Reflection	131
3.6.6	Symmetry Groups	139
3.7	Crystal Systems and Bravais Lattices in Two Dimensions	144
3.8	Crystallographic Classification of Netplanes and Monolayers	149
3.8.1	Oblique Netplanes	151
3.8.2	Primitive Rectangular Netplanes	151
3.8.3	Centered Rectangular Netplanes	155
3.8.4	Square Netplanes	157
3.8.5	Hexagonal Netplanes	158
3.8.6	Classification Overview	163
3.9	Exercises	164
4	**Ideal Single Crystal Surfaces**	**169**
4.1	Basic Definition, Termination	169
4.2	Morphology of Surfaces, Stepped and Kinked Surfaces	175
4.3	Miller Index Decomposition	178
4.4	Chiral and Achiral Surfaces	192
4.5	Exercises	204
5	**Real Crystal Surfaces**	**209**
5.1	Surface Relaxation	209
5.2	Surface Reconstruction	210
5.3	Growth Processes	222
5.4	Faceting	226
5.5	Exercises	231
6	**Adsorbate Layers**	**235**
6.1	Definition and Classification	235
6.2	Adsorbate Sites	241
6.3	Wood Notation of Surface Structure	251
6.4	High-Order Commensurate (HOC) Overlayers	258
6.5	Interference Lattices	263
6.5.1	Basic Formalism	264
6.5.2	Interference and Wood Notation	272
6.5.3	Anisotropic Scaling, Stretching, and Shifting	279

6.6	Symmetry and Domain Formation	*283*
6.7	Adsorption at Surfaces and Chirality	*293*
6.8	Exercises	*299*

7	**Experimental Analysis of Real Crystal Surfaces**	*305*
7.1	Experimental Methods	*305*
7.2	Surface Structure Compilations	*306*
7.3	Database Formats for Surface and Nanostructures	*311*
7.4	Exercises	*313*

8	**Nanotubes**	*315*
8.1	Basic Definition	*315*
8.2	Nanotubes and Symmetry	*319*
8.3	Complex Nanotubes	*323*
8.4	Exercises	*326*

	Appendix A: Sketches of High-Symmetry Adsorbate Sites	*329*
A.1	Face-Centered Cubic (fcc) Surface Sites	*330*
A.2	Body-Centered Cubic (bcc) Surface Sites	*338*
A.3	Hexagonal Close-Packed (hcp) Surface Sites	*342*
A.4	Diamond Surface Sites	*346*
A.5	Zincblende Surface Sites	*349*

Appendix B: Parameter Tables of Crystals *351*

	Appendix C: Mathematics of the Wood Notation	*355*
C.1	Basic Formalism and Examples	*355*
C.2	Wood-Representability	*361*

Appendix D: Mathematics of the Minkowski Reduction *367*

	Appendix E: Details of Number Theory	*371*
E.1	Basic Definitions and Functions	*371*
E.2	Euclid's Algorithm	*376*
E.3	Linear Diophantine Equations	*377*
E.4	Quadratic Diophantine Equations	*380*
E.5	Number Theory and 2×2 Matrices	*386*

Appendix F: Details of Vector Calculus and Linear Algebra *391*

Appendix G: Details of Fourier Theory *395*

Appendix H: List of Surface Web Sites *399*

Appendix I: List of Surface Structures *401*

Glossary and Abbreviations *403*

References *417*

Index *425*

Preface to the Second Edition

As a result of feedback from readers of the first edition of this book and from colleagues, this second edition is a major revision that goes beyond mere error correction and minor clarification. While the book has been modified and extended greatly, the initial concept of a combined tutorial introduction into surface crystallography, with bulk crystallography as a basis, and an overview of modern subjects for the advanced researcher has been conserved. Many sections have been updated for completeness and have been extended to include recent developments due to the advent of new and more refined measuring techniques.

The second edition is also targeted at researchers working on graphene and other weakly adsorbing overlayers that form large size moiré patterns observed by scanning tunneling and electron microscopy. They might appreciate the new section on moiré lattice formation, which until now has not been available in any textbook format. Nanoparticle physicists and materials scientists who are interested in structure information of very small particles and seek to connect, for example, electronic and magnetic properties with structural data, may benefit from the sections on nanoparticles, crystal spheres, nanotubes, as well as faceting. This might also interest catalytic chemists trying to interpret chemical behavior, such as reactivity, by structural information of small particles.

Specific items that have been newly added or revised include
- nanoclusters and crystallites, giving a basic overview on structure details;
- incommensurate and quasicrystals, being treated on a common basis;
- basics of epitaxy and crystal growth;
- further details on chiral surfaces and adsorbates;
- the theoretical treatment of high-order commensurate (HOC) overlayers;
- the theory of interference lattices and moiré patterns;
- the geometric structure of high-symmetry adsorbate sites;
- more detailed computational algorithms in the appendices; and
- structure database formats, documenting measured surface structures.

Furthermore, the list of references to original publications and books on specific subjects has been revised and extended to account for more recent experimental and theoretical developments. The set of exercises that conclude each section has been substantially enlarged following suggestions by readers of the first book edition. All structure graphics in this book have been created using the interactive software *Balsac* (Build and Analyze Lattices, Surfaces, And Clusters) developed by the author ((C) K. Hermann, Fritz-Haber-Institut, Berlin, 1990–2016).

Michel A. Van Hove and Wolfgang Moritz have once again lent invaluable support through their constructive criticism and detailed suggestions. I am particularly indebted to Michel for our fruitful discussions during various extended research visits to the Institute of Computational and Theoretical Studies (ICTS) at Hong Kong Baptist University, which have helped to improve the revised text in innumerable ways. Advice from other colleagues on subjects specific to the second edition has likewise been instrumental in improving this edition, as also the suggestions on interference lattices by Michael S. Altman, on growth mechanisms by Ernst G. Bauer, on quasicrystals by Renee D. Diehl, and on chirality by Andrew J. Gellman and Rasmita Raval. Critical reading of the final manuscript by Travis Jones and color design advice on figures by Liudmyla Masliuk are greatly acknowledged.

Finally, unsurpassed support and overwhelming patience by my wife Hanna has again proven essential for the completion of this book project.

Fritz-Haber-Institut, Berlin   *Klaus Hermann*
Autumn 2015

Preface to the First Edition

The objective of this book is to provide students and researchers with the crystallographic foundations necessary to understand the structure and symmetry of surfaces and the interfaces of crystalline materials. This includes macroscopic single crystals as well as crystalline nanoparticles. Knowledge of their geometric properties is a prerequisite for the interpretation of corresponding experimental and theoretical results, which explain both their physical and chemical behavior. In particular, surface and interface structure is of vital importance not only for the study of properties near single crystal surfaces, but also for research on thin films at solid substrates. Here, technological applications range from semiconductor devices and magnetic storage disks to heterogeneous catalysts.

Crystalline nanoparticles, such as nanotubes, nanowires, or compact particles of finite size have recently attracted considerable interest due to their novel chemical and physical properties. Examples are carbon nanotubes, silicon nanowires, nanosize quantum dots at semiconductor surfaces, or catalytically active crystallites. These particles are of finite size in one or more dimensions, but their local atom arrangement can still be close to that of extended bulk crystals. In addition, their surfaces and interfaces with other material can be described analogously to those found for single crystal surfaces. Thus, surface crystallography, covered in this book, can also be applied for the analysis of structural properties of nanoparticle surfaces.

While treatises on three-dimensional crystallography are abundant, there are only few chapters on surface crystallography that are available in specialized surface science reviews. In particular, comprehensive textbooks on surface structure have not yet been published. Nevertheless, students and researchers entering the field need to obtain a thorough overview of surface structure and geometry, which includes all relevant basic crystallographic methods required for theoretical and experimental analyses. This book tries to serve this purpose. It is primarily meant for graduate and PhD students in physics, chemistry, and crystallography, but will also help researchers who want to learn more details about geometric structure at surfaces of single crystals or nanoparticles.

The book is written by a theoretical surface scientist. Therefore, the discussion of methods and approaches in the text is frequently adapted to surfaces and differs in some places from traditional treatment of crystallography. As an example, number theoretical methods are used to derive appropriate transformations between equivalent lattice descriptions. Further, some of the conventional surface structure concepts are looked at from a different perspective and go beyond the standard treatment that is practiced inside the surface science community. Examples are the introduction of Miller indices based on netplane-adapted lattices and a thorough mathematical treatment of symmetry, which results in the 17 two-dimensional space groups. Therefore, the text can also be used as a resource that is complementary to the standard surface science literature.

This book started as a manuscript of a series of lectures on surface crystallography, given by the author at several international workshops and in universities as well as in research institutions where surface science and catalysis groups were engaged in research on the structural properties of surfaces. Questions and discussions during the lectures were often the source for more detailed work on different sections of the manuscript and thus helped to improve its presentation. Furthermore, research visits to various surface science groups raised the author's awareness of new or incompletely treated issues that had to be dealt with. The author is indebted to all those who contributed with their scientific curiosity and criticism. The text has benefited from numerous discussions with surface scientists, crystallographers, and mathematicians of whom only a few are mentioned in alphabetic order: Gerhard Ertl, Klaus Heinz, Bernhard Hornfeck, Klaus Müller, John B. Pendry, Gabor A. Somorjai, D. Phil Woodruff. Wolfgang Moritz served as an extremely valuable sparring partner in the world of crystallography. Very special thanks go to Michel A. Van Hove whose constructive criticism, rich ideas, and continuous support during the writing phase were unmatched. Without him the book would not exist in its present form.

Finally, I am greatly indebted to my wife Hanna for her patience and loving care throughout the time it took to finish this book and beyond.

Fritz-Haber-Institut, Berlin *Klaus Hermann*
Summer 2010

1
Introduction

Research in many areas of materials science requires detailed knowledge about crystalline solids on an atomic scale. These systems may represent real materials such as complex semiconductors, or may act as meaningful models, for example, by simulating reactive sites of catalysts. Here, physical and chemical insight depends very much on the details of the geometric structure of local environments of atoms and of the possible periodic atom arrangements inside the crystal as well as at its surface. As examples we mention the following:

- **Chemical binding** between atoms inside a crystal but also at its surfaces depends, apart from atomic parameters, strongly on local geometry [1, 2]. This is very often expressed by local **coordination** that describes the number and arrangement of nearest neighbor atoms with respect to the binding atom. As an example, metal atoms inside a bulk metal crystal are usually characterized by a large number of nearest neighbors, 8 or 12, yielding metallic binding. At surfaces, the change in chemical binding due to different coordination, compared with that inside the bulk, is tightly connected with local structure, which can be expressed by relaxation and reconstruction. Further, atoms or molecules can adsorb at specific sites of crystalline substrates, where the adsorption geometry is essential for understanding their local binding behavior.
- **Electronic properties** at surfaces of single crystals can differ substantially from those of the corresponding bulk. For example, the existence of a surface can induce additional electronic states, so-called surface states, that have been found in experiments and studied theoretically for some time [3]. Here, the detailed surface structure determines the existence as well as the energetic behavior of the states. Further, electronic interband transitions in silicon nanowires and nanodots are found to cause photoluminescence that does not occur in silicon bulk crystals [4]. The difference is explained by both the spatial confinement of the nanoparticles and also by changes in geometric properties of their atom arrangement. Finally, it has been claimed from experiments that semiconducting bulk silicon shows metallicity at its (7×7) reconstructed (1 1 1) surface [5], and metallicity is also found in theoretical studies on silicon nanowires [6].

- **Magnetism** of crystalline bulk material as well as of its surfaces depends on the crystal structure and local coordination. For example, vanadium sesquioxide, V_2O_3, in its monoclinic crystal structure is antiferromagnetic at low temperatures, whereas its high-temperature phase, described by a trigonal corundum lattice, is paramagnetic [7]. Vanadium crystals with a body-centered cubic lattice are found to be paramagnetic in their bulk volume but ferromagnetic at their surfaces [8]. Other examples are thin iron films grown on top of copper single crystal surfaces where, as a function of film thickness, their crystal structure changes and, as a consequence, their magnetic properties [9].
- **Anisotropic electrical conductivity** is often connected with dense atom packing along specific directions inside a crystal. An example is given by trigonal $LiCoO_2$ crystals that form the most common lithium storage material for rechargeable batteries. Here, the electrical conductivity is greatly enhanced along densely packed Co and Li planes while it is much smaller perpendicular to the planes [10].
- **Catalytic surface reactions** depend crucially on structural properties of the surfaces of crystalline catalyst materials at an atomic scale [11, 12]. The atomic surface structure determines possible adsorption and reaction sites for molecules, which can support specific catalytic reactions but can also exclude others known as *structure–reactivity relationships* [11]. For example, catalytic CO oxidation happens at single crystal surfaces of platinum with different efficiency depending on the surface orientation [13], where the surface structure determines the type and density of reactive sites.

In addition to bulk crystals and their surfaces, studies on crystalline **nanoparticles** [14, 15] have become an exciting field of research. This includes nanotubes [16], nanowires [14], or compact particles of finite size, such as atom clusters [17], fullerenes [18], or quantum dots [19], which show novel physical and chemical properties deviating from those of the corresponding bulk material. Examples are carbon nanotubes providing substrate material to yield new active catalysts [20] or silicon nanowires whose visible photoluminescence is determined by their size [21]. Further, nanosize quantum dots at semiconductor surfaces are found to yield quite powerful light emitting diodes (LEDs) of technological relevance [19].

These nanoparticle systems are described as **atom aggregates** of finite size in one or more dimensions, where their local geometric arrangement can still be close to that of extended bulk crystals. Likewise, their spatial confinement with corresponding surfaces and interfaces can be considered analogous to that appearing at bulk crystal surfaces. Therefore, surface crystallography, initially developed to describe structural properties at single crystal surfaces, also forms a sound basis by which the structure of nanoparticle surfaces can be characterized. This is particularly interesting since the relative number of atoms positioned at nanoparticle surfaces compared with those of their inner volume is always larger than that of extended macroscopic single crystals. Thus, the relative importance of atoms at **nanoparticle surfaces** in determining physical properties is expected to be greater than that of atoms at single crystal surfaces. In addition, nanoparticles

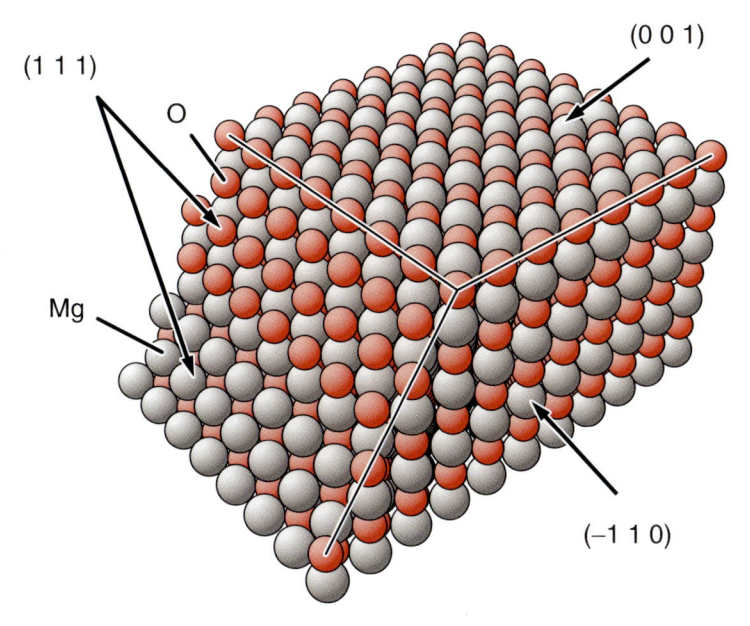

Figure 1.1 Section of an MgO crystal (NaCl lattice). The atoms are shown as shaded balls of different color and labeled accordingly. The section is enclosed by nonpolar (0 0 1), (−1 1 0), and by polar (1 1 1) oriented surfaces.

can possess symmetry and geometric properties that do not appear in single crystals or at their surfaces. Examples are icosahedral clusters or curved nanoparticle surfaces that originate from bending single crystal sections, where in this book **nanotubes** will be discussed as examples.

In many experimental and theoretical studies real crystalline systems are, for the sake of simplicity, approximately described as **ideal single crystals** with a well-defined atomic composition and an unperturbed three-dimensional periodicity. In addition, planar surfaces of single crystals are often assumed to be bulk-terminated and of unperturbed two-dimensional periodicity. With this approximation in mind a rigorous mathematical description of all structural parameters becomes possible and is one of the basic subjects of classical crystallography. As an illustration, Figure 1.1 shows the structure of a section of an ideal single crystal of magnesium oxide, MgO, with its perfect three-dimensional periodic arrangement of atoms. Here, sections of ideal planar surfaces, originating from bulk truncation, become visible and demonstrate the variety of surface types for the same crystal depending on the crystal cut.

In the following chapters of this book we will discuss the basic elements as well as the **mathematical methods** used in crystallography to evaluate structural parameters of single crystals with particular emphasis on their surfaces. We start with ideal bulk crystals of three-dimensional periodicity, where classical bulk crystallography provides a quantitative description. Then we introduce ideal two-dimensional surfaces as a result of bulk truncation along specific directions including high density, vicinal, stepped, kinked, and chiral surfaces. We give a detailed account of their two-dimensional symmetry behavior following the crystallographic classification scheme of Bravais lattices and two-dimensional space groups. Next, we discuss details of the deviation of atomic structures at surfaces due to changes in surface binding compared with that in the bulk. This is usually described by surface relaxation and reconstruction, where we consider different schemes. In addition, structural behavior during growth processes is discussed. Then we deal with the crystallographic aspects of commensurate, high-order commensurate, and incommensurate adsorbate systems as special cases of surface reconstruction. Here also the different structure notations used in the literature will be described. The discussion of surface structure will be completed by an overview of the surfaces that have been analyzed quantitatively at an atomic level in scattering, diffraction, imaging, or spectroscopic experiments. Further, formal requirements of complete quantitative surface structure databases will be considered. Finally, we describe the theoretical aspects and structural details of nanotubes of different element composition as special cases of rolled sections of crystal monolayers. These nanotubes are examples of a larger class of crystalline materials, nanoparticles mentioned above, and demonstrate that crystallographic methods can also be applied to these systems in order to account for their structural properties. Finally, the book concludes with appendices spelling out further details of the mathematical methods used in the different sections, with tabulations of typical surface sites, and with compilations of structural parameters of crystals.

The theoretical concepts treated in this book will be illustrated by example applications for further understanding, which include results from **measured** real single crystal surfaces documented in the Surface Structure Database (**SSD**) [22–24] or its earlier version **SCIS** (Surface Crystallographic Information Service) [25]. In addition, each chapter of the book concludes with a set of **exercises**. These exercises are of varying difficulty, ranging from simple questions to small research projects, and are meant to stimulate the discussion on the different subjects and to contribute to their clarification. Some of the exercises may require **visualization tools** for crystals, such as **Balsac** [26] or **Survis** [27] or the like.

In the theoretical treatment of some structural properties of ideal single crystals we will apply **number theoretical methods,** dealing with relations between integers. While this approach is not commonly used in textbooks on surface science or crystallography it can simplify the formal treatment considerably. Examples are solutions of linear and quadratic Diophantine equations that facilitate the discussion of monolayers or of atom neighbor shells in crystals. Therefore, number theoretical methods will be introduced briefly as required and further details are found in Appendix E.

A few illustrations are included as **stereo pictures** for an enhanced three-dimensional impression. These pictures may be viewed either by using optical stereo glasses (available separately) or by cross-eyed viewing without glasses.

In the latter case, viewing for an extended time may overstrain the eyes and should be avoided.

Clearly, the present book cannot cover all aspects of the field and may, in some cases, be quite brief. Further, the selection of topics as well as their presentation is, to some degree, determined by the author's personal preferences. However, the interested reader is referred to the extensive crystallographic literature, see for example, [28–33], to the surface science literature, see for example, [34–39], or to the solid state physics literature, see for example, [1, 2, 40], to explore additional details.

2
Bulk Crystals: Three-Dimensional Lattices

This section deals with the geometric properties of three-dimensional **bulk crystals**, which are described, in their perfect structure, by atom arrangements that are periodic in three dimensions. As an example, Figure 2.1 shows a section of a tetragonal $YBa_2Cu_3O_7$ crystal, where vectors $\underline{R}_1$, $\underline{R}_2$, $\underline{R}_3$ (lattice vectors) indicate the mutually perpendicular directions of periodicity. Further, the basis of the crystal structure consists of 13 atoms (1 × yttrium, 2 × barium, 3 × copper, 7 × oxygen) inside a rectangular block (unit cell) that is repeated periodically inside the crystal. The building unit is shown to the left of the figure.

In this section, all **basic definitions** used for a quantitative description of structural properties of perfect three-dimensionally periodic crystals will be provided. Here, the crystals are considered not only in terms of their translational symmetry, that is, periodicity, but also by their different point symmetry elements, such as inversion points, mirror planes, or rotation axes, which characterize the positions of all atoms inside a crystal. While the definitions and general properties are rather abstract and **mathematical,** they can be quite relevant for theoretical studies of real three-dimensional crystals. As an example, lattice representations of crystals are required as input to any electronic structure calculation for solid crystalline material. Further, the theoretical treatment of three-dimensional crystals serves as a foundation to study the surfaces of single crystals, as will be discussed in Chapters 4, 5, and 6.

2.1
Basic Definition

The basic definition of a perfect three-dimensional bulk crystal becomes quite clear by considering a simple example. Figure 2.2a shows a section of the cubic CsCl crystal, which is periodic in three perpendicular directions. Thus, its periodicity can be described by orthogonal vectors $\underline{R}_1$, $\underline{R}_2$, $\underline{R}_3$ (**lattice vectors**), as indicated in Figure 2.2b, whose lengths define the corresponding periodicity lengths. The lattice vectors span a cubic cell (**morphological unit cell**) that contains one cesium and one chlorine atom each at positions given by vectors $\underline{r}_1$ (Cs) and $\underline{r}_2$ (Cl) (**lattice basis vectors**), see Figure 2.2b. A periodic repetition

2 Bulk Crystals: Three-Dimensional Lattices

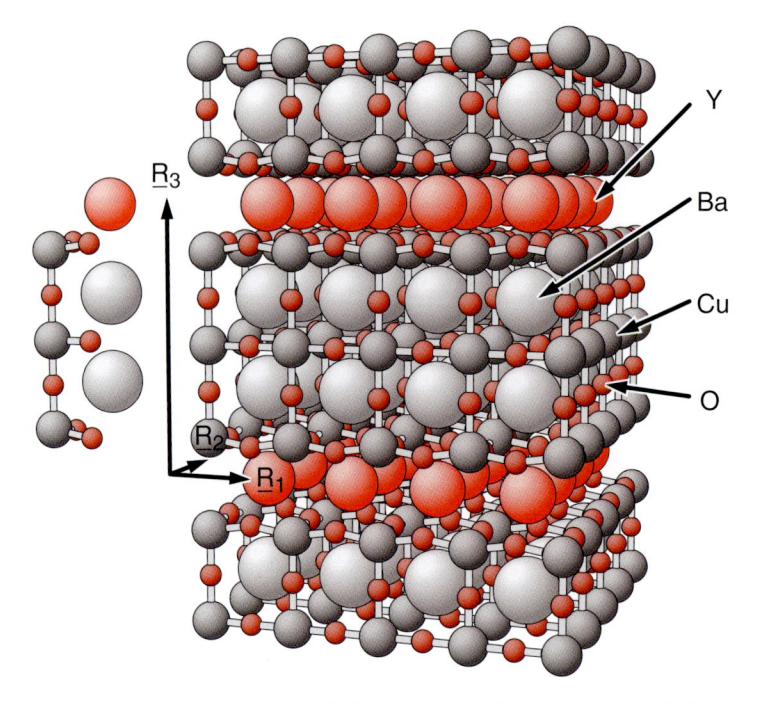

Figure 2.1 Section of a tetragonal $YBa_2Cu_3O_7$ crystal. The atoms are labeled accordingly. In addition, the basis of 13 atoms inside a rectangular cell and lattice vectors $\underline{R}_1$, $\underline{R}_2$, $\underline{R}_3$ are included to the left.

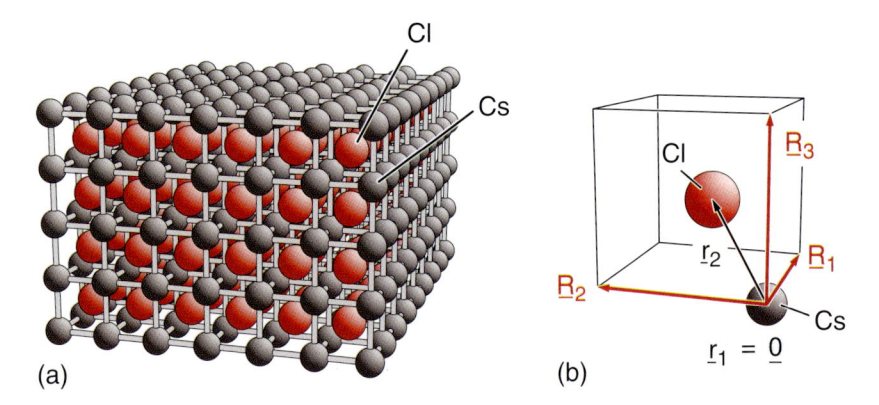

Figure 2.2 (a) Section of a cubic CsCl crystal. Sticks connect neighboring Cs atoms to indicate the crystal structure. (b) Primitive morphological unit cell with two atoms, Cs and Cl, inside. The lattice vectors $\underline{R}_1$, $\underline{R}_2$, $\underline{R}_3$ as well as lattice basis vectors, $\underline{r}_1 = \underline{0}$ for Cs and $\underline{r}_2$ for Cl, are shown and labeled accordingly.

of the unit cell along $\underline{R}_1$, $\underline{R}_2$, $\underline{R}_3$ can then be used to build the complete infinite crystal.

In the general case, the formal definition of a perfect three-dimensional bulk **crystal** starts from a three-dimensionally periodic arrangement of atoms. Here, the crystal periodicity is described by a **lattice** with **lattice vectors** $\underline{R}_1$, $\underline{R}_2$, $\underline{R}_3$. Thus, the lattice forms an infinite and periodic array of **lattice points** reached from a common origin by vectors $\underline{R}$ with

$$\underline{R} = n_1 \underline{R}_1 + n_2 \underline{R}_2 + n_3 \underline{R}_3 \tag{2.1}$$

where the coefficients n_1, n_2, n_3 can assume any integer value. This means, in particular, that each lattice point experiences the same environment created by all other points.

The lattice vectors can be given in different ways, where the choice depends on the type of application. While for numerical calculations it may be preferable to define $\underline{R}_1$, $\underline{R}_2$, $\underline{R}_3$ with respect to an absolute **Cartesian coordinate system** as

$$\underline{R}_i = (x_i, y_i, z_i), \quad i = 1, 2, 3 \tag{2.2}$$

it is common in the crystallographic literature to define these vectors by **lattice parameters** describing their lengths (**lattice constants**) a, b, c and by their mutual **angles** α, β, γ, as sketched in Figure 2.3, where

$$a = |\underline{R}_1|, \quad b = |\underline{R}_2|, \quad c = |\underline{R}_3|$$
$$(\underline{R}_1 \underline{R}_2) = ab \cos(\gamma), \quad (\underline{R}_1 \underline{R}_3) = ac \cos(\beta), \quad (\underline{R}_2 \underline{R}_3) = bc \cos(\alpha) \tag{2.3}$$

Examples are given by lattices denoted as

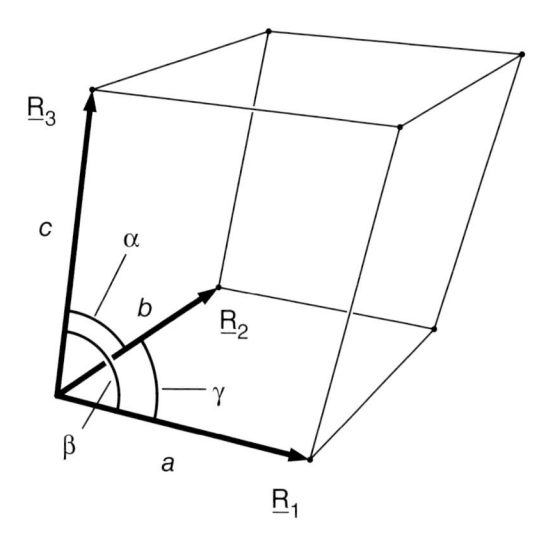

$$\text{simple cubic} \quad \text{where} \quad a = b = c, \quad \alpha = \beta = \gamma = 90° \tag{2.4}$$

Figure 2.3 Definition of crystallographic lattice parameters $a, b, c, \alpha, \beta, \gamma$ in a perspective view.

hexagonal   where $a = b \neq c$, $\alpha = \beta = 90°, \gamma = 120°$ (2.5)

Relations (2.3) can be **converted** to yield lattice vectors in Cartesian coordinates starting from the six parameters, a, b, c, and α, β, γ, given in Eq. (2.3), where one possible conversion is

$$\underline{R}_1 = a\,(1, 0, 0), \quad \underline{R}_2 = b\,(\cos(\gamma), \sin(\gamma), 0)$$
$$\underline{R}_3 = c\,(\cos(\beta), (\cos(\alpha) - \cos(\beta)\cos(\gamma))/\sin(\gamma), v_3/\sin(\gamma)) \quad (2.6a)$$

with

$$v_3 = \{(\cos(\beta - \gamma) - \cos(\alpha))(\cos(\alpha) - \cos(\beta + \gamma))\}^{1/2} \quad (2.6b)$$

This yields for simple cubic (sc) lattices with Eq. (2.4)

$$\underline{R}_1 = a\,(1, 0, 0), \quad \underline{R}_2 = a\,(0, 1, 0), \quad \underline{R}_3 = a\,(0, 0, 1) \quad (2.7)$$

and for hexagonal lattices with Eq. (2.5)

$$\underline{R}_1 = a\,(1, 0, 0), \quad \underline{R}_2 = a\,(-1/2, \sqrt{3}/2, 0), \quad \underline{R}_3 = c\,(0, 0, 1) \quad (2.8)$$

The lattice vectors $\underline{R}_1$, $\underline{R}_2$, $\underline{R}_3$ span a six-faced polyhedron (so-called parallelepiped), defining the **morphological unit cell**, often referred to as the **unit cell**, whose edges are parallel to $\underline{R}_1$, $\underline{R}_2$, $\underline{R}_3$ and whose volume V_{el} is given by

$$V_{el} = |(\underline{R}_1 \times \underline{R}_2)\,\underline{R}_3| \quad (2.9)$$

The unit cell is called a **primitive unit cell** if its volume is the smallest of all possible unit cells in the crystal. This is equivalent to requiring that there is no additional lattice point, described by vector $\underline{R}'$ with

$$\underline{R}' = \kappa_1\underline{R}_1 + \kappa_2\underline{R}_2 + \kappa_3\underline{R}_3, \quad 0 \leq \kappa_i < 1 \quad (2.10)$$

inside the morphological unit cell of the lattice. Otherwise, the cell is **non-primitive** and there must be one or more additional lattice points $\underline{R}'$ inside the unit cell. Analogously, lattice vectors $\underline{R}_1$, $\underline{R}_2$, $\underline{R}_3$ whose morphological unit cell is primitive are called **primitive lattice vectors**, otherwise **non-primitive**. As an example, the cubic unit cell of CsCl as well as the corresponding lattice vectors, shown in Figure 2.2, are primitive. On the other hand, replacing all cesium and chlorine atoms in Figure 2.2 by one atom type, for example, iron, yields a body-centered cubic (bcc) crystal. Here, the lattice vectors $\underline{R}_1$, $\underline{R}_2$, $\underline{R}_3$, shown in the figure, are non-primitive, since vector $\underline{r}_2$ now becomes a lattice vector inside the morphological unit cell.

In a crystal, the morphological unit cell contains, in general, p atoms at positions given by vectors $\underline{r}_1, \ldots, \underline{r}_p$ (**lattice basis vectors**), which form the **basis** of the crystal structure (the basis is sometimes also called the **structure**). Each atom at $\underline{r}_i$ carries a **label** characterizing its properties, such as its nuclear charge or element name. These labels, usually omitted in the following, will be attached to each lattice basis vector if needed. For example, a definition $\underline{r}_3^{Cl}$ would refer to a chlorine atom placed at a position given by the third lattice basis vector. All lattice basis vectors $\underline{r}_i$ inside the morphological unit cell can be written as **linear combinations** of the lattice vectors $\underline{R}_1, \underline{R}_2, \underline{R}_3$ according to

$$\underline{r}_i = x_i \underline{R}_1 + y_i \underline{R}_2 + z_i \underline{R}_3, \quad i = 1 \ldots p \tag{2.11}$$

where x_i, y_i, and z_i are real-valued coefficients with $|x_i| < 1$, $|y_i| < 1$, $|z_i| < 1$. This use of **relative coordinates** x_i, y_i, z_i to describe atoms inside the unit cell is common practice in the crystallographic literature [28–33]. Note that, according to definition (2.11) the coefficients x_i, y_i, z_i are generally not connected with the Cartesian coordinate system but with coordinate axes given by the lattice vectors $\underline{R}_1, \underline{R}_2, \underline{R}_3$.

The **origin** of the morphological unit cell inside a crystal can always be chosen freely since the complete infinite crystal consists of a periodic arrangement of unit cells in three dimensions. In particular, the origin does not need to coincide with a specific atom position, as considered in the example of CsCl discussed above. However, it is usually chosen to coincide with the location of the largest number of point symmetry elements, such as inversion points or origins of mirror planes and rotation axes, which are given by the lattice vectors $\underline{R}_1, \underline{R}_2, \underline{R}_3$ together with the lattice basis vectors $\underline{r}_1, \ldots, \underline{r}_p$. This will be discussed in greater detail in Section 2.4.

Altogether, a **crystal** is characterized uniquely by its **lattice** defined by lattice vectors $\underline{R}_1, \underline{R}_2, \underline{R}_3$ and its **basis** defined by lattice basis vectors $\underline{r}_1, \ldots, \underline{r}_p$. Thus, general atom positions inside the crystal can be given by

$$\underline{r} = n_1 \underline{R}_1 + n_2 \underline{R}_2 + n_3 \underline{R}_3 + \underline{r}_i \tag{2.12}$$

where the coefficients n_1, n_2, n_3 can assume any integer value and index $i = 1, \ldots, p$ counts the number of atoms in the unit cell. Here, the lattice and the basis can be treated as **separate** elements of a crystal structure (which are only connected by the symmetry elements as will be discussed in Section 2.4).

2.2 Representation of Bulk Crystals

There is one important aspect that characterizes all formal descriptions of crystal structures, the fact that mathematical descriptions of crystals are **not unique.** This means that, for a given definition of a crystal, one can always find an infinite number of alternatives that describe the same crystal. While this ambiguity may be considered a drawback at first glance, it allows choosing crystal representations according to additional constraints, for example, those given by symmetry, physical, or chemical properties. Here, one can distinguish between alternative descriptions that affect the crystal basis but not its lattice representation and those where both the lattice representation and the basis are affected.

2.2.1
Alternative Descriptions Conserving the Lattice Representation

Examples of alternative crystal descriptions that do not affect the crystal lattice are given by elemental or compound **decompositions** of a crystal. Here, the basic idea is to decompose the basis inside the unit cell of a complex crystal into components and consider (fictitious) crystals of these components with the same periodicity as that of the initial crystal, given by its lattice. This decomposition is of didactic value but may also help to understand details of chemical binding inside the crystal, for example, discriminating between intra- and inter-molecular binding in molecular crystals. In the simplest case, a crystal with p atoms in its primitive unit cell can be considered alternatively as a combination of p crystals with the same lattice but with only one atom in their primitive unit cells. The origins of the corresponding p crystals can be set at positions given by the lattice basis vectors $\underline{r}_i$ of the complete non-primitive crystal.

As a very simple example, the cubic **cesium chloride**, CsCl, crystal, shown in Figure 2.2, is defined by a simple cubic (sc) lattice with lattice vectors $\underline{R}_1, \underline{R}_2, \underline{R}_3$ given by Eq. (2.7). Further, its basis includes two atoms, Cs and Cl, which can be positioned at

$$\underline{r}_1 = a\,(0,0,0) \text{ for Cs}, \quad \underline{r}_2 = a\,(1/2, 1/2, 1/2) \text{ for Cl} \qquad (2.13)$$

with a denoting the lattice constant of CsCl. Thus, the crystal can be considered as a combination of two sc monoatomic crystals, one for cesium and one for chlorine, where their origins are shifted by $\underline{r}_o = \underline{r}_2 - \underline{r}_1 = a\,(1/2, 1/2, 1/2)$ with respect to each other.

A more complex example is the tetragonal **YBa$_2$Cu$_3$O$_7$** crystal, shown in Figure 2.1. Here, the lattice vectors can be written in Cartesian coordinates as

$$\underline{R}_1 = a\,(1,0,0), \quad \underline{R}_2 = a\,(0,1,0), \quad \underline{R}_3 = c\,(0,0,1) \qquad (2.14a)$$

and the morphological unit cell contains 13 atoms resulting in 13 lattice basis vectors $\underline{r}_i$ with

Y atom : $\underline{r}_1 = (1/2, 1/2, 5/6)$

Ba atoms : $\underline{r}_2 = (1/2, 1/2, 1/6), \quad \underline{r}_3 = (1/2, 1/2, 1/2)$

Cu atoms : $\underline{r}_4 = (0, 0, 0), \quad \underline{r}_5 = (0, 0, 1/3), \quad \underline{r}_6 = (0, 0, 2/3)$

O atoms : $\underline{r}_7 = (1/2, 0, -\varepsilon), \quad \underline{r}_8 = (0, 1/2, -\varepsilon), \quad \underline{r}_9 = (0, 0, 1/6)$

$\underline{r}_{10} = (0, 1/2, 1/3), \quad \underline{r}_{11} = (0, 0, 1/2)$

$\underline{r}_{12} = (1/2, 0, 2/3 + \varepsilon), \quad \underline{r}_{13} = (0, 1/2, 2/3 + \varepsilon) \qquad (2.14b)$

using relative coordinates (2.11) where experiments yield a relative position shift $\varepsilon = 0.026$ of four oxygen atoms. This crystal can be decomposed conceptually into 13 monoatomic (tetragonal) crystals, one yttrium, two barium, three copper, and seven oxygen crystals.

Alternatively, one can decompose the YBa$_2$Cu$_3$O$_7$ crystal into physically more meaningful subunits that include **several** of the atoms of the initial unit cell. As an

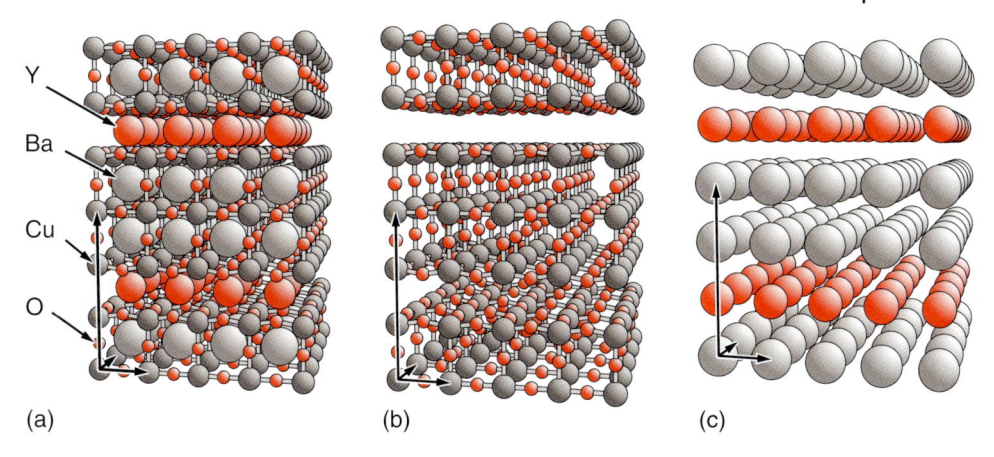

Figure 2.4 Decomposition of the $YBa_2Cu_3O_7$ crystal (a) into its copper oxide, Cu_3O_7 (b) and heavy metal, YBa_2, components (c). Atoms are shown as colored balls and labeled accordingly. In addition, the lattice vectors $\underline{R}_1$, $\underline{R}_2$, $\underline{R}_3$ are indicated by arrows.

example, Figure 2.4 illustrates a decomposition of the $YBa_2Cu_3O_7$ crystal into its copper oxide and its heavy metal components, namely, Cu_3O_7 and YBa_2, respectively. Here, the unit cells of the component crystals contain 10 and 3 atoms each, where the Cu_3O_7 component is believed to contribute to the high-temperature superconductivity of $YBa_2Cu_3O_7$.

A very illustrative example of crystal decomposition is given by the **diamond** crystal, shown in Figure 2.5. Its lattice can be defined as a cubic lattice where the lattice vectors are given by Eq. (2.7). The basis of the crystal includes eight carbon atoms in tetrahedral arrangements resulting in eight lattice basis vectors $\underline{r}_i$ with

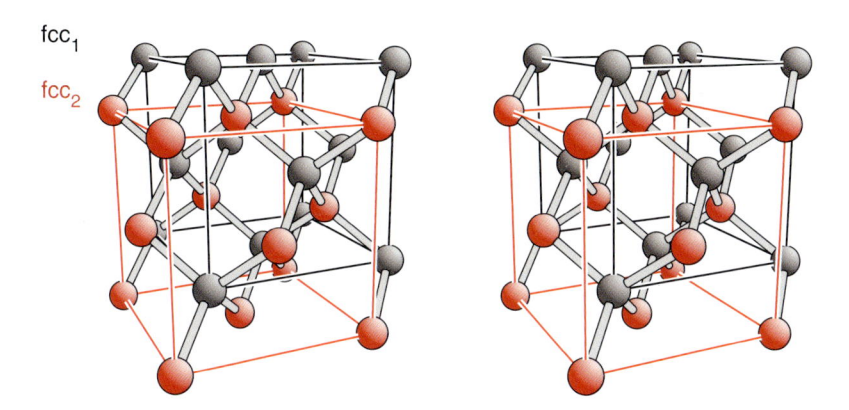

Figure 2.5 Decomposition of the diamond crystal into two (shifted) face-centered cubic crystals, denoted fcc_1 (gray balls, black lines) and fcc_2 (red balls and red lines). The crystal is shown in a stereo view where the visual three-dimensional impression is obtained by cross-eyed viewing.

$$\underline{r}_1 = (0,0,0), \qquad \underline{r}_2 = (0,1/2,1/2), \qquad \underline{r}_3 = (1/2,0,1/2)$$
$$\underline{r}_4 = (1/2,1/2,0), \qquad \underline{r}_5 = (1/4,1/4,1/4), \qquad \underline{r}_6 = (1/4,3/4,3/4)$$
$$\underline{r}_7 = (3/4,1/4,3/4), \qquad \underline{r}_8 = (3/4,3/4,1/4) \tag{2.15}$$

in relative coordinates (2.11). This shows, first, that the diamond crystal can be decomposed into eight sc crystals, each with one carbon in the primitive unit cell. Further, the lattice basis vectors $\underline{r}_5, \underline{r}_6, \underline{r}_7, \underline{r}_8$ arise from $\underline{r}_1, \underline{r}_2, \underline{r}_3, \underline{r}_4$ by identical shifts with

$$\underline{r}_{i+4} = \underline{r}_i + 1/4\,(1,1,1), \quad i = 1,2,3,4 \tag{2.16}$$

This suggests that the diamond crystal can also be decomposed into two identical cubic crystals, each with four atoms in their unit cells, where the origins of the two crystals are shifted by a vector $a/4$ (1, 1, 1) with respect to each other. The lattices of the two component crystals will be shown in Section 2.2.2.1 to be identical to face-centered cubic (fcc) lattices. Thus, the diamond crystal can be alternatively described by a superposition of two fcc crystals which becomes clear by an inspection of Figure 2.5.

2.2.2
Alternative Descriptions Affecting the Lattice Representation

There are many possibilities of providing alternative descriptions of crystals where their lattices are represented differently. These alternatives may be preferred because of **conceptual** convenience but may also be required due to **computational** necessity. Examples are symmetry-adapted lattices combining translational with point symmetry properties or surface-adapted lattices facilitating the definition of atom coordinates in surface studies.

Crystallographers have defined a set of constraints on lattice vectors $\underline{R}_1, \underline{R}_2, \underline{R}_3$ to yield a unique description of a lattice according to Niggli [41], which allows an easy distinction between the different types of three-dimensional Bravais lattices discussed in Section 2.4. First, the lattice vectors are chosen such that they form a right-handed vector triplet, which can be expressed mathematically by the constraint

$$(\underline{R}_1 \times \underline{R}_2)\,\underline{R}_3 > 0 \tag{2.17}$$

Further, they are assumed to reflect three smallest periodicity lengths along different directions in the crystal and are arranged such that

$$|\underline{R}_1| \leq |\underline{R}_2| \leq |\underline{R}_3| \tag{2.18}$$

In addition, all lattices are grouped according to their scalar products $s_{ij} = (\underline{R}_i\,\underline{R}_j)$ into two classes,

$$s_{12} \geq 0, \quad s_{13} \geq 0, \quad s_{23} \geq 0 \quad (\text{type 1}, \textbf{acute}) \tag{2.19a}$$

$$s_{12} \leq 0, \quad s_{13} \leq 0, \quad s_{23} \leq 0, \quad \text{with at least one } s_{ij} < 0 \quad (\text{type 2}, \textbf{obtuse}) \tag{2.19b}$$

where lattices with other s_{ij} combinations can be easily converted to one of the two classes by inverting two of the lattice vectors $\underline{R}_i, \underline{R}_j$ to yield $-\underline{R}_i, -\underline{R}_j$. Further, simple iterative algorithms have been developed [42, 43] to reduce a general vector set $\underline{R}_1, \underline{R}_2, \underline{R}_3$ of type 1 or 2 to a unique description with $\underline{R}_1, \underline{R}_2, \underline{R}_3$ referring to vectors of smallest length in the lattice. This **reduced lattice vector set** fulfills, apart from Eqs. (2.17) and (2.18), the inequalities

$$-\min(\underline{R}_i^2, \underline{R}_j^2) \leq 2\,(\underline{R}_i\,\underline{R}_j) < \min(\underline{R}_i^2, \underline{R}_j^2), \quad i \neq j, \quad i, j = 1, 2, 3 \qquad (2.20)$$

which can be used to identify and classify reduced unique vector sets $\underline{R}_1, \underline{R}_2, \underline{R}_3$. Type 2 lattices require an additional constraint which reads

$$2\,(|\underline{R}_1\,\underline{R}_2| + |\underline{R}_1\,\underline{R}_3| + |\underline{R}_2\,\underline{R}_3|) \leq \underline{R}_1^2 + \underline{R}_2^2 \qquad (2.21)$$

to yield a unique description [42].

The application of the above constraints to two-dimensional lattices described by lattice vectors $\underline{R}_1, \underline{R}_2$ is straightforward. Here, the two vectors are required to yield the smallest periodicity lengths along different directions in the lattice and are ordered according to

$$|\underline{R}_1| \leq |\underline{R}_2| \qquad (2.22)$$

This allows, as in the three-dimensional case, two lattice classes differing by the scalar product $s_{12} = (\underline{R}_1\,\underline{R}_2)$,

$$s_{12} \geq 0 \quad (\text{type 1, \textbf{acute}}) \quad \text{and} \quad s_{12} < 0 \quad (\text{type 2, \textbf{obtuse}}) \qquad (2.23)$$

The Minkowski reduction, see Section 3.3 and Appendix D, can be used to reduce a general vector set $\underline{R}_1, \underline{R}_2$ of type 1 or 2 to a unique description referring to vectors of smallest length in the lattice. This reduced vector set fulfills, apart from Eq. (2.22), the inequality

$$-\min(\underline{R}_1^2, \underline{R}_2^2) \leq 2\,(\underline{R}_1\,\underline{R}_2) < \min(\underline{R}_1^2, \underline{R}_2^2) \qquad (2.24)$$

which can be used to test whether a vector set $\underline{R}_1, \underline{R}_2$ is reduced or not. The constraints (2.22) and (2.24) yield a unique description that allows a simple distinction between the different types of two-dimensional Bravais lattices discussed in Section 3.7. They can also serve as a basis for a more general classification scheme proposed in the literature [44]. In two dimensions, obtuse lattice descriptions can always be converted to acute descriptions, which is preferred by many surface scientists, by swapping the lattice vectors and replacing one of the two, for example, $\underline{R}_1$, by its negative, $-\underline{R}_1$, where, however, one of the two representations may violate constraint (2.22).

Many researchers in the surface science community (and not only there) find it convenient to think in Cartesian coordinates, using orthogonal unit vectors in three-dimensional space. Therefore, they prefer to characterize lattices, if possible, by orthogonal lattice vectors $\underline{R}_1, \underline{R}_2, \underline{R}_3$ even at the expense of having to consider corresponding crystal bases with a larger number of atoms. This will be discussed for face- and body-centered cubic lattices in Section 2.2.2.1.

Theoretical studies on extended geometric perturbations inside a crystal, such as those originating from periodic imperfections or distortions, often require considering unit cells with lattice vectors $\underline{R}'_1, \underline{R}'_2, \underline{R}'_3$ which are larger than those given by $\underline{R}_1, \underline{R}_2, \underline{R}_3$ of the unperturbed crystal. Here, a direct computational comparison of results for the perturbed crystal with those for the unperturbed crystal is often facilitated by applying the same (enlarged) lattice vectors $\underline{R}'_1, \underline{R}'_2, \underline{R}'_3$ to both systems. As a result, the unperturbed crystal is described by a lattice with a larger than primitive unit cell and an appropriately increased number of atoms. This is the basic idea behind so-called **superlattice** methods, which will be discussed in Section 2.2.2.2.

Ideal single crystal surfaces that originate from bulk truncation yielding two-dimensional periodicity at the surface, will be treated in great detail in Chapter 4. Here, the analysis of structural properties at the surface can be facilitated greatly by using so-called **netplane-adapted** lattice vectors $\underline{R}'_1, \underline{R}'_2, \underline{R}'_3$. These are given by linear transformations of the initial bulk lattice vectors, where the shape of the morphological unit cell may change but not its volume nor the number of atoms inside the cell. Differently oriented surfaces require different sets of netplane-adapted lattice vectors leading to many alternative descriptions of the bulk lattice, as discussed in Section 2.2.2.3.

2.2.2.1 Cubic, Hexagonal, and Trigonal Lattices

The family of cubic lattices, simple, body-, and face-centered, are closely connected with each other, which is why many scientists use the simplest of the three, the simple cubic lattice as their reference. This lattice, also called **cubic-P** and often abbreviated by **sc** is described in Cartesian coordinates by lattice vectors

$$\underline{R}_1^{sc} = a\,(1,0,0), \quad \underline{R}_2^{sc} = a\,(0,1,0), \quad \underline{R}_3^{sc} = a\,(0,0,1) \tag{2.25}$$

which are three mutually orthogonal vectors of equal length, given by the lattice constant a.

The **body-centered cubic** lattice, also called **I-centered** or **cubic-I** and often abbreviated by **bcc**, see Figure 2.6, can be defined in Cartesian coordinates by lattice vectors

$$\underline{R}_1 = a/2\,(-1,1,1), \quad \underline{R}_2 = a/2\,(1,-1,1), \quad \underline{R}_3 = a/2\,(1,1,-1) \tag{2.26}$$

Here, the three vectors are of the same length

$$|\underline{R}_1| = |\underline{R}_2| = |\underline{R}_3| = (\sqrt{3}/2)a \tag{2.27}$$

but they are not orthogonal to each other, forming angles $\alpha = \beta = \gamma = 109.47°$ ($\cos \alpha = -1/3$) according to Eq. (2.3). **General lattice points** of the bcc lattice are given in Cartesian coordinates by vectors

$$\begin{aligned}\underline{R} &= n_1\underline{R}_1 + n_2\underline{R}_2 + n_3\underline{R}_3 \\ &= a/2\,(-n_1 + n_2 + n_3, n_1 - n_2 + n_3, n_1 + n_2 - n_3) \\ &= a/2\,(N_1, N_2, N_3), \quad n_1, n_2, n_3, N_1, N_2, N_3 \text{ integer}\end{aligned} \tag{2.28}$$

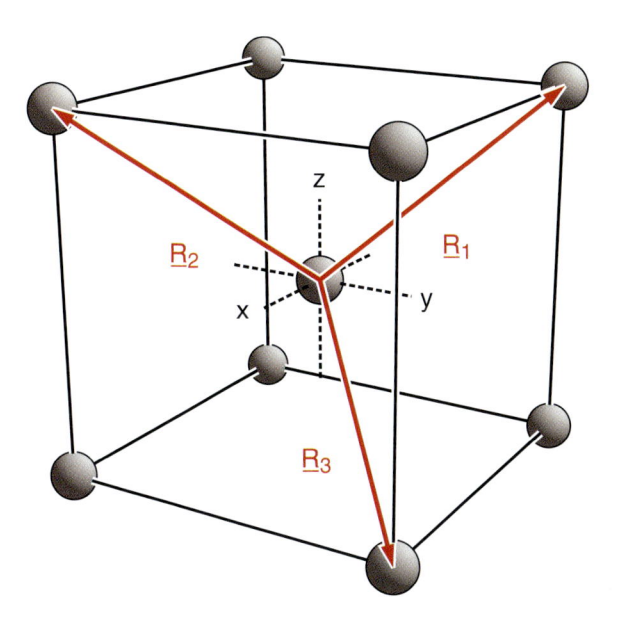

Figure 2.6 Lattice vectors $\underline{R}_1, \underline{R}_2, \underline{R}_3$ of the body-centered cubic (bcc) lattice sketched inside a cubic frame with Cartesian coordinates, x, y, and z, indicated. Atoms of the corresponding bcc crystal are shown as balls.

where the integers n_1, n_2, n_3 and N_1, N_2, N_3 are connected by

$$N_1 = -n_1 + n_2 + n_3, \quad N_2 = n_1 - n_2 + n_3, \quad N_3 = n_1 + n_2 - n_3 \quad (2.29)$$

Relation (2.28) together with the definition of the sc lattice vectors can be written as

$$\underline{R} = n_1 \underline{R}_1 + n_2 \underline{R}_2 + n_3 \underline{R}_3 = 1/2\, (N_1 \underline{R}_1^{sc} + N_2 \underline{R}_2^{sc} + N_3 \underline{R}_3^{sc}) \quad (2.30)$$

which demonstrates the connection between the bcc and the sc lattices. While the integer coefficients n_1, n_2, n_3 can be chosen freely the integer coefficients N_1, N_2, N_3 are not independent. Relations (2.29) yield

$$N_2 = N_1 + 2\,(n_1 - n_2), \quad N_3 = N_1 + 2\,(n_1 - n_3) \quad (2.31)$$

Hence, the integers N_1, N_2, N_3 can only be all odd or all even for any choice of n_1, n_2, n_3.

If N_1, N_2, N_3 in Eq. (2.28) are **all even,** that is, they can be represented by

$$N_i = 2m_i, \quad i = 1, 2, 3 \quad \text{for any integer } m_i \quad (2.32)$$

then relation (2.30) together with Eq. (2.32) leads to

$$\underline{R} = m_1 \underline{R}_1^{sc} + m_2 \underline{R}_2^{sc} + m_3 \underline{R}_3^{sc} \quad m_1, m_2, m_3 \text{ integer} \quad (2.33)$$

which describes an sc lattice as one subset of the bcc lattice.

If, on the other hand, N_1, N_2, N_3 in Eq. (2.28) are **all odd,** that is, they can be represented by

$$N_i = 2m_i + 1, \quad i = 1, 2, 3 \quad \text{for any integer } m_i \quad (2.34)$$

then relation (2.30) together with Eq. (2.34) leads to

$$\underline{R} = m_1 \underline{R}_1^{sc} + m_2 \underline{R}_2^{sc} + m_3 \underline{R}_3^{sc} + \underline{v} \quad m_1, m_2, m_3 \text{ integer} \tag{2.35}$$

with

$$\underline{v} = 1/2\,(\underline{R}_1^{sc} + \underline{R}_2^{sc} + \underline{R}_3^{sc}) \tag{2.36}$$

This also describes an sc lattice as the second subset of the bcc lattice, where the second sc lattice is, however, shifted by a vector $\underline{v}$ with respect to the first. Thus, the constraints for N_1, N_2, N_3 in Eq. (2.29) yield a decomposition of the bcc lattice into **two** identical sc lattices that are shifted with respect to each other by vector $\underline{v}$ given by Eq. (2.36). The two sc lattices are sketched in Figure 2.7 and denoted "sc_1", "sc_2" in the figure.

As a consequence, any crystal with a bcc lattice given by lattice vectors (2.26) can be alternatively described by a crystal with an sc lattice with lattice vectors (2.25), where the unit cell of the sc lattice contains twice as many atoms with atom pairs separated by vector $\underline{v}$. Further, the lattice vectors $\underline{R}_1^{sc}, \underline{R}_2^{sc}, \underline{R}_3^{sc}$ of the sc lattice representation are **non-primitive** since vector

$$\underline{v} = 1/2\,(\underline{R}_1^{sc} + \underline{R}_2^{sc} + \underline{R}_3^{sc}) = \underline{R}_1 + \underline{R}_2 + \underline{R}_3 \tag{2.37}$$

according to Eq. (2.26) is a true lattice vector.

The **face-centered cubic** lattice, also called **F-centered** or **cubic-F** and often abbreviated by **fcc**, see Figure 2.8, can be defined in Cartesian coordinates by lattice vectors

$$\underline{R}_1 = a/2\,(0, 1, 1), \quad \underline{R}_2 = a/2\,(1, 0, 1), \quad \underline{R}_3 = a/2\,(1, 1, 0) \tag{2.38}$$

As for the bcc lattice, the three vectors are of the same length

$$|\underline{R}_1| = |\underline{R}_2| = |\underline{R}_3| = a/\sqrt{2} \tag{2.39}$$

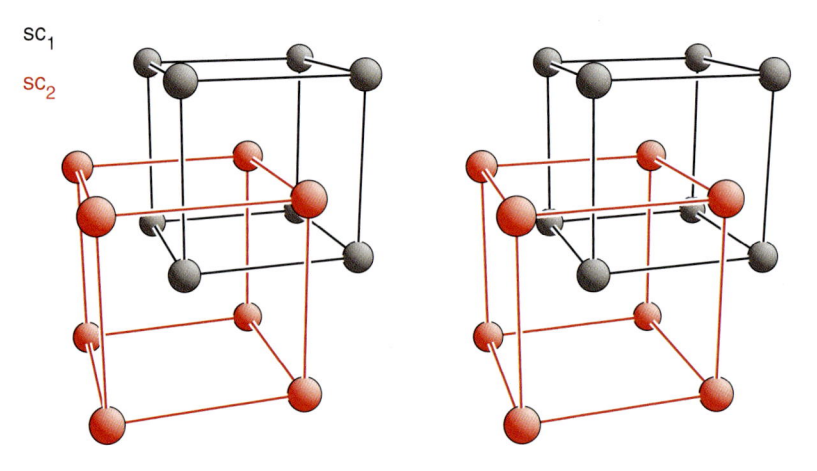

Figure 2.7 Visual decomposition of the bcc crystal into two (shifted) sc crystals, denoted sc_1 (gray balls, black lines) and sc_2 (red balls and lines). The crystal is shown in a stereo view where the visual three-dimensional impression is obtained by cross-eyed viewing.

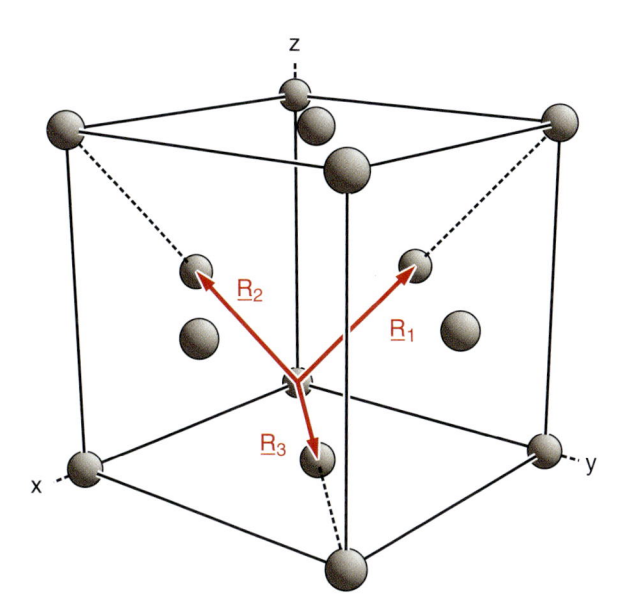

Figure 2.8 Lattice vectors $\underline{R}_1$, $\underline{R}_2$, $\underline{R}_3$ of the fcc lattice sketched inside a cubic frame and labeled accordingly. Atoms of the corresponding fcc crystal are shown as balls. The dashed lines are meant to assist the visual orientation inside the figure.

but are not orthogonal to each other, forming angles $\alpha = \beta = \gamma = 60°$ ($\cos \alpha = 1/2$) according to Eq. (2.3). **General lattice points** of the fcc lattice are given in Cartesian coordinates by vectors

$$\underline{R} = n_1 \underline{R}_1 + n_2 \underline{R}_2 + n_3 \underline{R}_3 = a/2 \left(n_2 + n_3, n_1 + n_3, n_1 + n_2\right)$$
$$= a/2 \left(N_1, N_2, N_3\right), \quad n_1, n_2, n_3, N_1, N_2, N_3 \text{ integer} \quad (2.40)$$

where the integers n_1, n_2, n_3 and N_1, N_2, N_3 are connected by

$$N_1 = n_2 + n_3, \quad N_2 = n_1 + n_3, \quad N_3 = n_1 + n_2 \quad (2.41)$$

Relation (2.40) together with the definition of the sc lattice vectors can be written as

$$\underline{R} = n_1 \underline{R}_1 + n_2 \underline{R}_2 + n_3 \underline{R}_3 = 1/2 \left(N_1 \underline{R}_1^{sc} + N_2 \underline{R}_2^{sc} + N_3 \underline{R}_3^{sc}\right) \quad (2.42)$$

which shows the connection between the fcc and the sc lattice. As in the bcc case, the integer coefficients N_1, N_2, N_3 are not independent. Even- and odd-valued combinations of the initial coefficients n_1, n_2, n_3 yield eight cases as shown in Table 2.1.

As a result, integers N_1, N_2, N_3 reduce to **four** different types of **even/odd combinations**,

1) $N_i = 2m_i$, $i = 1, 2, 3$, (cases 1, 2 in Table 2.1) which results, according to Eq. (2.42), in

$$\underline{R} = a/2 \left(N_1, N_2, N_3\right) = a \left(m_1, m_2, m_3\right), \quad m_1, m_2, m_3 \text{ integer} \quad (2.43a)$$

Table 2.1 List of all possible even/odd integer combinations N_1, N_2, N_3 following from even/odd integer combinations n_1, n_2, n_3 according to Eq. (2.41).

Case	n_1	n_2	n_3	N_1	N_2	N_3
1	e	e	e	e	e	e
2	o	o	o	e	e	e
3	o	e	e	e	o	o
4	e	o	o	e	o	o
5	e	o	e	o	e	o
6	o	e	o	o	e	o
7	e	e	o	o	o	e
8	o	o	e	o	o	e

Characters "e" and "o" stand for even and odd integers, respectively.

describing the sc lattice given by Eq. (2.25) with its origin coinciding with that of the fcc lattice, corresponding to an origin shift $\underline{v}_1 = \underline{0}$, see below.

2) $N_1 = 2m_1$, $N_2 = 2m_2 + 1$, $N_3 = 2m_3 + 1$, (cases 3, 4) resulting in

$$\underline{R} = a/2 \, (N_1, N_2, N_3) = a \, (m_1, m_2, m_3) + \underline{v}_2$$
$$\underline{v}_2 = 1/2 \, (\underline{R}_2^{sc} + \underline{R}_3^{sc}) \tag{2.43b}$$

describing the sc lattice with an origin shift $\underline{v}_2$.

3) $N_1 = 2m_1 + 1$, $N_2 = 2m_2$, $N_3 = 2m_3 + 1$, (cases 5, 6) resulting in

$$\underline{R} = a/2 \, (N_1, N_2, N_3) = a \, (m_1, m_2, m_3) + \underline{v}_3$$
$$\underline{v}_3 = 1/2 \, (\underline{R}_1^{sc} + \underline{R}_3^{sc}) \tag{2.43c}$$

describing the sc lattice with an origin shift $\underline{v}_3$.

4) $N_1 = 2m_1 + 1$, $N_2 = 2m_2 + 1$, $N_3 = 2m_3$, (cases 7, 8) resulting in

$$\underline{R} = a/2 \, (N_1, N_2, N_3) = a \, (m_1, m_2, m_3) + \underline{v}_4$$
$$\underline{v}_4 = 1/2 \, (\underline{R}_1^{sc} + \underline{R}_2^{sc}) \tag{2.43d}$$

describing the sc lattice with an origin shift $\underline{v}_4$.

Therefore, the constraints for N_1, N_2, N_3 in Eq. (2.41) yield a decomposition of the fcc lattice into **four** identical sc lattices that are shifted with respect to each other according to their origins at $\underline{v}_1, \underline{v}_2, \underline{v}_3, \underline{v}_4$, given by Eqs. (2.43a–2.43d). The four sc lattices are sketched in Figure 2.9 and denoted "sc_1" to "sc_4" in the figure.

As a consequence, any crystal with an fcc lattice given by lattice vectors Eq. (2.38) can be alternatively described by a crystal with an sc lattice with lattice vectors Eq. (2.25), where the unit cell of the sc lattice contains four times as many atoms with atom pairs separated by vectors $\underline{v}_i - \underline{v}_j$, $i, j = 1, \ldots, 4$. Further, the lattice vectors $\underline{R}_1^{sc}, \underline{R}_2^{sc}, \underline{R}_3^{sc}$ of the sc lattice representation are **non-primitive** since

sc₁

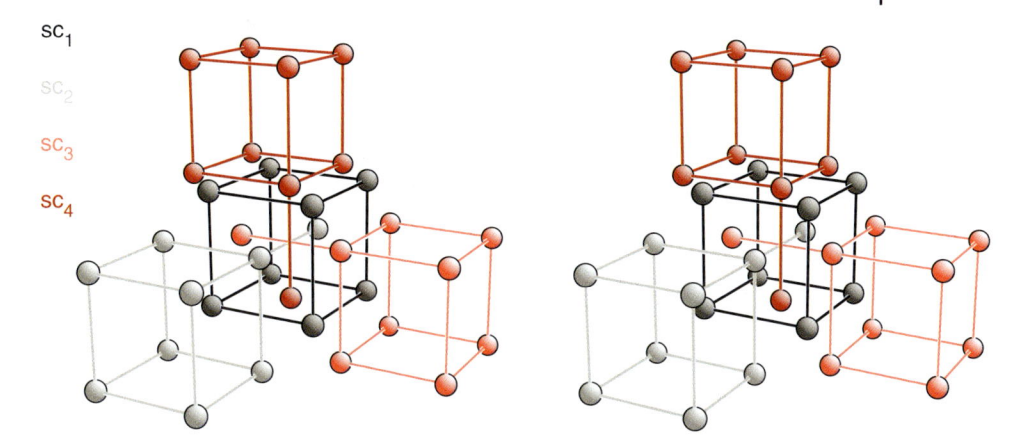

sc₂

sc₃

sc₄

Figure 2.9 Visual decomposition of the fcc crystal into four (shifted) sc crystals, denoted sc$_1$ (dark gray balls, black lines), sc$_2$ (dark red balls and lines), sc$_3$ (light gray balls and lines), and sc$_4$ (light red balls and lines). The crystal is shown in a stereo view where the visual three-dimensional impression is obtained by cross-eyed viewing.

the four vectors $\underline{v}_i$

$$\underline{v}_1 = \underline{0} \qquad \underline{v}_2 = 1/2 \left(\underline{R}_2^{sc} + \underline{R}_3^{sc} \right) = \underline{R}_1$$
$$\underline{v}_3 = 1/2 \left(\underline{R}_1^{sc} + \underline{R}_3^{sc} \right) = \underline{R}_2 \quad \underline{v}_4 = 1/2 \left(\underline{R}_1^{sc} + \underline{R}_2^{sc} \right) = \underline{R}_3 \qquad (2.44)$$

according to Eq. (2.38) are true lattice vectors.

The **hexagonal** lattice, also called **hexagonal-P** and often abbreviated by **hex**, is described by two lattice vectors $\underline{R}_1^{hex}$, $\underline{R}_2^{hex}$ of equal length a, forming an angle of either 120° (**obtuse representation**) or 60° (**acute representation**) between them. A third lattice vector $\underline{R}_3^{hex}$ of length c, is perpendicular to both $\underline{R}_1^{hex}$ and $\underline{R}_2^{hex}$. Thus, the vectors of the obtuse representation can be described in Cartesian coordinates by

$$\underline{R}_1^{hex} = a\,(1,0,0), \quad \underline{R}_2^{hex} = a\,(-1/2, \sqrt{3}/2, 0), \quad \underline{R}_3^{hex} = c\,(0,0,1) \qquad (2.45a)$$

and those of the acute representation by

$$\underline{R}_1^{hex} = a\,(1,0,0), \quad \underline{R}_2^{hex} = a\,(1/2, \sqrt{3}/2, 0), \quad \underline{R}_3^{hex} = c\,(0,0,1) \qquad (2.45b)$$

where a and c are the lattice constants of the hexagonal lattice. While the two representations are equivalent, the **obtuse** representation of crystal lattices is often **preferred** over the acute one and will be used in the following discussions.

There is a special type of crystal structure with hexagonal lattice, the so-called **hexagonal close-packed (hcp)** crystal structure, illustrated by Figure 2.10 and called **hex (hcp)** in the following it is defined by a hexagonal lattice, given in obtuse representation by Eq. (2.45a) with a lattice constant ratio c/a of $\sqrt{(8/3)} = 1.63299$.

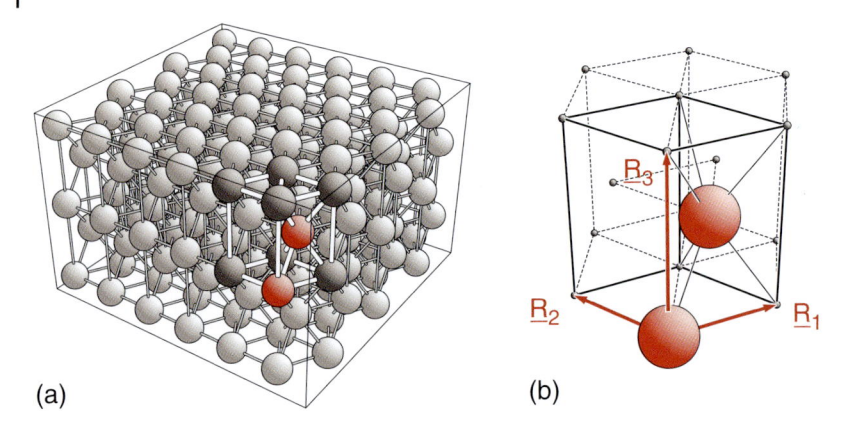

Figure 2.10 (a) Section of a hexagonal crystal with close-packed geometry (hcp). Sticks connect atoms with nearest and second nearest neighbors to indicate the crystal structure. (b) Primitive morphological unit cell with two atoms inside. The lattice vectors $\underline{R}_1, \underline{R}_2, \underline{R}_3$ (obtuse representation) are labeled accordingly. The unit cell is embedded in a hexagonal environment (dashed lines) to indicate its symmetry.

Further, the hexagonal unit cell contains two identical atoms at positions

$$\underline{r}_1^{hcp} = a\,(0,0,0), \quad \underline{r}_2^{hcp} = (a/2, a/\sqrt{12}, c/2) = a\,(1/2, 1/\sqrt{12}, \sqrt{(2/3)}) \quad (2.45c)$$

see Figure 2.10b. The c/a ratio and the atom positions are chosen such that each atom is surrounded by 12 nearest neighbor atoms at equal distance (equal to lattice constant a), achieving the same atom density as crystals with a corresponding fcc lattice. While hcp crystals in their rigorous mathematical definition do not exist in nature, they occur, to a good approximation, that is, with ratios c/a close to $\sqrt{(8/3)}$, for many single crystals of metals, such as beryllium, magnesium, titanium, cobalt, ruthenium, and cadmium, see Table B.3.

Analogous to the family of cubic lattices, there is also a close connection between trigonal and hexagonal lattices, where scientists often prefer hexagonal over trigonal lattice descriptions. The **trigonal** lattice, also called **trigonal-R** or **rhombohedral**, is described by three lattice vectors $\underline{R}_1, \underline{R}_2, \underline{R}_3$ of equal length a, which form identical angles $\alpha = \beta = \gamma$. Thus, the lattice vectors can be thought of as arising from each other by a 120° rotation about a common axis given by $(\underline{R}_1 + \underline{R}_2 + \underline{R}_3)$, see Figure 2.11a. Assuming the rotation axis as the z axis of a Cartesian coordinate system, the vectors can be described in Cartesian coordinates by

$$\underline{R}_1 = a\,(c_1, 0, c_2), \quad \underline{R}_2 = a\,(-1/2\,c_1, \sqrt{3}/2\,c_1, c_2),$$
$$\underline{R}_3 = a\,(-1/2\,c_1, -\sqrt{3}/2\,c_1, c_2), \quad c_1 = \cos(\varphi), \quad c_2 = \sin(\varphi) \quad (2.46)$$

where φ denotes the angle between each of the three lattice vectors and the xy plane, see Figure 2.11a, and is determined by

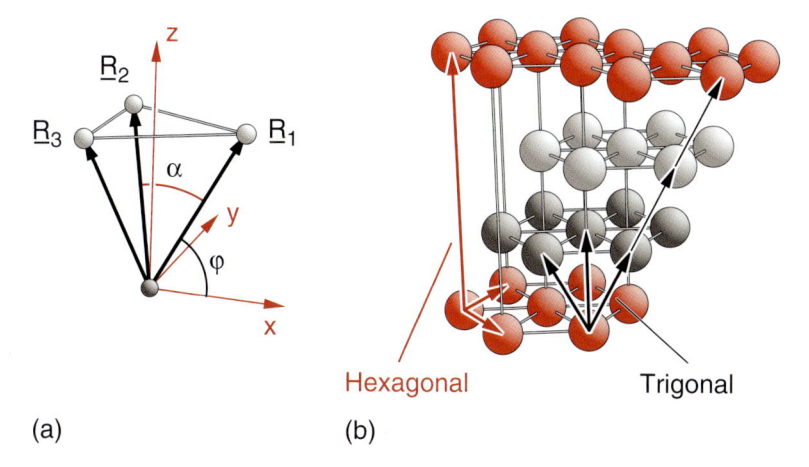

Figure 2.11 (a) Lattice vectors $\underline{R}_1$, $\underline{R}_2$, $\underline{R}_3$ of the trigonal (rhombohedral) lattice with definitions of the Cartesian coordinate system and of angles φ, α, see text. (b) Three trigonal lattices combining to form a non-primitive hexagonal lattice. Lattice vectors are shown by arrows, black for trigonal and red for hexagonal. The visual correlation between the two lattices is indicated by thin gray sticks connecting hexagonal lattice points.

$$\cos(\alpha) = \cos(\beta) = \cos(\gamma) = 1/4\,\{1 - 3\cos(2\varphi)\} \tag{2.47}$$

Thus, the three vectors $\underline{R}'_1$, $\underline{R}'_2$, $\underline{R}'_3$ with

$$\underline{R}'_1 = \underline{R}_1 - \underline{R}_2 = \sqrt{3}c_1 a\,(\sqrt{3}/2, -1/2, 0)$$

$$\underline{R}'_2 = \underline{R}_2 - \underline{R}_3 = \sqrt{3}c_1 a\,(0, 1, 0)$$

$$\underline{R}'_3 = \underline{R}_1 + \underline{R}_2 + \underline{R}_3 = 3c_2 a\,(0, 0, 1) \tag{2.48}$$

form a **hexagonal sublattice** (**obtuse** representation) of the trigonal lattice since

$$|\underline{R}'_1|^2 = |\underline{R}'_2|^2 = 3a^2\cos^2(\varphi), \quad |\underline{R}'_3| = 9a^2\sin^2(\varphi)$$

$$\angle(\underline{R}'_1, \underline{R}'_2) = 120°, \quad \angle(\underline{R}'_1, \underline{R}'_3) = \angle(\underline{R}'_2, \underline{R}'_3) = 90° \tag{2.49}$$

(Actually, lattice vectors (2.48) can be easily shown to coincide with definition (2.45a) of a hexagonal lattice by applying a rotation by 30° about the axis through $\underline{R}'_3$ and a scaling of the lattice constants where constants a and c in Eq. (2.45a) correspond to ($\sqrt{3}\,c_1\,a$) and ($3\,c_2\,a$) in Eq. (2.48).)

General lattice points of the hexagonal sublattice are given according to Eqs. (2.46) and (2.48) by vectors

$$\underline{R} = n_1 \underline{R}'_1 + n_2 \underline{R}'_2 + n_3 \underline{R}'_3$$

$$= (n_1 + n_3)\,\underline{R}_1 + (n_2 - n_1 + n_3)\,\underline{R}_2 + (n_3 - n_2)\,\underline{R}_3$$

$$= m_1 \underline{R}_1 + m_2 \underline{R}_2 + m_3 \underline{R}_3 \tag{2.50}$$

where the coefficients m_i and n_i are connected by linear transformations written in matrix form as

$$\begin{pmatrix} m_1 \\ m_2 \\ m_3 \end{pmatrix} = \begin{pmatrix} 1 & 0 & 1 \\ -1 & 1 & 1 \\ 0 & -1 & 1 \end{pmatrix} \cdot \begin{pmatrix} n_1 \\ n_2 \\ n_3 \end{pmatrix} \qquad (2.51a)$$

and

$$\begin{pmatrix} n_1 \\ n_2 \\ n_3 \end{pmatrix} = \frac{1}{3} \begin{pmatrix} 2 & -1 & -1 \\ 1 & 1 & -2 \\ 1 & 1 & 1 \end{pmatrix} \cdot \begin{pmatrix} m_1 \\ m_2 \\ m_3 \end{pmatrix} \qquad (2.51b)$$

According to Eq. (2.51b), the hexagonal sublattice is described by integer values n_1, n_2, n_3 only if the corresponding trigonal coefficients m_1, m_2, m_3 fulfill the **three** conditions

$$2m_1 - m_2 - m_3 = 3g, \quad m_1 + m_2 - 2m_3 = 3g', \quad m_1 + m_2 + m_3 = 3g'' \quad (2.52)$$

where g, g', g'' are integers. Since

$$m_1 + m_2 + m_3 = (m_1 + m_2 - 2m_3) + 3m_3 = -(2m_1 - m_2 - m_3) + 3m_1 \qquad (2.53)$$

fulfilling one of the three conditions (2.52) will automatically satisfy the other two. Considering the complete trigonal lattice, all sets of coefficients m_1, m_2, m_3 can be grouped according to one of the three categories,

$$m_1 + m_2 + m_3 = 3g \qquad (2.54a)$$

$$m_1 + m_2 + m_3 = 3g + 1 \quad \text{or} \quad (m_1 - 1) + m_2 + m_3 = 3g \qquad (2.54b)$$

$$m_1 + m_2 + m_3 = 3g + 2 \quad \text{or} \quad (m_1 - 2) + m_2 + m_3 = 3g \qquad (2.54c)$$

Here the condition (2.54a) was shown to result in a hexagonal lattice whose origin coincides with that of the trigonal lattice. The conditions of Eq. (2.54b) also lead to a hexagonal lattice. However, its origin is shifted with respect to that of the trigonal lattice by a trigonal lattice vector $\underline{R}_1$ (or $\underline{R}_2$ or $\underline{R}_3$). Analogously, the conditions of Eq. (2.54c) lead to an identical hexagonal lattice with its origin shifted by a trigonal lattice vector $2\underline{R}_1$ (or any combination of two trigonal lattice vectors). Since all lattice points of the trigonal lattice satisfy one of the three conditions (2.54) the **trigonal** lattice can be decomposed into **three** identical **hexagonal** lattices that are shifted with respect to each other as sketched by the thinner arrows in Figure 2.11b.

Altogether, any crystal with a trigonal lattice, given by lattice vectors (2.46), can be alternatively described by a crystal with a non-primitive hexagonal lattice, with lattice vectors (2.48), where the unit cell of the hexagonal lattice contains three times as many atoms compared with that of the trigonal lattice. Further, the lattice vectors $\underline{R}_1^{hex}, \underline{R}_2^{hex}, \underline{R}_3^{hex}$ of the hexagonal lattice representation are **non-primitive**.

2.2.2.2 Superlattices and Repeated Slabs

As mentioned earlier, theoretical studies on the physical or chemical parameters inside a crystal often require considering a unit cell with lattice vectors $\underline{R}_1$, $\underline{R}_2$, $\underline{R}_3$, which is larger than the primitive cell of the lattice given by $\underline{R}_{o1}$, $\underline{R}_{o2}$, $\underline{R}_{o3}$. Examples of this so-called **supercell** or **superlattice** concept include spin alignment in antiferromagnetic crystals [45], where the magnetic lattice, defined by positions of the different spins, differs from the geometric lattice of the crystal. In addition, local perturbations, such as vacancies, added atoms, or substituted atoms in alloy formation [46], of otherwise perfect crystals have been examined theoretically [47] applying supercell concepts. Here, single perturbations are simulated by those in an artificial crystal with large supercells such that distances between periodic copies of the perturbations are large enough to avoid physical coupling. Further, small distortions of lattice positions, which can result in periodicity with large supercells have been considered in so-called frozen phonon calculations [48]. Finally, we mention the use of supercell geometry in calculations of physical and chemical properties of single crystal surfaces. These calculations are often based on the so-called **repeated slab geometry** [48], where the surface region is approximated by a slab of finite thickness and a vacuum gap repeated periodically such that overall a three-dimensional periodicity with a large supercell is achieved.

The basic **mathematical** idea behind **conventional** supercell descriptions relies on the fact that any crystal with a lattice described by primitive lattice vectors $\underline{R}_{o1}$, $\underline{R}_{o2}$, $\underline{R}_{o3}$ and an atom basis can be represented by an alternative (non-primitive) lattice with (larger) lattice vectors $\underline{R}_1$, $\underline{R}_2$, $\underline{R}_3$ and an appropriately modified basis. The alternative vectors are connected with those of the initial lattice by a linear transformation that must be integer-valued if the global three-dimensional periodicity is to be conserved. This can be expressed mathematically by a **transformation matrix** $\underline{\underline{T}}$ with

$$\begin{pmatrix} \underline{R}_1 \\ \underline{R}_2 \\ \underline{R}_3 \end{pmatrix} = \begin{pmatrix} t_{11} & t_{12} & t_{13} \\ t_{21} & t_{22} & t_{23} \\ t_{31} & t_{32} & t_{33} \end{pmatrix} \cdot \begin{pmatrix} \underline{R}_{o1} \\ \underline{R}_{o2} \\ \underline{R}_{o3} \end{pmatrix} = \underline{\underline{T}} \cdot \begin{pmatrix} \underline{R}_{o1} \\ \underline{R}_{o2} \\ \underline{R}_{o3} \end{pmatrix} \qquad (2.55)$$

where the elements t_{ij} of matrix $\underline{\underline{T}}$ are integers. As a consequence, the volumes V_{el} and V_{el}^o of the unit cells of the two lattices, defined by Eq. (2.9), are connected by

$$V_{el} = |(\underline{R}_1 \times \underline{R}_2)\underline{R}_3| = |\det(\underline{\underline{T}})| \, |(\underline{R}_{o1} \times \underline{R}_{o2})\underline{R}_{o3}| = |\det(\underline{\underline{T}})| \, V_{el}^o \qquad (2.56)$$

where Eq. (2.55) together with vector relation (F.9) of Appendix F is applied. This means, in particular, that the volume V_{el} of the supercell must be an integer multiple of volume V_{el}^o of the initial unit cell.

In the simplest case, the superlattice description results from simple **scaling** of the initial lattice vectors, corresponding to a transformation

$$\underline{R}_1 = m_1 \underline{R}_{o1}, \quad \underline{R}_2 = m_2 \underline{R}_{o2}, \quad \underline{R}_3 = m_3 \underline{R}_{o3} \qquad (2.57)$$

with integer-valued m_1, m_2, m_3. Thus, the transformation matrix $\underline{\underline{T}}$ becomes diagonal, that is,

$$\underline{\underline{T}} = \begin{pmatrix} m_1 & 0 & 0 \\ 0 & m_2 & 0 \\ 0 & 0 & m_3 \end{pmatrix} \qquad (2.58)$$

As an **illustration,** we consider a fictitious sc crystal with ferromagnetic and antiferromagnetic ordering of its atoms, where the antiferromagnetism introduces a doubling of the lattice vectors in two dimensions, as sketched in Figure 2.12. Thus, the lattice vectors of the antiferromagnetic crystal, $\underline{R}_1, \underline{R}_2, \underline{R}_3$, can be connected with those of the ferromagnetic crystal, $\underline{R}_{o1}, \underline{R}_{o2}, \underline{R}_{o3}$, by

$$\underline{R}_1 = 2\,\underline{R}_{o1}, \quad \underline{R}_2 = 2\,\underline{R}_{o2}, \quad \underline{R}_3 = \underline{R}_{o3} \qquad (2.59)$$

Theoretical studies of the antiferromagnetic crystal must be based on a lattice description given by $\underline{R}_1, \underline{R}_2, \underline{R}_3$ while studies of the ferromagnetic crystal allow the use of the smaller lattice vectors $\underline{R}_{o1}, \underline{R}_{o2}, \underline{R}_{o3}$. However, a direct comparison of physical properties of the two crystals with different spin alignments can be simplified by using identical lattice parameters, which suggests applying the superlattice vectors $\underline{R}_1, \underline{R}_2, \underline{R}_3$ also for the ferromagnetic crystal.

Incidentally, Figure 2.12 shows that, for the present sc crystal with its antiferromagnetic spin alignment, alternative lattice vectors $\underline{R}'_1, \underline{R}'_2, \underline{R}'_3$ with

$$\underline{R}'_1 = 1/2\,(\underline{R}_1 + \underline{R}_2), \quad \underline{R}'_2 = 1/2\,(\underline{R}_2 - \underline{R}_1), \quad \underline{R}'_3 = \underline{R}_3 \qquad (2.60)$$

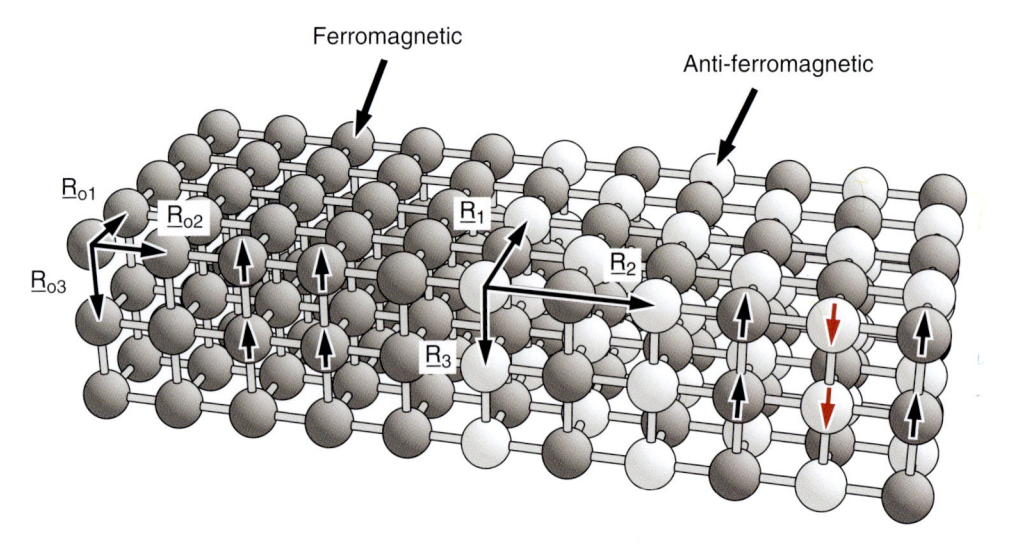

Figure 2.12 Fictitious sc crystal with ferromagnetic (left) and anti-ferromagnetic ordering (right). Atoms are shown as dark (spin up) and light (spin down) balls with their spin orientation indicated by black and red arrows. The superlattice vectors $\underline{R}_1, \underline{R}_2, \underline{R}_3$ and primitive lattice vectors $\underline{R}_{o1}, \underline{R}_{o2}, \underline{R}_{o3}$ are labeled accordingly.

could also be chosen, yielding a smaller morphological unit cell than that given by Eq. (2.59). This vector set can also be used to describe a superlattice of the ferromagnetic crystal.

As mentioned earlier, computational studies of the physical and chemical properties of single crystal surfaces are often based on the so-called **repeated slab geometry** [48], which can be considered a **modified** supercell concept. Within this concept a single crystal with a confining planar surface of given orientation and periodicity is described approximately by a two-dimensionally periodic solid layer of finite thickness (**slab**) cut out of the bulk crystal. Here, two bulk lattice vectors, $\underline{R}_1$ and $\underline{R}_2$, characterize the two-dimensional periodicity of the surface (and that of the slab). In addition, the slab is repeated periodically along the direction of its surface normal with a vacuum gap between adjacent slabs where the periodicity vector $\underline{R}_3$ is chosen appropriately. This procedure creates an altogether three-dimensionally periodic crystal system with a **fictitious** superlattice

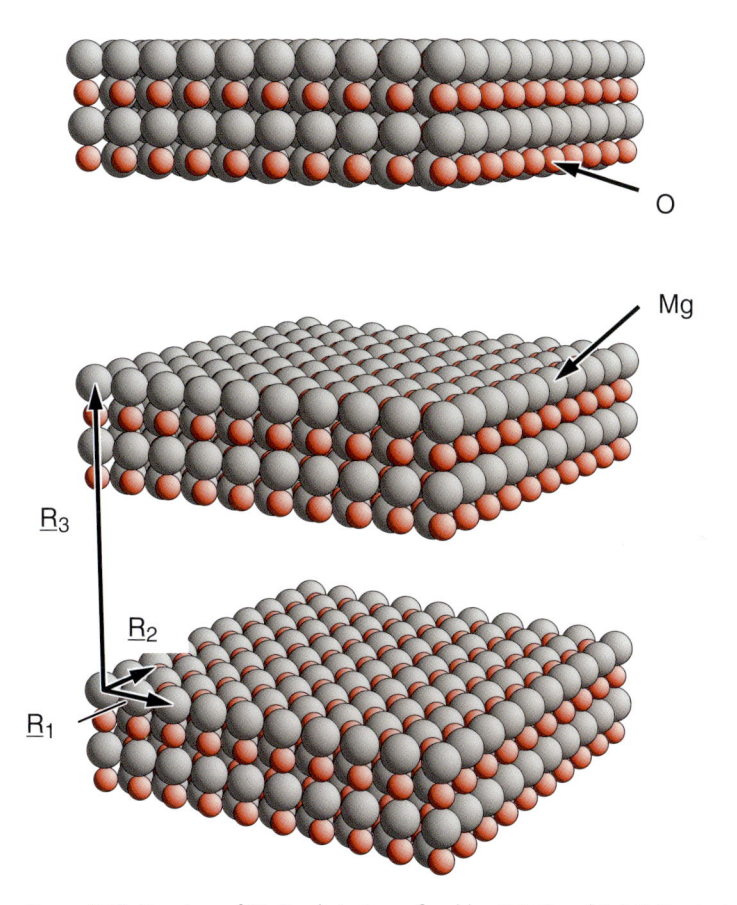

Figure 2.13 Structure of MgO substrate confined by (1 0 0) and (−1 0 0) oriented surfaces in repeated slab geometry (three slabs). The superlattice vectors $\underline{R}_1$, $\underline{R}_2$, $\underline{R}_3$ are labeled accordingly.

$\underline{R}_1$, $\underline{R}_2$, $\underline{R}_3$, which is connected with the initial crystal lattice only by vectors $\underline{R}_1$ and $\underline{R}_2$. As a result, matrix $\underline{T}$ of Eq. (2.55) contains integer-valued elements in its first and second rows while its third row may be real-valued. Within the **repeated slab** concept, the physical and chemical parameters of crystalline surfaces can be evaluated by well-established computational methods developed a long time ago for three-dimensionally periodic bulk crystals in solid state physics. For this approach to be meaningful the slabs must be sufficiently thick so that the surfaces of their upper and lower sides are electronically decoupled. Further, the vacuum distance between neighboring slabs must be sufficiently large such that they do not influence each other electronically.

As an illustration, Figure 2.13 shows three slabs of a magnesium oxide crystal confined by (1 0 0) and (−1 0 0) oriented surfaces at their tops and bottoms consisting of four MgO layers each with a vacuum separation of about three times the slab thickness. The appropriate superlattice vectors $\underline{R}_1$, $\underline{R}_2$, $\underline{R}_3$, referring to a 2×2 supercell laterally, that is, along $\underline{R}_1$ and $\underline{R}_2$, are sketched and labeled accordingly. The size of the supercell is much larger than that of the bulk crystal and the number of atoms in the cell, $4 \times 8 = 32$ in the present model structure, is rather large compared with that of the primitive bulk containing two atoms. Therefore, computational studies applying the repeated slab geometry are usually much more demanding than those of the corresponding bulk crystal.

As another illustration, Figure 2.14 shows a more complicated structure of three slabs of an fcc silver crystal confined by kinked surfaces (denoted (12 11 7) and (−12 −11 −7), respectively) at their top and bottom with a vacuum separation corresponding to four times the slab thickness. Again, the appropriate superlattice vectors $\underline{R}_1$, $\underline{R}_2$, $\underline{R}_3$ show that the size of the supercell with 25 atoms is much larger than the primitive cell of the bulk crystal with one atom only, demonstrating the difference in computational effort between slab and bulk calculations.

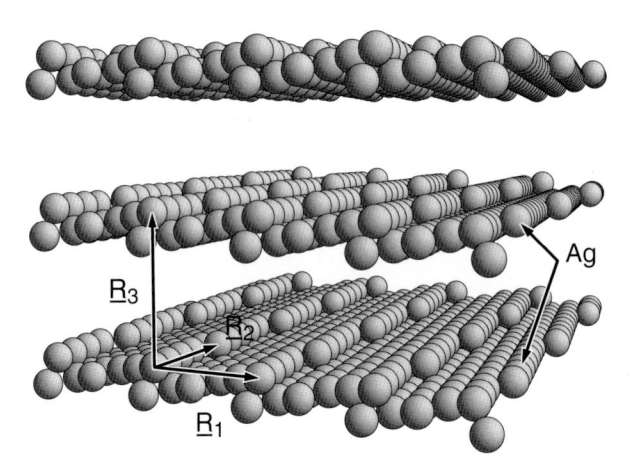

Figure 2.14 Structure of silver substrate confined by (12 11 7) and (−12 −11 −7) oriented surfaces in repeated slab geometry (three slabs). The superlattice vectors $\underline{R}_1$, $\underline{R}_2$, $\underline{R}_3$ are labeled accordingly.

2.2.2.3 Linear Transformations of Lattice Vectors

One group of alternative descriptions of crystal lattices is given by those where the alternative lattice vectors $\underline{R}_1$, $\underline{R}_2$, $\underline{R}_3$ are **linear combinations** of their initial counterparts $\underline{R}_{o1}$, $\underline{R}_{o2}$, $\underline{R}_{o3}$ with integer coefficients. This was already discussed in connection with the superlattice concept, and the basic linear transformation was defined by Eq. (2.55). Among these alternatives, there are lattice descriptions, whose morphological unit cells change their shape but not their volume, when compared with those of the initial lattice.

The latter alternatives can be used in practical cases to **adapt** the lattice description of a single crystal to additional **geometric constraints**, in particular those introduced by the existence of a single crystal surface. Therefore, these alternative descriptions are important for a crystallographic characterization of single crystal surfaces, as will become more evident in Chapters 4 and 5. In addition, they can be used to adapt lattice descriptions such that symmetry elements of the lattice become easily visible. As a simple example in two dimensions, Figure 2.15 shows two alternative descriptions of the square lattice by lattice vectors $\underline{R}_{o1}$, $\underline{R}_{o2}$ and $\underline{R}_1$, $\underline{R}_2$, respectively, where the two sets are connected by a linear transformation

$$\underline{R}_1 = -\underline{R}_{o1} - \underline{R}_{o2}, \quad \underline{R}_2 = 2\underline{R}_{o1} + \underline{R}_{o2} \tag{2.61}$$

Both vector sets, $\underline{R}_{o1}$, $\underline{R}_{o2}$ and $\underline{R}_1$, $\underline{R}_2$, provide mathematically exact descriptions of the square lattice and form morphological unit cells of the same volume. However, lattice vectors $\underline{R}_{o1}$, $\underline{R}_{o2}$ are of the same length and perpendicular to each other. Thus, their unit cell reveals additional symmetry properties of the lattice, such as mirror and rotational symmetry.

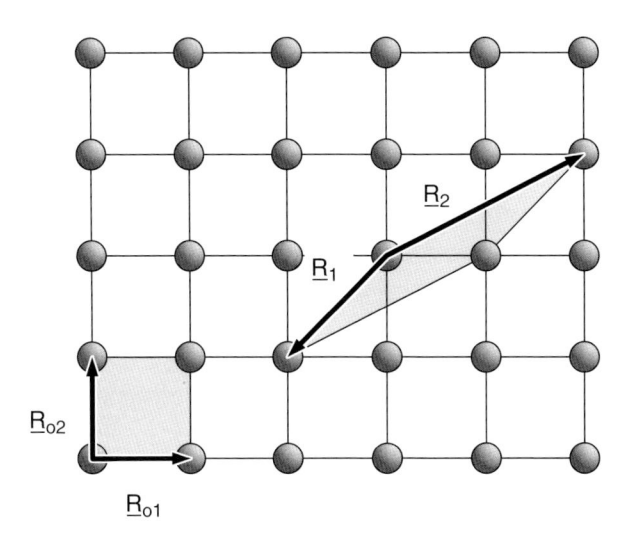

Figure 2.15 Alternative description of the square lattice by lattice vectors $\underline{R}_{o1}$, $\underline{R}_{o2}$ and $\underline{R}_1$, $\underline{R}_2$, respectively. The morphological unit cells of the two descriptions are emphasized by gray painting.

In the **general case,** we consider lattice vectors $\underline{R}_1, \underline{R}_2, \underline{R}_3$ of an alternative lattice description as a result of a linear transformation applied to an initial set of lattice vectors $\underline{R}_{o1}, \underline{R}_{o2}, \underline{R}_{o3}$, which according to Eq. (2.55) can be written in matrix form as

$$\begin{pmatrix} \underline{R}_1 \\ \underline{R}_2 \\ \underline{R}_3 \end{pmatrix} = \begin{pmatrix} t_{11} & t_{12} & t_{13} \\ t_{21} & t_{22} & t_{23} \\ t_{31} & t_{32} & t_{33} \end{pmatrix} \cdot \begin{pmatrix} \underline{R}_{o1} \\ \underline{R}_{o2} \\ \underline{R}_{o3} \end{pmatrix} = \underline{\underline{T}} \cdot \begin{pmatrix} \underline{R}_{o1} \\ \underline{R}_{o2} \\ \underline{R}_{o3} \end{pmatrix} \quad (2.62)$$

If the lattice vectors $\underline{R}_1, \underline{R}_2, \underline{R}_3$ are to describe the same set of lattice points as vectors $\underline{R}_{o1}, \underline{R}_{o2}, \underline{R}_{o3}$, then a general lattice point at $\underline{R}$ must be representable by an integer-valued linear combination of both sets of lattice vectors, that is,

$$\underline{R} = n_{o1}\underline{R}_{o1} + n_{o2}\underline{R}_{o2} + n_{o3}\underline{R}_{o3} = n_1\underline{R}_1 + n_2\underline{R}_2 + n_3\underline{R}_3, \quad n_{oi}, n_i \text{ integer} \quad (2.63)$$

Thus, any triplet of integers n_1, n_2, n_3 corresponds to another integer triplet n_{o1}, n_{o2}, n_{o3} and vice versa. This means, in particular, that the transformation matrix $\underline{\underline{T}} = (t_{ij})$ in Eq. (2.62) must be **integer-valued.** Further, transformation (2.62) can be inverted to yield

$$\begin{pmatrix} \underline{R}_{o1} \\ \underline{R}_{o2} \\ \underline{R}_{o3} \end{pmatrix} = \begin{pmatrix} t'_{11} & t'_{12} & t'_{13} \\ t'_{21} & t'_{22} & t'_{23} \\ t'_{31} & t'_{32} & t'_{33} \end{pmatrix} \cdot \begin{pmatrix} \underline{R}_1 \\ \underline{R}_2 \\ \underline{R}_3 \end{pmatrix} = \underline{\underline{T}}^{-1} \cdot \begin{pmatrix} \underline{R}_1 \\ \underline{R}_2 \\ \underline{R}_3 \end{pmatrix} \quad (2.64)$$

where the matrix elements t'_{ij} of the inverse matrix $\underline{\underline{T}}^{-1}$ also must be integers. Since all elements of $\underline{\underline{T}}$ are integers the determinant of matrix $\underline{\underline{T}}$, given by

$$\det(\underline{\underline{T}}) = t_{11}\left(t_{22}t_{33} - t_{23}t_{32}\right) + t_{12}\left(t_{23}t_{31} - t_{21}t_{33}\right) + t_{13}\left(t_{21}t_{32} - t_{22}t_{31}\right) \quad (2.65)$$

must be integer-valued. The same must be true for the inverse matrix $\underline{\underline{T}}^{-1}$. From linear algebra, we know that

$$\det(\underline{\underline{T}}^{-1}) = 1/\det(\underline{\underline{T}}) \quad (2.66)$$

Thus, both determinant values must be non-zero integers, that is, $|\det(\underline{\underline{T}})| \geq 1$ and $|\det(\underline{\underline{T}}^{-1})| \geq 1$, which according to Eq. (2.66) can only be possible if

$$\det(\underline{\underline{T}}) = \det(\underline{\underline{T}}^{-1}) = \pm 1 \quad (2.67)$$

Here the determinant value -1 can be safely ignored since it affects only the sequence in which the lattice vectors appear in the transformation (connected with handedness of the vector set). Any transformation Eq. (2.62) with $\det(\underline{\underline{T}}) = -1$ can be modified to yield $\det(\underline{\underline{T}}) = 1$ by exchanging one vector pair $\underline{R}_i, \underline{R}_j$ in the transformation.

Relation (2.67) imposes a **constraint** to possible transformation matrices $\underline{\underline{T}}$. Combining Eq. (2.67) with Eq. (2.65) one can write

$$\det(\underline{\underline{T}}) = a_1 t_{11} + a_2 t_{12} + a_3 t_{13} = 1 \quad (2.68)$$

with integer-valued coefficients a_i where

$$a_1 = t_{22}t_{33} - t_{23}t_{32}$$
$$a_2 = t_{23}t_{31} - t_{21}t_{33}$$
$$a_3 = t_{21}t_{32} - t_{22}t_{31} \tag{2.69}$$

Equation (2.68) represents a linear Diophantine equation containing only integers as parameters and variables. As shown in Appendix E.3, this equation has integer solutions a_1, a_2, a_3 for given t_{11}, t_{12}, t_{13} only if the latter three numbers have no common divisor greater than 1. Thus, the transformed lattice vector

$$\underline{R}_1 = t_{11}\underline{R}_{o1} + t_{12}\underline{R}_{o2} + t_{13}\underline{R}_{o3} \tag{2.70}$$

is of **smallest length** along its direction in the lattice. Rearranging the components in the determinant (2.65) we can easily derive analogous relations

$$\det(\underline{T}) = b_1 t_{21} + b_2 t_{22} + b_3 t_{23} = 1 \tag{2.71}$$

$$\det(\underline{T}) = c_1 t_{31} + c_2 t_{32} + c_3 t_{33} = 1 \tag{2.72}$$

with integer-valued coefficients b_i, c_i, where

$$b_1 = t_{32}t_{13} - t_{12}t_{33}, \quad c_1 = t_{12}t_{23} - t_{13}t_{22}$$
$$b_2 = t_{33}t_{11} - t_{13}t_{31}, \quad c_2 = t_{13}t_{21} - t_{11}t_{23}$$
$$b_3 = t_{31}t_{12} - t_{11}t_{32}, \quad c_3 = t_{11}t_{22} - t_{12}t_{21} \tag{2.73}$$

Then the corresponding linear Diophantine equations (2.71) and (2.72) have integer solutions b_1, b_2, b_3 for given t_{21}, t_{22}, t_{23} (and c_1, c_2, c_3 for given t_{31}, t_{32}, t_{33}) only if the latter three numbers have no common divisor greater than 1. Thus, the transformed lattice vectors

$$\underline{R}_2 = t_{21}\underline{R}_{o1} + t_{22}\underline{R}_{o2} + t_{23}\underline{R}_{o3} \tag{2.74}$$

$$\underline{R}_3 = t_{31}\underline{R}_{o1} + t_{32}\underline{R}_{o2} + t_{33}\underline{R}_{o3} \tag{2.75}$$

are also of **smallest length** along their direction in the lattice.

2.2.3
Centered Lattices

In Section 2.2.2.1 it was shown that the bcc lattice, characterizing, for example, iron single crystals, see Figure 2.16a, can be described by non-primitive lattice vectors $\underline{R}_1$, $\underline{R}_2$, $\underline{R}_3$ that form an sc lattice. However, there is an additional lattice vector $\underline{R}'$ inside the morphological unit cell, spanned by $\underline{R}_1$, $\underline{R}_2$, $\underline{R}_3$, which points to the center of the cubic unit cell, as illustrated in Figure 2.16b. This is an example of a more general property of non-primitive lattice representations, commonly denoted as **centering** and discussed in the following.

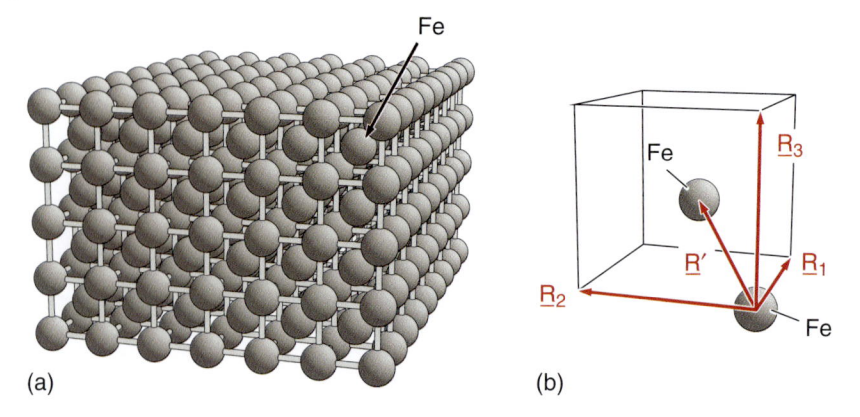

Figure 2.16 (a) Section of a bcc iron (Fe) crystal. Sticks between atom balls indicate the crystal structure. (b) Non-primitive sc morphological unit cell with two iron atoms inside. Lattice vectors $\underline{R}_1$, $\underline{R}_2$, $\underline{R}_3$, and $\underline{R}'$ are labeled accordingly.

First, we consider possible **lattice vectors** $\underline{R}'$ at the **faces** of the morphological unit cell of a lattice represented by **non-primitive** lattice vectors $\underline{R}_1$, $\underline{R}_2$, $\underline{R}_3$. For this we assume that each of the vectors $\underline{R}_1$, $\underline{R}_2$, $\underline{R}_3$ is of smallest length compared with all general lattice vectors along the same direction. Then an additional lattice point at a face of the morphological unit cell (excluding cell edges) can be described by a vector $\underline{R}'$ given by

$$\underline{R}' = \kappa_i\,\underline{R}_i + \kappa_j\,\underline{R}_j, \quad 0 < \kappa_i, \kappa_j < 1, \quad (i, j) = (1, 2), (1, 3), (2, 3) \tag{2.76}$$

and for general values κ_i, κ_j there is always a second lattice point at the face of the cell with vector $\underline{R}''$ given by

$$\underline{R}'' = \underline{R}_i + \underline{R}_j - \underline{R}' = (1 - \kappa_i)\underline{R}_i + (1 - \kappa_j)\underline{R}_j \tag{2.77}$$

If, however, the face is assumed to contain **only one** additional lattice point then the vectors $\underline{R}'$ and $\underline{R}''$ must coincide, that is,

$$\underline{R}'' - \underline{R}' = (1 - 2\kappa_i)\underline{R}_i + (1 - 2\kappa_j)\underline{R}_j = \underline{0} \tag{2.78}$$

Here the expressions in brackets must both be zero since the vectors $\underline{R}_i$, $\underline{R}_j$ are linearly independent. This leads to

$$\kappa_i = \kappa_j = 1/2, \quad \underline{R}' = 1/2\,(\underline{R}_i + \underline{R}_j) \tag{2.79}$$

yielding a lattice vector $\underline{R}'$ in the center of the cell face spanned by $\underline{R}_i$, $\underline{R}_j$ (**face centering**). Relation (2.79) written as

$$\underline{R}_i = 2\underline{R}' - \underline{R}_j, \quad \underline{R}_j = 2\underline{R}' - \underline{R}_i \tag{2.80}$$

means in particular that $(\underline{R}', \underline{R}_j)$ and $(\underline{R}_i, \underline{R}')$ can be used as alternative lattice vector sets to represent the set $(\underline{R}_i, \underline{R}_j)$.

Next, we consider possible **lattice vectors** $\underline{R}'$ **inside** the morphological unit cell defined by lattice vectors $\underline{R}_1$, $\underline{R}_2$, $\underline{R}_3$. An additional lattice point inside the cell (excluding cell faces and edges) can be described by a vector $\underline{R}'$ given by

$$\underline{R}' = \kappa_1 \underline{R}_1 + \kappa_2 \underline{R}_2 + \kappa_3 \underline{R}_3, \quad 0 < \kappa_i < 1, \quad i = 1, 2, 3 \quad (2.81)$$

and for general values $\kappa_1, \kappa_2, \kappa_3$ there is always a second lattice point at the face of the cell with vector $\underline{R}''$ given by

$$\underline{R}'' = \underline{R}_1 + \underline{R}_2 + \underline{R}_3 - \underline{R}' = (1 - \kappa_1)\underline{R}_1 + (1 - \kappa_2)\underline{R}_2 + (1 - \kappa_3)\underline{R}_3 \quad (2.82)$$

If, however, the cell is assumed to contain **only one** additional lattice point then the vectors $\underline{R}'$ and $\underline{R}''$ must coincide, that is,

$$\underline{R}'' - \underline{R}' = (1 - 2\kappa_1)\underline{R}_1 + (1 - 2\kappa_2)\underline{R}_2 + (1 - 2\kappa_3)\underline{R}_3 = \underline{0} \quad (2.83)$$

Again the expressions in brackets must all be zero since the vectors $\underline{R}_1, \underline{R}_2, \underline{R}_3$ are linearly independent. This leads to

$$\kappa_1 = \kappa_2 = \kappa_3 = 1/2, \quad \underline{R}' = 1/2(\underline{R}_1 + \underline{R}_2 + \underline{R}_3) \quad (2.84)$$

yielding a lattice vector $\underline{R}'$ in the center of the cell spanned by $\underline{R}_1, \underline{R}_2, \underline{R}_3$ (**body centering**). Relation (2.84) written as

$$\underline{R}_i = 2\underline{R}' - \underline{R}_j - \underline{R}_k, \quad (i, j, k) = (1, 2, 3), (2, 3, 1), (3, 1, 2) \quad (2.85)$$

means in particular that $(\underline{R}', \underline{R}_2, \underline{R}_3)$, $(\underline{R}_1, \underline{R}', \underline{R}_3)$, and $(\underline{R}_1, \underline{R}_2, \underline{R}')$ can be used as alternative lattice vector sets to represent the set $(\underline{R}_1, \underline{R}_2, \underline{R}_3)$.

Altogether, additional lattice vectors $\underline{R}' = \kappa_1 \underline{R}_1 + \kappa_2 \underline{R}_2 + \kappa_3 \underline{R}_3$ inside the morphological unit cell of a lattice, represented by non-primitive lattice vectors $\underline{R}_1, \underline{R}_2, \underline{R}_3$, allow **four** choices of centering

(a) body centering $\kappa_1 = \kappa_2 = \kappa_3 = 1/2$, $\underline{R}_a = 1/2\,(\underline{R}_1 + \underline{R}_2 + \underline{R}_3)$
(b) face centering $\kappa_2 = \kappa_3 = 1/2, \kappa_1 = 0$, $\underline{R}_b = 1/2\,(\underline{R}_2 + \underline{R}_3)$
(c) face centering $\kappa_1 = \kappa_3 = 1/2, \kappa_2 = 0$, $\underline{R}_c = 1/2\,(\underline{R}_1 + \underline{R}_3)$
(d) face centering $\kappa_1 = \kappa_2 = 1/2, \kappa_3 = 0$, $\underline{R}_d = 1/2\,(\underline{R}_1 + \underline{R}_2)$
$$(2.86)$$

As an illustration, Figure 2.17 shows the lattice vectors $\underline{R}_a, \underline{R}_b, \underline{R}_c, \underline{R}_d$, referring to the four choices.

If several of the vector choices (a–d) for $\underline{R}'$ in Eq. (2.86) appear simultaneously in the unit cell there are additional **compatibility restrictions**. First, a lattice point (a) in the center of the unit cell excludes the appearance of any of the additional lattice points (b–d), and vice versa, since this would also result in additional lattice points $\underline{R}' = (1/2)\underline{R}_i$ at the edges of the cell, which contradicts the assumption of vectors $\underline{R}_i$ being of smallest length along their direction. Thus, lattices with non-primitive lattice vectors and additional lattice points (a) form a separate group of lattices, called **I-centered lattices** ("I" = "Inner," preferred by crystallographers) or **body-centered lattices**. Examples of this lattice type are the bcc lattices describing many metal crystal structures, such as Cr, V, Mo, W, and Fe, the latter shown in Figure 2.16.

Second, lattices with only additional lattice points (b) form their own group, called **A-centered lattices** ("A" refers to the first lattice vector $\underline{R}_1$, determining

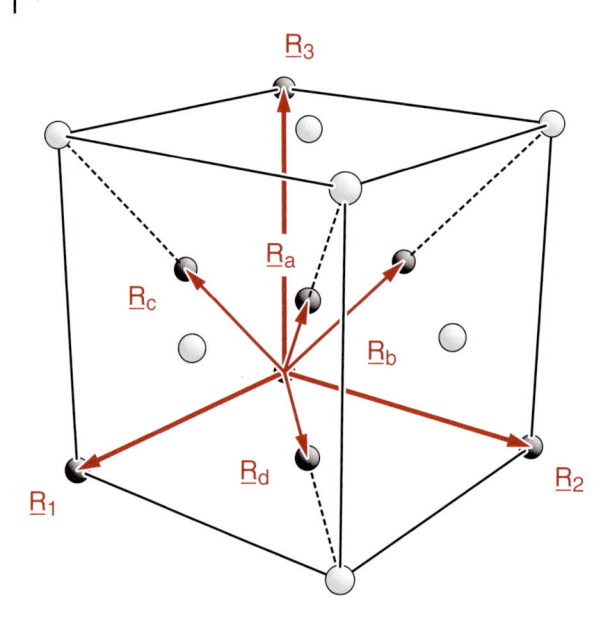

Figure 2.17 Non-primitive morphological unit cell with choices (a–d) for possible lattice vectors inside the cell. The possible lattice vectors $\underline{R}_a$, $\underline{R}_b$, $\underline{R}_c$, $\underline{R}_d$, as well as the lattice vectors $\underline{R}_1$, $\underline{R}_2$, $\underline{R}_3$ defining the morphological unit cell are labeled accordingly. The dashed lines are meant to assist the visual orientation inside the figure.

the stacking of the additional lattice points). Those with only additional lattice points (c) form a group, called **B-centered lattices** ("B" refers to the second lattice vector $\underline{R}_2$), and those with only additional lattice points (d) form a group, called **C-centered lattices** ("C" refers to the third lattice vector $\underline{R}_3$). The appearance of two additional lattice points of choices (b–d) in Eq. (2.86) leads immediately to the appearance of a third lattice point as can be shown quite easily. As an example, the existence of lattice vectors $\underline{R}_b$ and $\underline{R}_c$ implies a lattice vector $\underline{R} = \underline{R}_b + \underline{R}_c - \underline{R}_3$, which equals $\underline{R}_d$. Thus, the existence of more than one additional lattice point of choices (b–d) in Eq. (2.86) is only possible if all three types of lattice points exist at the same time. This group of lattices is called **F-centered lattices** ("F" = "Face," preferred by crystallographers) or **face-centered lattices**. Examples of this lattice type are the fcc lattices describing many metal crystal structures, such as Ni, Cu, Pt, and Ag.

Altogether, additional lattice points inside the morphological unit cell of a lattice with non-primitive lattice vectors resulting in a centered unit cell, allow **five choices** as shown in Figure 2.18. An additional lattice point in the cell center defines **I-centered** lattices. Further, additional points in the centers of the cell faces parallel to $\underline{R}_2$ and $\underline{R}_3$ (with no other additional lattice points) define **A-centered** lattices. Analogously, centering of cell faces parallel to $\underline{R}_1$ and $\underline{R}_3$ defines **B-centered** lattices, and centering of cell faces parallel to $\underline{R}_1$ and $\underline{R}_2$ defines **C-centered** lattices. Finally, **F-centered** lattices have additional lattice points at centers of all six faces of the unit cell.

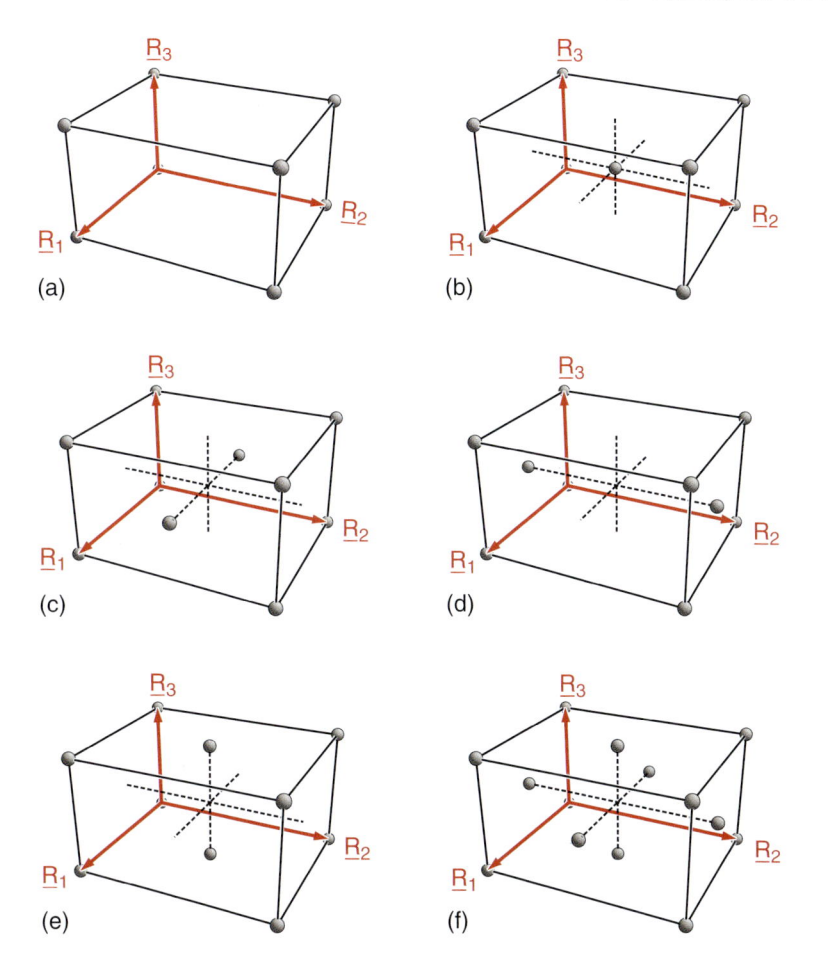

Figure 2.18 Different centering of morphological unit cells. (a) Primitive, (b) I-centered, (c) A-centered, (d) B-centered, (e) C-centered, and (f) F-centered cell. The lattice vectors $\underline{R}_1$, $\underline{R}_2$, $\underline{R}_3$ defining the morphological unit cell are labeled accordingly. The dashed lines are meant to assist the visual orientation inside the cells.

2.3
Periodicity Cells of Lattices

In Section 2.1, the **morphological unit cell** of a lattice with lattice vectors $\underline{R}_1$, $\underline{R}_2$, $\underline{R}_3$ was defined as a six-faced polyhedron spanned by $\underline{R}_1$, $\underline{R}_2$, $\underline{R}_3$ with its edges parallel to the lattice vectors and a volume V_{el} given by Eq. (2.9). If the lattice vectors are of smallest length in the lattice the corresponding polyhedral unit cells are also called **Buerger cells** [49]. If, in addition, the lattice vectors result from a reduction according to Niggli [41], providing a unique description in a crystallographical sense, see Section 2.2.2, then the unit cells are referred to as **Niggli cells**. As examples, the three **cubic** lattices discussed in Section 2.2.2.1 with lattice vectors defined by Eqs. (2.25), (2.26), and (2.38) yield cell volumes according to Eq. (2.9)

$$V_{el} = a^3 \quad \text{for sc lattices}$$
$$V_{el} = 1/2\, a^3 \quad \text{for bcc lattices}$$
$$V_{el} = 1/4\, a^3 \quad \text{for fcc lattices} \tag{2.87}$$

In general, the morphological unit cell contains all p atoms that form the basis of a crystal structure. A continued repetition of the cell in the three directions along $\underline{R}_1, \underline{R}_2, \underline{R}_3$ fills the complete three-dimensional space and describes the complete infinite crystal.

Assuming primitive lattice vectors $\underline{R}_1, \underline{R}_2, \underline{R}_3$, the **volume** of the morphological unit cell, given by Eq. (2.9) and connected with the atom density of the crystal, is unique whereas the cell **shape** is not. The shape is only determined by the requirement that a continued repetition of the cell in the three directions along $\underline{R}_1, \underline{R}_2, \underline{R}_3$ fills the complete three-dimensional space without holes and overlaps. This can be achieved by very differently shaped **alternative unit cells.** As an illustration in two dimensions, Figure 2.19 shows a section of the square lattice, where, apart from the square-shaped morphological unit cell, spanned by lattice vectors $\underline{R}_1$ and $\underline{R}_2$ (bottom left), two alternative (polygonal) unit cells are included.

The arbitrariness in the shape of alternative unit cells can be removed by additional constraints such as symmetry requirements, where the unit cell is assumed to reflect all point symmetry elements of the lattice. An additional constraint is compactness where all points inside the unit cell are assumed to be as close as possible to a lattice point. This leads to the definition of **Voronoi** or **Wigner–Seitz cells** (WSCs) that are commonly known in solid state physics [1, 2].

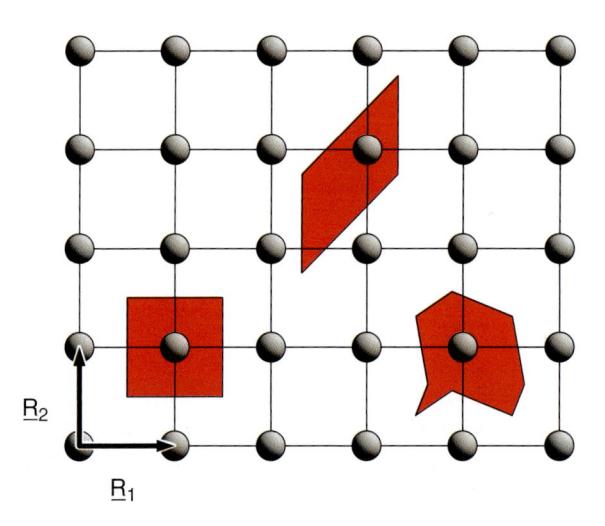

Figure 2.19 Alternative unit cells of the square lattice. The different unit cells are emphasized by red filling. The lattice vectors $\underline{R}_1$ and $\underline{R}_2$ are sketched at lower left corner.

2.3 Periodicity Cells of Lattices

The **formal definition** of WSCs considers for each point at position $\underline{r}$ its distances $d = |\underline{r} - \underline{R}|$ with respect to any lattice point $\underline{R} = n_1 \underline{R}_1 + n_2 \underline{R}_2 + n_3 \underline{R}_3$. Then position $\underline{r}$ can be assigned to a lattice point $\underline{R}$ by requiring that its distance d with respect to this lattice point is the smallest of all possible distances. The collection of all points $\underline{r}$ assigned to a given lattice point defines its WSC. There will always be points $\underline{r}$ whose distances with respect to two (or more, up to four [50]) lattice points are identical. These points define the **boundaries** separating WSCs of adjacent lattice points. More precisely, if $\underline{R}_a$ and $\underline{R}_b$ denote two adjacent lattice points, then points $\underline{r}$ of equal distance with respect to $\underline{R}_a$ and $\underline{R}_b$ satisfy relation

$$|\underline{r} - \underline{R}_a|^2 = \underline{r}^2 - 2(\underline{r}\,\underline{R}_a) + \underline{R}_a^2 = |\underline{r} - \underline{R}_b|^2 = \underline{r}^2 - 2(\underline{r}\,\underline{R}_b) + \underline{R}_b^2 \tag{2.88}$$

leading to

$$(\underline{R}_a - \underline{R}_b)\,\underline{r} = 1/2\,(\underline{R}_a^2 - \underline{R}_b^2) = 1/2\,(\underline{R}_a - \underline{R}_b)(\underline{R}_a + \underline{R}_b)$$
$$(\underline{R}_a - \underline{R}_b)\left[\underline{r} - 1/2\,(\underline{R}_a + \underline{R}_b)\right] = 0 \tag{2.89}$$

Equation (2.89) defines all points $\underline{r}$ on a plane perpendicular to $(\underline{R}_a - \underline{R}_b)$ and bisecting the line connecting $\underline{R}_a$ with $\underline{R}_b$. Thus, boundaries separating adjacent WSCs must be sections of planes and WSCs must be **polyhedral** in shape. As an illustration, Figures 2.20 and 2.21 compare morphological unit cells (Niggli cells) with WSCs for the bcc- and fcc lattices. Both polygonal WSCs are highly symmetric, which reflects the large number of point symmetry elements of the cubic lattice. This applies also to the WSC of the sc lattice which forms a cube and is identical in shape with the morphological unit cell. A complete set of WSCs for all 14 Bravais lattices can be found in [32, 50].

It is interesting to note that there is a continuous transition, the so-called **Bain path** [51], from bcc to fcc lattices, where the intermediate lattice type is centered tetragonal (ct) characterized by lattice vectors

$$\underline{R}_1 = (a, 0, 0),\ \underline{R}_2 = (0, a, 0),\ \underline{R}_3 = 1/2\,(a, a, c) \tag{2.90}$$

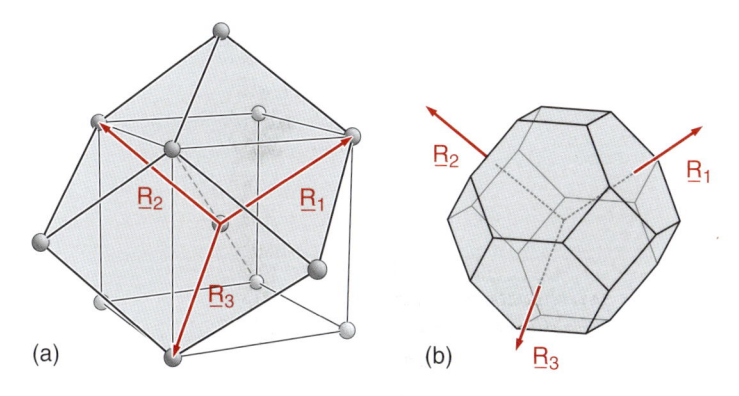

Figure 2.20 (a) Morphological unit cell (Niggli cell) and (b) Wigner–Seitz cell of the bcc lattice. The polygonal cells are shaded in gray with lattice vectors labeled accordingly. The dashed lines are meant to assist the visual orientation inside the figures.

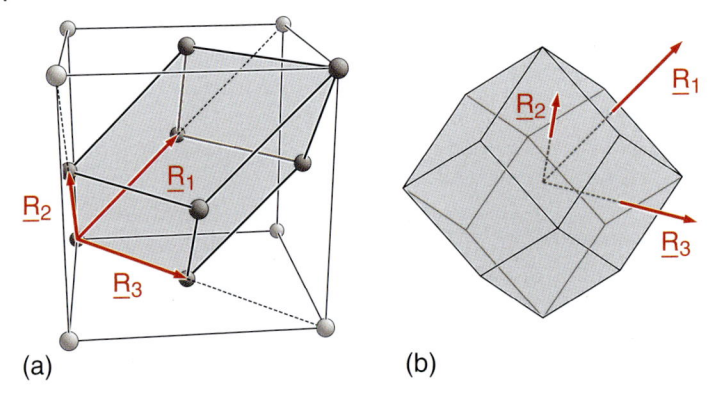

Figure 2.21 (a) Morphological unit cell (Niggli cell) and (b) Wigner–Seitz cell of the fcc lattice. The polygonal cells are shaded in gray with lattice vectors labeled accordingly. The dashed lines are meant to assist the visual orientation inside the figures.

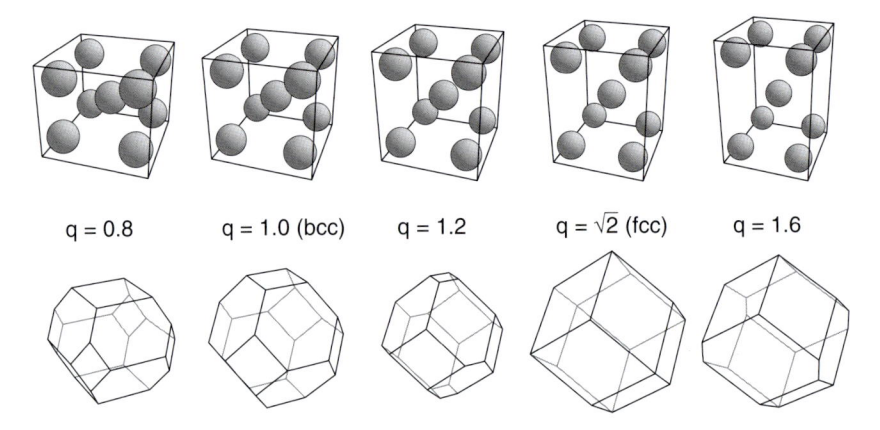

$q = 0.8$ $q = 1.0$ (bcc) $q = 1.2$ $q = \sqrt{2}$ (fcc) $q = 1.6$

Figure 2.22 Shape of Wigner–Seitz cells of the centered tetragonal (ct) lattice for different ratios $q = c/a$. Here $q = 1$ and $q = \sqrt{2}$ represent the bcc and fcc lattice, respectively. The top figures refer to rectangular blocks of a fictitious monoatomic ct crystal illustrating the lattice structure for corresponding ratios q.

The ratio $q = c/a$ of the two lattice constants determines the actual lattice type, where $q = 1$ reflects the bcc and $q = \sqrt{2}$ yields the fcc lattice. The WSCs must also transform continuously along the Bain path. This is illustrated in Figure 2.22, where WSCs of the ct lattice are shown for different ratios q between 0.8 and 1.6.

2.4
Lattice Symmetry

A wide area of crystallography concerns the **classification** of all possible types of crystal structure based on their symmetry behavior. This subject will be discussed

2.4 Lattice Symmetry

extensively for two-dimensional lattices (netplanes) in Sections 3.7 and 3.8, due to its importance for the characterization of single crystal surfaces. The present section deals with the symmetry of crystal structures in three dimensions. However, only **basic results** will be briefly discussed and the reader is referred to the literature [28–33] for more detailed information.

Based on its initial definition, every lattice, given by lattice vectors $\underline{R}_1, \underline{R}_2, \underline{R}_3$, has **translational symmetry** along any direction described by **general lattice vectors**

$$\underline{R} = n_1 \underline{R}_1 + n_2 \underline{R}_2 + n_3 \underline{R}_3, \quad n_i \text{ integer} \tag{2.91}$$

Thus, a lattice viewed from two points $\underline{r}$ and $\underline{r}'$, which are separated by $\underline{R}$, looks identical. This can be expressed mathematically using a **translation operation** $t(\underline{R})$ which acts on vector $\underline{r}$ to yield a shifted vector $\underline{r}'$ with

$$\underline{r}' = t(\underline{R})\underline{r} = \underline{r} + \underline{R} \tag{2.92}$$

Then translational symmetry of the lattice means that the lattice does not change geometrically when a translation operation (2.92) is applied.

In addition, lattices may exhibit **point symmetry** with respect to given points $\underline{r}_o$ of the lattice space (symmetry origins), which do not need to coincide with general lattice points defined by vectors (2.91). A lattice is considered symmetric with respect to a point symmetry operation P if it does not change geometrically when the operation is applied. A **point symmetry operation** can be written formally as

$$\underline{r} \to \underline{r}' = P\underline{r} \tag{2.93}$$

where in three dimensions five different types of operations P are generally available. These are

- $i(\underline{r}_o)$: **inversion** with respect to symmetry origin $\underline{r}_o$,
- $C_\varphi(\underline{r}_o, \underline{e})$: **rotation** by an angle φ about an axis along vector $\underline{e}$ through symmetry origin $\underline{r}_o$,
- $\sigma(\underline{r}_o, \underline{e})$: **mirroring (reflection)** with respect to a plane with normal vector $\underline{e}$ through symmetry origin $\underline{r}_o$,
- $S_\varphi(\underline{r}_o, \underline{e})$: **rotoinversion** by an angle φ about an axis along $\underline{e}$ through $\underline{r}_o$. This combines a rotation $C_\varphi(\underline{r}_o, \underline{e})$ with an inversion $i(\underline{r}_o)$,
- $S'_\varphi(\underline{r}_o, \underline{e})$: **rotoreflection** by an angle φ about an axis along $\underline{e}$ through $\underline{r}_o$. This combines a rotation $C_\varphi(\underline{r}_o, \underline{e})$ with a mirror operation $\sigma(\underline{r}_o, \underline{e})$, where direction vector $\underline{e}$ of the rotation axis coincides with the normal vector of the mirror plane.

The latter two symmetry operations combine two of the other operations, a rotation with inversion or with reflection. This means, in particular, that a lattice may be symmetric with respect to one of the two combined symmetry operations but may not exhibit the symmetry of the two component operations. Rotoinversion and rotoreflection are connected with each other by a rotation by 180°, which can be formally written as

$$S'_\varphi(\underline{r}_o, \underline{e}) = S_\varphi(\underline{r}_o, \underline{e})C_{180}(\underline{r}_o, \underline{e}) = C_{180}(\underline{r}_o, \underline{e})S_\varphi(\underline{r}_o, \underline{e}) \tag{2.94}$$

Thus, the two symmetry operations can be used equivalently in a symmetry classification of three-dimensional crystal lattices. Here, the **Hermann–Mauguin** or **international notation**, which forms the basis of the International Tables of Crystallography [33] and is by now the standard among crystallographers, considers rotoinversion as the standard symmetry operation while the **Schönflies notation**, practiced by many physicists, uses rotoreflection.

There are two additional symmetry operations that can appear in three-dimensional crystal lattices, namely

- $T_\varphi(\underline{r}_o, \underline{e}, t)$: **rototranslation (screw operation)** by an angle φ about an axis along $\underline{e}$ through $\underline{r}_o$ and subsequent translation by vector $t\underline{e}$.
- $g(\underline{r}_o, \underline{e}, \underline{g})$: **glide reflection**, combining a reflection $\sigma(\underline{r}_o, \underline{e})$ with a translation by vector $\underline{g}$, where vectors $\underline{g}$ and $\underline{e}$ are perpendicular to each other.

Both operations are not true point symmetry operation since they contain a translational component. However, rototranslations are required to describe the symmetry of crystals that contain screw axes. Glide reflections appear as symmetry elements in centered three-dimensional crystal lattices.

Translational and point symmetry elements of a lattice are subject to **compatibility constraints**. These constraints limit the number of possible point symmetry operations available for a characterization of different types of lattices. Examples are constraints on possible angles φ of rotation operations and of the direction $\underline{e}$ of their axes, of mirror plane orientations, or positions of inversion centers. These constraints for three-dimensional lattices will not be detailed in this book. However, they will be discussed extensively for the case of two-dimensional lattices (netplanes) in Section 3.6. Using these constraints together with group theoretical methods provides the basis of a **general classification scheme** of all crystal lattices, which is documented in the International Tables of Crystallography [33]. As an illustration of the compatibility constraints, possible rotation axes inside lattices will be considered. This can already serve for a rough classification of all possible lattice types into 7 **crystal systems** and 14 different types of lattices, the so-called **three-dimensional Bravais lattices**.

In Section 3.6.3 it will be shown that the compatibility of rotational and translational symmetry in two-dimensional lattices restricts rotation angles φ to integer multiples of $(360°/n)$, where only values n = 2, 3, 4, and 6 are allowed. The corresponding mathematical proof is also valid for three-dimensional lattices. If a lattice transforms into itself by a rotation by $\varphi = (360°/n)$, it will also do so for all rotations by $\varphi' = p\,(360°/n)$, p = 1, …, n. This property can be used to characterize rotation axes by their "foldedness" n. An **n-fold rotation axis** in a lattice allows rotations by all integer multiples of the angle $360°/n$ about its axis, where the rotated images coincide with the initial lattice. Thus, lattices allow only two-, three-, four-, and sixfold rotation axes as rotational point symmetry elements.

Table 2.2 Naming conventions (Hermann–Mauguin, Schönflies) for the simple point symmetry elements of three-dimensional lattices.

Symmetry	Hermann–Mauguin	Schönflies
Twofold rotation axis	2	C_2
Threefold rotation axis	3	C_3
Fourfold rotation axis	4	C_4
Sixfold rotation axis	6	C_6
Mirror plane	m	σ
Inversion	$\bar{1}$ or −1	i

There are two different naming conventions for rotation axes in crystal lattices. The so-called **Hermann–Mauguin** or **International notation**, is preferred by crystallographers and used in the International Tables of Crystallography [33]. In this notation an n-fold rotation axis is denoted by its foldedness as **n**. In contrast, the so-called **Schönflies** notation uses the symbol C_n for an n-fold rotation axis. Table 2.2 lists all possible n-fold rotation axes of three-dimensional lattices together with their Hermann–Mauguin and Schönflies names. It also includes the corresponding names for mirror symmetry planes and inversion centers.

The four different rotation axes can **distinguish** between the different types of three-dimensional lattices. First, there are lattices that do not possess any rotational axis. They form the most general type of **Bravais lattices** and will be called **triclinic-P**. Any centering of a triclinic-P lattice according to Section 2.2.3 will lead to another triclinic-P lattice, that is, will not create any new lattice type. Thus, the triclinic-P lattice, shown in Figure 2.23, is the only member of the **triclinic crystal system**.

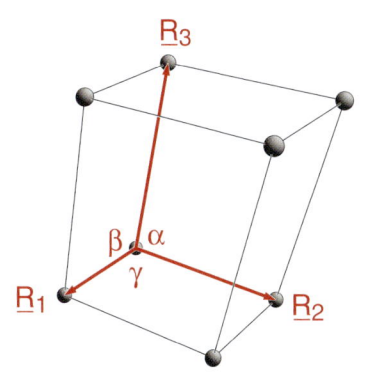

Figure 2.23 Morphological unit cell of the triclinic-P lattice with lattice vectors $\underline{R}_1$, $\underline{R}_2$, $\underline{R}_3$ and angles α, β, γ labeled accordingly.

Next, we assume that a lattice possesses a **twofold rotation axis,** where the origin of the lattice can always be set to lie on the axis. Then, we consider two lattice points, given by a general lattice vector $\underline{R}^{(0)}$ (not on the axis), and its image $\underline{R}^{(1)}$, which arises from rotating $\underline{R}^{(0)}$ about the axis by 180°. The sum of the two general lattice vectors, $\underline{R}^c = (\underline{R}^{(0)} + \underline{R}^{(1)})$, is a general lattice vector pointing along the rotation axis. Therefore, as a result of translational symmetry, there are an infinite number of lattice points on the rotation axis. Of these lattice points, the one nearest to the origin can be used to define lattice vector $\underline{R}_3$ of the lattice. On the other hand, the difference vector $\underline{R}^a = (\underline{R}^{(1)} - \underline{R}^{(0)})$ is a general lattice vector perpendicular to the rotation axis suggesting infinitely many lattice points along its direction. Of these, again the one nearest to the origin can be used to define lattice vector $\underline{R}_1$ of the lattice. The same procedure can be applied to a different general lattice vector $\underline{R}^{(2)}$ and its rotational image $\underline{R}^{(3)}$, where the difference vector $\underline{R}^b = (\underline{R}^{(3)} - \underline{R}^{(2)})$ is also perpendicular to the rotation axis. Then the smallest lattice vector along $\underline{R}^b$ can be used to define lattice vector $\underline{R}_2$ of the lattice. Vectors $\underline{R}_1$ and $\underline{R}_2$ may have to be exchanged to guarantee a right-handed system, but, altogether, the vector triplet $\underline{R}_1, \underline{R}_2, \underline{R}_3$ provides an appropriate set of lattice vectors describing the lattice with its twofold rotation axis. For the following discussion, these lattice vectors will be described by their lengths a, b, c (lattice constants) and mutual angles α, β, γ, according to Eq. (2.3) and sketched in Figure 2.3. This means, in particular, for the present symmetry and choice of lattice vectors that $\alpha = \beta = 90°$.

The lattice vectors $\underline{R}_1, \underline{R}_2, \underline{R}_3$ where the angle γ assumes any value different from 60°, 90°, and 120° define the **monoclinic crystal system**. If the morphological unit cell of $\underline{R}_1, \underline{R}_2, \underline{R}_3$ is primitive the corresponding Bravais lattice will be called **monoclinic-P**. Centering a monoclinic-P lattice according to Section 2.2.3 can create different lattices depending on the type of centering. Here C-centering will only modify the lattice vectors $\underline{R}_1$ and $\underline{R}_2$ but will keep the monoclinic-P lattice. In contrast, A-centering leads to a new lattice type, **monoclinic-A**, which cannot be described by a monoclinic-P lattice. Likewise, B-centering, creates a new lattice type, **monoclinic-B**, different from monoclinic-P. However, monoclinic-A and monoclinic-B lattices are morphologically equivalent since they differ only by an interchange of lattice vectors $\underline{R}_1$ and $\underline{R}_2$. Thus, it is sufficient to consider one of the two lattice types, where crystallographers prefer **monoclinic-B** over monoclinic-A. Further, F- and I-centering can be shown to also be equivalent to B-centering by appropriate origin shifts and lattice vector modifications. Thus, the monoclinic crystal system can be represented by **two** unique Bravais lattices, monoclinic-P and monoclinic-B, shown in Figure 2.24.

It should be noted that crystallographers often describe the monoclinic crystal system by using lattice vector $\underline{R}_2$ (**B axis**) to define the direction of the twofold rotation axis of a crystal. This geometry, referred to in the International Tables of Crystallography [33] as the "**first setting**" (as opposed to the "second setting" discussed above), corresponds to lattice vector angles $\alpha = \gamma = 90°, \beta \neq 60°, 90°, 120°$. The interchange of crystal axes between the two settings does not affect the discussion of possible crystal types, except that the first setting considers **monoclinic-P** and centered **monoclinic-C** as the unique monoclinic lattices.

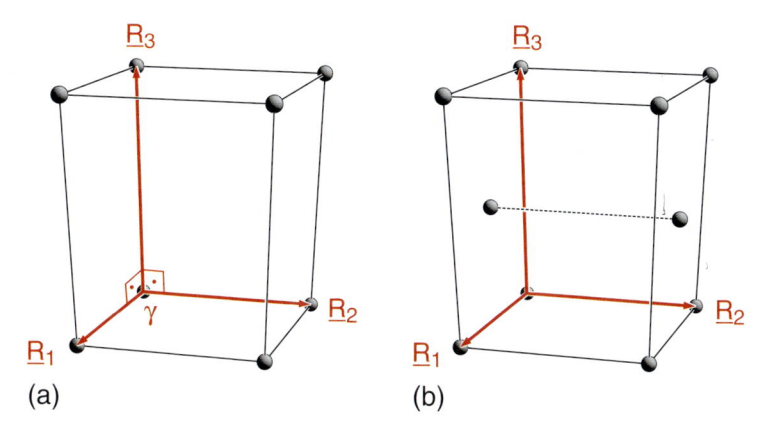

Figure 2.24 Morphological unit cells of the monoclinic crystal system, (a) monoclinic-P and (b) monoclinic-B lattices with lattice vectors $\underline{R}_1, \underline{R}_2, \underline{R}_3$ and angles $\alpha = \beta = 90°$, γ labeled accordingly (angles only for monoclinic-P). Angles of 90° are indicated by small rectangles filled with a dot. The dashed line connecting opposing lattice points is meant to guide the eye.

Lattice vectors $\underline{R}_1, \underline{R}_2, \underline{R}_3$ with the angle $\gamma = 90°$, that is, for $\alpha = \beta = \gamma = 90°$, but with different vector lengths, $a \neq b$, $a \neq c$, $b \neq c$, define the **orthorhombic crystal system**. If the morphological unit cell of $\underline{R}_1, \underline{R}_2, \underline{R}_3$ is primitive the corresponding Bravais lattice will be called **orthorhombic-P**. Centering an orthorhombic-P lattice according to Section 2.2.3 will always create different lattices. Here A-, B-, and C-centerings lead to new lattice types, **orthorhombic-A, -B, and -C**, respectively, which cannot be described by an orthorhombic-P lattice. However, these three centered lattices are morphologically equivalent and differ only by an interchange of corresponding lattice vectors. Thus, only one of these lattice types needs to be considered, where crystallographers often prefer **orthorhombic-C** over the other two. In addition, I- and F-centerings yield new lattice types, **orthorhombic-I** and **orthorhombic-F**. Therefore, the orthorhombic crystal system can be represented by **four** unique Bravais lattices, orthorhombic-P, orthorhombic-C, orthorhombic-I, and orthorhombic-F, shown in Figure 2.25.

Lattice vectors $\underline{R}_1, \underline{R}_2, \underline{R}_3$ with angles $\alpha = \beta = \gamma = 90°$ but with vector lengths, $a = b \neq c$, define the **tetragonal crystal system**. If the morphological unit cell of $\underline{R}_1, \underline{R}_2, \underline{R}_3$ is primitive the corresponding Bravais lattice will be called **tetragonal-P**. Here the two constraints, $a = b$ and $\gamma = 90°$ mean, in particular, that the twofold rotation axis along $\underline{R}_3$ is also a **fourfold** rotation axis. Therefore, A- and B-centering of a tetragonal-P lattice is not possible. Further, C-centering will only modify the lattice vectors $\underline{R}_1$ and $\underline{R}_2$ but will keep the tetragonal-P lattice. However, I-centering results in a new lattice type, **tetragonal-I**, which cannot be represented by a tetragonal-P lattice. In addition, F-centering can be shown to result in a tetragonal-I lattice by appropriate origin shifts and lattice vector modifications. Thus, the tetragonal crystal system can be represented

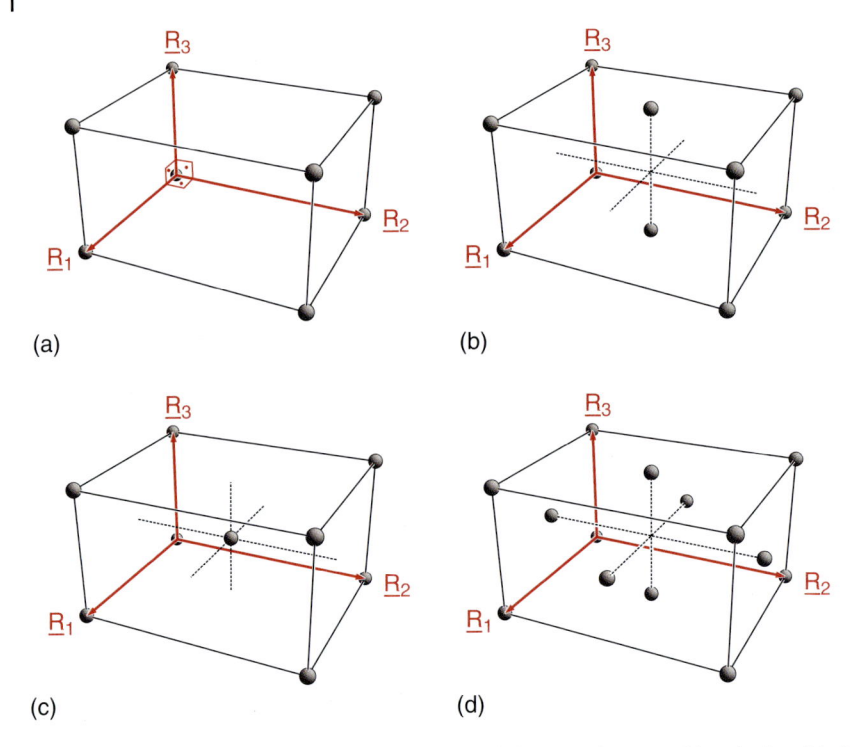

Figure 2.25 Morphological unit cells of the orthorhombic crystal system, (a) orthorhombic-P, (b) orthorhombic-C, (c) orthorhombic-I, and (d) orthorhombic-F, lattices with lattice vectors $\underline{R}_1$, $\underline{R}_2$, $\underline{R}_3$ and angles $\alpha = \beta = \gamma = 90°$ labeled accordingly (angles only for orthorhombic-P indicated by small rectangles filled with a dot). The dashed lines connecting lattice points are meant to guide the eye.

by **two** unique Bravais lattices, tetragonal-P and tetragonal-I, shown in Figure 2.26.

Lattice vectors $\underline{R}_1$, $\underline{R}_2$, $\underline{R}_3$ with angles $\alpha = \beta = \gamma = 90°$ but with three equal vector lengths, $a = b = c$, define the **cubic crystal system**. If the morphological unit cell of $\underline{R}_1$, $\underline{R}_2$, $\underline{R}_3$ is primitive the corresponding Bravais lattice will be called **cubic-P** or **sc**. For this crystal system the twofold rotation axis along $\underline{R}_3$ is also a **fourfold** rotation axis, analogous to the tetragonal case. In addition, there are fourfold rotation axes along $\underline{R}_1$ and $\underline{R}_2$. As a consequence, neither A- nor B- nor C-centering of a cubic-P lattice is possible. However, I- and F-centerings yield new lattice types, **cubic-I** or **bcc** and **cubic-F** or **fcc**. As a consequence, the cubic crystal system includes **three** unique Bravais lattices, cubic-P, cubic-I, and cubic-F, shown in Figure 2.27.

So far, all lattice vector sets $\underline{R}_1$, $\underline{R}_2$, $\underline{R}_3$ with vector $\underline{R}_3$ pointing along a two- and fourfold rotation axis have been considered. In addition, $\underline{R}_3$ can define the direction of a **sixfold** rotation axis. This corresponds to $\alpha = \beta = 90°$, $\gamma = 60°$, $a = b$, and defines the **hexagonal crystal system**. In this system, the lattice vectors $\underline{R}_1$ and $\underline{R}_2$ can be represented in two ways. The initial definition is based on the sixfold

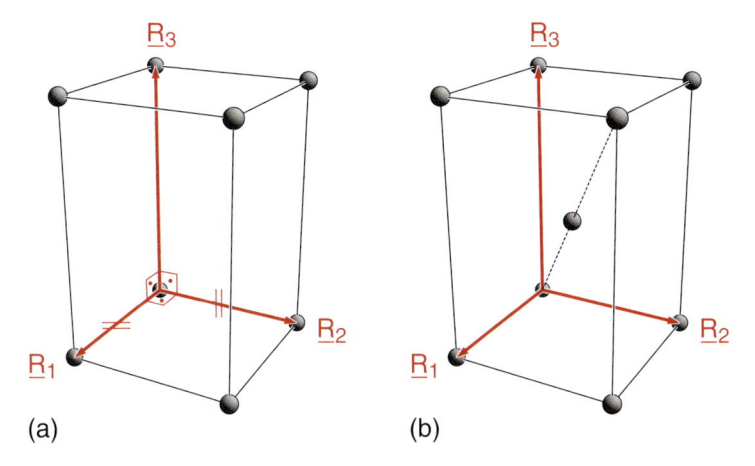

Figure 2.26 Morphological unit cells of the tetragonal crystal system, (a) tetragonal-P and (b) tetragonal-I lattices with lattice vectors $\underline{R}_1$, $\underline{R}_2$, $\underline{R}_3$ and angles $\alpha = \beta = \gamma = 90°$ labeled accordingly (angles only for tetragonal-P indicated by small rectangles filled with a dot). Parallel pairs of short lines indicate vectors of equal length. The dashed line connecting opposing lattice points is meant to guide the eye.

rotation axis and uses an angle $\gamma = 60°$ beween $\underline{R}_1$ and $\underline{R}_2$ (**acute representation**). The alternative definition uses an angle $\gamma = 120°$ beween $\underline{R}_1$ and $\underline{R}_2$ (**obtuse representation**) and is often preferred by crystallographers. The latter representation emphasizes the threefold rotation axis along $\underline{R}_3$, which is, however, combined with a twofold axis to form the sixfold rotation axis. If the morphological unit cell of the hexagonal lattice is primitive the corresponding Bravais lattice will be called **hexagonal-P**. Any centering of a hexagonal-P lattice according to Section 2.2.3 will destroy the hexagonal symmetry. Thus, the hexagonal crystal system includes only the hexagonal-P Bravais lattice, shown in Figure 2.28.

The present lattice classification is based on lattices where lattice vector $\underline{R}_3$ points along a twofold rotation axis. Therefore, it cannot immediately be used to describe lattices with a pure threefold rotation axis along $\underline{R}_3$ since the combination of coinciding two- and threefold rotation axes leads to a sixfold rotation axis and, thus, to the hexagonal crystal system discussed previously. However, we can start from a threefold rotation axis through a lattice point and consider three other lattice points. These are given by a general lattice vector $\underline{R}^{(0)}$ from the lattice point and its two images $\underline{R}^{(1)}$, $\underline{R}^{(2)}$, which arise from rotating $\underline{R}^{(0)}$ about the axis by 120° and 240°, respectively. If these vectors are of the smallest length along their direction they can be used as lattice vectors $\underline{R}_1$, $\underline{R}_2$, $\underline{R}_3$. By construction these lattice vectors are of identical length and form identical angles with each other, that is, $a = b = c$ and $\alpha = \beta = \gamma$, which defines the **trigonal** or **rhombohedral crystal system**. This crystal system also includes only one Bravais lattice, the **trigonal-R** lattice ("R" = "Rhombohedral" reminds of the alternative name), shown in Figure 2.29. The trigonal and hexagonal lattices are closely connected with each other. In Section 2.2.2.1 it is shown that a trigonal-R lattice can be alternatively described by a hexagonal lattice with non-primitive lattice

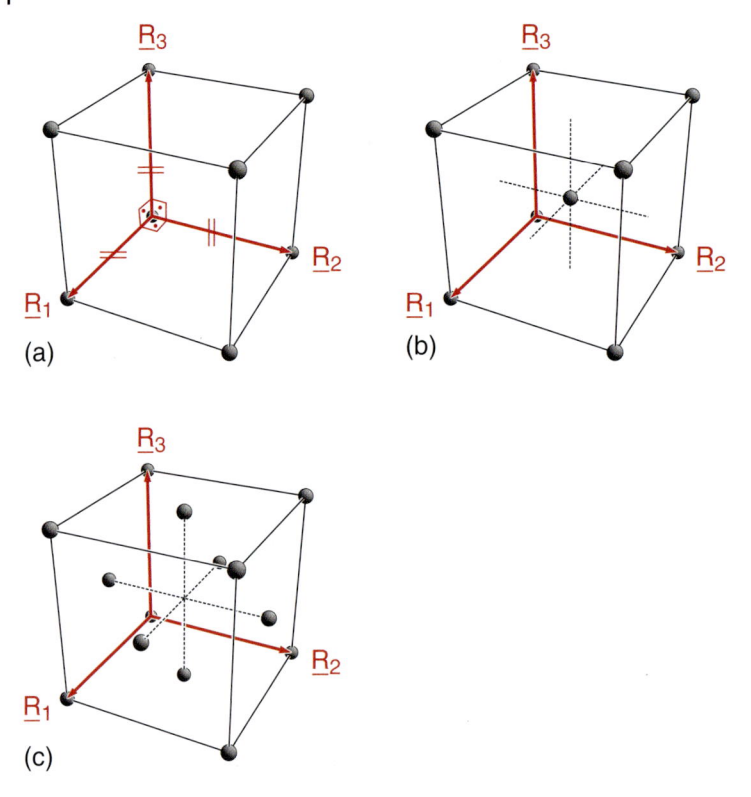

Figure 2.27 Morphological unit cells of the cubic crystal system, (a) cubic-P, (b) cubic-I, and (c) cubic-F, lattices with lattice vectors $\underline{R}_1$, $\underline{R}_2$, $\underline{R}_3$ and angles $\alpha = \beta = \gamma = 90°$ labeled accordingly (angles only for cubic-P indicated by small rectangles filled with a dot). Parallel pairs of short lines indicate vectors of equal length. The dashed lines connecting opposing lattice points are meant to guide the eye.

vectors. Further, the trigonal-R lattice yields for $\alpha = \beta = \gamma = 90°$ a cubic-P lattice, for $\alpha = \beta = \gamma = 60°$ a cubic-F lattice, and for $\alpha = \beta = \gamma = 109.47°$ $(\cos(\alpha) = -1/3)$ a cubic-I lattice. Thus, the cubic lattices may also be defined by their (four different) threefold rotation axes rather than by two- and fourfold rotation axes.

Altogether, the existence of rotation axes inside lattices allows a first classification of all lattice types yielding the **7 crystal systems** and **14 different Bravais lattices** listed in Table 2.3 and sketched in Figure 2.30.

A complete classification of all possible lattice types must also take into account point symmetry elements other than rotations. However, this will not affect the basic family of crystal systems and Bravais lattices obtained so far. In fact, Bravais lattices and their morphological units cells are always found to exhibit the largest number of symmetry elements of all lattices of a given crystal system. Atom positions, defining the basis inside the morphological unit cell of a crystal, may result in lower symmetry than suggested by the shape of the unit cell (given by

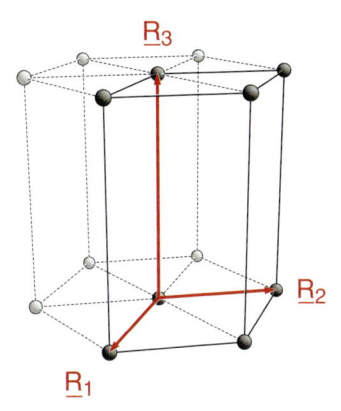

Figure 2.28 Morphological unit cell of the hexagonal-P lattice with lattice vectors $\underline{R}_1$, $\underline{R}_2$, $\underline{R}_3$ labeled accordingly. The dashed lines connecting lattice points in hexagonal arrangements at top and bottom planes are meant to guide the eye.

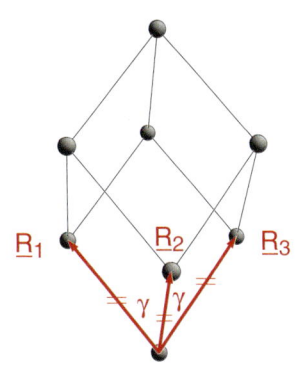

Figure 2.29 Morphological unit cell of the trigonal-R (rhombohedral) lattice with lattice vectors $\underline{R}_1$, $\underline{R}_2$, $\underline{R}_3$ and angles γ (given for the two front sides) labeled accordingly. Parallel pairs of short lines indicate vectors of equal length.

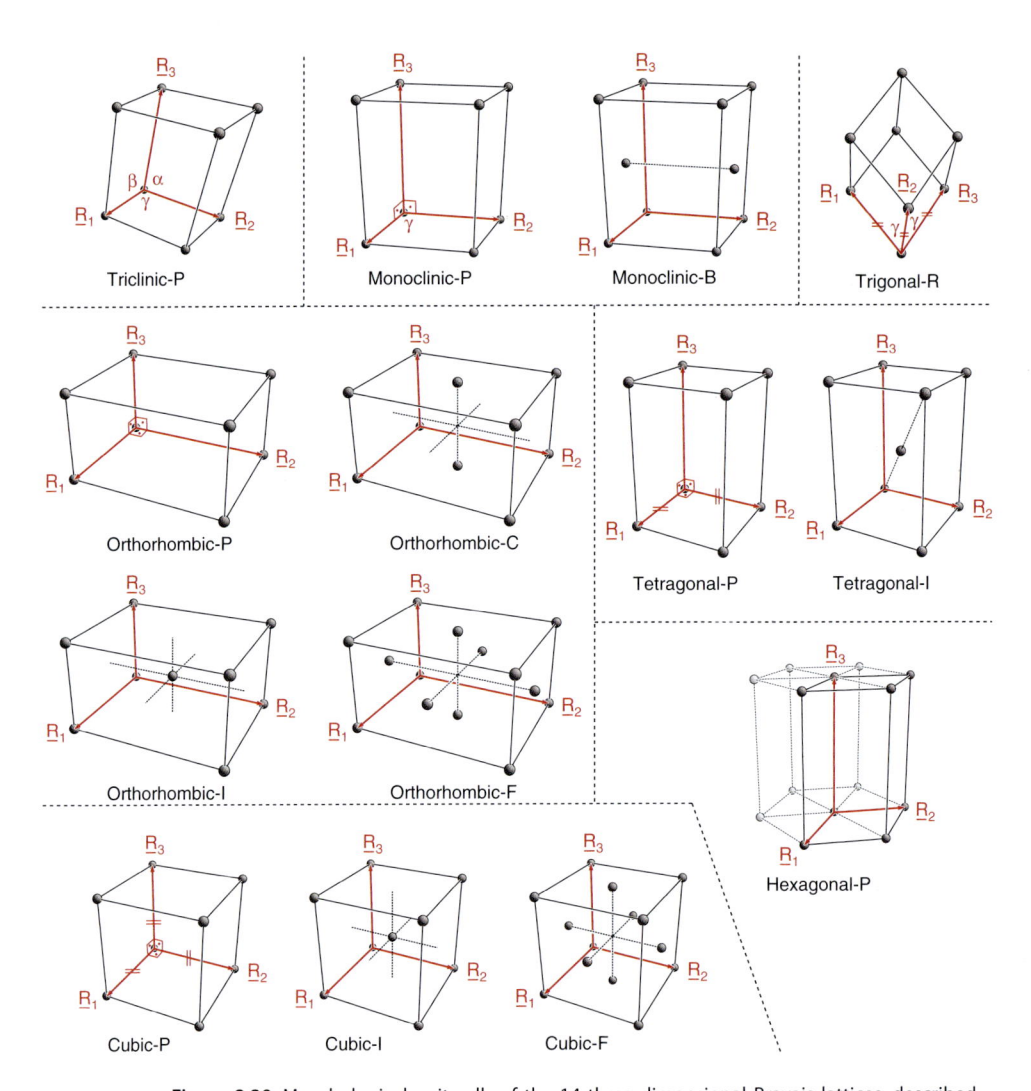

Figure 2.30 Morphological unit cells of the 14 three-dimensional Bravais lattices, described in Table 2.3, with lattice vectors $\underline{R}_1$, $\underline{R}_2$, $\underline{R}_3$ and angles α, β, γ labeled accordingly. Angles of 90° are indicated by small rectangles filled with a dot. Parallel pairs of short lines indicate vectors of equal length. The dashed lines connecting opposing lattice points are meant to guide the eye.

Table 2.3 List of the seven crystal systems with their Bravais lattice members described by lattice constants a, b, c, and angles α, β, γ.

Crystal system	Lattice constants	Bravais lattices	Symmetry
Triclinic	$a \neq b \neq c$, α, β, γ	-P	C_i
Monoclinic	$a \neq b \neq c$, $\alpha = 90°$, $\beta = 90°$, γ	-P, -B	C_{2h}
Orthorhombic	$a \neq b \neq c$, $\alpha = \beta = \gamma = 90°$	-P, -C, -I, -F	D_{2h}
Tetragonal	$a = b \neq c$, $\alpha = \beta = \gamma = 90°$	-P, -I	D_{4h}
Hexagonal	$a = b \neq c$, $\alpha = \beta = 90°$, $\gamma = 60°$ (acute), $120°$ (obtuse)	-P	D_{6h}
Trigonal, Rhombohedral	$a = b = c$, $\alpha = \beta = \gamma$	-R	D_{3d}
Cubic	$a = b = c$, $\alpha = \beta = \gamma = 90°$	-P, -I, -F	O_h

Angles quoted without specific values are assumed to differ from 60°, 90°, and 120°. For each crystal system, the corresponding Bravais lattices with crystallographic labels as well as with the highest point symmetry group in Schoenflies notation are included.

the crystal system and the corresponding Bravais lattice). This allows for lattices with identical lattice vectors but different point symmetry properties. Applying group theoretical methods, it can be shown that, altogether, there are 230 different ways to combine symmetry with lattices, as described by the **230 different three-dimensional space groups,** tabulated in the International Tables of Crystallography [33]. For a full discussion see [32] and references therein.

2.5
Reciprocal Lattice

In addition to the initial lattice vectors $\underline{R}_1, \underline{R}_2, \underline{R}_3$ that describe a lattice in real space the definition of a second set of vectors, $\underline{G}_1, \underline{G}_2, \underline{G}_3$, given by vector products

$$\underline{G}_1 = \beta \, (\underline{R}_2 \times \underline{R}_3), \quad \underline{G}_2 = \beta \, (\underline{R}_3 \times \underline{R}_1), \quad \underline{G}_3 = \beta \, (\underline{R}_1 \times \underline{R}_2)$$
$$\beta = (2\pi)/[(\underline{R}_1 \times \underline{R}_2)\underline{R}_3] = (2\pi)/V_{el} \tag{2.95}$$

has proven to be quite useful for various applications discussed in this book. Examples are the calculation of distances between adjacent monolayers or of atom densities of monolayers described in Section 3.1. The reciprocal lattice defined by Eq. (2.95) is also central to wave diffraction by lattices and to electronic band structure theory of crystals. Since the scaling factor β in Eq. (2.95) is of dimension [length^{-3}] vectors $\underline{G}_i$ are of dimension [length^{-1}]. They are closely related to the initial lattice vectors and can be used to define a complementary lattice in reciprocal space, the **reciprocal lattice.** Therefore, these vectors are called **reciprocal lattice vectors**. They have a number of interesting properties of which we mention only a few in the following.

1) The reciprocal lattice vectors $\underline{G}_i$ fulfill **orthogonality relations**

$$(\underline{G}_i \underline{R}_i) = 2\pi \text{ for } i = 1, 2, 3; \quad (\underline{G}_i \underline{R}_j) = 0 \quad \text{for } i \neq j \tag{2.96}$$

which is clear from definitions (2.95) and basic properties of vector products.

2) The **volume** of the **unit cell** of the reciprocal lattice is **inverse** to that of the real space lattice. According to definition (2.9) and using a property of the vector product of three vectors $\underline{a}, \underline{b}, \underline{c}$

$$\underline{a} \times (\underline{b} \times \underline{c}) = (\underline{a}\,\underline{c})\underline{b} - (\underline{a}\,\underline{b})\underline{c} \tag{2.97}$$

we obtain

$$\begin{aligned}V_G &= (\underline{G}_1 \times \underline{G}_2)\,\underline{G}_3 = \beta^3\{(\underline{R}_2 \times \underline{R}_3) \times (\underline{R}_3 \times \underline{R}_1)\}(\underline{R}_1 \times \underline{R}_2) \\ &= \beta^3\{((\underline{R}_2 \times \underline{R}_3)\,\underline{R}_1)\,\underline{R}_3 - ((\underline{R}_2 \times \underline{R}_3)\,\underline{R}_3)\,\underline{R}_1\}(\underline{R}_1 \times \underline{R}_2) \\ &= \beta^3\{((\underline{R}_2 \times \underline{R}_3)\,\underline{R}_1)\,\underline{R}_3\}(\underline{R}_1 \times \underline{R}_2) \\ &= \beta^3\{\underline{R}_3\,(\underline{R}_1 \times \underline{R}_2)\}^2 = (2\pi)^3/[(\underline{R}_1 \times \underline{R}_2)\,\underline{R}_3] = (2\pi)^3/V_{el}\end{aligned} \tag{2.98}$$

and, thus,

$$V_G V_{el} = (2\pi)^3 \tag{2.99}$$

3) The **reciprocal** lattice of a real space lattice is identical to the **real** space lattice. This can be proven by simple vector calculus using relations (2.95) and (2.97) as discussed in Appendix F. Thus, we can write formally

$$\{\underline{R}_1, \underline{R}_2, \underline{R}_3\}^{-1} = \{\underline{G}_1, \underline{G}_2, \underline{G}_3\}, \quad \{\underline{G}_1, \underline{G}_2, \underline{G}_3\}^{-1} = \{\underline{R}_1, \underline{R}_2, \underline{R}_3\} \tag{2.100}$$

Explicit examples of reciprocal lattices are

- the **sc lattice** whose reciprocal lattice also defines an **sc** lattice, that is,

$$\begin{aligned}\underline{R}_1^{sc} &= a\,(1,0,0) & \underline{G}_1^{sc} &= 2\pi/a\,(1,0,0) \\ \underline{R}_2^{sc} &= a\,(0,1,0) & \underline{G}_2^{sc} &= 2\pi/a\,(0,1,0) \\ \underline{R}_3^{sc} &= a\,(0,0,1) & \underline{G}_3^{sc} &= 2\pi/a\,(0,0,1)\end{aligned} \tag{2.101}$$

- the **fcc lattice** whose reciprocal lattice defines a **bcc** lattice, that is,

$$\begin{aligned}\underline{R}_1^{fcc} &= a/2\,(0,1,1) & \underline{G}_1^{fcc} &= 2\pi/a\,(-1,1,1) \\ \underline{R}_2^{fcc} &= a/2\,(1,0,1) & \underline{G}_2^{fcc} &= 2\pi/a\,(1,-1,1) \\ \underline{R}_3^{fcc} &= a/2\,(1,1,0) & \underline{G}_3^{fcc} &= 2\pi/a\,(1,1,-1)\end{aligned} \tag{2.102}$$

- the **bcc lattice** whose reciprocal lattice defines an **fcc** lattice, that is,

$$\begin{aligned}\underline{R}_1^{bcc} &= a/2\,(-1,1,1) & \underline{G}_1^{bcc} &= 2\pi/a\,(0,1,1) \\ \underline{R}_2^{bcc} &= a/2\,(1,-1,1) & \underline{G}_2^{bcc} &= 2\pi/a\,(1,0,1) \\ \underline{R}_3^{bcc} &= a/2\,(1,1,-1) & \underline{G}_3^{bcc} &= 2\pi/a\,(1,1,0)\end{aligned} \tag{2.103}$$

In general, lattice types and their symmetry properties in reciprocal space, defined by reciprocal lattice vectors $\underline{G}_1$, $\underline{G}_2$, $\underline{G}_3$, can be related with those of the corresponding real space lattices, lattice vectors $\underline{R}_1$, $\underline{R}_2$, $\underline{R}_3$. This is clear from Table 2.4 which lists the corresponding reciprocal Bravais lattice for each of the 14 three-dimensional Bravais lattices discussed in Section 2.4.

Reciprocal lattice vectors are also useful for the decomposition of real space coordinates $\underline{r}$ into multiples of lattice vectors $\underline{R}_1$, $\underline{R}_2$, $\underline{R}_3$. If a spatial coordinate $\underline{r}$ is written as

$$\underline{r} = x_1 \underline{R}_1 + x_2 \underline{R}_2 + x_3 \underline{R}_3, \quad x_i \text{ real} \tag{2.104}$$

then the orthogonality theorem (2.96) yields for $i = 1, 2, 3$

$$\underline{G}_i \underline{r} = x_1 (\underline{G}_i \underline{R}_1) + x_2 (\underline{G}_i \underline{R}_2) + x_3 (\underline{G}_i \underline{R}_3) = 2\pi x_i \tag{2.105}$$

Thus, the mixing coefficients x_i can be calculated as scalar products involving reciprocal lattice vectors.

Many physical properties of a perfect three-dimensional bulk crystal with its periodicity described by lattice vectors $\underline{R}_1$, $\underline{R}_2$, $\underline{R}_3$ are characterized by functions $f(\underline{r})$ that are periodic in space where the periodicity coincides with that of the crystal lattice, that is,

$$f(\underline{r} + \underline{R}) = f(\underline{r}) \quad \text{with} \quad \underline{R} = n_1 \underline{R}_1 + n_2 \underline{R}_2 + n_3 \underline{R}_3, \quad n_i \text{ integer} \tag{2.106}$$

Examples are the electron density $\rho(\underline{r})$ or the electrostatic potential $V(\underline{r})$ in a perfect crystal. In a harmonic analysis (**Fourier analysis**) these periodic functions $f(\underline{r})$ are represented by a (generally infinite) series of harmonic functions written as exponentials $\exp(i\,\underline{G}\,\underline{r})$ of imaginary argument (i being the imaginary unit number), that is,

$$f(\underline{r}) = \sum_{\underline{G}} c_{\underline{G}} \exp(i\,\underline{G}\,\underline{r}) \quad \text{with} \quad \underline{G} = k_1 \underline{G}_1 + k_2 \underline{G}_2 + k_3 \underline{G}_3, \quad k_i \text{ integer} \tag{2.107}$$

Table 2.4 List of real and corresponding reciprocal Bravais lattices.

	Real space lattice	Reciprocal space lattice
1	Triclinic-P	Triclinic-P
2	Monoclinic-P	Monoclinic-P
3	Monoclinic-A, -B	Monoclinic-A, -B
4	Orthorhombic-P	Orthorhombic-P
5	Orthorhombic-A, -B, -C	Orthorhombic-A, -B, -C
6	Orthorhombic-I	Orthorhombic-F
7	Orthorhombic-F	Orthorhombic-I
8	Tetragonal-P	Tetragonal-P
9	Tetragonal-I	Tetragonal-I
10	Hexagonal-P	Hexagonal-P
11	Trigonal-R	Trigonal-R
12	Cubic-P (sc)	Cubic-P (sc)
13	Cubic-I (bcc)	Cubic-F (fcc)
14	Cubic-F (fcc)	Cubic-I (bcc)

Vectors $\underline{G}$ in this expansion are linear combinations of reciprocal lattice vectors with integer-valued mixing coefficients k_i and $c_{\underline{G}}$ are expansion coefficients of the series determined by integrals

$$c_{\underline{G}} = \frac{1}{V_{el}} \iiint_{V_{el}} f(\underline{r}) \exp(-i \underline{G}\, \underline{r}) d^3 r \qquad (2.108)$$

where the three-dimensional integration is carried out over the elementary cell V_{el} of the real space lattice, for further mathematical details see Appendix G. Thus, the Fourier expansion (2.107) is based on a summation of terms that are determined by reciprocal lattice vectors. This is justified by real and reciprocal lattice vectors obeying the orthogonality theorem (2.96). According to

$$\begin{aligned} \underline{G}\, \underline{R} &= (k_1 \underline{G}_1 + k_2 \underline{G}_2 + k_3 \underline{G}_3)(n_1 \underline{R}_1 + n_2 \underline{R}_2 + n_3 \underline{R}_3) \\ &= 2\pi (k_1 n_1 + k_2 n_2 + k_3 n_3) = 2\pi N \end{aligned} \qquad (2.109)$$

and

$$\exp(i\, 2\pi N) = 1, \quad N \text{ integer} \qquad (2.110)$$

we obtain with Eq. (2.107)

$$f(\underline{r} + \underline{R}) = \sum_{\underline{G}} c_{\underline{G}} \exp(i\, \underline{G}\, (\underline{r} + \underline{R})) = \sum_{\underline{G}} c_{\underline{G}} \exp(i\, \underline{G}\, \underline{r}) \exp(i\, \underline{G}\, \underline{R}) = f(\underline{r}) \qquad (2.111)$$

Thus, the Fourier series reproduces the periodicity (2.106).

2.6 Neighbor Shells

Geometric parameters of periodic crystals are fully described by their translational and point symmetries. However, in some cases physical and chemical properties may be represented more appropriately by considering local atom neighborhoods and relationships between atoms in spherical environments. This leads to the concept of **neighbor shells**. Neighbor shells start from an atom of the crystal and characterize its environment by surrounding atoms. Here, all atoms at a given **distance** or distance **range** with respect to the central atom and irrespective of their direction are collected to form a **shell**. The neighbor shells are ordered according to their **shell radii** (given by the interatomic distances) where the closest shell with the smallest radius is sometimes called the (first) **coordination shell**. Then the atoms of the different shells span the complete crystal, if all radii up to infinity are considered.

Physical applications of the neighbor shell concept include using shell models to describe lattice vibrations (balls-and-springs approach to phonons [52]), tight-binding methods [53] to describe electronic properties of crystals, or for electrostatic potential calculations based on point charges in ionic crystals [54]. Figure 2.31 illustrates the neighbor shell concept by sketching the six smallest shells in a crystal with an fcc lattice. Further, Table 2.5 lists radii and numbers of

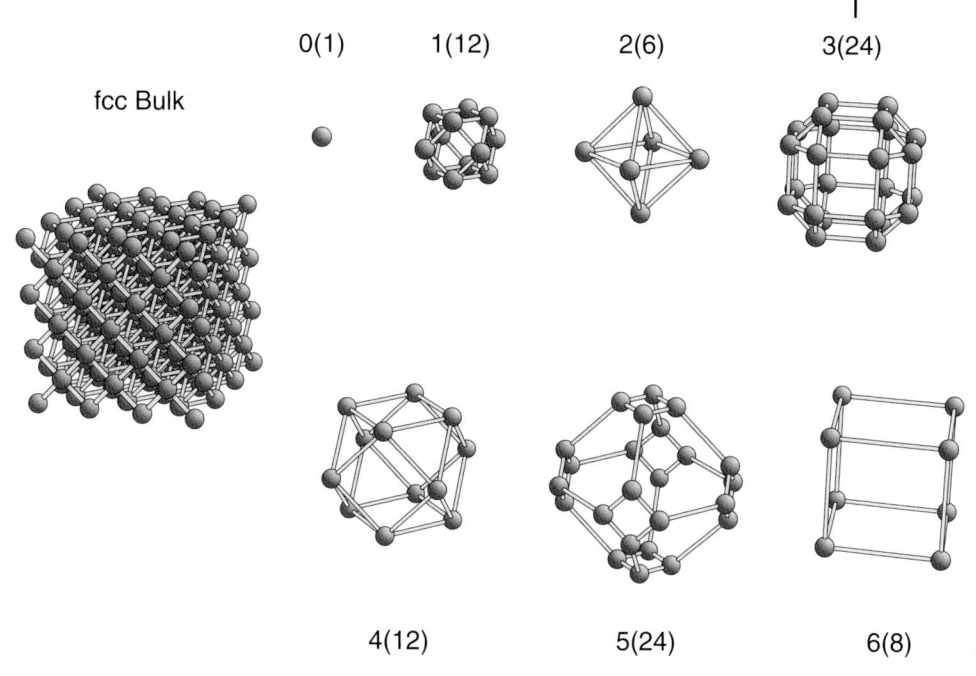

Figure 2.31 Neighbor shells of a crystal with an fcc lattice. The labels "i(M_i)" combine the shell index i (0 for central atom, 1–6) with the corresponding shell multiplicity M_i (1–24). A section of the fcc bulk crystal (labeled "fcc bulk") is included to the left.

atom members M_i (**shell multiplicities**) of the smallest six neighbor shells of crystals with cubic (sc, bcc, fcc) and hcp lattices. Sometimes, the multiplicity of the smallest neighbor shell is also called the **coordination number** denoting the number of closest atom neighbors that can form direct bonds with the central atom.

The **formal definition** of a neighbor shell starts from a crystal with lattice vectors $\underline{R}_1, \underline{R}_2, \underline{R}_3$ and a basis, given by atom positions $r_1, \ldots, r_p$. Then the ith neighbor shell $S_i(\underline{R}_c, D_i, \varepsilon_i)$ inside a crystal is defined as a collection of crystal atoms surrounding a **shell center** $\underline{R}_c$, which may or may not coincide with the position of a crystal atom. The shell includes all atoms, at general positions $\underline{R} = n_1 \underline{R}_1 + n_2 \underline{R}_2 + n_3 \underline{R}_3 + \underline{r}_i$, whose distances $D = |\underline{R} - \underline{R}_c|$ lie within the **shell range**,

$$(D_i - \varepsilon_i/2) \leq D \leq (D_i + \varepsilon_i/2) \qquad (2.112)$$

where D_i defines the **shell radius** and ε_i the **shell thickness**. The number M_i of atoms belonging to a neighbor shell, also called the **shell multiplicity**, is determined by the position of the shell center $\underline{R}_c$ in the crystal, the geometry of the crystal lattice, and its basis. Here, monoatomic crystals with lattices of high symmetry, providing many atom pairs of identical distance, are expected to result in shells with large shell multiplicities M_i even for vanishing ε_i. On the other hand, crystals with lattices of low symmetry may lead to sets of shells, where, even for $\varepsilon_i > 0$ ("fuzzy shells"), each shell contains only few atoms such that the shell concept may not be useful. The determination of the **complete set** of **shells** for a

Table 2.5 Radii D_i and shell multiplicities M_i of the smallest six neighbor shells of crystals with (a) simple (sc), (b) body-centered cubic (bcc), (c) face-centered (fcc), and (d) hexagonal close-packed (hcp) lattice.

i	M_i	D_i/a	M_i	D_i/a
	(a) sc		(b) bcc	
1	6	$1 = 1.0000$	8	$\sqrt{(3/4)} = 0.8660$
2	12	$\sqrt{2} = 1.4142$	6	$1 = 1.0000$
3	8	$\sqrt{3} = 1.7321$	12	$\sqrt{2} = 1.4142$
4	6	$2 = 2.0000$	24	$\sqrt{(11/4)} = 1.6583$
5	24	$\sqrt{5} = 2.2361$	8	$\sqrt{3} = 1.7321$
6	24	$\sqrt{6} = 2.4495$	6	$2 = 2.0000$
	(c) fcc		(d) hcp	
1	12	$1/\sqrt{2} = 0.7071$	12	$1 = 1.0000$
2	6	$1 = 1.0000$	6	$\sqrt{2} = 1.4142$
3	24	$\sqrt{(3/2)} = 1.2247$	2	$\sqrt{(8/3)} = 1.6330$
4	12	$\sqrt{2} = 1.4142$	18	$\sqrt{3} = 1.7321$
5	24	$\sqrt{(5/2)} = 1.5811$	12	$\sqrt{(11/3)} = 1.9149$
6	8	$\sqrt{3} = 1.7321$	6	$2 = 2.0000$

The radii D_i are given with respect to the corresponding lattice constant a (with $c/a = \sqrt{(8/3)}$ for hcp).

perfect single crystal seems straightforward since all atom positions are defined mathematically. However, the actual computation can be quite tedious as will be discussed in the following.

All atom positions inside a monoatomic crystal with an sc lattice can be represented by vectors

$$\underline{R} = n_1 \underline{R}_1 + n_2 \underline{R}_2 + n_3 \underline{R}_3 = a\,(n_1, n_2, n_3), \quad n_i \text{ integer} \qquad (2.113)$$

in Cartesian coordinates with a denoting the lattice constant. Then, neighbor shells about the origin $\underline{R}_c = (0, 0, 0)$ with shell radius D_i and a range $\varepsilon_i = 0$ can be defined, according to Eq. (2.112), by

$$D_i^2 = |\underline{R} - \underline{R}_c|^2 = R^2 = a^2\,(n_1^2 + n_2^2 + n_3^2) \qquad (2.114)$$

Thus, neighbor shells for radii D_i are determined by

$$D_i = a\sqrt{N_i} \qquad (2.115)$$

where

$$N_i = n_1^2 + n_2^2 + n_3^2 \qquad (2.116)$$

and $N_i, n_1, n_2,$ and n_3 are integer-valued. (**Shell indices** i count the shells according to the size of their radii.) Equation (2.116) forms a **quadratic Diophantine** equation for given N_i with possible solutions n_1, n_2, n_3. Depending on the specific values n_k of a solution, there are always alternative solutions and, thus, other shell members, which reflect the symmetry of the cubic lattice and determine the number of symmetry related shell members, also called the **symmetry related shell multiplicity** M_i^{sym}. Here, we can distinguish six different cases, defined by constraints for the solutions n_k, as given in Table 2.6.

In addition to the symmetry-related shell multiplicities M_i^{sym} of the different shells there may be **accidental shell multiplicities** M_i^{acc}. They arise from the fact that Eq. (2.116) may have different solutions n_1, n_2, n_3 where the actual absolute values n_k differ. This leads to neighbor shells with increased total shell multiplicity. Examples are listed in Table 2.7. In fact, accidental shell multiplicities M_i^{acc} of neighbor shells are responsible for the fact that **total shell multiplicities** $M_i^{tot} = M_i^{sym} + M_i^{acc}$ do not have an upper limit when shell radii increase to an arbitrary size. In addition, total shell multiplicities M_i^{tot} depend on the shell radius in a chaotic fashion. While M_i^{tot} values increase on the average with shell radii there are always shells of very small total shell multiplicity. As examples we mention the 54th neighbor shell ($N_i = 64$, $M_i^{tot} = 6$) and the 107th neighbor shell ($N_i = 128$, $M_i^{tot} = 12$). Figure 2.32a illustrates the chaotic behavior of M_i^{tot} with shell index for the first 100 neighbor shells of a crystal with an sc lattice (shell

Table 2.6 Alternative solutions of Diophantine equation (2.116) and symmetry-related shell multiplicities M_i^{sym}.

Case	Constraints on n_k	Alternatives (n_1, n_2, n_3)	M_i^{sym}
1	$n_1 = n > 0, n_2 = n_3 = 0$	$(\pm n, 0, 0), (0, \pm n, 0), (0, 0, \pm n)$	6
2	$n_1 = n_2 = n > 0, n_3 = 0$	$(\pm n, \pm n, 0), (\pm n, 0, \pm n), (0, \pm n, \pm n)$	12
3	$n_1 = n_2 = n_3 = n > 0$	$(\pm n, \pm n, \pm n)$	8
4	$n_1 > 0, n_2 > 0, n_1 \neq n_2, n_3 = 0$	$(\pm n_1, \pm n_2, 0), (\pm n_1, 0, \pm n_2), (0, \pm n_1, \pm n_2),$ $(\pm n_2, \pm n_1, 0), (\pm n_2, 0, \pm n_1), (0, \pm n_2, \pm n_1)$	24
5	$n_1 > 0, n_1 \neq n, n_2 = n_3 = n > 0$	$(\pm n_1, \pm n, \pm n), (\pm n, \pm n_1, \pm n), (\pm n, \pm n, \pm n_1)$	24
6	$n_1 > 0, n_2 > 0, n_3 > 0, n_1 \neq n_2,$ $n_1 \neq n_3, n_2 \neq n_3$	$(\pm n_1, \pm n_2, \pm n_3), (\pm n_1, \pm n_3, \pm n_2),$ $(\pm n_2, \pm n_1, \pm n_3), (\pm n_2, \pm n_3, \pm n_1),$ $(\pm n_3, \pm n_1, \pm n_2), (\pm n_3, \pm n_2, \pm n_1)$	48

Table 2.7 Examples of alternative solutions for Diophantine equation (2.116).

N_i	(n_1, n_2, n_3)	M_i^{tot}
9	(3, 0, 0), (2, 2, 1)	6 + 24 = 30
25	(5, 0, 0), (4, 3, 0)	6 + 24 = 30
74	(8, 3, 1), (7, 5, 0), (7, 4, 3)	48 + 24 + 48 = 120
101	(10, 1, 0), (9, 4, 2), (8, 6, 1), (7, 6, 4)	24 + 48 + 48 + 48 = 168

Parameter $M_i^{tot} = M_i^{sym} + M_i^{acc}$ denotes the total number of alternative solutions for each value of N_i including symmetry-related and accidental shell multiplicities.

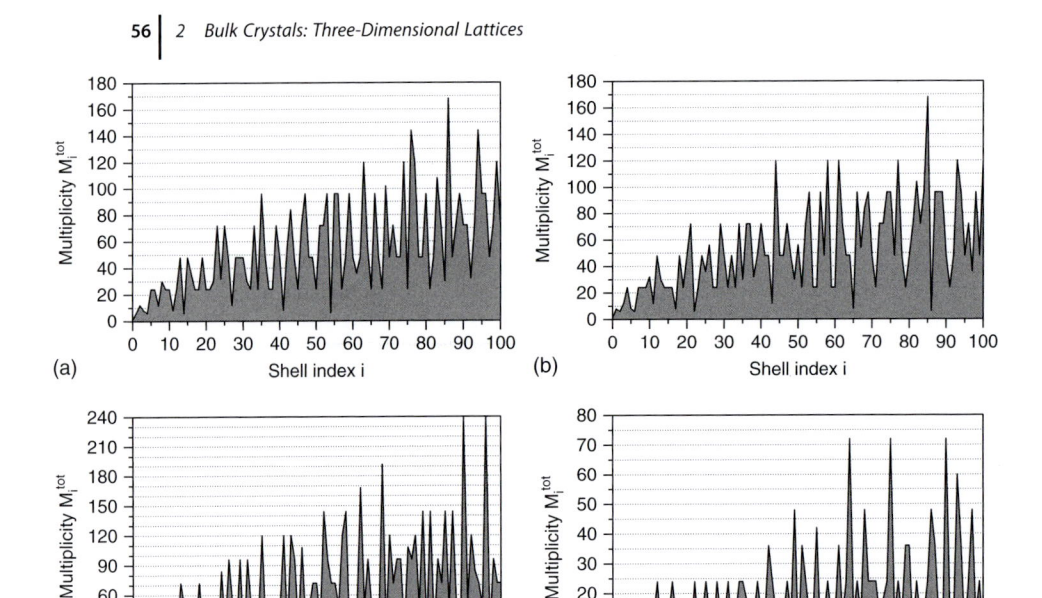

Figure 2.32 Shell multiplicity M_i^{tot} as a function of the shell index i up to the 100th shell for crystals with an (a) sc, (b) bcc, (c) fcc, and (d) hcp lattice. For the range of shell radii, see Figure 2.33.

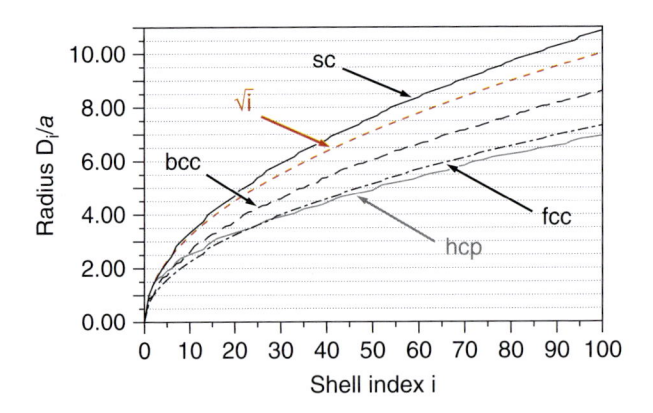

Figure 2.33 Relative shell radii D_i/a as a function of the shell index i up to the 100th shell for crystals with an sc, bcc, fcc, and hcp lattice.

no. 100 corresponds to a radius of 10.816 in units of the lattice constant *a*, see Figure 2.33).

Note that Eq. (2.116) does not yield solutions for integer values N_i with $N_i = 4^p (8q + 7)$, where p and q are positive integers as shown in Appendix E.4. This means, in particular, that N_i according to Eq. (2.116) is larger than i for i > 6.

2.6 Neighbor Shells

As a consequence, shell radii D_i of the sc lattice given by Eq. (2.115) do not scale with $\sqrt{i}$. This is also clear from Figure 2.33, which shows the relative shell radii D_i/a as a function of the shell index i and demonstrates the deviation of D_i/a from the $\sqrt{i}$ dependence for the sc lattice.

The results obtained for crystals with an sc lattice can be applied to crystals with **centered cubic** lattices. In Section 2.2.3 it was shown that the **bcc** lattice can be represented by non-primitive sc lattice vectors $\underline{R}_1, \underline{R}_2, \underline{R}_3$ with an additional lattice vector $\underline{R}'$ pointing to the center of the morphological unit cell. Therefore, general lattice points can be described by **two sets** of lattice vectors, that is, by

$$\underline{R} = n_1 \underline{R}_1 + n_2 \underline{R}_2 + n_3 \underline{R}_3 = a\,(n_1, n_2, n_3), \quad n_i \text{ integer} \tag{2.117}$$

and by

$$\underline{R} = \underline{R}' + n_1 \underline{R}_1 + n_2 \underline{R}_2 + n_3 \underline{R}_3 = a\,(n_1 + 1/2, n_2 + 1/2, n_3 + 1/2), \quad n_i \text{ integer} \tag{2.118}$$

in Cartesian coordinates. As a consequence, for a crystal with a bcc lattice, neighbor shells about the origin $\underline{R}_c = (0, 0, 0)$ with shell radius D_i and a range $\varepsilon_i = 0$ can be defined, according to Eq. (2.112), by two sets, by

$$D_i^2 = a^2\,(n_1^2 + n_2^2 + n_3^2) = a^2/4\,\{(2n_1)^2 + (2n_2)^2 + (2n_3)^2\} \tag{2.119}$$

referring to vectors (2.117) and by

$$D_i^2 = a^2\left\{(n_1 + 1/2)^2 + (n_2 + 1/2)^2 + (n_3 + 1/2)^2\right\}$$
$$= a^2/4\left\{(2n_1 + 1)^2 + (2n_2 + 1)^2 + (2n_3 + 1)^2\right\} \tag{2.120}$$

referring to vectors (2.118). Thus, in both cases neighbor shells for shell radii D_i are determined by

$$D_i = a/2\,\sqrt{N_i} \tag{2.121}$$

which agrees with the result for crystals with an sc lattice. However, the integers N_i are determined by **two** different quadratic Diophantine equations

$$N_i = (2n_1)^2 + (2n_2)^2 + (2n_3)^2 = 4\,\{n_1^2 + n_2^2 + n_3^2\} \tag{2.122a}$$

and

$$N_i = (2n_1 + 1)^2 + (2n_2 + 1)^2 + (2n_3 + 1)^2 = 4\,\{n_1^2 + n_2^2 + n_3^2 + n_1 + n_2 + n_3\} + 3 \tag{2.122b}$$

with possible solutions n_1, n_2, n_3. The N_i values of the two equations cannot coincide since N_i of Eq. (2.122a) must be even, $N_i = 4P$ (P integer), to yield integers n_1, n_2, n_3 while N_i of Eq. (2.122b) must be odd, $N_i = 4P + 3$ (P integer). Further, Eqs. (2.122a) and (2.122b) can be understood as special cases of the quadratic Diophantine equation (2.116) defining neighbor shells of a crystal with an sc lattice, however, with two separate sets of N_i values. Altogether, the complete set of neighbor shells of a crystal with a bcc lattice can be decomposed into **two**

disjoint sets, corresponding to $N_i = 4P$ and $N_i = 4P + 3$, which are each described by **selected shells** of a crystal with an sc lattice and a lattice constant $a/2$ according to Eq. (2.121).

As to the **symmetry-related shell multiplicities** M_i^{sym} of the different shells, the first shell set, defined by Eq. (2.122a), allows all cases 1–6 of Table 2.6 while the second set, determined by Eq. (2.122b), is restricted to cases 3, 5, 6 (cases 1, 2, 4 apply to even numbers n_i). Analogous to crystals with an sc lattice, the **total shell multiplicities** M_i^{tot} of crystals with a bcc lattice depend on the shell radius in a chaotic fashion, with M_i^{tot} values increasing with shell radii on the average but also with shells of very small total shell multiplicity in between. As examples we mention the 65th neighbor shell ($N_i = 48$, $M_i^{tot} = 8$) and the 86th neighbor shell ($N_i = 64$, $M_i^{tot} = 6$). Figure 2.32b shows the chaotic behavior of M_i^{tot} with shell index for the first 100 neighbor shells of a crystal with a bcc lattice (shell no. 100 corresponds to a radius of 8.602 in units of the lattice constant a, see Figure 2.33).

The results for crystals with a bcc lattice can also be applied analogously to the **cesium chloride** (CsCl) crystal. This crystal, shown in Figure 2.2a, is described by an sc lattice and a basis of two atoms, Cs and Cl, where the primitive morphological unit cell contains Cs atoms at its corners and a Cl atom at its center (or vice versa), see Figure 2.2b. Thus, if the element types were ignored, the lattice would be described as a bcc lattice. Therefore, according to Eqs. (2.119)–(2.122) the neighbor shell arrangement in the CsCl crystal is given by the **two sets** of shells found for crystals with a bcc lattice. Each set contains only one element type, **Cs** or **Cl**. This means, in particular, that for this ionic crystal neighbor shells contain ions of only one kind, of positive (Cs$^+$) or of negative (Cl$^-$) charge. However, the shells do not strictly alternate between positive and negative charge with increasing shell radius. As an illustration, Table 2.8 lists the charge sequence for the first 20 neighbor shells starting at the center of a Cs atom (listed as shell no. 0).

The results obtained for crystals with an sc lattice can also be applied to crystals with an fcc lattice. In Section 2.2.3 it was shown that the fcc lattice can be represented by non-primitive sc lattice vectors $\underline{R}_1$, $\underline{R}_2$, $\underline{R}_3$ with three additional lattice vectors $\underline{R}^{(1)}$, $\underline{R}^{(2)}$, $\underline{R}^{(3)}$, pointing to the centers of the three unique faces of the morphological unit cell. Therefore, general lattice points can be described by

Table 2.8 Charge sequence of the first 20 neighbor shells of the CsCl crystal starting at the center of a Cs atom (shell no. 0).

Shell no.	0	1	2	3	4	5	6	7	8	9	10
(element)	(Cs)	(Cl)	(Cs)	(Cs)	(Cl)	(Cs)	(Cs)	(Cl)	(Cs)	(Cs)	(Cl)
Charge	+1	−8	+6	+12	−24	+8	+6	−24	+24	+24	−32

Shell no.	11	12	13	14	15	16	17	18	19	20
(element)	(Cs)	(Cl)	(Cs)	(Cs)	(Cl)	(Cs)	(Cs)	(Cl)	(Cs)	(Cs)
Charge	+12	−48	+30	+24	−24	+24	+8	−48	+24	+48

2.6 Neighbor Shells

four sets of lattice vectors, that is, by

$$\underline{R} = \underline{r}_c + n_1 \underline{R}_1 + n_2 \underline{R}_2 + n_3 \underline{R}_3 \ , \ \underline{r}_c = \underline{R}^{(0)} = \underline{0}, \quad \underline{R}^{(1)}, \quad \underline{R}^{(2)}, \quad \underline{R}^{(3)} \tag{2.123}$$

yielding in Cartesian coordinates

$$\underline{R}^{(0)} = (0,0,0), \qquad \underline{R} = a/2\,(2n_1, 2n_2, 2n_3),$$
$$\underline{R}^{(1)} = a/2\,(0,1,1), \quad \underline{R} = a/2\,(2n_1, 2n_2+1, 2n_3+1),$$
$$\underline{R}^{(2)} = a/2\,(1,0,1), \quad \underline{R} = a/2\,(2n_1+1, 2n_2, 2n_3+1),$$
$$\underline{R}^{(3)} = a/2\,(1,1,0), \quad \underline{R} = a/2\,(2n_1+1, 2n_2+1, 2n_3), \quad n_i \text{ integer} \tag{2.124}$$

As a consequence, for a crystal with an fcc lattice, neighbor shells about the origin $\underline{R}_c = (0, 0, 0)$ with shell radius D_i and a range $\varepsilon_i = 0$ can be defined, according to Eq. (2.112), by four sets, by

$$D_i^2 = a^2/4\,\left\{(2n_1)^2 + (2n_2)^2 + (2n_3)^2\right\},$$
$$D_i^2 = a^2/4\,\left\{(2n_1)^2 + (2n_2+1)^2 + (2n_3+1)^2\right\},$$
$$D_i^2 = a^2/4\,\left\{(2n_1+1)^2 + (2n_2)^2 + (2n_3+1)^2\right\},$$
$$D_i^2 = a^2/4\,\{(2n_1+1)^2 + (2n_2+1)^2 + (2n_3)^2\}, \quad n_i \text{ integer} \tag{2.125}$$

referring to vectors $\underline{R}$ according to Eq. (2.124). Thus, in all four cases neighbor shells for shell radii D_i are determined by

$$D_i = a/2\sqrt{N_i} \tag{2.126}$$

which agrees with the result for crystals with sc and bcc lattices. However, the integers N_i are now determined by **four** different quadratic Diophantine equations

$$N_i = (2n_1)^2 + (2n_2)^2 + (2n_3)^2 = 4\,\{n_1^2 + n_2^2 + n_3^2\} \tag{2.127a}$$

$$N_i = (2n_1)^2 + (2n_2+1)^2 + (2n_3+1)^2 = 4\,\{n_1^2 + n_2^2 + n_3^2 + n_2 + n_3\} + 2 \tag{2.127b}$$

$$N_i = (2n_1+1)^2 + (2n_2)^2 + (2n_3+1)^2 = 4\,\{n_1^2 + n_2^2 + n_3^2 + n_1 + n_3\} + 2 \tag{2.127c}$$

$$N_i = (2n_1+1)^2 + (2n_2+1)^2 + (2n_3)^2 = 4\,\{n_1^2 + n_2^2 + n_3^2 + n_1 + n_2\} + 2 \tag{2.127d}$$

with possible solutions n_1, n_2, n_3. Here, Eq. (2.127a) yields integers n_1, n_2, n_3 only for $N_i = 4P$ (P integer), whereas Eqs. (2.127b)–(2.127d) yield integers n_1, n_2, n_3 for $N_i = 4P + 2$ (P integer). Thus, N_i values of the three Eqs. (2.127b)–(2.127d) can coincide whereas they do not coincide with N_i values of Eq. (2.127a). Further, Eqs. (2.127a)–(2.127d) can be considered as special cases of the quadratic

Diophantine equation (2.116) for an sc lattice, however, with two separate sets of N_i values. Altogether, the complete set of neighbor shells of a crystal with an fcc lattice can be decomposed into **two disjoint sets,** corresponding to $N_i = 4P$ and $N_i = 4P + 2$, which are described each by **selected shells** of a crystal with an sc lattice and a lattice constant $a/2$ according to Eq. (2.121).

As to **symmetry-related shell multiplicities** M_i^{sym} of the different shells, the first shell set, defined by Eq. (2.127a), refers to all cases 1–6 of Table 2.6. The second set, determined by Eqs. (2.127b)–(2.127d), is restricted to cases 2, 4, 5, 6 (cases 1, 3 do not allow one even and two odd numbers n_k). Analogous to crystals with sc and bcc lattices, the **total shell multiplicities** of crystals with an fcc lattice depend on the shell radius in a chaotic fashion with M_i^{tot} values increasing with shell radii on the average and also with shells of very small total shell multiplicity in between. As examples, we mention the 30th neighbor shell ($N_i = 16$, $M_i^{tot} = 6$) and the 90th neighbor shell ($N_i = 48$, $M_i^{tot} = 8$). Figure 2.32c illustrates the chaotic behavior of M_i^{tot} with shell index for the first 100 neighbor shells of a crystal with an fcc lattice (shell No. 100 corresponds to a radius of 7.316 in units of the lattice constant a, see Figure 2.33).

The results for crystals with sc and fcc lattices can also be applied to the **sodium chloride** (NaCl) crystal. This crystal, shown in Figure 2.34a can be defined by an sc lattice and a basis of eight atoms, four Na and Cl each. The primitive morphological unit cell contains Na atoms at its corners as well as at the centers of the three unique faces, while Cl atoms reside in the cell center and at midpoints of all unique edges, see Figure 2.34b. (Na and Cl atoms can be interchanged in the definition.) In fact the two elemental parts of the crystal can both be characterized by fcc lattices shifted with respect to each other. Further, if the element types were ignored, the lattice would be described by an sc lattice of lattice constant $a/2$. Therefore, according to the previous discussion, the neighbor shell arrangement of the NaCl crystal is given by the **two sets** of shells for the **Na** part, described by Eqs. (2.127a)–(2.127d) for an fcc crystal, where $N_i = 4P$ or

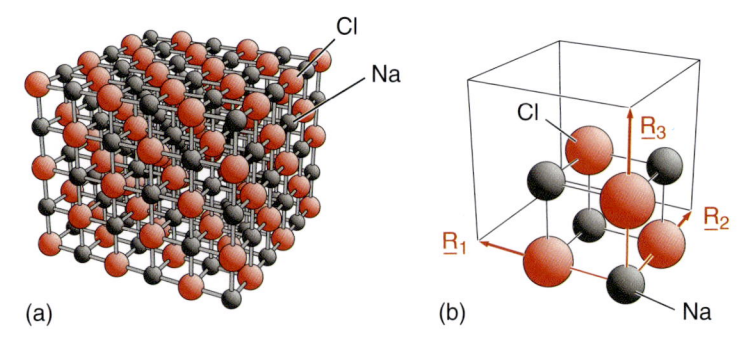

Figure 2.34 (a) Section of a cubic NaCl crystal. Sticks connect Na with neighboring Cl atoms to indicate the crystal structure. (b) Primitive morphological unit cell with eight atoms, $4 \times$ Na and $4 \times$ Cl, inside. The lattice vectors $\underline{R}_1$, $\underline{R}_2$, $\underline{R}_3$ are labeled accordingly.

$N_i = 4P + 2$. Further, the shells of the Cl part can be shown to be described by

$$N_i = (2n_1 + 1)^2 + (2n_2 + 1)^2 + (2n_3 + 1)^2$$
$$= 4\{n_1^2 + n_2^2 + n_3^2 + n_1 + n_2 + n_3\} + 3$$
$$N_i = (2n_1 + 1)^2 + (2n_2)^2 + (2n_3)^2 = 4\{n_1^2 + n_2^2 + n_3^2 + n_1\} + 1$$
$$N_i = (2n_1)^2 + (2n_2 + 1)^2 + (2n_3)^2 = 4\{n_1^2 + n_2^2 + n_3^2 + n_2\} + 1$$
$$N_i = (2n_1)^2 + (2n_2)^2 + (2n_3 + 1)^2 = 4\{n_1^2 + n_2^2 + n_3^2 + n_3\} + 1 \quad (2.128)$$

leading to $N_i = 4P + 1$ or $N_i = 4P + 3$. This also defines **two sets** of shells of the **Cl** part. Thus, the shell arrangement in the NaCl lattice is given by **four sets** of shells, two each for Na and Cl, which are disjoint and contain only one element type. Therefore, neighbor shells of ionic crystals of the NaCl type contain ions of only one kind, of positive (Na$^+$) or of negative (Cl$^-$) charge, analogous to CsCl discussed above. Further, the ionic Na$^+$ and Cl$^-$ shells do not strictly alternate with increasing shell radius, also in analogy with CsCl. As an illustration, Table 2.9 lists the charge sequence for the first 20 neighbor shells starting at the center of a Na atom (listed as shell no. 0).

Crystals with a **hexagonal** lattice and close-packed (**hcp**) structure (i.e., with $c/a = \sqrt{(8/3)}$ and two identical atoms in the primitive unit cell, see Figure 2.10) are described by neighbor shells surrounding atoms, which have, at larger distances, less atom members in each shell compared to the fcc lattice of equal atom density. This is confirmed by Figure 2.32d, which shows the chaotic behavior of M_i^{tot} with increasing shell radius for the first 100 neighbor shells (shell No. 100 corresponds to a radius of $6.931a$, where a is the lattice constant, see Figure 2.33).

There are **no explicit** formulas to determine properties, such as radius or shell multiplicity, of the nth neighbor shell of a given lattice. However, there is a simple **strategy** to evaluate neighbor shells up to a maximum radius D_{max} in a crystal with its lattice described by any lattice vectors $\underline{R}_1, \underline{R}_2, \underline{R}_3$ and with a corresponding basis. For the sake of simplicity, we confine ourselves to crystals with one atom in the primitive unit cell and primitive lattice vectors $\underline{R}_1, \underline{R}_2, \underline{R}_3$. Then, after selecting

Table 2.9 Charge sequence of the first 20 neighbor shells of the NaCl lattice starting at the center of a Na atom (denoted as shell No. 0).

Shell no.	0	1	2	3	4	5	6	7	8	9	10
(element)	(Na)	(Cl)	(Na)	(Cl)	(Na)	(Cl)	(Na)	(Na)	(Cl)	(Na)	(Cl)
Charge	+1	−6	+12	−8	+6	−24	+24	+12	−30	+24	−24
Shell no.	11	12	13	14	15	16	17	18	19	20	
(element)	(Na)	(Cl)	(Na)	(Na)	(Cl)	(Na)	(Cl)	(Na)	(Cl)	(Na)	
Charge	+8	−24	+48	+6	−48	+36	−24	+24	−48	+24	

a shell center $\underline{R}_c$ inside the morphological unit cell we build a polyhedral cell around $\underline{R}_c$ including all atom positions $\underline{R}$ with respect to the shell center, that is,

$$\underline{R} = n_1 \underline{R}_1 + n_2 \underline{R}_2 + n_3 \underline{R}_3 - \underline{R}_c \quad \text{with} \quad -N_k \leq n_k \leq N_k, \quad k = 1, 2, 3 \quad (2.129)$$

The inscribed sphere of this polyhedral cell, centered at $\underline{R}_c$, has a radius D_{max} which is given by the smallest of three lengths, that is, by

$$\begin{aligned} D_{max} &= \min\left(\left|N_1 \underline{R}_1 \frac{(\underline{R}_2 \times \underline{R}_3)}{|\underline{R}_2 \times \underline{R}_3|}\right|, \left|N_2 \underline{R}_2 \frac{(\underline{R}_3 \times \underline{R}_1)}{|\underline{R}_3 \times \underline{R}_1|}\right|, \left|N_3 \underline{R}_3 \frac{(\underline{R}_1 \times \underline{R}_2)}{|\underline{R}_1 \times \underline{R}_2|}\right|\right) \\ &= V_{el} \min\left(\frac{N_1}{|\underline{R}_2 \times \underline{R}_3|}, \frac{N_2}{|\underline{R}_3 \times \underline{R}_1|}, \frac{N_3}{|\underline{R}_1 \times \underline{R}_2|}\right) \end{aligned} \quad (2.130)$$

Figure 2.35 illustrates the inscribed sphere for a polyhedral cell of a crystal with triclinic-P lattice with $N_1 = N_2 = N_3 = 2$.

As a next step we evaluate all atom positions of the polyhedral cell according to Eq. (2.129) together with their distances D with respect to the center $\underline{R}_c$ given by

$$D = |\underline{R}| = |n_1 \underline{R}_1 + n_2 \underline{R}_2 + n_3 \underline{R}_3 - \underline{R}_c| \quad (2.131)$$

Sorting these atom positions according to their D values in increasing order and grouping those with equal (or very similar) distances yields neighbor shells with respect to $\underline{R}_c$ where, however, only those shells with $D \leq D_{max}$ are guaranteed to be complete. Determining shells of radii larger than D_{max} is achieved by increasing the ranges N_i in Eq. (2.130) and going through the same procedure.

The concept of neighbor shells also becomes important for bulk crystals with a **surface**. There the truncation of the perfect three-dimensional crystal, yielding the crystal substrate below and vacuum above, creates atom environments near the surface that are incomplete if compared with perfect bulk environments. This will be discussed in more detail in Section 4.1.

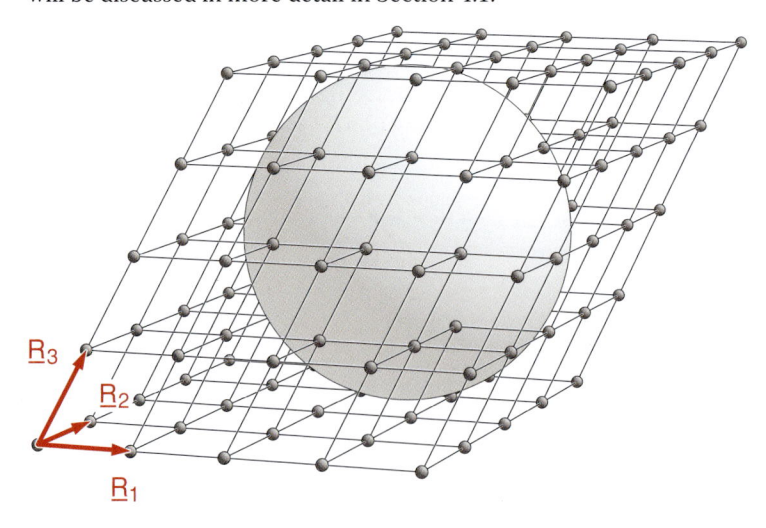

Figure 2.35 Polyhedral cell of a crystal with triclinic-P lattice with inscribed sphere. The lattice vectors are labeled accordingly.

2.7
Nanoparticles and Crystallites

Finite size particles are characterized by aggregates of atoms that can be of the same element type but are exposed to different local environments depending on their location inside the particle or at its surface. Atoms close to the particle surface are coordinated with fewer atom neighbors compared to those near the particle center, which influences their interatomic binding and affects the particle structure. This is different from the atom arrangement inside a perfect monoatomic crystal with its three-dimensional periodicity, which results in equivalent atom centers and, thus, leads to identical binding environments and atom coordination. The inhomogeneity of atom environments in finite particles depends strongly on the particle size since the relative number of surface atoms compared with those of the inner particle core becomes smaller with increasing size. Thus, one might expect that deviations from a crystalline bulk structure become less important for most of the atoms as the particle size increases.

In many cases, structural properties of **clusters** with only a few atoms, typically 2–200, do not reflect those of corresponding bulk crystals. Here, the structural details depend on the specific cluster and there are no general guidelines as to interatomic distances or angles or as to symmetry. As an example, density-functional theory studies on silver clusters with up to 12 atoms, Ag_n, n = 2, ..., 12, [55] have identified equilibrium structures that differ substantially from those of sections of the fcc crystal found for bulk silver. This is illustrated by Figure 2.36, which shows two energetically very close isomers of the Ag_7 cluster, (a) a tricapped tetrahedron (also verified by experiment) and (b) a pentagonal bipyramid where in both isomers interatomic distances d_{Ag-Ag} close to 2.7 Å on the average are obtained. This is considerably smaller than the nearest neighbor value $d_{Ag-Ag} = 2.89$ Å in the fcc silver crystal. Further, the isomer (b) includes a fivefold rotation symmetry axis, which is forbidden in bulk crystals as discussed in Section 2.4.

Larger metal clusters are also found to exhibit symmetry properties that are not compatible with those of bulk crystals. As examples, many alkaline earth (Be, Mg, Ca, Sr) and transition metal clusters (Ni, Co) in gas phase, the former with up to 5000 atoms [56], are believed to form compact particles with **icosahedral**

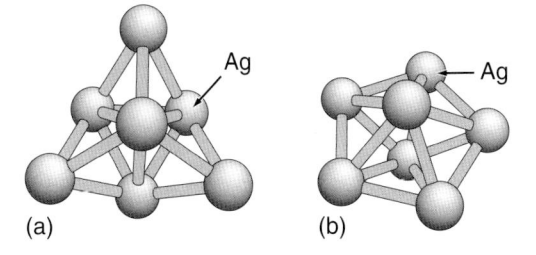

Figure 2.36 Balls-and-sticks models of two isomers of the Ag_7 cluster, (a) tricapped tetrahedron and (b) pentagonal bipyramid.

symmetry for selected atom numbers N, so-called **magic numbers**, for which particles Me_N are found to occur in preferred abundance. However, icosahedral symmetry cannot appear in perfect bulk crystals since it includes fivefold rotational axes.

The geometric definition of an icosahedral cluster is based on the concept of **polyhedral atom shells** of increasing size about a central atom where the shells are of icosahedral symmetry I_h. These compact closed shells consist of planar sections of triangular shape where the equilateral triangles are filled by close-packed (hexagonal) arrays of atoms with n atoms at each triangle side. Each triangle shares its three edges with adjacent triangles where five triangles meet at corners that are centers of fivefold rotational axes. This yields, altogether, shells with 20 triangles sharing 30 identical edges and 12 corners. As an illustration, Figure 2.37 shows icosahedral shells for n = 2, 4, and 9 atoms per edge, which form the outer shells of icosahedral clusters of N = 13, 147, and 2057 atoms, respectively.

If the shell edges contain n atoms each, the total number of atoms $n_s(n)$ of the nth icosahedral shell (n > 1) amounts to

$$n_s(n) = 20\left(3\frac{1}{5} + 3(n-2)\frac{1}{2} + \frac{1}{2}(n-3)(n-2)\right) = 10(n-1)^2 + 2 \quad (2.132)$$

and the normalized distance $r_a(n)$ of the corner atoms from the shell center (neglecting any surface relaxations of the atomic positions) is, after some trigonometric calculus, given by

$$r_a(n) = \frac{r(n)}{a} = \left(\frac{1}{2}\sqrt{1 + \frac{2}{3-\sqrt{5}}}\right)(n-1) = 0.95106(n-1) \quad (2.133)$$

where a is the distance between neighboring atoms on the shell. Further, the combination of n concentric icosahedral shells to form an icosahedral cluster yields N(n) atoms where

$$N(n) = 1 + \sum_{i=2}^{n} n_s(i) = \frac{1}{3}(2n-1)[5n(n-1)+3] \quad (2.134)$$

Numbers N(n) are usually called **magic numbers** and have been discussed in mathematical detail in the literature [57]. In the larger icosahedral clusters, n > 4,

(a) (b) (c)

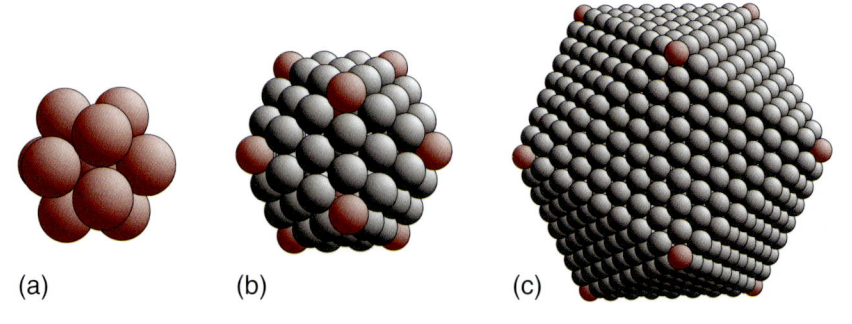

Figure 2.37 Atom ball models of icosahedral shells for (a) n = 2, (b) n = 4, (c) n = 9. Corner atoms are emphasized by red shading.

one can distinguish between atoms inside the cluster ("bulk" atoms) which experience a full nearest neighbor environment of 12 atoms (6 at distances $d = 0.951\,a$ and 6 at $d = a$, respectively) and atoms of the outer shell ("surface" atoms) where the environment consists of 6 (1 at $d = 0.951\,a$ and 5 at $d = a$, corner), 8 (2 at $d = 0.951\,a$ and 6 at $d = a$, edge), or 9 (3 at $d = 0.951\,a$ and 6 at $d = a$, facet inside triangle) atoms. The ratio of the number of surface and bulk atoms $\rho(n)$ given by

$$\rho(n) = \frac{n_s(n)}{(N(n) - n_s(n))} \quad (2.135)$$

decreases with increasing cluster size converging to $\rho(n) \approx 3/n$ for $n \gg 1$. Table 2.10 lists shell and cluster sizes as well as relative radii of corner atoms for icosahedral clusters up to 3000 atoms.

Large metal clusters have also been found to exhibit symmetry, which phenomenon can be associated with finite sections of cubic bulk crystal structures, both fcc and bcc. Here, examples are aluminum and indium clusters between 1000 and 10 000 atoms in gas phase [56] which are suggested to form compact particles with **cubic symmetry,** reflecting sections of the fcc bulk crystal, where for selected atom numbers N (**magic numbers**) the metal clusters are found to occur in preferred abundance. Among these, **cuboctahedral** cluster shapes have been discussed [56].

As in the icosahedral case, a cuboctahedral cluster can be constructed by packing polyhedral atom shells of increasing size about a central atom where, the shells are of cubic symmetry O_h. They consist of planar sections of both triangular and square shape. The equilateral triangles are filled by close-packed (hexagonal) arrays of atoms with n atoms at each triangle side while the squares reflect n × n square arrays of atoms where the nearest neighbor distances agree with those of the triangular sections. This yields a lower atom packing inside the

Table 2.10 Shell sizes $n_s(n)$, cluster sizes N(n), normalized radii $r_a(n)$ of corner atoms, and ratio of surface to bulk atoms $\rho(n)$ for icosahedral and cuboctahedral clusters with up to 3000 atoms.

N	$n_s(n)$	N(n)	$r_a(n)$	$\rho(n)$
1	1	1	—	—
2	12	13	0.951 (1.000)	12.000
3	42	55	1.902 (2.000)	3.231
4	92	147	2.853 (3.000)	1.673
5	162	309	3.804 (4.000)	1.102
6	252	561	4.755 (5.000)	0.816
7	362	923	5.706 (6.000)	0.645
8	492	1415	6.657 (7.000)	0.533
9	642	2057	7.609 (8.000)	0.454
10	812	2869	8.560 (9.000)	0.395

The values of $n_s(n)$, N(n), and $\rho(n)$ are valid for both cluster types while $r_a(n)$ values differ with the results for cuboctahedral clusters added in parentheses.

squares compared with the triangles. Each triangle shares its three edges with adjacent squares and squares are connected only with triangles at their edges. Further, two triangles and two squares join at each corner of the cluster. This yields altogether shells with 8 triangles and 6 squares sharing 24 identical edges and 12 corners. Figure 2.38 shows cuboctahedral shells for n = 2, 4, and 9 atoms per edge, which form the outer shells of cuboctahedral clusters of N = 13, 147, and 2057 atoms, respectively.

If the shell edges contain n atoms each, the eight triangular sections account for

$$n_s^{tria}(n) = 8\left(3\frac{1}{4} + 3(n-2)\frac{1}{2} + \frac{1}{2}(n-3)(n-2)\right) = 4(n-1)^2 + 2 \quad (2.136)$$

atoms and the six square sections for

$$n_s^{sqr}(n) = 6\left(4\frac{1}{4} + 4(n-2)\frac{1}{2} + (n-2)^2\right) = 6(n-1)^2 \quad (2.137)$$

atoms, which yields a total number of atoms $n_s(n)$ of the nth cuboctahedral shell (n > 1) with

$$n_s(n) = n_s^{tria}(n) + n_s^{sqr}(n) = 10(n-1)^2 + 2 \quad (2.138)$$

which, according to Eq. (2.132), agrees with the atom count of the nth icosahedral shell. The normalized distance $r_a(n)$ of the corner atoms from the shell center (shell radius) is given by

$$r_a(n) = \frac{r(n)}{a} = (n-1) \quad (2.139)$$

where a is the distance between neighboring atoms on the shell. Since the atom count of cuboctahedral shells (2.138) is identical to that of icosahedral shells the total numbers of atoms N(n) in a cuboctahedral cluster agrees with the corresponding icosahedral value given by Eq. (2.134). Thus, the magic numbers N(n) are identical for cuboctahedral and icosahedral clusters [57] while the atom packing differs between the two cluster types. The distinction between atoms inside the cluster ("bulk" atoms) and those of the outer shell ("surface" atoms) for larger cuboctahedral clusters, n > 3, is analogous to that for icosahedral clusters.

(a) (b) (c)

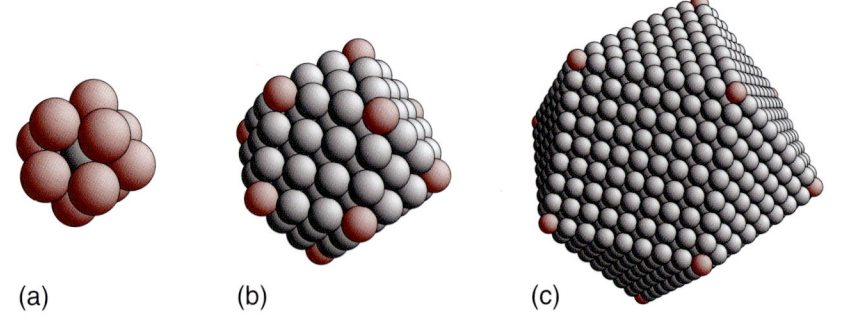

Figure 2.38 Atom ball models of cuboctahedral shells for (a) n = 2, (b) n = 4, (c) n = 9. Corner atoms are emphasized by red shading.

Here, bulk atoms experience the nearest neighbor environment of 12 atoms (at distances $d = a$), which is identical to the geometry inside the fcc bulk crystal. In contrast, the environment of surface atoms consists of five (corner), seven (edge), eight (inside square), or nine (inside triangle) atoms with identical distances $d = a$. In addition, the ratio of the number of surface and bulk atoms $\rho(n)$ given by Eq. (2.135) yields values that agree for corresponding cuboctahedral and icosahedral clusters.

Clusters of cuboctahedral shape and moderate size, up to 1000 atoms, represent a fairly good approximation to spherical particles since the atoms of the different planar sections do not vary too much in their distance from the cluster center. This is illustrated in Figure 2.39, which compares the cuboctahedral cluster for $n = 4$ (147 atoms) with a "spherical" cluster that contains all atoms inside a sphere whose radius is equal to the radius r_a of the corner atoms of the cuboctahedral cluster. The spherical cluster, containing 177 atoms, differs from the cuboctahedral cluster only by five additional atoms (painted red in Figure 2.39b) sitting on top of each square section and roughening the cluster surface. Clearly, the difference between cuboctahedral and the corresponding spherical clusters will become larger with increasing cluster size when more atoms can fill the void between the planar shell sections and the spherical boundary which results in smooth transitions between planar shell sections of high packing density.

Other high-symmetry sections of **fcc** bulk crystals have been suggested in the literature [56, 57] as possible structures of metal clusters where we mention only **simple octahedral** clusters whose closed shells consist of 8 planar sections of equilateral triangular shape filled by close-packed (hexagonal) arrays of atoms resulting in 12 identical edges with n atoms each and 6 corners. This leads to closed shell clusters with $N(n)$ atoms (magic numbers), $n_s(n)$ atoms per shell, and normalized shell radii $r_a(n)$ given by

$$N(n) = \frac{1}{3} n (2n^2 + 1), \quad n_s(n) = 4(n-1)^2 + 2, \quad r_a(n) = \frac{1}{\sqrt{2}} (n-1) \quad (2.140)$$

where clusters with odd n contain an atom in the cluster center while for even n the cluster center is void. As an illustration, Figure 2.40a shows a simple octahedral

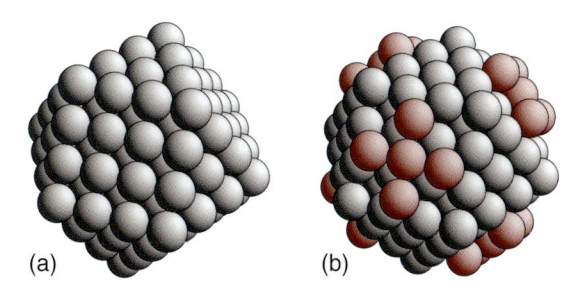

Figure 2.39 Comparison of atom ball models of (a) the cuboctahedral cluster ($n = 4$, see Figure 2.38b) with (b) a cluster of similar size with spherical constraints. Atom balls outside the cuboctahedron are emphasized by red shading.

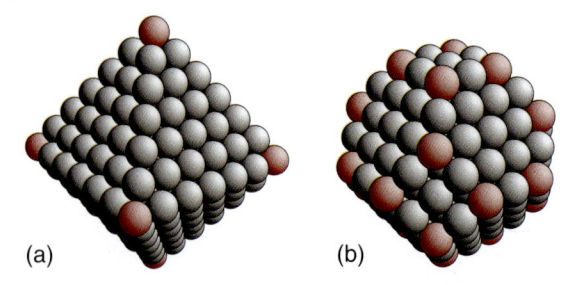

Figure 2.40 Atom ball models of clusters (a) simple octahedral with n = 7 (231 atoms) and (b) truncated octahedral with n = 3 (201 atoms). Corner atoms are emphasized by red shading.

cluster for n = 7 containing 231 atoms. In addition, more complex cubic cluster structures with polyhedral shells representing mixtures of square and hexagonal sections have been considered. Here, we mention only the **truncated octahedral** type, shown with 201 atoms in Figure 2.40b, which is obtained from a simple octahedral cluster by cutting off the six corners symmetrically. This cluster shape is reminiscent of the Wigner-Seitz cell of a bcc crystal, see Section 2.3. The corresponding clusters consist of 6 square and 8 hexagonal sections forming 36 identical edges with n atoms each and 24 corners. The closed shell clusters are characterized by

$$N(n) = 16n^3 - 33n^2 + 24n - 6, \quad n_s(n) = 30(n-1)^2 + 2, \quad r_a(n) = \frac{\sqrt{10}}{2}(n-1)$$
(2.141)

with $n_s(n)$ denoting the number of atoms at the cluster surface.

Further, metal atom clusters of high symmetry described by sections of **bcc** bulk crystals have been discussed in the literature [56] where we mention only **dodecahedral** and **truncated dodecahedral** clusters, shown in Figure 2.41. The dodecahedral shells consist of 12 identical rhombohedral sections each, which are filled with close-packed arrays of atoms (distorted hexagonal, reflecting the densest planar packing inside the bcc crystal). The cluster shape, see Figure 2.41a, is

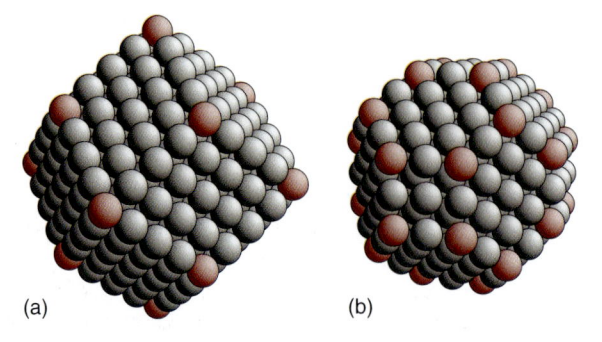

Figure 2.41 Atom ball models of clusters (a) dodecahedral with n = 5 and (b) truncated dodecahedral with n = 3. Corner atoms are emphasized by red shading.

reminiscent of the Wigner-Seitz cell of an fcc crystal, see Section 2.3. Truncated dodecahedral clusters, see Figure 2.41b, are obtained from dodecahedral clusters by cutting off six corners symmetrically. The resulting cluster shells contain 6 square and 12 distorted hexagonal sections each and are similar to the Wigner-Seitz cell of a bcc crystal, see Section 2.3.

Very large clusters above 100 000 atoms up to macroscopic sizes, usually referred to as **crystallites** or **grains**, can be assumed to form a crystalline structure inside their inner core with periodic arrangement of atoms within a finite volume. Under equilibrium conditions their exterior shape is often determined by planar surface sections (**facets**) of densely packed atom arrangements where there are transition regions between the facets. The geometric structure of the planar sections has been found to be very close to that of high-density crystal layers discussed in Section 4.2 with transitions involving more open layers with steps and kinks.

The global shape of crystallites has been discussed applying **quasi-continuum** models [58, 59] to polyhedral particles whose facets are considered as planar sections with normal directions along those of high-density crystal layers. In these models each polygonal facet is characterized by its size and a **surface free energy** per area where the latter depends on the atom density as well as on local binding properties of the facet layer. The size of the different facets can then be determined in an optimization of the total surface free energy of the particle for a given particle volume [60, 61] neglecting differences in binding energy at corners and edges. This yields the equilibrium shape of the crystallite. The procedure based on the **Gibbs–Wulff theorem** [58, 59] is also known as *Wulff construction*.

In its simplest version the **Wulff construction** starts from a crystallite center and a continuous energy function $\gamma(\underline{n})$ that depends only on the crystallographic direction vector $\underline{n}$ from the center and defines the surface free energy per area of a crystal layer whose layer normal vector equals vector $\underline{n}$. Then the total surface free energy of the polyhedral crystallite with N surface facets is given by

$$\gamma_{tot} = \sum_{i=1}^{N} \gamma(\underline{n}_i) \, a_i \qquad (2.142)$$

where $\underline{n}_i$ denotes the normal vector of the ith facet and a_i is the polygon area of the facet. To obtain the equilibrium shape of the crystallite for a given volume, function γ_{tot} needs to be minimized with respect to the number N of facets, the facet orientations $\underline{n}_i$, and their polygonal areas a_i. The Gibbs–Wulff theorem [58–60] states, in its simplest version, that the shape of a polyhedral crystallite in its thermodynamic equilibrium is achieved if all facets lie at distances d_i from the crystallite center, which are proportional to the corresponding surface free energies per area $\gamma(\underline{n}_i)$. As a consequence, a polyhedral crystallite can be constructed from a surface free energy function $\gamma(\underline{n})$ by considering a set of discrete planes $S(\underline{n}_i)$ with normal vectors $\underline{n}_i$ at distances $d_i = \lambda \gamma(\underline{n}_i)$ from the crystallite center enclosing the center. (Here λ is a global scaling factor to be adjusted later.) Then the equilibrium surface of the crystallite is described by all points on the planes $S(\underline{n}_i)$ which, for a given direction $\underline{n}$, are closest to the crystallite center. This yields polygonal areas (facets) forming the shape of the polyhedral crystallite where, in a subsequent step, the scaling constant λ is adjusted such that the required crystallite volume is obtained. In practical applications, the selection of planes $S(\underline{n}_i)$ is restricted to sets of high-density crystal layers (perpendicular to $\underline{n}_i$) and respective free energy values $\gamma(\underline{n}_i)$ are taken from experimental data for the corresponding surfaces of high atom density or from theoretical surface energies.

As an illustration, Figure 2.42 shows sketches of Wulff polyhedra of a cubic crystallite where 26 planes $S(\underline{n}_i)$ perpendicular to high-symmetry directions (x y z) are allowed to contribute to the shape of the crystallite. Further, free energy values $\gamma(\underline{n}_i)$ are assumed to be equal for symmetry equivalent directions $\underline{n}_i$ resulting in a parameter set

$$\begin{aligned}\underline{n}_i = (x, y, z) &= (\pm 1, \pm 1, \pm 1) \quad \text{with } \gamma_{(1\,1\,1)} \\ &= (\pm 1, 0, 0), (0, \pm 1, 0), (0, 0, \pm 1) \quad \text{with } \gamma_{(0\,0\,1)} \\ &= (\pm 1, \pm 1, 0), (\pm 1, 0, \pm 1), (0, \pm 1, \pm 1) \quad \text{with } \gamma_{(0\,1\,1)}\end{aligned} \quad (2.143)$$

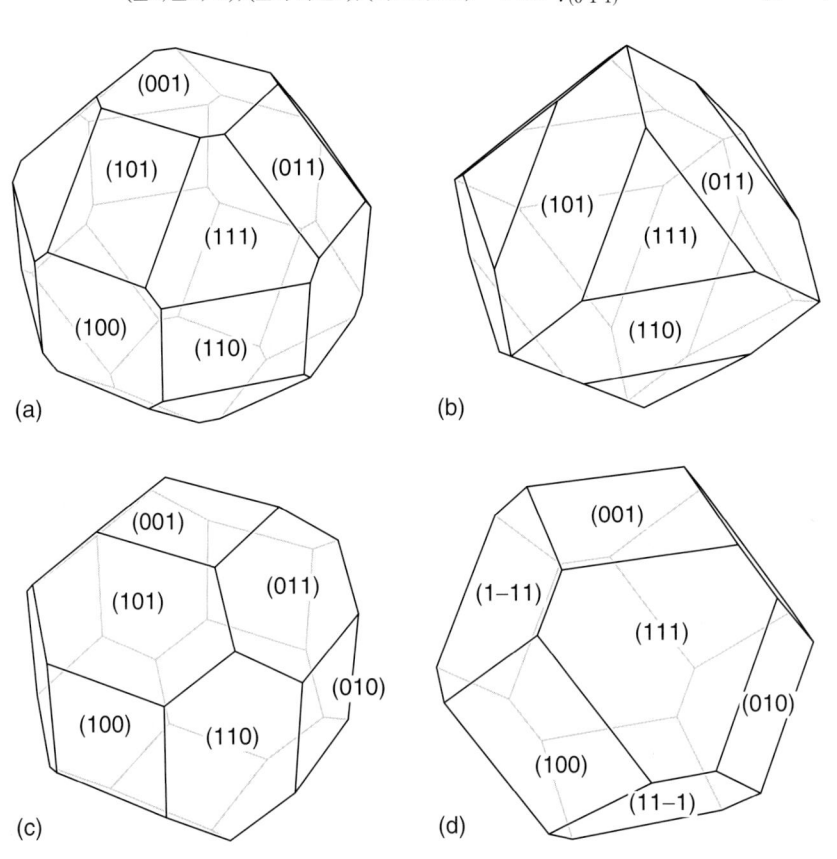

Figure 2.42 Sketches of Wulff polyhedra of a cubic crystallite for different surface free energy scenarios. (a) $\gamma_{(1\,1\,1)} = \gamma_{(0\,0\,1)} = \gamma_{(0\,1\,1)} = 1.0$; (b) $\gamma_{(1\,1\,1)} = \gamma_{(0\,1\,1)} = 1.0$, $\gamma_{(0\,0\,1)} = 1.5$; (c) $\gamma_{(0\,0\,1)} = \gamma_{(0\,1\,1)} = 1.0$, $\gamma_{(1\,1\,1)} = 1.5$; and (d) $\gamma_{(1\,1\,1)} = \gamma_{(0\,0\,1)} = 1.0$, $\gamma_{(0\,1\,1)} = 1.5$. Directions of selected polygon surfaces, (x y z), are labeled accordingly.

Figure 2.42a shows the Wulff polyhedron for surface free energy values $\gamma_{(1\,1\,1)} = \gamma_{(0\,0\,1)} = \gamma_{(0\,1\,1)} = 1.0$ in relative units. If one of the three values, for example, $\gamma_{(0\,0\,1)}$, is increased, the corresponding high-symmetry planes are less preferred and do not appear or contribute smaller polygon areas to the Wulff polyhedron, which reflects the minimum energy shape. This is evident in Figure 2.42b where $\gamma_{(0\,0\,1)}$ is increased to $\gamma_{(0\,0\,1)} = 1.5$ and high-symmetry planes corresponding to normal vectors (0 0 1) and their symmetry equivalents do not appear at the polygon surface. Figures 2.42c and d show the analogous effect for increased values $\gamma_{(1\,1\,1)} = 1.5$ and $\gamma_{(0\,1\,1)} = 1.5$, respectively.

The Wulff construction can be generalized [61] formally to smoothed polygonal shapes of crystallites with continuous transitions between planar areas which, on an atomic scale, account for open atom arrangements with steps and kinks between high-density atom arrays. Even a formal treatment of particles with generally curved surfaces has been considered. These generalizations are achieved by replacing the summation in Eq. (2.142) by a surface integration which, however, complicates the optimization procedure considerably. Further details that go beyond the scope of this book can be found in [61]. Finally, it should be emphasized that the shape of real crystallites observed by experiment is not exclusively determined by surface free energy considerations, which are the basis of the Wulff construction. Depending on preparation, kinetics, and local impurities, growth conditions of the different crystal planes can influence the shape of crystallites and may lead to metastable shapes that differ considerably from those of a Wulff construction.

2.8
Incommensurate Crystals and Quasicrystals

There are solid materials that exhibit **long-range atomic order** as well as **local symmetry** behavior. However, they are not periodic in three dimensions and are usually referred to as **aperiodic crystals**. They are different in their structural properties from amorphous materials, which show only some short-range order in their atom arrangement. As a result, the International Union of Crystallography extended its definition of a crystal to a solid producing discrete X-ray diffraction patterns where its ordering can be either periodic or aperiodic [62]. This includes three different groups of aperiodic crystals [63], those with modulated structures, incommensurate composite crystals, and quasicrystals which will be briefly discussed in the following.

2.8.1
Modulated Structures

According to Section 2.1, ideal periodic crystals are defined by lattice vectors $\underline{R}_{o1}, \underline{R}_{o2}, \underline{R}_{o3}$ describing their basic periodicity where

$$\underline{R}_o = n_1 \underline{R}_{o1} + n_2 \underline{R}_{o2} + n_3 \underline{R}_{o3}, \quad n_i \text{ integer} \qquad (2.144)$$

denotes all lattice positions which, for primitive lattices, may be set to coincide with atom sites. In crystals with modulated structures, the basic periodicity is modulated by adding a modulation vector $\underline{u}\,(\underline{r})$ to each atom position where $\underline{u}\,(\underline{r})$ is a periodic function in space with its periodicity described by a second lattice given by modulation vectors $\underline{U}_1, \underline{U}_2, \underline{U}_3$ such that

$$\underline{u}\,(\underline{r} + m_1 \underline{U}_1 + m_2 \underline{U}_2 + m_3 \underline{U}_3) = \underline{u}\,(\underline{r}), \quad m_i \text{ integer} \tag{2.145}$$

Thus, general atom positions $\underline{R}$ in a crystal with a modulated structure are given by

$$\underline{R} = \underline{R}_o + \underline{u}\,(\underline{R}_o) \tag{2.146}$$

The basic and modulation lattices are connected by a linear transformation written as

$$\begin{pmatrix} \underline{U}_1 \\ \underline{U}_2 \\ \underline{U}_3 \end{pmatrix} = \begin{pmatrix} d_{11} & d_{12} & d_{13} \\ d_{21} & d_{22} & d_{23} \\ d_{31} & d_{32} & d_{33} \end{pmatrix} \cdot \begin{pmatrix} \underline{R}_{o1} \\ \underline{R}_{o2} \\ \underline{R}_{o3} \end{pmatrix} = \underline{\underline{D}} \cdot \begin{pmatrix} \underline{R}_{o1} \\ \underline{R}_{o2} \\ \underline{R}_{o3} \end{pmatrix} \tag{2.147}$$

where one can distinguish three cases,

1) Matrix $\underline{\underline{D}}$ or its inverse $\underline{\underline{D}}^{-1}$ contain only integer-valued elements d_{ij}. Then the two lattices are commensurate with the joint periodicity determined by lattice vectors $\underline{U}_1, \underline{U}_2, \underline{U}_3$ or $\underline{R}_{o1}, \underline{R}_{o2}, \underline{R}_{o3}$. Thus the modulated structure describes a strictly **periodic** atom arrangement. This is illustrated in Figure 2.43a sketching a square lattice with lattice constant a modulated by a longitudinal wave displacement along vector $\underline{L}_{wave}$ whose length (modulation wavelength) equals $4a$.

2) Matrix $\underline{\underline{D}}$ contains a mixture of integer or fractional-valued elements d_{ij}. Then matrix $\underline{\underline{D}}$ can be written as a product of two matrices, $\underline{\underline{D}} = \underline{\underline{B}}^{-1}\underline{\underline{A}}$ where $\underline{\underline{A}}, \underline{\underline{B}}$ are integer-valued such that there is also a joint periodicity. The lattices are high-order commensurate, as discussed in detail for two-dimensional periodicities at surfaces in Section 6.4, and the modulated structure describes again a strictly **periodic** atom arrangement.

3) Matrix $\underline{\underline{D}}$ contains real-valued elements d_{ij}. Then the two lattices are incommensurate and the resulting atom arrangement (2.146) is **aperiodic**.

Altogether, an aperiodic primitive crystal with a modulated structure is defined by atom positions (2.146) with a displacement function $\underline{u}(\underline{r})$ whose periodicity is incommensurate with that of the basic lattice. This corresponds formally to the requirement that the transformation matrix $\underline{\underline{D}}$ in Eq. (2.147) contains real-valued elements d_{ij}. Modulated crystal structures have been observed, for example, for monoclinic sodium carbonate, $\gamma\text{-Na}_2\text{CO}_3$, [64] or $\text{KSm}(\text{MoO}_4)_2$ [65].

Modulation vectors $\underline{u}(\underline{r})$ can also refer to one- and two-dimensionally periodic functions, reflecting crystals with preferred modulation directions. As an illustration, Figure 2.43b shows a simple two-dimensional example of an aperiodic lattice. Here, the square lattice with lattice constant a is modulated by a transverse

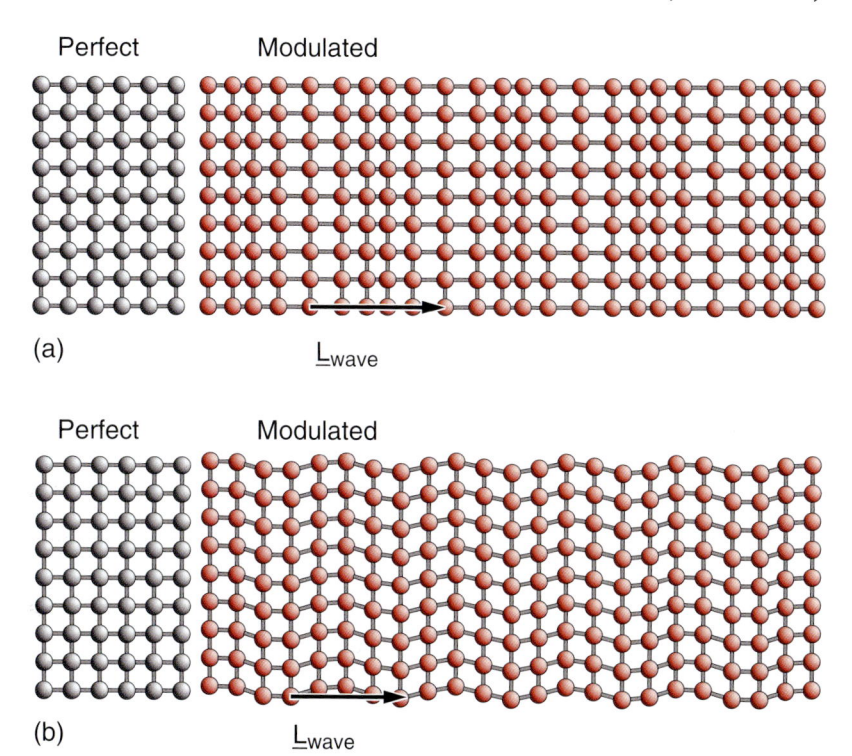

2.8 Incommensurate Crystals and Quasicrystals

Figure 2.43 Commensurate and incommensurate modulation of a square lattice; (a) periodic lattice with longitudinal wave modulation and (b) aperiodic lattice with transverse wave modulation. The left sections ("Perfect") show the perfect square lattice with lattice constant a, while the modulated lattices ("Modulated") are shown to the right. Vectors $\underline{L}_{wave}$ denote the directions and wave lengths of the modulations with $L_{wave} = 4a$ in (a), $(3\sqrt{2})a$ in (b).

wave displacement perpendicular to vector $\underline{L}_{wave}$ whose length (modulation wavelength) is given by $(3\sqrt{2})a$.

2.8.2
Incommensurate Composite Crystals

These crystals consist, in the simplest case, of two crystal components, S_1 and S_2, which themselves form periodic structures with lattice vectors $\underline{R}_{11}$, $\underline{R}_{12}$, $\underline{R}_{13}$ and $\underline{R}_{21}$, $\underline{R}_{22}$, $\underline{R}_{23}$, respectively, where the two sublattices are mutually incommensurate yielding an altogether aperiodic crystal structure with long-range order. This requires the two sublattices to be sufficiently open such that crystal component S_1 can accommodate all atoms of component S_2 or vice versa. Examples are layer type crystals where layers of atoms from component S_1 alternate with those from component S_2 yielding strict periodicity perpendicular to the layers. By contrast, the two-dimensional periodicity of the two components parallel to the layers is incommensurate. In general, this yields a linear transformation between the two sublattices where

$$\begin{pmatrix} \underline{R}_{21} \\ \underline{R}_{22} \\ \underline{R}_{23} \end{pmatrix} = \begin{pmatrix} d_{11} & d_{12} & 0 \\ d_{21} & d_{22} & 0 \\ 0 & 0 & m \end{pmatrix} \cdot \begin{pmatrix} \underline{R}_{11} \\ \underline{R}_{12} \\ \underline{R}_{13} \end{pmatrix} = \underline{D} \cdot \begin{pmatrix} \underline{R}_{11} \\ \underline{R}_{12} \\ \underline{R}_{13} \end{pmatrix}, \quad d_{ij} \text{ real, m integer or fractional}$$

(2.148)

assuming that the lattice vectors $\underline{R}_{1i}$, $\underline{R}_{2i}$, i = 1, 2, point parallel to the layers and $\underline{R}_{13}$, $\underline{R}_{23}$ connect between (e.g., are perpendicular to) adjacent layers. (If all elements d_{ij} in Eq. (2.148) would be integer- or fractional-valued the resulting lattice combination and thus the crystal would be strictly periodic in three dimensions.)

As an example, the crystal of the misfit layer compound $(LaS)_{1.13}TaS_2$ has been found [66] to form a layer type structure with alternating LaS and TaS_2 layers, see Figure 2.44. Here, the two lattice vectors perpendicular to the layers, $\underline{R}_{13}$ and $\underline{R}_{23}$, as well as one pair, $\underline{R}_{12}$ and $\underline{R}_{22}$, parallel to the layers coincide, while the lattice vectors $\underline{R}_{11}$ and $\underline{R}_{21}$ along the layers are different in length and incommensurate as indicated in Figure 2.44. In experiment, the LaS layer is found to buckle and both layers also exhibit incommensurate modulations [66], which are ignored for the present purpose and not included in Figure 2.44.

The different periodicities of incommensurate composite layer crystals along their layers, expressed by transformation Eq. (2.148) between $\underline{R}_{11}$, $\underline{R}_{12}$ and $\underline{R}_{21}$, $\underline{R}_{22}$ are formally equivalent with the periodic arrangement of **incommensurate adsorbate overlayers** at single crystal surfaces, discussed in Section 6.1.

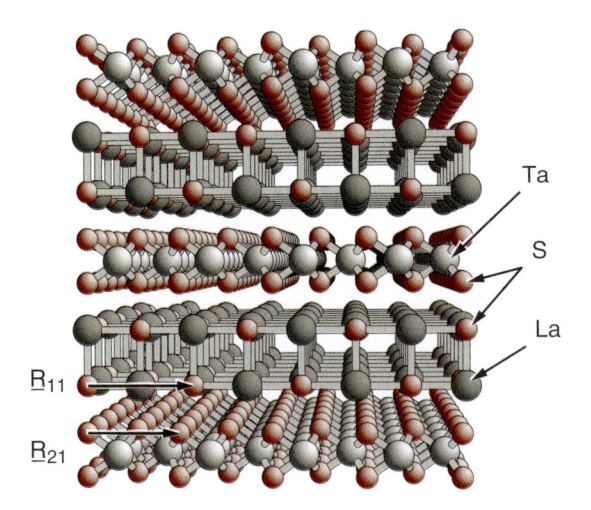

Figure 2.44 (a) Schematic view of a section of the layer type $(LaS)_{1.13}TaS_2$ crystal. Sticks connect neighboring atoms of the LaS and TaS_2 layers with atoms labeled accordingly. The incommensurate lattice vectors $\underline{R}_{11}$ (LaS layer) and $\underline{R}_{21}$ (TaS_2 layer) are indicated by arrows connecting sulfur atoms.

Therefore, the mathematical treatment of the two types of systems can be performed on the same basis.

Other groups of incommensurate composite crystals include two-component materials where component S_1 forms a framework with open channels. Then the atoms of component S_2 arrange in one-dimensionally periodic chains inside the channels where the chain periodicity is incommensurate with the periodicity of the host along the channels. This yields, in general, a linear transformation between the two sublattices where

$$\begin{pmatrix}\underline{R}_{21}\\ \underline{R}_{22}\\ \underline{R}_{23}\end{pmatrix}=\begin{pmatrix}m_{11}&m_{12}&0\\ m_{21}&m_{22}&0\\ 0&0&d\end{pmatrix}\cdot\begin{pmatrix}\underline{R}_{11}\\ \underline{R}_{12}\\ \underline{R}_{13}\end{pmatrix}=\underline{\underline{D}}\cdot\begin{pmatrix}\underline{R}_{11}\\ \underline{R}_{12}\\ \underline{R}_{13}\end{pmatrix}, \quad m_{ij} \text{ integer or fractional, d real}$$

(2.149)

assuming that the lattice vectors $\underline{R}_{1i}$, $\underline{R}_{2i}$, $i=1, 2$, define a common two-dimensional lattice of the crystal (e.g., perpendicular to the channels) and $\underline{R}_{13}$, $\underline{R}_{23}$ point along the channels.

An example, the compound $Rb_{1.37}MnO_2$ (derived from $Rb_{15}Mn_{11}O_{22}$) [67] has been found to consist of an open hexagonal host lattice of rubidium atoms where chains of connected MnO_4 tetrahedra form inside the hexagonal Rb channels. Here, the two lattice vectors along the hexagonal channels, $\underline{R}_{13}$ and $\underline{R}_{23}$, are incommensurate and the periodicity along the MnO_4 chains is further perturbed by modulation. In contrast, the two components, Rb and MnO_4, share the two-dimensional periodicity perpendicular to the channels. as shown in Figure 2.45. In experiment, the connected MnO_4 units defining the inner chains

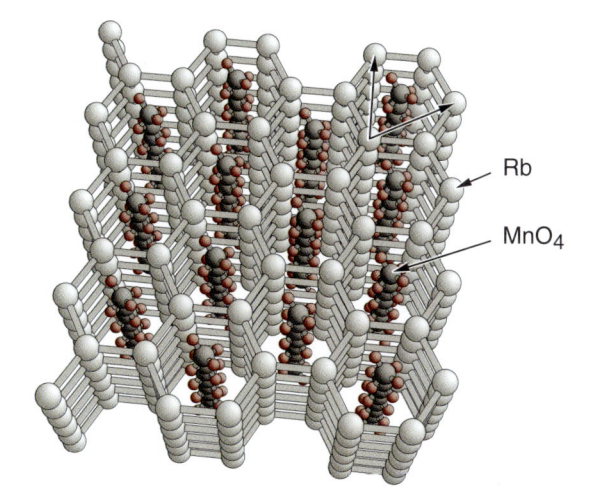

Figure 2.45 Schematic view of a section of the $Rb_{1.37}MnO_2$ crystal. Sticks connect neighboring atoms of the Rb host lattice and the MnO_4 chains with atoms labeled accordingly. The periodicity perpendicular to the hexagonal channels is indicated by arrows connecting rubidium atoms.

are found to distort and form spiral columns [67], which are ignored for the present purpose and not shown in Figure 2.45.

2.8.3
Quasicrystals

There is a third class of crystals that exhibit long-range order but are not periodic in three dimensions. They were first suggested by experiment [68] and are referred to as *quasiperiodic crystals* or *quasicrystals*. These materials, consisting in many cases of aluminum-rich metal alloys, exhibit numerous exciting physical and chemical properties [69–75]. Examples are high mechanical hardness as well as relatively low electrical and thermal conductance, which make quasicrystalline materials good candidates for surface coatings. Surfaces of quasicrystals are known for their low friction and adhesion, as well as for their good oxidation resistance [76, 77]. Aperiodic crystals including quasicrystals have also become a common playground for mathematicians [70, 71, 78]. As a result, there is a wealth of publications and textbooks dealing with quasicrystals, going far beyond the scope of the present book. References [28, 69–74, 78, 79] represent only a few examples. Here, we confine ourselves to some aspects of structure and symmetry of quasicrystals by simple examples illustrating general issues.

As a first example of quasicrystalline order, a two-dimensional model of a quasicrystal will be considered. This model is connected with the mathematical theory of **Penrose tiling** [78, 80]. The basic subject of this theory is to cover a plane completely without holes or overlaps using tiles of a finite set of different polygons, so-called **prototiles**. Here we consider only **rhombic** prototiles of two different shapes, a "**fat**" rhombus with its smallest vertex angle at $\alpha = 360°/5 = 72°$ and a "**thin**" rhombus with its smallest vertex angle at $\beta = 36°$, where both rhombuses have edges of the same length a, as shown in Figure 2.46. As a consequence of

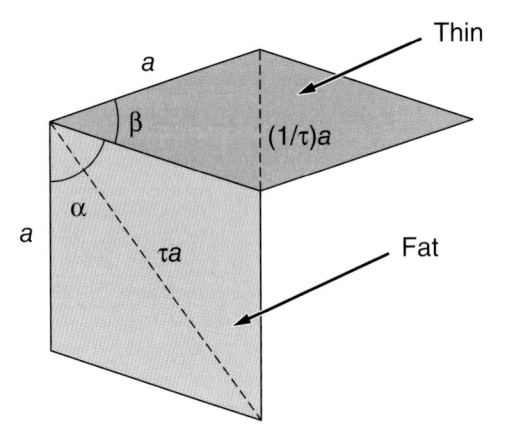

Figure 2.46 Rhombic Penrose prototiles; fat (light gray) and thin (dark gray) tiles with edge lengths a and with smallest vertex angles of $\alpha = 72°$ and $\beta = 36°$, respectively.

the choice of angles (whose values will become more evident when rotational symmetry of the tiling pattern is considered), the long diagonal of the fat rhombus has a length of τa, while the short diagonal of the thin rhombus has a length of $(1/\tau)a$ (diagonals as dashed lines in Figure 2.46), where $\tau = (1+\sqrt{5})/2 = 1.618034$ denotes the **golden ratio** or **golden mean.**

Sets of the two prototiles can be arranged such that they form a two-dimensionally **periodic pattern,** shown in Figure 2.47. In this figure, atom balls are placed at the centers of all rhombuses, one large for each fat rhombus and one small for each thin rhombus. This leads to a planar atom arrangement that describes a two-dimensionally **periodic crystal** with primitive lattice vectors $\underline{R}_1$ and $\underline{R}_2$ and a basis of three atoms, two large, and one small, as shown in Figure 2.47. The corresponding lattice is characterized as centered rectangular by its symmetry, see Section 3.8.6.

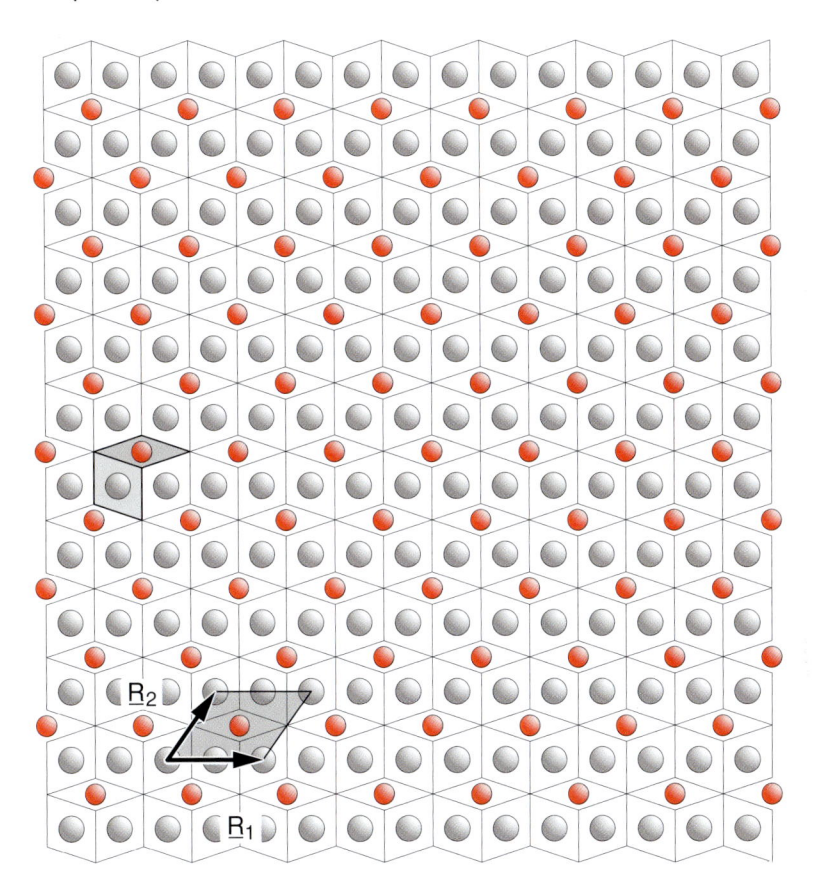

Figure 2.47 Rhombic Penrose prototiles arranged to form a two-dimensionally periodic pattern. Two different prototiles are emphasized by light and dark background, respectively. Atom balls are added at the centers of the tiles to form a two-dimensional crystal. Lattice vectors are labeled accordingly with the morphological unit cell emphasized by a gray background.

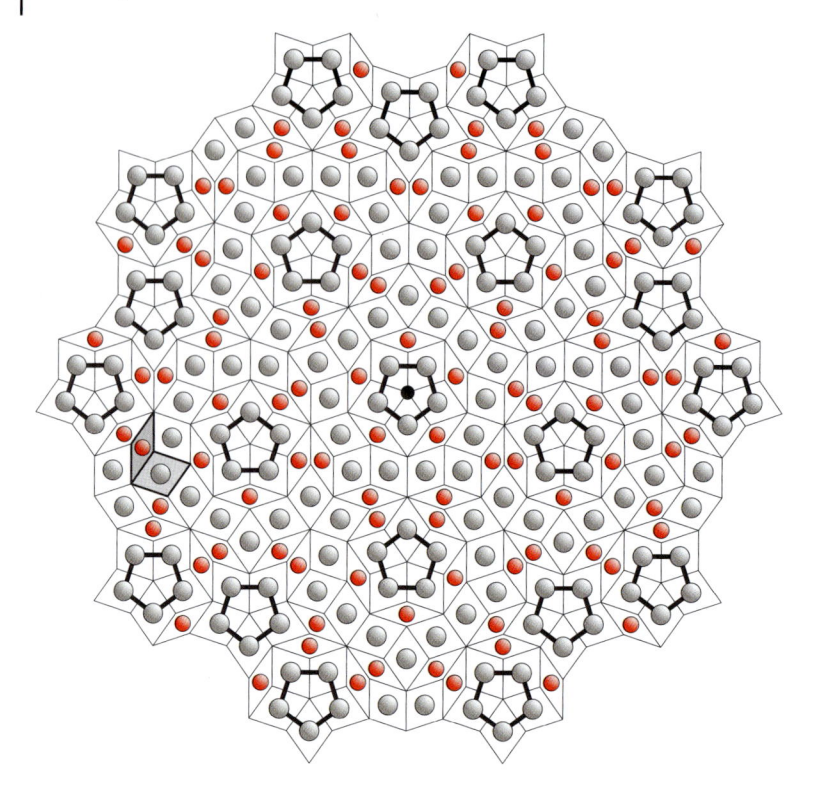

Figure 2.48 Aperiodic (quasicrystal) arrangement of rhombic Penrose prototiles illustrating fivefold symmetry. Two different prototiles are emphasized by light and dark background, respectively. Atom balls are added at the centers of all tiles to form a two-dimensional quasicrystal.

On the other hand, sets of the two rhombuses can also be positioned such that they cover the plane completely but do not exhibit periodicity in any direction as indicated in Figure 2.48. A closer inspection of the tiling evidences **global fivefold symmetry** for exactly one vertex at the center (labeled by a black dot in Figure 2.48). In addition, there are many smaller regions (indicated by black sticks connecting thick neighbor atoms), which exhibit **local fivefold symmetry** without being periodic repeat units. Global fivefold symmetry is not compatible with translational symmetry as shown in Section 3.6.3. Thus, placing atoms at the centers of all rhombuses, as indicated in Figure 2.48, does not yield a two-dimensionally periodic crystal. (Atomic positions in a real quasicrystal are slightly more complicated, as will be shown below.) However, the existence of local symmetry suggests some order which justifies calling the set of atoms a **quasicrystal**. There is an additional geometric aspect that becomes clear from an analysis of Figure 2.48. The local regions with fivefold symmetry are not completely random in their distribution but follow the global fivefold symmetry such that similar pentagonal regions with higher complexity are formed. This property is connected with **self-similarity** and is also encountered in models of three-dimensional quasicrystals.

Further, a detailed analysis of the atom arrangement in Figure 2.48 evidences that many atoms are positioned along **(infinite) rows** that are **parallel** to each other and point in different directions. This becomes even clearer in Figure 2.49, where some of these rows are emphasized by lines to guide the eye. The sequences of separations between specific parallel lines follow a pattern, where only two different distances appear, labeled L (large) and s (small) in Figure 2.49. The distance sequence, indicated by "sLLsLsL" from the top of the figure, seems to be random at first sight. However, it can be shown to be associated with sequences appearing in the mathematical theory of **Fibonacci numbers** [78], which will not be discussed further, see Exercise 2.32.

A more complex mathematical treatment of Penrose tiling [78] shows that its structural properties can be obtained by considering periodic lattices in **higher**

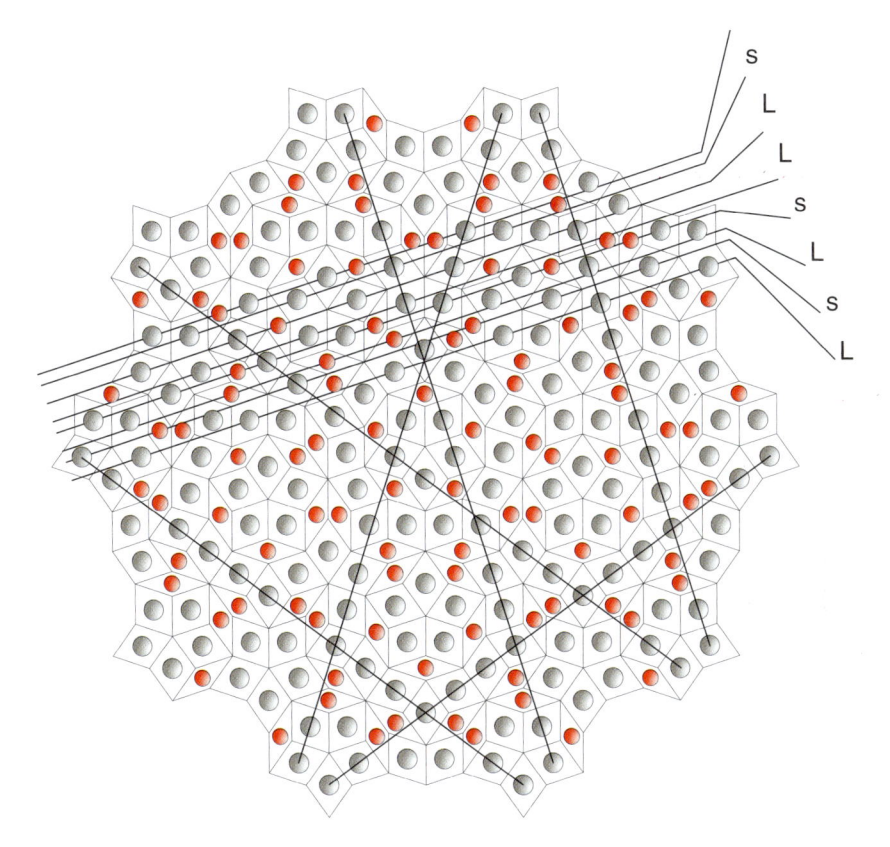

Figure 2.49 Aperiodic (quasicrystal) arrangement of rhombic Penrose prototiles illustrating linear atom rows. The rows are emphasized by black lines. In addition, parallel atom rows, separated by two distinct distances, large (L) and small (s), are shown with their sequence labeled accordingly.

dimensions, five-dimensional in the case of two-dimensional Penrose tiling. Then, projections of an aperiodic section of the five-dimensional lattice can provide the structure of the two-dimensional aperiodic quasicrystal. This result is more general and also applies to three-dimensional quasicrystals, where the corresponding higher dimensional lattices are six-dimensional. In this approach, the atom composition of the aperiodic sections in six dimensions determines the composition of the three-dimensional quasicrystal. The approach leads to a classification of quasicrystals into **two types** [81]. First, **icosahedral** quasicrystals do not exhibit periodicity in any direction but allow one global and many local fivefold rotation axes. Second, **polygonal (dihedral)** quasicrystals contain one global 8-, 10-, or 12-fold rotation axis (octagonal, decagonal, and dodecagonal quasicrystals) and are periodic along this axis. However, these quasicrystals are aperiodic in planes perpendicular to the rotation axis (quasiperiodic ordering). Further details concerning the mathematics behind possible projections go far beyond the scope of this book and can be found in the literature cited above, see, for example, [78].

As an illustration of projecting a periodic lattice of higher dimension to yield a lower-dimensional aperiodic quasicrystal we consider the **cut-and-project (CAP)** method applied to a two-dimensional square lattice of lattice constant a as shown in Figure 2.50. The square unit cell of the lattice, outlined

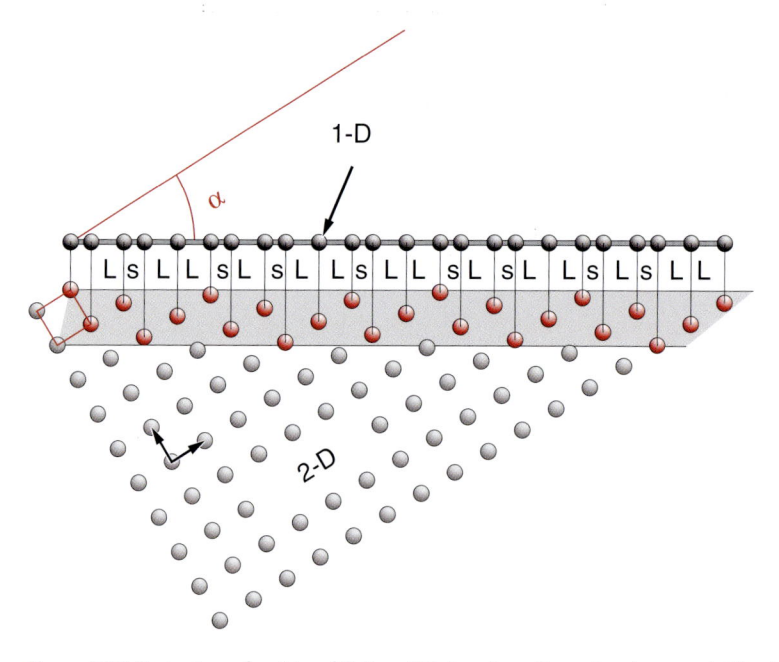

Figure 2.50 Projection of a strip of finite width in a two-dimensional square lattice (cut-and-project procedure) to yield a Fibonacci chain. The atoms inside the strip are painted red and those of the Fibonacci chain are dark gray. Long and short distances between chain atoms are labeled L and s, respectively. The unit cell of the square lattice is outlined in red and lattice vectors indicated by arrows.

in red in the figure, serves as a starting point to construct a strip between two parallel lines through two diagonal corners where the lines form an angle $\alpha = \arctan(1/\tau) = 31.717°$ with the corresponding sides of the cell (τ denoting the golden ratio) and are, thus, separated by a distance $w = a\{\cos(\alpha) + \sin(\alpha)\}$, see Figure 2.50. The area between the parallel lines includes atoms of the square lattice (painted red in the figure) whose coordinates are projected onto one of the parallel lines. This defines a linear string of locations where atoms are placed (the string is shifted upward in the figure for better visibility). This array of atoms is **aperiodic** with two distinct interatomic distances, a long and a short distance, L and s, respectively. The ratio of the distances is given by $L/s = \tau$ and the sequence of distances, LsLLsLsLLsLLsLsLLsLsLL ... , reflects the Fibonacci series which defines a one-dimensional quasicrystal, also known as a **Fibonacci chain**. This sequence of long and short distances was discussed before for Penrose tiling, see Figure 2.49.

The structural features discussed for purely mathematical models, like Penrose quasicrystals and Fibonacci chains can be also found in real three-dimensional quasicrystals and at their surfaces. As an illustration, Figures 2.51 and 2.52 show

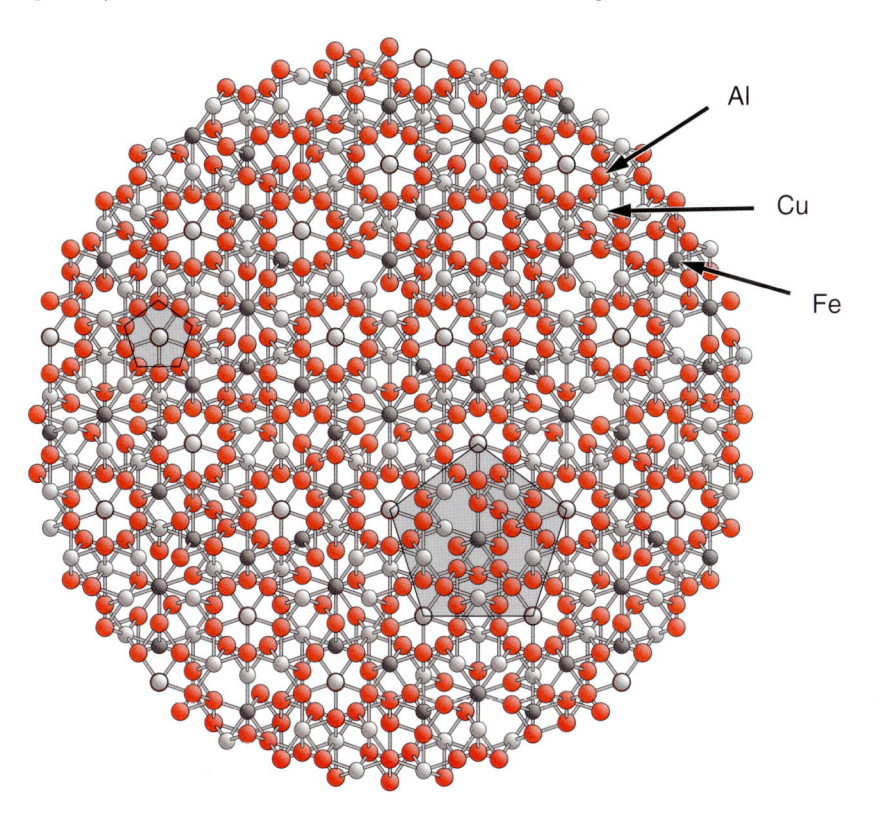

Figure 2.51 Structural model of the surface of an icosahedral i-AlCuFe quasicrystal in a top view. The different atom types, Al, Cu, Fe, are labeled accordingly. Two local environments of different size with fivefold symmetry are emphasized by gray background.

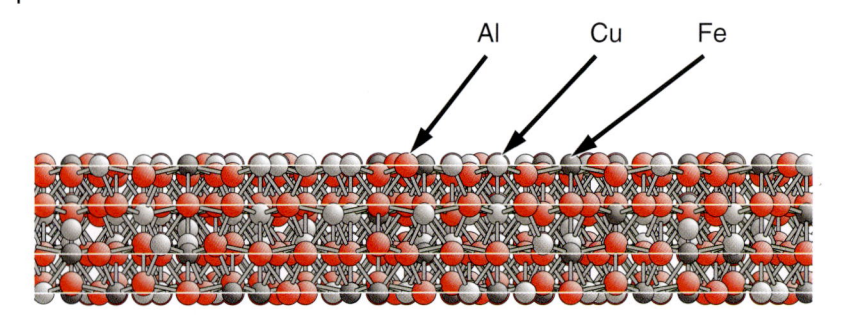

Figure 2.52 Structural model of a surface of an icosahedral i-AlCuFe quasicrystal in a side view. The different atom types, Al, Cu, Fe, are labeled accordingly. The approximate layer structure is indicated by white horizontal lines.

a geometric model of the surface of an icosahedral i-AlCuFe quasicrystal (characterized in its chemical composition as $Al_{65}Cu_{20}Fe_{15}$ [82]). This model was used to analyze the surface structure of i-AlCuFe by low-energy electron diffraction (LEED) [83].

The top view, given in Figure 2.51, appears less ordered than in Figure 2.48 because it does not contain a global fivefold axis. It has many rings of mostly 10 Al atoms, each of which form two pentagonal rings of five atoms in two different planes and, thus, have local fivefold symmetry. Either a copper or an iron atom resides in the symmetry centers. In addition, there are larger atom environments of more complex structure, such as pentagons consisting of five-rings of 10 Al atoms each, which also exhibit local fivefold symmetry: see the larger pentagon emphasized in Figure 2.51, which is blown up relative to the smaller emphasized pentagon. This can be taken as a first indication of self-similarity of the surface. Figure 2.52 shows the same quasicrystal surface in a side view, that is, along the surface. Here, an additional structural feature becomes clear. The i-AlCuFe quasicrystal forms approximate layers parallel to the surface, similar to those found for periodic crystals. This is emphasized in Figure 2.52 by white horizontal lines with a similar distance with respect to each other. While there are few atoms between the layers, most of the atoms are positioned close to the planes indicated by the lines. (It should be noted that the structure shown in Figures 2.51 and 2.52 is not an "open" low-density network, but a nearly close-packed structure. The atomic spheres in the figures are reduced in size to help visibility.)

2.9
Exercises

2.1 A crystal lattice is given by lattice vectors $\underline{R}_1, \underline{R}_2, \underline{R}_3$ with lattice constants a, b, c and mutual angles α, β, γ according to Eq. (2.3). Prove that the volume of the morphological unit cell is given by

$$V_{el} = abc\,\{1 - \cos^2(\alpha) - \cos^2(\beta) - \cos^2(\gamma) + 2\,\cos(\alpha)\,\cos(\beta)\,\cos(\gamma)\}^{1/2}$$

2.2 A crystal is described by lattice vectors $\underline{R}_1, \underline{R}_2, \underline{R}_3$ and a monoatomic basis. Replacing the atoms by hard balls of equal radii such that the balls are the largest without overlapping fills the crystal space partly, leaving empty space in between. The volume ratio of the space filled by balls and that of the complete crystal defines the packing ratio q_{pack}. Determine the packing ratio q_{pack} for crystals with (a) sc, (b) fcc, (c) bcc, (d) hex (hcp) lattice.

2.3 Characterize visually and formally (primitive) sublattices of the ions inside the NaCl, CsCl, and diamond crystal. For lattice vectors, see Eqs. (2.25), (2.26), (2.38), (2.13), and (2.15).

2.4 Analyze the centered tetragonal (ct) lattice with

$$\underline{R}_1 = (a, 0, 0), \quad \underline{R}_2 = (0, a, 0), \quad \underline{R}_3 = 1/2\,(a, a, c)$$

and show that fcc- and bcc lattices are special cases. Determine the value of c/a for these cases.

2.5 Discuss the structural phase transition bcc → ct → fcc (Bain path) based on the lattice definition of Exercise 2.4 and visualize respective morphological and WSCs.

2.6 Consider an alternative Bain path characterized by lattice vectors

$$\underline{R}_1 = a\,(1, 0, 0), \quad \underline{R}_2 = a\,(1/2, \sqrt{3}/2, 0), \quad \underline{R}_3 = a\,(1/2, 1/\sqrt{12}, x)$$

for $0.1 < x < 1.0$ and show that bcc, sc, and fcc lattices are special cases with $x_{bcc} = 1/\sqrt{24}$, $x_{sc} = 2/\sqrt{24}$, $x_{fcc} = 4/\sqrt{24}$. Which lattice symmetries appear along the continuous path $0.1 < x < 1.0$? Visualize respective morphological unit and WSCs.

2.7 The hexagonal graphite crystal can be defined by lattice vectors $\underline{R}_1, \underline{R}_2, \underline{R}_3$ and a basis of four C atoms where

$$\underline{R}_1 = a\,(1, 0, 0), \quad \underline{R}_2 = a\,(1/2, \sqrt{3}/2, 0), \quad \underline{R}_3 = c\,(0, 0, 1)$$
$$\underline{r}_1 = (0, 0, 0), \quad \underline{r}_2 = 1/3\,(1, 1, 0), \quad \underline{r}_3 = 1/2\,(0, 0, 1)$$
$$\underline{r}_4 = (2/3, 2/3, 1/2), \quad c/a = 2.72$$

with $\underline{R}_i$, $i = 1–3$, in Cartesian coordinates and $\underline{r}_k$, $k = 1–4$, in relative coordinates, see Eq. (2.11). Show that the crystal structure is hexagonal layer-type and can be built by stacking honeycomb-structured planes of atoms. What is the stacking direction?

2.8 The rhombohedral graphite crystal can be defined by lattice vectors $\underline{R}_1, \underline{R}_2, \underline{R}_3$ and a basis of two C atoms where

$$\underline{R}_1 = a\,(\sqrt{3}/2, -1/2, c/a), \quad \underline{R}_2 = a\,(0, 1, c/a),$$
$$\underline{R}_3 = a\,(-\sqrt{3}/2, -1/2, c/a)$$
$$\underline{r}_1 = (1/6, 1/6, 1/6), \quad \underline{r}_2 = (-1/6, -1/6, -1/6), \quad c/a = 2.36$$

with $\underline{R}_i$, $i = 1–3$, given in Cartesian coordinates and $\underline{r}_k$, $k = 1, 2$, in relative coordinates, see Eq. (2.11). Show that this crystal structure is also hexagonal layer-type and can be built by stacking honeycomb-structured monolayers of atoms. Prove that the hexagonal structure description

$$\underline{R}_1 = a\,(1, 0, 0), \quad \underline{R}_2 = a\,(1/2, \sqrt{3}/2, 0), \quad \underline{R}_3 = c\,(0, 0, 1)$$
$$\underline{r}_1 = (0, 0, 0), \quad \underline{r}_2 = 1/3\,(1, 1, 0), \quad \underline{r}_3 = 1/3\,(1, 1, 1)$$
$$\underline{r}_4 = 1/3\,(2, 2, 1), \quad \underline{r}_5 = 1/3\,(2, 2, 2), \quad \underline{r}_6 = 1/3\,(0, 0, 2)$$
$$c/a = 4.08$$

is equivalent to the rhombohedral description above.

2.9 Compare the crystal structures of rhombohedral and hexagonal graphite. Show that hexagonal graphite results in stacking of honeycomb-structured planes of atoms according to … A B A B … as opposed to … A B C A B C … for rhombohedral graphite. Here A, B, C denotes the planes with those of the same label positioned directly above each other (shifted by $\underline{R}_3$ of the hexagonal lattice).

2.10 Consider a hexagonal lattice with lattice vectors $\underline{R}_{o1}$, $\underline{R}_{o2}$, $\underline{R}_{o3}$ given in obtuse representation where $a = |\underline{R}_{o1}| = |\underline{R}_{o2}|$ and $\underline{R}_{o1}\underline{R}_{o2} = -1/2\,a^2$. Show that there are, altogether, 12 equivalent representations (6 obtuse and 6 acute) with vectors $\underline{R}_1$ and $\underline{R}_2$ of the same length as $|\underline{R}_{o1}|$ and identical $\underline{R}_3 = \underline{R}_{o3}$.

2.11 Show that there is a continuous structural transition from rhombohedral graphite to diamond and discuss the structural elements.

2.12 Consider initial lattice vectors $\underline{R}_{o1}$, $\underline{R}_{o2}$, $\underline{R}_{o3}$ of a three-dimensional crystal and a corresponding superlattice representation by lattice vectors $\underline{R}_1, \underline{R}_2, \underline{R}_3$ where

$$\underline{R}_1 = \kappa_{11}\,\underline{R}_{o1} + \kappa_{12}\,\underline{R}_{o2} + \underline{R}_{o3}, \quad \underline{R}_2 = \kappa_{21}\,\underline{R}_{o1} + \kappa_{22}\,\underline{R}_{o2} + \underline{R}_{o3}, \quad \underline{R}_3 = \underline{R}_{o3}$$

Determine the constraints on κ_{ij} such that the two representations result in identical volumes of their morphological unit cells.

2.13 Consider a three-dimensional crystal with non-primitive lattice vectors $\underline{R}_1, \underline{R}_2, \underline{R}_3$ and a lattice point inside the morphological unit cell described by vector

$$\underline{R}' = \kappa_1\,\underline{R}_1 + \kappa_2\,\underline{R}_2 + \kappa_3\,\underline{R}_3, \quad 0 \le \kappa_i < 1$$

Which parameter combinations $\kappa_1, \kappa_2, \kappa_3$ must be excluded and why?

2.14 Consider a three-dimensional crystal with non-primitive lattice vectors $\underline{R}_1, \underline{R}_2, \underline{R}_3$ which are of smallest length compared with all general lattice vectors pointing along the same directions. Assume a lattice vector $\underline{R}'$ inside the morphological unit cell, that is,

$$\underline{R}' = \kappa_1\,\underline{R}_1 + \kappa_2\,\underline{R}_2 + \kappa_3\,\underline{R}_3, \quad 0 < \kappa_i < 1$$

where $(\underline{R}', \underline{R}_2, \underline{R}_3)$, $(\underline{R}_1, \underline{R}', \underline{R}_3)$, and $(\underline{R}_1, \underline{R}_2, \underline{R}')$ form alternative lattice vector sets to describe the initial lattice. Show that this can be achieved only for

$$\underline{R}' = (\underline{R}_1 + \underline{R}_2 + \underline{R}_3)/p, \quad p \text{ integer}$$

2.15 Consider a two-dimensional set of non-primitive lattice vectors, $\underline{R}_1$ and $\underline{R}_2$, which are of smallest length compared with all general lattice vectors pointing along the same directions. Assume a lattice vector $\underline{R}'$ inside the morphological unit cell, that is,

$$\underline{R}' = \kappa_1 \underline{R}_1 + \kappa_2 \underline{R}_2, \quad 0 < \kappa_i < 1$$

where $(\underline{R}', \underline{R}_2)$ and $(\underline{R}_1, \underline{R}')$ form alternative lattice vector sets to describe $\underline{R}_1$ and $\underline{R}_2$. Show that this can be achieved only for

$$\underline{R}' = (\underline{R}_1 + \underline{R}_2)/p, \quad p \text{ integer}$$

2.16 Consider the Wigner-Seitz cells (WSCs) of a (a) sc, (b) fcc, (c) bcc lattice with lattice constant a. Determine for each WSC the volume of the largest inscribed sphere V_{sph} and compare it with the WSC volume V_{el}. How do the ratios $q_{WSC} = V_{sph}/V_{el}$ compare with the packing ratios q_{pack} determined in Exercise 2.2?

2.17 A crystal lattice is described by lattice vectors $\underline{R}_1, \underline{R}_2, \underline{R}_3$ where the morphological unit cell contains threefold rotation axes along directions

$$\underline{e}_1 = (1, 1, 1), \quad \underline{e}_2 = (1, -1, 1), \quad \underline{e}_3 = (-1, 1, 1), \quad \underline{e}_4 = (-1, -1, 1)$$

in Cartesian coordinates.

(a) Determine the 3×3 matrices $\underline{T}$ of Cartesian coordinate transformations

$$\begin{pmatrix} x' \\ y' \\ z' \end{pmatrix} = \begin{pmatrix} t_{11} & t_{12} & t_{13} \\ t_{21} & t_{22} & t_{23} \\ t_{31} & t_{32} & t_{33} \end{pmatrix} \cdot \begin{pmatrix} x \\ y \\ z \end{pmatrix} = \underline{T} \cdot \begin{pmatrix} x \\ y \\ z \end{pmatrix}$$

referring to the corresponding eight rotation operations.

(b) Determine all point symmetry elements of the morphological unit cell given by $\underline{R}_1, \underline{R}_2, \underline{R}_3$.

2.18 The morphological unit cell of a crystal, described by orthogonal lattice vectors $\underline{R}_1, \underline{R}_2, \underline{R}_3$, contains three twofold rotation axes along directions

$$\underline{e}_1 = (1, 0, 0), \quad \underline{e}_2 = (0, 1, 0), \quad \underline{e}_3 = (0, 0, 1) \quad \text{in Cartesian coordinates}$$

(a) Determine all possible symmetry elements of the unit cell originating from the twofold rotation axes.
(b) Fill the unit cell with atoms such that the cell allows only the symmetry elements found in (a).

2.19 A crystal is described by cubic lattice vectors $\underline{R}_1, \underline{R}_2, \underline{R}_3$ and a basis of two different atoms A, B, located at

$$\underline{r}_1^A = (0,0,0), \quad \underline{r}_2^B = \kappa_1 \underline{R}_1 + \kappa_2 \underline{R}_2 + \kappa_3 \underline{R}_3, \quad |\kappa_i| < 1, \quad i = 1, 2, 3$$

in relative coordinates, see Eq. (2.11). Find values $\kappa_1, \kappa_2, \kappa_3$ where neighbor shells with respect to atom A contain both types of atoms in the same shell.

2.20 Determine neighbor shells (distances, shell multiplicities) of monoatomic crystals with (a) fcc, (b) bcc, (c) hex (hcp), and (d) diamond lattice up to fifth nearest neighbors. Which shells are identical for these lattices assuming the same value of lattice constant a.

2.21 A crystal of the semiconductor gallium arsenite, GaAs, can be described by an fcc lattice and a two-atom basis with

$$\underline{R}_1 = a/2\,(0,1,1), \quad \underline{R}_2 = a/2\,(1,0,1), \quad \underline{R}_3 = a/2\,(1,1,0)$$
$$\underline{r}_1 = (0,0,0), \quad \underline{r}_2 = 1/4\,(1,1,1)$$

with $\underline{R}_i$, $i = 1-3$, in Cartesian coordinates and $\underline{r}_k$, $k = 1, 2$, in relative coordinates, see Eq. (2.11). (This is also the crystal structure of cubic zincblende, ZnS.) Show that neighbor shells with respect to Ga or As centers contain only atoms of the same type.

2.22 Consider a monoatomic crystal described by sc lattice vectors $\underline{R}_1, \underline{R}_2, \underline{R}_3$ and determine neighbor shells with respect to the midpoint between two adjacent atoms in the crystal up to the 10th shell. Calculate the number of shell multiplicities M_i and corresponding cluster radii D_i as functions of the shell index i.

2.23 The titanium dioxide, TiO_2, crystal with rutile structure can be described by a tetragonal lattice and a six-atom basis with

$$\underline{R}_1 = a\,(1,0,0), \quad \underline{R}_2 = a\,(0,1,0), \quad \underline{R}_3 = c\,(0,0,1)$$
$$\underline{r}_1^{Ti} = (0,0,0), \quad \underline{r}_2^{Ti} = (1/2, 1/2, 1/2), \quad \underline{r}_3^O = (x, x, 0)$$
$$\underline{r}_4^O = (1-x, 1-x, 0), \quad \underline{r}_5^O = (1/2+x, 1/2-x, 1/2),$$
$$\underline{r}_6^O = (1/2-x, 1/2+x, 1/2)$$
$$a = 4.593\ \text{Å}, \quad c = 2.958\ \text{Å}, \quad x = 0.3053$$

with $\underline{R}_i$, $i = 1-3$, in Cartesian coordinates and $\underline{r}_k$, $k = 1-6$, in relative coordinates.

(a) Show that each titanium atom has four oxygen atoms in its first and two in its second neighbor shell. Determine Ti–O distances of the two oxygen shells with respect to the central titanium.
(b) The six oxygen atoms of the two neighbor shells form edges of a polyhedron. Determine its shape.
(c) Determine point symmetry elements of the crystal.

2.24 Consider atom clusters A_N with N atoms originating from a central atom A and its n neighbor shells in a crystal (n < 10) with an (a) fcc, (b) bcc, (c) hex (hcp) lattice. Determine the total number of atoms N(n) and corresponding cluster radii R(n) as a function of the shell index n.

2.25 Build radially symmetric C_{60} ("Buckminster ball"), C_{24}, and C_{12} clusters. Here, equilateral carbon hexagons join with each other and with pentagons, squares, and triangles, respectively to form a polyhedral structure of "spherical" shape, that is, all atom centers lie on a sphere with respect to a common center. Determine the sphere radius of each cluster as a function of the interatomic C–C distance d_{C-C} assuming equal distances in all cases ($d_{C-C} = 1.4$ Å in experiment).
Hints:
  (a) C_{60} combines 20 hexagons with 12 pentagons. Each pentagon is adjacent only to hexagons. The cluster resembles a competition soccer ball.
  (b) C_{24} combines eight hexagons with six squares. Each square is adjacent only to hexagons. The cluster resembles the WSC of a bcc crystal.
  (c) C_{12} combines four hexagons with four triangles. Each triangle is adjacent only to hexagons. The cluster resembles a tetrahedron with edges cut off symmetrically.

2.26 Determine orientations and areas of all surfaces of the Wulff polyhedra for the cubic crystallites shown in Figure 2.42.

2.27 Consider a Wulff polyhedron of a cubic crystallite with low-energy planes perpendicular to $(\pm1, \pm1, \pm1)$ with surface energies $\gamma_{(1\,1\,1)}$ and perpendicular to $(\pm1, 0, 0), (0, \pm1, 0), (0, 0, \pm1)$ with surface energies $\gamma_{(0\,0\,1)}$ where (x, y, z) refers to vectors in Cartesian coordinates.

  (a) Determine the polyhedron volume and surface area as a function of $q = \gamma_{(0\,0\,1)}/\gamma_{(1\,1\,1)}$.
  (b) Assume surface energy ratios $q = \gamma_{(0\,0\,1)}/\gamma_{(1\,1\,1)}$ and show that the Wulff polyhedral are cubes for $q \geq \sqrt{3}$, capped cubes for $2/\sqrt{3} \leq q < \sqrt{3}$, capped octahedra for $1/\sqrt{3} \leq q < 2/\sqrt{3}$, and octahedra for $q < 1/\sqrt{3}$.
  (c) Determine the volume of the sphere surrounding the Wulff polyhedron for given ratio q where the polyhedron edges lie on the sphere surface.

2.28 Consider a Wulff polyhedron with low-energy planes of equal surface energy γ and perpendicular to $(1, 1, 1), (-1, -1, 1), (-1, 1, -1), (1, -1, -1)$ in Cartesian coordinates. Show that the shape is tetrahedral.

2.29 Consider a Wulff polyhedron to be defined by 12 planes through $\underline{R}_i$ with normal vectors along $\underline{R}_i$ in Cartesian coordinates where $\underline{R}_i = (x, y, z) = (\pm1, \pm1, \pm\sqrt{2}), (\pm2, 0, 0), (0, \pm2, 0)$. Show that the shape of the polyhedron is that of a WSC of the fcc lattice. Determine the relative surface energies $\gamma_{(x, y, z)}$ of the different planes.

2.30 Consider a Wulff polyhedron to be defined by 14 planes through $\underline{R}_i$ with normal vectors along $\underline{R}_i$ in Cartesian coordinates where $\underline{R}_i = (x, y, z) = (\pm 1, \pm 1, \pm 1)$, $(\pm 2, 0, 0)$, $(0, \pm 2, 0)$, $(0, 0, \pm 2)$. Show that the shape of the polyhedron is that of a WSC of the bcc lattice. Determine the relative surface energies $\gamma_{(x, y, z)}$ of the different planes.

2.31 Show that the Wulff polyhedron of a hexagonal crystal with its hexagonal axis along z in Cartesian coordinates and surface energies $\gamma_{(x, y, 0)} = \gamma_a$, $\gamma_{(0, 0, z)} = \gamma_c$ is a hexagonal prism independent of γ_a and γ_c.

2.32 The two-dimensional Penrose crystal, see Figure 2.49, includes parallel atom rows where adjacent rows are separated by two distinct distances, a small and a large distance, s and L, respectively. The sequence of these distances can be generated iteratively by the following procedure:
- start with a one-member sequence S_1 of distance s or L;
- generate the next sequence S_{k+1} by replacing in the present sequence S_k all distances s by L and all L by two distances s, L.

This generates sequences S_k with an increasing number N_k of distances, of which N_k^L counts large and N_k^s small distances. An example is the eight-distance sequence "L s L L s L s L."

(a) Show that the number of distances N_k in sequence S_k, starting with sequences $S_1 = s$, $S_2 = L$, can be determined iteratively by
$$N_k = N_{k-1} + N_{k-2}, \quad N_1 = N_2 = 1$$
yielding the Fibonacci number series 1, 1, 2, 3, 5, 8, 13, …

(b) Show that the starting N_k members of distances in sequence S_{k+1} are identical to those of S_k.

(c) Show that the ratio of the number of large distances, N_k^L, and that of small distances, N_k^s, of sequence S_k can be determined iteratively by
$$q_k = N_k^L/N_k^s: \quad q_k = 1 + 1/q_{k-1}$$
Show that ratio q_k converges for infinitely large sequences to the golden ratio
$$\tau = (1 + \sqrt{5})/2 = 1.618034.$$

(d) Show that the ratio $p_k = N_k/N_{k-1}$ of the number of distances N_k of two successive sequences converges for infinitely large sequences to the golden ratio τ.

2.33 A Fibonacci chain is created by projecting atoms inside a strip cut from a square lattice of lattice constant c where the cutting angle α is given by $\tan(\alpha) = 1/\tau$, see Figure 2.50.

(a) Show that the width w of the strip is given by $w = c(\tau + 1)(\tau^2 + 1)^{-1/2}$.

(b) Assume that the two different interatomic distances along the chain are L (long) and s (short). Prove that
$$L/s = \tau, \quad L = c\cos(\alpha) = c\tau(\tau^2+1)^{-1/2}, \quad s = c\sin(\alpha) = c(\tau^2+1)^{-1/2}$$
with $\tan(\alpha) = 1/\tau$.

2.34 In general, cut-and-project (CAP) chains can be created by projecting atoms inside a strip cut from a square lattice of lattice constant c where the cutting angle is α analogous to Figure 2.50.

(a) Show that CAP chains for $\tan(\alpha) = n/m$ (n, m integer) are periodic. For other angles α they are aperiodic with always only two different interatomic lengths L and s.
(b) Show that the width w of the strips is given by $w = c\{\sin(\alpha) + \cos(\alpha)\}$.
(c) Prove that $s/L = \tan(\alpha), \quad L = c\cos(\alpha), \quad s = c\sin(\alpha)$.

2.35 Consider a monoatomic crystal with modulated lattice structure where atom positions are given by
$$\underline{R}_o = n_1 \underline{R}_{o1} + n_2 \underline{R}_{o2} + n_3 \underline{R}_{o3} + \underline{u}(\underline{R}_o), \quad n_i \text{ integer}$$
$$\underline{u}(\underline{r}) = \underline{u}_o \sin(\underline{G}\,\underline{r}), \quad \underline{u}_o, \underline{G} \text{ fixed}$$
Determine vectors $\underline{G}$ which result in a truly periodic lattice.

3
Crystal Layers: Two-Dimensional Lattices

3.1
Basic Definition, Miller Indices

Understanding the concept of **monolayers** and **netplanes** in crystals is of central importance for the analysis of many structural properties of single crystals as well as of their surfaces. It can simplify conceptual thinking about crystal structure and is also essential for studying practical applications. For example, electron and photon diffraction from single crystal surfaces are often treated by theoretical methods that consider scattering from different crystal layers to build the complete diffraction image. Here the basic idea is that any three-dimensionally periodic crystal can be **decomposed** into planar two-dimensionally periodic monolayers that are **stacked** along the third dimension. This is illustrated in Figure 3.1, which shows different two-dimensionally periodic monolayers inside a monoatomic crystal with a face-centered cubic (fcc) lattice, where square, rectangular, and hexagonal layers are displayed. (The netplane nomenclature that has been used to label the monolayers in Figure 3.1, refers to Miller indices which will be explained below). Clearly, monolayers can assume quite different structures depending on how they are oriented in the crystal.

The **formal definition** of a monolayer starts from a perfect bulk crystal with its periodicity defined by lattice vectors $\underline{R}_{o1}$, $\underline{R}_{o2}$, and $\underline{R}_{o3}$ and its basis by lattice basis vectors $\underline{r}_{o1}, \ldots, \underline{r}_{op}$, where we focus first on the lattice. **General lattice vectors** $\underline{R}$ of the lattice can always be described by linear combinations

$$\underline{R} = n_{o1}\underline{R}_{o1} + n_{o2}\underline{R}_{o2} + n_{o3}\underline{R}_{o3}, \quad n_{oi} \text{ integer} \tag{3.1}$$

However, according to Section 2.2.2.3, there is an infinite number of alternative descriptions of the same lattice by other vector sets $\underline{R}_1$, $\underline{R}_2$, $\underline{R}_3$, where some may be more appropriate for describing specific monolayers. These alternative vector sets are connected with their initial counterparts, $\underline{R}_{o1}$, $\underline{R}_{o2}$, $\underline{R}_{o3}$, by linear transformations according to Eq. (2.62) with integer-valued transformation matrices $\underline{\underline{T}}$, where $\det(\underline{\underline{T}}) = 1$. (Matrix $\underline{\underline{T}}$ also transforms all lattice basis vectors $\underline{r}_{o1} \ldots \underline{r}_{op}$ to yield $\underline{r}_1 \ldots \underline{r}_p$.) If we consider one of these **transformed lattice vector sets** defined by an appropriate matrix $\underline{\underline{T}}$ then general lattice vectors $\underline{R}$ can be written,

Crystallography and Surface Structure: An Introduction for Surface Scientists and Nanoscientists,
Second Edition. Klaus Hermann.
© 2017 Wiley-VCH Verlag GmbH & Co. KGaA. Published 2017 by Wiley-VCH Verlag GmbH & Co. KGaA.

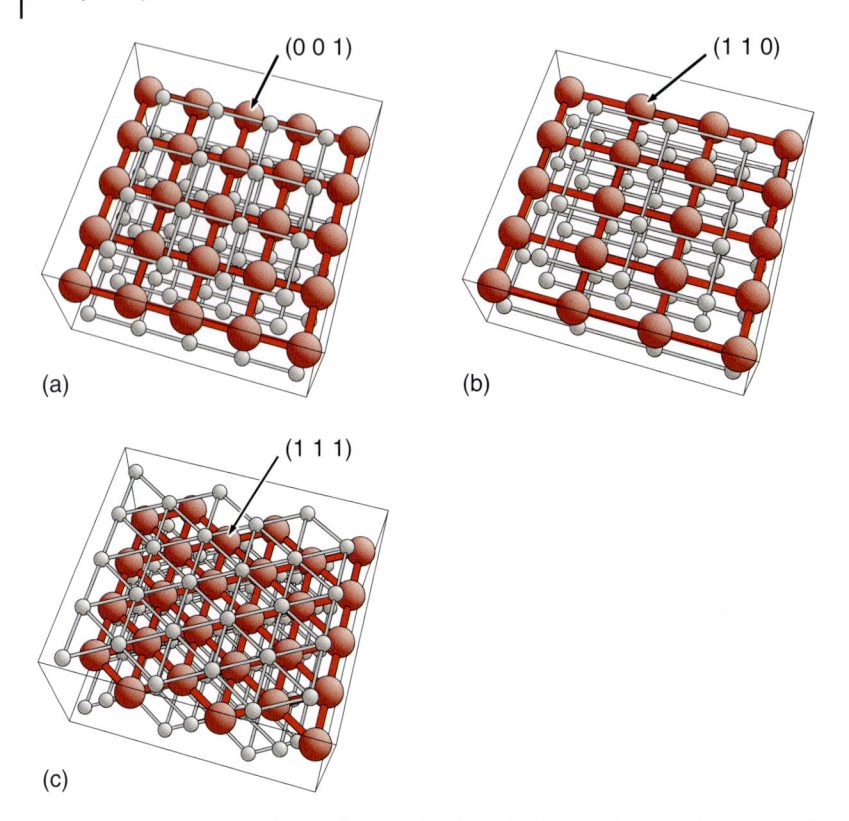

Figure 3.1 Different monolayers of a crystal with an fcc lattice, (a) square (0 0 1) monolayer, (b) rectangular (1 1 0) monolayer, and (c) hexagonal (1 1 1) monolayer. In each case the second monolayer from the top is emphasized by large red balls.

analogous to Eq. (3.1), as

$$\underline{R} = (n_1 \underline{R}_1 + n_2 \underline{R}_2) + n_3 \underline{R}_3, \quad n_i \text{ integer} \tag{3.2}$$

The parentheses in Eq. (3.2) emphasize that, for a fixed value n_3, the infinite set of vectors $\underline{R}$ for all integer values n_1, n_2 forms a two-dimensional lattice with lattice vectors $\underline{R}_1$ and $\underline{R}_2$. Thus, different values of n_3 provide a collection of parallel two-dimensional lattices that are identical in their periodicity defined by $\underline{R}_1$ and $\underline{R}_2$ and whose origins are separated from each other by vector $\underline{R}_3$. These sublattices of the bulk lattice are called **netplanes**, sometimes also as **crystallographic planes**, and the complete set of parallel netplanes spans the three-dimensional lattice. Further, the planar symmetry of the netplanes depends on the choice of the transformed lattice vectors $\underline{R}_1$, $\underline{R}_2$, and, therefore, on the corresponding transformation matrix $\underline{\underline{T}}$. (Additional symmetry issues will be discussed in greater detail in Sections 3.7 and 3.8.)

The concept of netplanes is intimately connected with the definition of monolayers. A **monolayer** describes a collection of atoms on a **plane** in a crystal. The atoms are located at **positions** given by a **lattice basis vector** $\underline{r}_i$, $i = 1 \ldots p$, and its periodic equivalents where the periodicity is defined by vectors $\underline{R}_1$ and $\underline{R}_2$ of a **netplane**. Then the complete set of parallel monolayers is obtained by shifting each of the origins $\underline{r}_i$, $i = 1, \ldots, p$, by any integer multiple of lattice vector $\underline{R}_3$. As examples, Figure 3.1 shows three different monolayers for corresponding choices of lattice vectors $\underline{R}_1$ and $\underline{R}_2$. The complete set of monolayers for all $\underline{r}_i$ fills the three-dimensional crystal. Further, depending on vectors $\underline{R}_1$ and $\underline{R}_2$, monolayers originating from different positions $\underline{r}_i$ can lie on the same plane, which leads to polyatomic monolayers. This can be seen already in Figure 1.1 for the MgO crystal where monolayers denoted by (1 1 1) refer to planes of single Mg or O atoms while those labeled (0 0 1) contain both Mg and O atoms in the same plane.

It should be noted that the **definition** of a **monolayer** given in this book is rather general and **deviates** from definitions used in numerous surface science publications. In the latter, the term "monolayer" often refers to adsorbate overlayers at single crystal bulk where the overlayer density corresponds to one adsorbate particle per substrate atom or unit cell (described as monolayer coverage). With the definition in this book, this would be an adsorbate monolayer with a specific coverage while those of different coverage would also be called monolayers.

The relationship between netplanes and monolayers forms the two-dimensional **analog** to that between lattices and crystals in the three-dimensional case. Thus, one can also consider a **morphological unit cell** of a netplane or monolayer, defined by the parallelogram spanned by lattice vectors $\underline{R}_1$ and $\underline{R}_2$ with cell area A_{el} given by

$$A_{el} = |\underline{R}_1 \times \underline{R}_2| \tag{3.3}$$

The unit cell will be called **primitive** if it is of the smallest possible area in the netplane. Further, a monolayer can be assigned a basis of atoms given by corresponding **lattice basis vectors** $\underline{r}_i$. Thus, in the example of a (0 0 1) monolayer of MgO one could define two lattice basis vectors, $\underline{r}_1^{Mg}$, $\underline{r}_2^{O}$, inside the corresponding morphological unit cell describing the unique positions of the two atoms Mg and O. It should be noted that surface scientists often use the word "netplane" to describe a monolayer. However, in a strict sense, a **netplane** can be considered only a **mathematical construct** to characterize the two-dimensional periodicity and symmetry properties of a monolayer.

The definition of a netplane is based on **transformed** lattice vectors $\underline{R}_1, \underline{R}_2, \underline{R}_3$ of a three-dimensional lattice, where the two vectors $\underline{R}_1$ and $\underline{R}_2$ determine the periodicity of the netplane. Thus, vectors $\underline{R}_1, \underline{R}_2, \underline{R}_3$ and the corresponding transformation **matrix** $\underline{T}$ can be considered to be **netplane-adapted** and matrix $\underline{T}$ may be used to characterize netplanes. The **normal direction** of a netplane can be viewed as the normal component of the stacking direction of the corresponding monolayers inside a crystal. It is given by a vector $\underline{n}$ where, using transformation $\underline{T}$ from Eq. (2.62),

$$\underline{n} = \alpha(\underline{R}_1 \times \underline{R}_2) = \alpha \sum_{i=1}^{3} \sum_{j=1}^{3} t_{1i} t_{2j} (\underline{R}_{oi} \times \underline{R}_{oj})$$
$$= \alpha\{h(\underline{R}_{o2} \times \underline{R}_{o3}) + k(\underline{R}_{o3} \times \underline{R}_{o1}) + l(\underline{R}_{o1} \times \underline{R}_{o2})\} \tag{3.4}$$

with coefficients

$$h = t_{12}t_{23} - t_{13}t_{22}, \quad k = t_{13}t_{21} - t_{11}t_{23}, \quad l = t_{11}t_{22} - t_{12}t_{21}$$

$$\alpha = |\underline{R}_1 \times \underline{R}_2|^{-1} \tag{3.5}$$

Here, α is only a normalization constant to guarantee that $|\underline{n}| = 1$. Since all elements of the transformation matrix $\underline{T}$ are integers, the coefficients h, k, and l must also be integer-valued and are commonly named **(generic) Miller indices**. This means in particular that, according to Eq. (3.4), normal directions of netplanes in a lattice are always discrete and generic Miller indices $(h\ k\ l)$ can be used to **characterize** sets of netplanes for a given **direction.** In this spirit transformation matrices $\underline{T}$ that are connected with netplane stacking directions will be labeled $\underline{T}^{(h\ k\ l)}$ in the following.

Miller indices can assume both positive and negative integer values where in crystallography negative indices are often written with the minus sign above the number, such as $(-4\ 1\ 2)$ being written as $(\overline{4}\ 1\ 2)$, which saves a character space. However, in this book we do not adopt this notation and all negative indices will be given with a minus sign in front.

According to Eq. (3.4) netplane normal vectors $\underline{n}$ are given as linear combinations of three vectors $(\underline{R}_{oi} \times \underline{R}_{oj})$ that arise from vector products of the initial lattice vectors. Thus, based on the definition (2.95) of reciprocal lattice vectors, $\underline{G}_{o1}$, $\underline{G}_{o2}$, $\underline{G}_{o3}$, in Section 2.5 any normal vector $\underline{n}$ points along a vector $\underline{G}_{(h\ k\ l)}$, where

$$\underline{G}_{(h\ k\ l)} = \beta\,(\underline{R}_1 \times \underline{R}_2) = h\,\underline{G}_{o1} + k\,\underline{G}_{o2} + l\,\underline{G}_{o3}$$

$$\beta = (2\pi)\,/\,[(\underline{R}_1 \times \underline{R}_2)\underline{R}_3] = (2\pi)/[(\underline{R}_{o1} \times \underline{R}_{o2})\underline{R}_{o3}] = (2\pi)/V_{el} \tag{3.6}$$

Thus, $\underline{G}_{(h\ k\ l)}$ is a **general reciprocal lattice** vector and is quite useful in describing the numerous properties of netplanes and monolayers.

As an example, the **distance** $d_{(h\ k\ l)}$ between two adjacent $(h\ k\ l)$ netplanes or monolayers that are connected by a lattice vector $\underline{R}_3$, can be written as

$$d_{(h\ k\ l)} = (\underline{n}\,\underline{R}_3) = (\underline{G}_{(h\ k\ l)}\underline{R}_3)/|\underline{G}_{(h\ k\ l)}| = (2\pi/V_{el})\,(\underline{R}_1 \times \underline{R}_2)\underline{R}_3/|\underline{G}_{(h\ k\ l)}|$$

$$= 2\pi/|\underline{G}_{(h\ k\ l)}| \tag{3.7}$$

Thus, if the length of $\underline{G}_{(h\ k\ l)}$, determined by the size of the generic Miller indices, becomes large, the distance between adjacent netplanes becomes small. Netplanes belonging to large Miller indices lie close together.

Further, the average **atom density** $\rho_{(h\ k\ l)}$ of a monolayer with an $(h\ k\ l)$ netplane is, according to Eqs. (3.6) and (3.7), given by

$$\rho_{(h\ k\ l)} = n_{(h\ k\ l)}/|\underline{R}_1 \times \underline{R}_2| = n_{(h\ k\ l)}\,(2\pi/V_{el})/|\underline{G}_{(h\ k\ l)}| = n_{(h\ k\ l)}\,d_{(h\ k\ l)}/V_{el} \tag{3.8}$$

where $n_{(h\ k\ l)}$ denotes the number of co-planar atoms in the unit cell of the monolayer. Thus, for large vectors $\underline{G}_{(h\ k\ l)}$ the atom density of corresponding monolayers becomes small while monolayers belonging to small Miller indices are the densest. As an illustration, Table 3.1 lists Miller indices $(h\ k\ l)$ and densities $\rho_{(h\ k\ l)}$ of the 10 densest monolayers with $(h\ k\ l)$ netplanes of the sc, fcc, bcc, and hcp lattices calculated using Eq. (3.8). All Miller

Table 3.1 Average atom densities $\rho_{(h\,k\,l)}$ of the 10 densest $(h\,k\,l)$ monolayers ($\{h\,k\,l\}$ monolayer families) of crystals with (a) sc, (b) fcc, (c) bcc, and (d) hcp lattice.

	$\{h\,k\,l\}$; (m)	$\rho_{(h\,k\,l)}/a^2$		$\{h\,k\,l\}$; (m)	$\rho_{(h\,k\,l)}/a^2$
(a) sc					
1	{1 0 0}; (6)	$1 = 1.000$	6	{2 2 1}; (24)	$1/\sqrt{9} = 0.333$
2	{1 1 0}; (12)	$1/\sqrt{2} = 0.707$	7	{3 1 0}; (24)	$1/\sqrt{10} = 0.316$
3	{1 1 1}; (8)	$1/\sqrt{3} = 0.577$	8	{3 1 1}; (24)	$1/\sqrt{11} = 0.302$
4	{2 1 0}; (24)	$1/\sqrt{5} = 0.447$	9	{3 2 0}; (24)	$1/\sqrt{13} = 0.277$
5	{2 1 1}; (24)	$1/\sqrt{6} = 0.408$	10	{3 2 1}; (24)	$1/\sqrt{14} = 0.267$
(b) fcc					
1	{1 1 1}; (8)	$4/\sqrt{3} = 2.309$	6	{4 2 0}; (24)	$4/\sqrt{20} = 0.894$
2	{2 0 0}; (6)	$4/\sqrt{4} = 2.000$	7	{4 2 2}; (24)	$4/\sqrt{24} = 0.816$
3	{2 2 0}; (12)	$4/\sqrt{8} = 1.414$	8	{5 1 1}; (24)	$4/\sqrt{27} = 0.770$
4	{3 1 1}; (24)	$4/\sqrt{11} = 1.206$	9	{5 3 1}; (48)	$4/\sqrt{35} = 0.676$
5	{3 3 1}; (24)	$4/\sqrt{19} = 0.918$	10	{4 4 2}; (24)	$4/\sqrt{36} = 0.667$
(c) bcc					
1	{1 1 0}; (12)	$2/\sqrt{2} = 1.414$	6	{3 2 1}; (48)	$2/\sqrt{14} = 0.535$
2	{2 0 0}; (6)	$2/\sqrt{4} = 1.000$	7	{4 1 1}; (24)	$2/\sqrt{18} = 0.471$
3	{2 1 1}; (24)	$2/\sqrt{6} = 0.816$	8	{4 2 0}; (24)	$2/\sqrt{20} = 0.447$
4	{3 1 0}; (24)	$2/\sqrt{10} = 0.632$	9	{3 3 2}; (24)	$2/\sqrt{22} = 0.426$
5	{2 2 2}; (8)	$2/\sqrt{12} = 0.577$	10	{5 1 1}; (24)	$2/\sqrt{27} = 0.385$

(d) hcp	$(h\,k\,l)$; (m)	$\rho_{(h\,k\,l)}/a^2$
1	(0 0 ±1); (2)	$2/\sqrt{3} = 1.155$
2°	±(1 1 0), ±(2 −1 0), ±(1 −2 0); (6)	$1/\sqrt{2} = 0.707$
3	±(1 0 0), ±(0 1 0), ±(1 −1 0); (6)	$\sqrt{(3/8)} = 0.612$
4°	±(1 1 ±2), ±(2 −1 ±2), ±(1 −2 ±2); (12)	$2/\sqrt{11} = 0.603$
5	±(1 0 ±1), ±(0 1 ±1), ±(1 −1 ±1); (12)	$\sqrt{(12/41)} = 0.541$
6	±(1 0 ±2), ±(0 1 ±2), ±(1 −1 ±2); (12)	$\sqrt{(3/17)} = 0.420$
7°	±(3 0 ±2), ±(0 3 ±2), ±(3 −3 ±2); (12)	$2/\sqrt{27} = 0.385$
8	±(1 1 ±1), ±(2 −1 ±1), ±(1 −2 ±1); (12)	$2/\sqrt{35} = 0.338$
9	±(2 0 ±1), ±(0 2 ±1), ±(1 −1 ±1); (12)	$\sqrt{(12/137)} = 0.296$
10	±(2 1 0), ±(1 2 0), ±(−1 3 0), ±(−2 3 0), ±(3 −2 0), ±(3 −1 0); (12)	$\sqrt{(3/56)} = 0.231$

The densities $\rho_{(h\,k\,l)}$ are defined with respect to the square of the corresponding lattice constant a (applying $c/a = \sqrt{(8/3)}$ for hcp).

indices of the cubic lattices are given in sc notation by symmetry-related families $\{h\,k\,l\}$; (m), with m denoting the number of family members, see below. Miller indices of the hcp lattice refer to the obtuse representation (i.e., $\angle(\underline{R}_{o1}, \underline{R}_{o2}) = 120°$, see Section 2.2.2.1) and are given in generic three-index notation. Note that for the hcp lattice the second, fourth and seventh densest monolayers, denoted by asterisks (*) in Table 3.1, contain two atoms in their morphological unit cells, that is, $n_{(h\,k\,l)} = 2$ while for all others $n_{(h\,k\,l)} = 1$.

If a lattice exhibits, in addition to translation symmetry, also **point symmetry** then geometrically identical netplanes may appear for different Miller index values. These equivalent netplanes are often grouped into **families**, where each family is characterized by Miller indices {*h k l*} written inside **curly** brackets. An example is given by the simple cubic lattice with the six equivalent netplanes, denoted by Miller indices ($\pm$1 0 0), (0 $\pm$1 0), (0 0 $\pm$1), forming a family described as {1 0 0}. This notation is also used in Table 3.1 for monolayers of cubic lattices.

In a generalization of Eq. (3.4), directions inside a lattice may also be defined by Miller indices *h, k, l*, which are, in general, non-integer- or integer-valued. These directions are usually written as [*h k l*] inside **square** brackets. In addition, lattices with point symmetry allow symmetry equivalent directions where the corresponding **direction families** are written as <*h k l*> inside **pointed** brackets. As an example, the simple cubic lattice includes eight equivalent directions $\pm$[1 1 1], $\pm$[−1 1 1], $\pm$[1 −1 1], $\pm$[1 1 −1], which form a direction family <1 1 1>.

3.2
Netplane-Adapted Lattice Vectors

The discussion in Section 3.1 showed that (*h k l*) netplanes can be associated with linear transformations $\underline{\underline{T}}^{(h\,k\,l)}$, connecting the initial lattice vectors $\underline{R}_{o1}, \underline{R}_{o2}, \underline{R}_{o3}$ with **netplane-adapted** lattice vectors $\underline{R}_1, \underline{R}_2, \underline{R}_3$ according to Eq. (2.62), where vectors $\underline{R}_1$ and $\underline{R}_2$ determine the netplane periodicity. Thus, transformations $\underline{\underline{T}}^{(h\,k\,l)}$ are essential for any **computation** of crystal properties, which requires quantitative information about netplanes by their explicit lattice vectors $\underline{R}_1$ and $\underline{R}_2$ derived from the initial lattice vectors $\underline{R}_{o1}, \underline{R}_{o2}, \underline{R}_{o3}$. As an example, we mention theoretical evaluations of elastic moduli along specific crystal directions. Transformed lattice vectors are also an essential ingredient for any quantitative theoretical treatment of ideal single crystal surfaces as will be discussed in detail in Chapter 4.

The close relationship between transformation matrices $\underline{\underline{T}}^{(h\,k\,l)}$ and Miller indices (*h k l*) becomes clear from the following relations obtained by using Eqs. (2.62), (2.96), and (3.6)

$$\underline{G}_{(h\,k\,l)}\,\underline{R}_1 = (h\,\underline{G}_{o1} + k\,\underline{G}_{o2} + l\,\underline{G}_{o3})\,(t_{11}\,\underline{R}_{o1} + t_{12}\,\underline{R}_{o2} + t_{13}\,\underline{R}_{o3})$$
$$= 2\pi\,(t_{11}\,h + t_{12}\,k + t_{13}\,l) = 0$$
$$\underline{G}_{(h\,k\,l)}\,\underline{R}_2 = (h\,\underline{G}_{o1} + k\,\underline{G}_{o2} + l\,\underline{G}_{o3})\,(t_{21}\,\underline{R}_{o1} + t_{22}\,\underline{R}_{o2} + t_{23}\,\underline{R}_{o3})$$
$$= 2\pi\,(t_{21}\,h + t_{22}\,k + t_{23}\,l) = 0$$
$$\underline{G}_{(h\,k\,l)}\,\underline{R}_3 = (h\,\underline{G}_{o1} + k\,\underline{G}_{o2} + l\,\underline{G}_{o3})\,(t_{31}\,\underline{R}_{o1} + t_{32}\,\underline{R}_{o2} + t_{33}\,\underline{R}_{o3})$$
$$= 2\pi\,(t_{31}\,h + t_{32}\,k + t_{33}\,l) = 2\pi \quad (3.9)$$

resulting in a set of three **linear Diophantine equations** that can be written in matrix form as

$$\begin{pmatrix} t_{11} & t_{12} & t_{13} \\ t_{21} & t_{22} & t_{23} \\ t_{31} & t_{32} & t_{33} \end{pmatrix} \cdot \begin{pmatrix} h \\ k \\ l \end{pmatrix} = \underline{\underline{T}}^{(hkl)} \cdot \begin{pmatrix} h \\ k \\ l \end{pmatrix} = \begin{pmatrix} 0 \\ 0 \\ 1 \end{pmatrix} \tag{3.10}$$

This shows that for any transformation matrix $\underline{\underline{T}}^{(hkl)}$, the corresponding Miller indices (h k l) can be obtained by solving the linear equations (3.10). In fact, the solutions are already given explicitly in Eq. (3.5).

On the other hand, for given (h k l) values example transformations $\underline{\underline{T}}^{(hkl)}$ can be evaluated from Eq. (3.10) using **number theoretical** methods as has been shown elsewhere [84]. Here we mention three **example solutions** which cover all possible h, k, l values.

1) Let integers a, b solve the linear Diophantine equation

$$ah + bk = 1 \tag{3.11a}$$

then

$$\underline{\underline{T}}^{(hkl)} = \begin{pmatrix} k & -h & 0 \\ la & lb & -1 \\ a & b & 0 \end{pmatrix}, \quad \left(\underline{\underline{T}}^{(hkl)}\right)^{-1} = \begin{pmatrix} b & 0 & h \\ -a & 0 & k \\ 0 & -1 & l \end{pmatrix} \tag{3.11b}$$

2) Let integers a, c solve the linear Diophantine equation

$$ah + cl = 1 \tag{3.12a}$$

then

$$\underline{\underline{T}}^{(hkl)} = \begin{pmatrix} -l & 0 & h \\ ka & -1 & kc \\ a & 0 & c \end{pmatrix}, \quad \left(\underline{\underline{T}}^{(hkl)}\right)^{-1} = \begin{pmatrix} -c & 0 & h \\ 0 & -1 & k \\ a & 0 & l \end{pmatrix} \tag{3.12b}$$

3) Let integers b, c solve the linear Diophantine equation

$$bk + cl = 1 \tag{3.13a}$$

then

$$\underline{\underline{T}}^{(hkl)} = \begin{pmatrix} 0 & l & -k \\ -1 & hb & hc \\ 0 & b & c \end{pmatrix}, \quad \left(\underline{\underline{T}}^{(hkl)}\right)^{-1} = \begin{pmatrix} 0 & -1 & h \\ c & 0 & k \\ -b & 0 & l \end{pmatrix} \tag{3.13b}$$

Thus, each of these solutions requires the computation of **one** linear Diophantine equation in two variables, Eqs. (3.11a), (3.12a), and (3.13a), where solutions may be guessed or determined numerically, using the algorithm discussed in Appendix E.3. Table 3.2 shows which of the three solutions can be used for any triplet (h k l). This table assumes that all non-zero Miller index values are

Table 3.2 Example transformations $\underline{T}^{(h\,k\,l)}$ for different Miller index values $(h\,k\,l)$ referring to solutions of Eqs. (3.11), (3.12), or (3.13).

Miller indices	Transformation
$(h\,0\,0) \equiv (1\,0\,0)$	(3.11) with a = 1, b = 0; (3.12) with a = 1, c = 0
$(0\,k\,0) \equiv (0\,1\,0)$	(3.11) with a = 0, b = 1; (3.13) with b = 1, c = 0
$(0\,0\,l) \equiv (0\,0\,1)$	(3.12) with a = 0, c = 1; (3.13) with b = 0, c = 1
$(h\,k\,0)$, $\gcd(|h|, |k|) = 1$	(3.11) compute a, b
$(h\,0\,l)$, $\gcd(|h|, |l|) = 1$	(3.12) compute a, c
$(0\,k\,l)$, $\gcd(|k|, |l|) = 1$	(3.13) compute b, c
$(h\,k\,l)$, $\gcd(|h|, |k|, |l|) = 1$	(3.11) compute a, b or (3.12) compute a, c or (3.13) compute b, c

normalized such that they do not have a common divisor greater than 1. Otherwise, all non-zero Miller indices have to be divided by gcd(x, y) or gcd(x, y, z) before transformations (3.11), (3.12), or (3.13) can be applied. (Functions gcd(x, y) and gcd(x, y, z) denote the greatest common divisor, see Appendix E.1.)

The calculation of transformation matrices $\underline{T}^{(h\,k\,l)}$ and, hence, of the corresponding netplane-adapted lattice vectors described in this section, does not make use of any specific lattice properties like lattice type or symmetry. Further, matrices $\underline{T}^{(h\,k\,l)}$ given by Eqs. (3.11b), (3.12b), and (3.13b) represent, in each case, an **infinite number of solutions**, since the generating Diophantine Eqs. (3.11a), (3.12a), and (3.13a) have infinitely many solutions, as shown in Appendix E.3. This reflects the fact that lattice descriptions are **not unique** and there is always an infinite number of alternatives. The next section shows how to select from this infinite manifold specific lattice descriptions, which can also reflect point symmetry properties of the $(h\,k\,l)$ netplane under consideration.

3.3
Symmetrically Appropriate Lattice Vectors: Minkowski Reduction

The netplane-adapted lattice vectors given by Eqs. (3.11), (3.12), and (3.13) yield **mathematically exact** lattice descriptions which may, however, **not** always be **intuitive.** As an example, we mention the simple cubic lattice, where (0 0 1) netplanes are of square shape, see Figure 2.15, which can be constructed by lattice vectors

$$\underline{R}_{o1} = a\,(1,\,0,\,0), \quad \underline{R}_{o2} = a\,(0,\,1,\,0) \qquad (3.14)$$

denoting a symmetrically appropriate vector set, since the two vectors are of equal length and are perpendicular to each other, suggesting a square pattern. However, there are alternative lattice vectors, for example,

$$\underline{R}_1 = a\,(-1,\,-1,\,0), \quad \underline{R}_2 = a\,(2,\,1,\,0) \qquad (3.15)$$

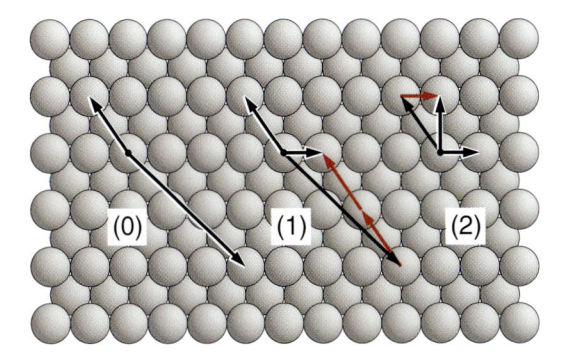

Figure 3.2 Minkowski reduction of a lattice vector set for the (1 1 0) netplane of an fcc monolayer. The underlying monolayer is added for orientation.

which do not give any idea of the square shape of the netplane. This problem can arise, in particular, when lattice vectors are generated numerically by a computer (i.e., without visual intuition). The rather hazy notion of "**symmetrically appropriate**" can be quantified by requiring that the lattice vectors $\underline{R}_1$ and $\underline{R}_2$ connect lattice points of **smallest** distance in the lattice with $|\underline{R}_1| \leq |\underline{R}_2|$. This requirement allows additional point symmetry elements to become visible already in the lattice vector representation of the corresponding netplane. Symmetrically appropriate vectors can be constructed iteratively following an algorithm due to Minkowski (**Minkowski reduction**) [85], see Appendix D for mathematical details. The iteration starts from any two lattice vectors $\underline{R}_1$ and $\underline{R}_2$ which are **reduced** successively in length by **linear mixing** until vectors of smallest length are obtained. This is illustrated in Figure 3.2 for the (1 1 0) netplane of an fcc monolayer. Here, the initial lattice vectors $\underline{R}_1$ and $\underline{R}_2$ (labeled "(0)" in the figure) are reduced in two iteration steps to yield Minkowski-reduced lattice vectors (labeled "(2)") where the vectors in red illustrate the reduction in each step.

The reduction yields always after a **finite** number of iterations two Minkowski-reduced lattice vectors $\underline{R}_1$ and $\underline{R}_2$ of lengths R_1 and R_2, which can be shown to satisfy the condition

$$-\min(R_1^2, R_2^2) \leq 2\,(\underline{R}_1\,\underline{R}_2) < \min(R_1^2, R_2^2) \tag{3.16}$$

This means that geometrically each of the two vectors $\underline{R}_i$ projected on to the other, $\underline{R}_j$, yields a vector along $\underline{R}_j$ of less than or equal to half the length of $\underline{R}_j$. As shown in Appendix D, relation (3.16) guarantees that at least one of the two lattice vectors, $\underline{R}_1$ or $\underline{R}_2$, connects lattice points of smallest distance in the lattice. If, as a result of the reduction, the vectors $\underline{R}_1$ and $\underline{R}_2$ are of the same length they must both be of smallest length. Further, relation (3.16) can be written as

$$-1 \leq -A \leq 2\cos(\gamma) = 2\,(\underline{R}_1\,\underline{R}_2)/(R_1 R_2) < A \leq 1$$
$$A = \min(R_1/R_2, R_2/R_1) \tag{3.17}$$

where γ is the angle between vectors $\underline{R}_1$ and $\underline{R}_2$. This proves that the angle between two Minkowski-reduced lattice vectors $\underline{R}_1$ and $\underline{R}_2$ must always lie between 60° and 120° where we distinguish between **acute**, $60° < \gamma \leq 90°$, and **obtuse** lattice vector sets, $90° < \gamma \leq 120°$.

Altogether, the Minkowski formalism leads to uniquely defined lattice vectors $\underline{R}_1$ and $\underline{R}_2$ for any two-dimensional lattice and allows easy classification of the different types of two-dimensional Bravais lattices discussed in Section 3.8. Centered rectangular and hexagonal lattices yield equivalent lattice vector sets for $\cos(\gamma) = A/2$ and $\cos(\gamma) = -A/2$ (with A according to Eq. (3.17)) corresponding to acute and obtuse vector sets that can both be considered to be Minkowski-reduced. Here, crystallographers prefer the obtuse representation referring to the strict definition given in Eq. (3.16). It should be mentioned that for three-dimensional bulk crystals analogous reduction schemes have been proposed [42, 43] to obtain unique lattice vector sets $\underline{R}_1, \underline{R}_2, \underline{R}_3$, see also Section 2.2.2. There, the simplification to two dimensions is equivalent to the Minkowski reduction.

3.4
Miller Indices for Cubic and Trigonal Lattices

Generic Miller indices $(h\,k\,l)$ are, by definition, based on reciprocal lattice vectors $\underline{G}_{o1}, \underline{G}_{o2}, \underline{G}_{o3}$, as given by Eq. (2.95), where lattice vectors $\underline{R}_{o1}, \underline{R}_{o2}, \underline{R}_{o3}$, referring to a primitive lattice representation, are the natural choice. However, in the case of **cubic** lattices, scientists very often use real space and reciprocal lattice vectors of the **simple cubic** lattice even in studies of crystals with fcc and bcc lattices. This is due to the **geometric simplicity** of the sc lattice with its three orthogonal lattice vectors of equal length where the reciprocal lattice is also of sc type and direction vectors in real space are parallel to those in reciprocal space. The choice of the sc lattice for fcc and bcc lattices affects the corresponding Miller index values, as will be discussed in the following.

According to Eq. (2.102), the **fcc** lattice is characterized by reciprocal lattice vectors $\underline{G}_{o1}^{fcc}, \underline{G}_{o2}^{fcc}, \underline{G}_{o3}^{fcc}$, which can be represented by those of the simple cubic lattice, using Eq. (2.101), where

$$\underline{G}_{o1}^{fcc} = -\underline{G}_{o1}^{sc} + \underline{G}_{o2}^{sc} + \underline{G}_{o3}^{sc}$$
$$\underline{G}_{o2}^{fcc} = \underline{G}_{o1}^{sc} - \underline{G}_{o2}^{sc} + \underline{G}_{o3}^{sc}$$
$$\underline{G}_{o3}^{fcc} = \underline{G}_{o1}^{sc} + \underline{G}_{o2}^{sc} - \underline{G}_{o3}^{sc} \tag{3.18}$$

As a consequence, netplane normal directions point along vectors

$$\begin{aligned}
\underline{G}_{(h\,k\,l)} &= h^{fcc}\,\underline{G}_{o1}^{fcc} + k^{fcc}\,\underline{G}_{o2}^{fcc} + l^{fcc}\,\underline{G}_{o3}^{fcc} \\
&= (-h^{fcc} + k^{fcc} + l^{fcc})\,\underline{G}_{o1}^{sc} \\
&\quad + (h^{fcc} - k^{fcc} + l^{fcc})\,\underline{G}_{o2}^{sc} + (h^{fcc} + k^{fcc} - l^{fcc})\,\underline{G}_{o3}^{sc} \\
&= h^{sc}\,\underline{G}_{o1}^{sc} + k^{sc}\,\underline{G}_{o2}^{sc} + l^{sc}\,\underline{G}_{o3}^{sc}
\end{aligned} \tag{3.19}$$

3.4 Miller Indices for Cubic and Trigonal Lattices

suggesting, in addition to the generic notation ($h^{fcc}\ k^{fcc}\ l^{fcc}$), a **simple cubic or sc notation** ($h^{sc}\ k^{sc}\ l^{sc}$) for Miller indices of the fcc lattice. According to Eq. (3.19), there is a linear transformation between the indices given by

$$\begin{pmatrix} h^{sc} \\ k^{sc} \\ l^{sc} \end{pmatrix} = \begin{pmatrix} -1 & 1 & 1 \\ 1 & -1 & 1 \\ 1 & 1 & -1 \end{pmatrix} \cdot \begin{pmatrix} h^{fcc} \\ k^{fcc} \\ l^{fcc} \end{pmatrix}, \tag{3.20a}$$

$$\begin{pmatrix} h^{fcc} \\ k^{fcc} \\ l^{fcc} \end{pmatrix} = \frac{1}{2}\begin{pmatrix} 0 & 1 & 1 \\ 1 & 0 & 1 \\ 1 & 1 & 0 \end{pmatrix} \cdot \begin{pmatrix} h^{sc} \\ k^{sc} \\ l^{sc} \end{pmatrix} \tag{3.20b}$$

Here the factor 1/2 in transformation (3.20b) restricts possible values of Miller indices in simple cubic notation. Transformation (3.20b) yields integer-valued Miller indices, $h^{fcc}, k^{fcc}, l^{fcc}$, only if the indices in simple cubic notation, h^{sc}, k^{sc}, l^{sc}, are **all even** or **all odd** integers.

Transformation (3.20a) can also be understood as the three basic Miller index triplets (−1 1 1), (1 −1 1), (1 1 −1) defining lattice vectors in the integer vector space, which describes all sc Miller indices ($h^{sc}\ k^{sc}\ l^{sc}$) of an fcc lattice in simple cubic notation. Thus, any valid Miller index triplet ($h^{sc}\ k^{sc}\ l^{sc}$) can be decomposed into contributions of basic Miller index triplets, where the above choice is not unique. Alternative basic lattice vectors ($h_1\ k_1\ l_1$), ($h_2\ k_2\ l_2$), ($h_3\ k_3\ l_3$) can be obtained by transformations

$$\begin{pmatrix} h_1 & k_1 & l_1 \\ h_2 & k_2 & l_2 \\ h_3 & k_3 & l_3 \end{pmatrix} = \underline{\underline{T}} \cdot \begin{pmatrix} -1 & 1 & 1 \\ 1 & -1 & 1 \\ 1 & 1 & -1 \end{pmatrix} \tag{3.21}$$

where $\underline{\underline{T}}$ is an integer-valued 3×3 matrix with det($\underline{\underline{T}}$) = 1. As an example, we mention

$$\underline{\underline{T}} = \begin{pmatrix} 1 & 1 & 1 \\ 0 & 1 & 1 \\ 0 & 0 & 1 \end{pmatrix}, \quad \begin{pmatrix} h_1 & k_1 & l_1 \\ h_2 & k_2 & l_2 \\ h_3 & k_3 & l_3 \end{pmatrix} = \begin{pmatrix} 1 & 1 & 1 \\ 2 & 0 & 0 \\ 1 & 1 & -1 \end{pmatrix} \tag{3.22}$$

The choice of appropriate basic lattice vectors in the vector space of integers describing all possible Miller indices will become important in connection with the decomposition of Miller indices characterizing stepped and kinked surfaces, see Section 4.3. Table 3.3 lists typical examples of lowest Miller index triplets ($h\ k\ l$) of the fcc lattice given both in simple cubic and in generic fcc notation.

The **bcc** lattice is, according to Eq. (2.103), characterized by reciprocal lattice vectors $\underline{G}_1^{bcc}, \underline{G}_2^{bcc}, \underline{G}_3^{bcc}$, which can be represented by those of the simple cubic lattice, using Eq. (2.101), where

$$\begin{aligned} \underline{G}_1^{bcc} &= \underline{G}_2^{sc} + \underline{G}_3^{sc} \\ \underline{G}_2^{bcc} &= \underline{G}_1^{sc} + \underline{G}_3^{sc} \\ \underline{G}_3^{bcc} &= \underline{G}_1^{sc} + \underline{G}_2^{sc} \end{aligned} \tag{3.23}$$

Table 3.3 Typical Miller index triplets of the fcc lattice given both in simple cubic notation, ($h^{sc}\ k^{sc}\ l^{sc}$), and in generic fcc notation ($h^{fcc}\ k^{fcc}\ l^{fcc}$).

($h^{sc}\ k^{sc}\ l^{sc}$)	($h^{fcc}\ k^{fcc}\ l^{fcc}$)
(1 1 1)	(1 1 1)
(2 0 0)	(0 1 1)
(2 2 0)	(1 1 2)
(3 1 1)	(1 2 2)
(3 3 1)	(2 2 3)
(4 2 0)	(1 2 3)
(4 2 2)	(2 3 3)

Therefore, netplane normal directions point along vectors

$$\begin{aligned} \underline{G}_{(hkl)} &= h^{bcc}\underline{G}_{o1}^{bcc} + k^{bcc}\underline{G}_{o2}^{bcc} + l^{bcc}\underline{G}_{o3}^{bcc} \\ &= (k^{fcc} + l^{fcc})\underline{G}_{o1}^{sc} + (h^{fcc} + l^{fcc})\underline{G}_{o2}^{sc} + (h^{fcc} + k^{fcc})\underline{G}_{o3}^{sc} \\ &= h^{sc}\underline{G}_{o1}^{sc} + k^{sc}\underline{G}_{o2}^{sc} + l^{sc}\underline{G}_{o3}^{sc} \end{aligned} \quad (3.24)$$

which suggests, in addition to the generic notation ($h^{bcc}\ k^{bcc}\ l^{bcc}$), a **simple cubic** or **sc notation** ($h^{sc}\ k^{sc}\ l^{sc}$) for Miller indices of the bcc lattice. According to Eq. (3.24), there is a linear transformation between the indices

$$\begin{pmatrix} h^{sc} \\ k^{sc} \\ l^{sc} \end{pmatrix} = \begin{pmatrix} 0 & 1 & 1 \\ 1 & 0 & 1 \\ 1 & 1 & 0 \end{pmatrix} \cdot \begin{pmatrix} h^{bcc} \\ k^{bcc} \\ l^{bcc} \end{pmatrix}, \quad (3.25a)$$

$$\begin{pmatrix} h^{bcc} \\ k^{bcc} \\ l^{bcc} \end{pmatrix} = \frac{1}{2}\begin{pmatrix} -1 & 1 & 1 \\ 1 & -1 & 1 \\ 1 & 1 & -1 \end{pmatrix} \cdot \begin{pmatrix} h^{sc} \\ k^{sc} \\ l^{sc} \end{pmatrix} \quad (3.25b)$$

Analogous to the fcc lattice, the factor 1/2 in transformation (3.25b) restricts possible values of Miller indices in simple cubic notation. Here, transformation (3.25b) yields integer-valued Miller indices, h^{bcc}, k^{bcc}, l^{bcc}, only for indices in simple cubic notation, h^{sc}, k^{sc}, l^{sc}, where the **sum** of all three values, that is, $g = h^{sc} + k^{sc} + l^{sc}$ is an **even** integer. This is achieved by either all indices being even or by one being even and two odd.

Transformation (3.25a) can, as before, be understood as the three basic Miller index triplets (0 1 1), (1 0 1), (1 1 0) defining lattice vectors in the integer vector space, which describes all sc Miller indices ($h^{sc}\ k^{sc}\ l^{sc}$) of a bcc lattice in simple cubic notation. Thus, any valid Miller index triplet ($h^{sc}\ k^{sc}\ l^{sc}$) can be decomposed into contributions of basic Miller index triplets, where the above choice is not unique. Alternative basic lattice vectors ($h_1\ k_1\ l_1$), ($h_2\ k_2\ l_2$), ($h_3\ k_3\ l_3$) can be obtained by transformations

3.4 Miller Indices for Cubic and Trigonal Lattices

Table 3.4 Typical Miller index triplets of the bcc lattice given both in simple cubic notation, (h^{sc} k^{sc} l^{sc}), and in generic bcc notation (h^{bcc} k^{bcc} l^{bcc}).

(h^{sc} k^{sc} l^{sc})	(h^{bcc} k^{bcc} l^{bcc})
(1 1 0)	(0 0 1)
(2 0 0)	(−1 1 1)
(2 1 1)	(0 1 1)
(3 1 0)	(−1 1 2)
(2 2 2)	(1 1 1)
(3 2 1)	(0 1 2)
(3 3 2)	(1 1 2)

$$\begin{pmatrix} h_1 & k_1 & l_1 \\ h_2 & k_2 & l_2 \\ h_3 & k_3 & l_3 \end{pmatrix} = \underline{\underline{T}} \cdot \begin{pmatrix} 0 & 1 & 1 \\ 1 & 0 & 1 \\ 1 & 1 & 0 \end{pmatrix} \quad (3.26)$$

where $\underline{\underline{T}}$ is an integer-valued 3×3 matrix with $\det(\underline{\underline{T}}) = 1$. As an example, we mention

$$\underline{\underline{T}} = \begin{pmatrix} 0 & 0 & 1 \\ 1 & -1 & 1 \\ 1 & 0 & 0 \end{pmatrix}, \quad \begin{pmatrix} h_1 & k_1 & l_1 \\ h_2 & k_2 & l_2 \\ h_3 & k_3 & l_3 \end{pmatrix} = \begin{pmatrix} 1 & 1 & 0 \\ 0 & 2 & 0 \\ 0 & 1 & 1 \end{pmatrix} \quad (3.27)$$

As for the fcc case, the choice of appropriate basic lattice vectors in the vector space of integers describing all possible Miller indices will become important in connection with the decomposition of Miller indices characterizing stepped and kinked surfaces, see Section 4.3. Table 3.4 lists typical examples of lowest Miller index triplets ($h\,k\,l$) of the bcc lattice given both in simple cubic and in generic bcc notation.

The numerical constraints on Miller indices, h^{sc}, k^{sc}, l^{sc}, in simple cubic notation for fcc or bcc lattices become important when Miller indices (and corresponding reciprocal lattice vectors $\underline{G}_{(h\,k\,l)}$) are used in **numerical calculus.** Examples are the evaluation of netplane distances $d_{(h\,k\,l)}$ or the decomposition of Miller indices discussed in see Section 4.3. As an illustration, distances $d_{(h\,k\,l)}$ between adjacent netplanes of an fcc lattice are, according to Eqs. (2.101), (2.102), and (3.7), given by

$$d_{(h\,k\,l)} = 2\pi / |G_{(h\,k\,l)}| = a / [3\{(h^{fcc})^2 + (k^{fcc})^2 + (l^{fcc})^2\}$$
$$-2\{h^{fcc}k^{fcc} + h^{fcc}l^{fcc} + k^{fcc}l^{fcc}\}]^{1/2}$$
$$= a / [(h^{sc})^2 + (k^{sc})^2 + (l^{sc})^2]^{1/2} \quad (3.28)$$

with the three sc Miller indices, h^{sc}, k^{sc}, l^{sc}, required to be either all even or all odd. Thus, netplanes with (1 1 2) orientation in simple cubic notation must use $h^{sc} = 2$, $k^{sc} = 2$, $l^{sc} = 4$ in the evaluation of Eq. (3.28). In general, Miller index triplets ($h\,k\,l$), given in simple cubic notation for fcc lattices and representing mixtures of even and odd numbers, have to be scaled by a factor of 2 for any quantitative calculus. For example, the triplet (1 2 3) needs to be replaced by (2 4 6). In analogy, Miller indices, given in simple cubic notation for bcc lattices with the sum $(h+k+l)$ being an odd number, must be scaled by a factor of 2. For example, the triplet (1 1 1) needs to be replaced by (2 2 2). However, when Miller indices $(h\,k\,l)$ are to be used only to denote **netplane directions** in the crystal, common integer factors in the indices can be omitted.

The correlations between generic fcc or bcc Miller indices and those referring to the simple cubic lattice are special cases of a more general behavior of Miller indices when lattice vectors are modified by transformations. If we assume that $\underline{R}_1'$, $\underline{R}_2'$, $\underline{R}_3'$ are lattice vectors arising from a linear transformation of initial lattice vectors $\underline{R}_1$, $\underline{R}_2$, $\underline{R}_3$ according to

$$\begin{pmatrix} \underline{R}_1' \\ \underline{R}_2' \\ \underline{R}_3' \end{pmatrix} = \underline{\underline{T}} \cdot \begin{pmatrix} \underline{R}_1 \\ \underline{R}_2 \\ \underline{R}_3 \end{pmatrix} \tag{3.29}$$

then the corresponding reciprocal lattice vectors $\underline{G}_1'$, $\underline{G}_2'$, $\underline{G}_3'$ and $\underline{G}_1$, $\underline{G}_2$, $\underline{G}_3$ are connected by

$$\begin{pmatrix} \underline{G}_1' \\ \underline{G}_2' \\ \underline{G}_3' \end{pmatrix} = (\underline{\underline{T}}^{-1})^+ \cdot \begin{pmatrix} \underline{G}_1 \\ \underline{G}_2 \\ \underline{G}_3 \end{pmatrix} \tag{3.30}$$

where the superscript "+" in $(\underline{\underline{T}}^{-1})^+$ denotes the transposed matrix of $\underline{\underline{T}}^{-1}$. Thus, a general reciprocal lattice vector $\underline{G}$ that can be written in both representations as

$$\underline{G} = h\underline{G}_1 + k\underline{G}_2 + l\underline{G}_3 = h'\underline{G}_1' + k'\underline{G}_2' + l'\underline{G}_3' \tag{3.31}$$

is given in matrix form together with Eq. (3.30) by

$$\underline{G} = (h'\,k'\,l') \cdot \begin{pmatrix} \underline{G}_1' \\ \underline{G}_2' \\ \underline{G}_3' \end{pmatrix} = (h'\,k'\,l') \cdot (\underline{\underline{T}}^{-1})^+ \cdot \begin{pmatrix} \underline{G}_1 \\ \underline{G}_2 \\ \underline{G}_3 \end{pmatrix} = (h\,k\,l) \cdot \begin{pmatrix} \underline{G}_1 \\ \underline{G}_2 \\ \underline{G}_3 \end{pmatrix} \tag{3.32}$$

This leads to a relation between the corresponding Miller indices where

$$(h'\,k'\,l') \cdot (\underline{\underline{T}}^{-1})^+ = (h\,k\,l), \quad (h'\,k'\,l') = (h\,k\,l) \cdot (\underline{\underline{T}})^+ \tag{3.33}$$

or

$$\begin{pmatrix} h' \\ k' \\ l' \end{pmatrix} = \underline{\underline{T}} \cdot \begin{pmatrix} h \\ k \\ l \end{pmatrix} \tag{3.34}$$

A comparison of Eq. (3.34) with Eq. (3.29) shows that the transformation between lattice vectors and that between the corresponding Miller indices uses the same transformation matrix $\underline{\underline{T}}$.

As an example, lattice vectors of trigonal crystals are often expressed by those of a related hexagonal lattice where the hexagonal vectors $\underline{R}_1^{hex}$, $\underline{R}_2^{hex}$, $\underline{R}_3^{hex}$ are connected with those of the trigonal lattice, $\underline{R}_1^{trg}$, $\underline{R}_2^{trg}$, $\underline{R}_3^{trg}$, for example, according to

$$\begin{pmatrix} R_1^{hex} \\ R_2^{hex} \\ R_3^{hex} \end{pmatrix} = \underline{\underline{T}} \cdot \begin{pmatrix} R_1^{trg} \\ R_2^{trg} \\ R_3^{trg} \end{pmatrix} = \begin{pmatrix} 1 & -1 & 0 \\ 0 & 1 & -1 \\ 1 & 1 & 1 \end{pmatrix} \cdot \begin{pmatrix} R_1^{trg} \\ R_2^{trg} \\ R_3^{trg} \end{pmatrix} \qquad (3.35)$$

see Eq. (2.48). (There are, altogether, six choices of obtuse and six of acute representations of the corresponding hexagonal lattice of which Eq. (3.35) is an obtuse representation.) Therefore, hexagonal Miller indices h^{hex}, k^{hex}, l^{hex} are connected with those of the corresponding trigonal lattice, h^{trg}, k^{trg}, l^{trg}, by

$$\begin{pmatrix} h^{hex} \\ k^{hex} \\ l^{hex} \end{pmatrix} = \underline{\underline{T}} \cdot \begin{pmatrix} h^{trg} \\ k^{trg} \\ l^{trg} \end{pmatrix} = \begin{pmatrix} 1 & -1 & 0 \\ 0 & 1 & -1 \\ 1 & 1 & 1 \end{pmatrix} \cdot \begin{pmatrix} h^{trg} \\ k^{trg} \\ l^{trg} \end{pmatrix} \qquad (3.36)$$

and hence

$$\begin{pmatrix} h^{trg} \\ k^{trg} \\ l^{trg} \end{pmatrix} = \frac{1}{3} \begin{pmatrix} 2 & 1 & 1 \\ -1 & 1 & 1 \\ -1 & -2 & 1 \end{pmatrix} \cdot \begin{pmatrix} h^{hex} \\ k^{hex} \\ l^{hex} \end{pmatrix} \qquad (3.37)$$

All Miller indices in Eq. (3.37) must be integer-valued which puts constraints on possible values of hexagonal Miller indices. These can be expressed as

$$-h^{hex} + k^{hex} + l^{hex} = 3g, \quad 2h^{hex} + k^{hex} + l^{hex} = 3g',$$
$$-h^{hex} - 2k^{hex} + l^{hex} = 3g'' \qquad (3.38)$$

where g, g', g'' are integers. Since

$$-h^{hex} + k^{hex} + l^{hex} = (2h^{hex} + k^{hex} + l^{hex}) - 3h^{hex}$$
$$= (-h^{hex} - 2k^{hex} + l^{hex}) + 3k^{hex} \qquad (3.39)$$

fulfilling one of the three conditions (3.38) will automatically satisfy the other two. Thus, the constraints of Eq. (3.38) can be replaced by one constraint,

$$-h^{hex} + k^{hex} + l^{hex} = 3g \qquad (3.40)$$

If all six possible obtuse representations of the hexagonal lattice are considered, three of them are subject to the constraint of Eq. (3.40) while the other three are subject to

$$h^{hex} - k^{hex} + l^{hex} = 3g \qquad (3.41)$$

Further, it can be shown that of the six different acute representation of the hexagonal lattice three are subject to the constraint

$$h^{hex} + k^{hex} - l^{hex} = 3g \qquad (3.42)$$

and three to

$$h^{hex} + k^{hex} + l^{hex} = 3g \qquad (3.43)$$

Table 3.5 lists typical examples of lowest Miller index triplets $(h\ k\ l)$ of the trigonal lattice given in both generic trigonal and in hexagonal notation according to Eq. (3.36). This table also includes the corresponding Miller–Bravais indices $(l^{hex}\ m^{hex}\ n^{hex}\ q^{hex})$ according to Eq. (3.53) as discussed in Section 3.5.

Table 3.5 Typical Miller index triplets of the trigonal lattice given both in generic trigonal ($h^{trg}\ k^{trg}\ l^{trg}$), in obtuse hexagonal notation, ($h^{hex}\ k^{hex}\ l^{hex}$) and in four-index notation ($l^{hex}\ m^{hex}\ n^{hex}\ q^{hex}$).

($h^{trg}\ k^{trg}\ l^{trg}$)	($h^{hex}\ k^{hex}\ l^{hex}$)	($l^{hex}\ m^{hex}\ n^{hex}\ q^{hex}$)
(1 0 0)	(1 0 1)	(1 0 −1 1)
(0 1 0)	(−1 1 1)	(−1 1 0 1)
(0 0 1)	(0 −1 1)	(0 −1 1 1)
(1 1 0)	(0 1 2)	(0 1 −1 2)
(1 0 1)	(1 −1 2)	(1 −1 0 2)
(0 1 1)	(−1 0 2)	(−1 0 1 2)
(1 1 1)	(0 0 3)	(0 0 0 3)

3.5
Alternative Definition of Miller Indices and Miller–Bravais Indices

There is an alternative way to define netplanes inside a lattice, which is usually preferred by crystallographers due to its seeming simplicity. Here, one considers two adjacent parallel $(h\ k\ l)$ netplanes in a lattice defined by lattice vectors $\underline{R}_{o1}, \underline{R}_{o2}, \underline{R}_{o3}$, where the lattice can always be positioned such that its origin coincides with that of one of the netplanes. Then the adjacent netplane will, in general, cross the lines along the three lattice vectors $\underline{R}_{o1}, \underline{R}_{o2}, \underline{R}_{o3}$ at crossing points $\underline{A}, \underline{B}, \underline{C}$ with

$$\underline{A} = \alpha_1 \underline{R}_{o1}, \quad \underline{B} = \alpha_2 \underline{R}_{o2}, \quad \text{and} \quad \underline{C} = \alpha_3 \underline{R}_{o3} \tag{3.44}$$

as shown in the Figure 3.3. Thus, the **intercept factors** $\alpha_1, \alpha_2, \alpha_3$ can be used to characterize the netplane uniquely. If vector $\underline{n}$ denotes the normal vector of the netplane, then the distance $d_{(h\ k\ l)}$ between the two adjacent netplanes is given by

$$d_{(h k l)} = \underline{A}\,\underline{n} = \alpha_1 (\underline{R}_{o1}\,\underline{n}) = \alpha_1 (\underline{R}_{o1}\,\underline{G}_{(h k l)}) / |\underline{G}_{(h k l)}|$$

$$= \alpha_1 \underline{R}_{o1} (h\,\underline{G}_{o1} + k\,\underline{G}_{o2} + l\,\underline{G}_{o3}) / G_{(h k l)}$$

$$= \alpha_1 h\, 2\pi / G_{(h k l)} = 2\pi / G_{(h k l)} \tag{3.45}$$

where Eqs. (3.6) and (3.7) together with the orthogonality relations (2.96) are used. This yields for the intercept factor α_1

$$\alpha_1 h = 1 \quad \text{or} \quad \alpha_1 = 1/h \tag{3.46}$$

In analogy, relations

$$d_{(h k l)} = \underline{B}\,\underline{n} \quad \text{and} \quad d_{(h k l)} = \underline{C}\,\underline{n} \tag{3.47}$$

result in

$$\alpha_2 = 1/k \quad \text{and} \quad \alpha_3 = 1/l \tag{3.48}$$

connecting, altogether, between inverse Miller indices h, k, l and the intercept factors α_i of the three lattice vectors cut by the netplane. Since Miller indices are all integer-valued, relations (3.46) and (3.48) show that for non-zero values of

3.5 Alternative Definition of Miller Indices and Miller–Bravais Indices

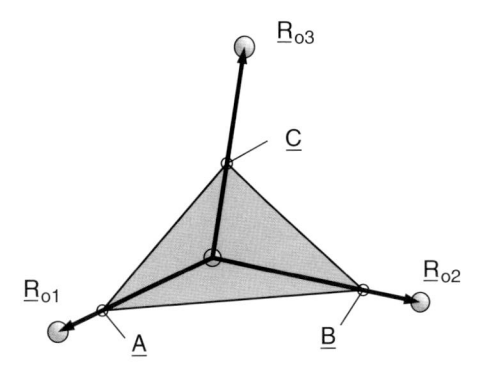

Figure 3.3 Netplane definition by its intercepts with the three lattice vectors $\underline{R}_{o1}, \underline{R}_{o2}, \underline{R}_{o3}$ at A, B, C. The lattice vectors and intercepts are sketched accordingly. The netplane is indicated by light gray.

h, k, l the corresponding intercept factors α_i are bound to $0 < |\alpha_i| \leq 1$. In addition, according to Eq. (3.46), $h = 0$ can be considered to be a result of the limiting case $\alpha_1 \to \infty$ such that the corresponding netplanes lie parallel to the lattice vector $\underline{R}_{o1}$. Analogously, $k = 0$ and $l = 0$ refer to netplanes parallel to vectors $\underline{R}_{o2}$ and $\underline{R}_{o3}$, respectively. If two Miller index values equal zero then the corresponding netplane must be parallel to two lattice vectors. For example, the (1 0 0) netplane cuts the $\underline{R}_{o1}$ axis at a lattice point ($\alpha_1 = 1$) and extends parallel to the plane spanned by $\underline{R}_{o2}$ and $\underline{R}_{o3}$. Relations (3.46) and (3.48) can be inverted to read

$$h = 1/\alpha_1, \quad k = 1/\alpha_2, \quad l = 1/\alpha_3 \quad \text{or} \quad (h\,k\,l) = (1/\alpha_1\,1/\alpha_2\,1/\alpha_3) \quad (3.49)$$

which shows that the **inverse intercept factors** are equivalent to **Miller indices** and can, thus, be alternatively used to characterize the orientation of a netplane inside a lattice.

There is a special variant of the alternative definition (3.49), which applies only to **hexagonal lattices.** These lattices are described by two lattice vectors, $\underline{R}_{o1}$ and $\underline{R}_{o2}$, forming a two-dimensional hexagonal lattice with angles $\angle(\underline{R}_{o1}, \underline{R}_{o2}) = 120°$ or $= 60°$ (obtuse or acute representation) while $\underline{R}_{o3}$ is perpendicular to both $\underline{R}_{o1}$ and $\underline{R}_{o2}$. Assuming an **obtuse representation**, the threefold symmetry of the planar sublattice given by $\underline{R}_{o1}$ and $\underline{R}_{o2}$ induces a third vector $\underline{R}'_{o2} = -\underline{R}_{o1} - \underline{R}_{o2}$, which forms an angle $\angle(\underline{R}_{o1}, \underline{R}'_{o2}) = \angle(\underline{R}_{o2}, \underline{R}'_{o2}) = 120°$ with respect to the initial lattice vectors and is of equal length, see Figure 3.4. Thus, the vector triplet $\underline{R}_{o1}, \underline{R}_{o2}, \underline{R}'_{o2}$ may be considered an equivalent set and each pair of vectors from this triplet can be used to describe the periodicity of the hexagonal netplane. Crystallographers treat the three lattice vectors $\underline{R}_{o1}, \underline{R}_{o2}, \underline{R}'_{o2}$ on equal footing and characterize netplanes by intercepts of the three lattice vectors $\underline{R}_{o1}, \underline{R}_{o2}, \underline{R}_{o3}$ and of vector $\underline{R}'_{o2}$, that is, by

$$A = \alpha_1 \underline{R}_{o1}, \quad B = \alpha_2 \underline{R}_{o2}, \quad C = \alpha_3 \underline{R}_{o3}, \quad \text{and} \quad D = \alpha'_2 \underline{R}'_{o2} \quad (3.50)$$

as shown in the Figure 3.4, where simple algebra yields

$$1/\alpha'_2 = -1/\alpha_1 - 1/\alpha_2 \quad (3.51)$$

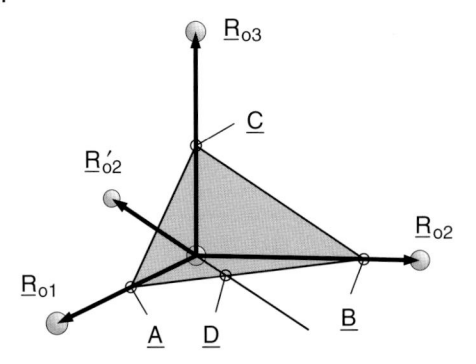

Figure 3.4 Netplane definition (hexagonal lattices) by its intercepts with the four lattice vectors $\underline{R}_{o1}$, $\underline{R}_{o2}$, $\underline{R}'_{o2}$, and $\underline{R}_{o3}$ at A, B, C, D. The lattice vectors and intercepts are sketched accordingly. The netplane is indicated by light gray.

Table 3.6 Examples of Miller indices (h k l) and corresponding Miller–Bravais indices (l m n q) based on an obtuse representation of the hexagonal lattice.

(h k l)	(l m n q)
(1 0 0)	(1 0 −1 0)
(0 1 0)	(0 1 −1 0)
(0 0 1)	(0 0 0 1)
(1 1 0)	(1 1 −2 0)
(1 0 1)	(1 0 −1 1)
(0 1 1)	(0 1 −1 1)
(1 1 1)	(1 1 −2 1)

This is the basis of the so-called **four-index notation** of the Miller indices, also referred to as **Miller–Bravais indices**, where the initial definition

$$(h\ k\ l) = (1/\alpha_1\ 1/\alpha_2\ 1/\alpha_3) \tag{3.52}$$

is, with the help of Eq. (3.51), replaced by

$$(l\ m\ n\ q) = (1/\alpha_1\ \ 1/\alpha_2\ \ 1/\alpha'_2\ \ 1/\alpha_3) = (h\ \ k\ \ (-h-k)\ \ l) \tag{3.53}$$

The quadruple $(l\ m\ n\ q)$ is sometimes also termed $(h\ k\ i\ l)$. Examples of corresponding Miller and Miller–Bravais indices are listed in Table 3.6.

The **geometric equivalence** of the three lattice vectors $\underline{R}_{o1}$, $\underline{R}_{o2}$, $\underline{R}'_{o2}$ in the obtuse representation is also visible in a symmetry property of the corresponding Miller–Bravais indices. A rotation of the hexagonal lattice by 120° (anticlockwise) about its symmetry axis along $\underline{R}_{o3}$, reproduces the lattice and leads to a transformation of its lattice vectors

$$\underline{R}_{o1} \rightarrow \underline{R}_{o2}, \quad \underline{R}_{o2} \rightarrow \underline{R}'_{o2}, \quad \underline{R}'_{o2} \rightarrow \underline{R}_{o1}, \quad \underline{R}_{o3} \rightarrow \underline{R}_{o3} \tag{3.54}$$

Table 3.7 Examples of Miller indices (h k l) and corresponding Miller–Bravais indices (l m n q) based on an acute representation of the hexagonal lattice.

(h k l)	(l m n q)
(1 0 0)	(1 −1 0 0)
(0 1 0)	(0 1 −1 0)
(0 0 1)	(0 0 0 1)
(1 1 0)	(1 0 −1 0)
(1 0 1)	(1 −1 0 1)
(0 1 1)	(0 1 −1 1)
(1 1 1)	(1 0 −1 1)

This also affects the intercepts, used for the definition of Miller–Bravais indices, see Figure 3.4, where the transformation yields

$$(l\ m\ n\ q) = (1/\alpha_1 \quad 1/\alpha_2 \quad 1/\alpha'_2 \quad 1/\alpha_3)$$
$$\rightarrow (1/\alpha_2 \quad 1/\alpha'_2 \quad 1/\alpha_1 \quad 1/\alpha_3) = (m\ n\ l\ q) \quad (3.55)$$

resulting in symmetry equivalent Miller index quadruplets. Thus, Miller–Bravais indices $(l\ m\ n\ q)$, $(m\ n\ l\ q)$, and $(n\ l\ m\ q)$ are symmetry equivalent and lead to netplanes of identical structure.

Assuming an **acute representation** of the hexagonal lattice the four points of Figure 3.4 are given by

$$\underline{A} = \alpha_1\,\underline{R}_1, \quad \underline{B} = \alpha'_2\,(\underline{R}_2 - \underline{R}_1), \quad \underline{C} = \alpha_3\,\underline{R}_3, \quad \text{and} \quad \underline{D} = \alpha_2\,(-\underline{R}_2) \quad (3.56)$$

where

$$1/\alpha'_2 = 1/\alpha_2 - 1/\alpha_1 \quad (3.57)$$

In this representation the initial definition

$$(h\ k\ l) = (1/\alpha_1\ 1/\alpha_2\ 1/\alpha_3) \quad (3.58)$$

is, with the help of Eq. (3.57), replaced by

$$(l\,m\ n\ q) = (1/\alpha_1 \quad 1/\alpha'_2 \quad -1/\alpha_2 \quad 1/\alpha_3) = (h\quad (k-h)\quad -k\quad l) \quad (3.59)$$

Examples of corresponding Miller and Miller–Bravais indices are listed in Table 3.7.

3.6
Symmetry Properties of Netplanes

Netplanes can also be analyzed in terms of their symmetry behavior, that is, by their translational symmetry and the corresponding point symmetry elements. This is analogous to the symmetry analyses of three-dimensional lattices discussed in Section 2.4. A comparison with the three-dimensional symmetry operations, available for lattices and listed in Section 2.4, shows immediately that there are only three different types of true point symmetry operations which qualify for two-dimensional netplanes. These are

- $i(\underline{r}_0)$: **inversion** with respect to symmetry origin $\underline{r}_0$, equivalent to a **180° rotation** about $\underline{r}_0$, as discussed below,
- $C_\varphi(\underline{r}_0)$: **rotation** by an angle φ about the symmetry origin $\underline{r}_0$,
- $\sigma(\underline{r}_0, \underline{e})$: **mirroring (reflection)** with respect to a line through symmetry origin $\underline{r}_0$ with the line direction defined by its normal vector $\underline{e}$.

In addition, netplanes may be symmetric with respect to a mixed mirror and translational symmetry operation which is known as

- $g(\underline{r}_0, \underline{g})$: **glide reflection**, combining a reflection $\sigma(\underline{r}_0, \underline{e})$ with a translation by vector $\underline{g}$, where vectors $\underline{g}$ and $\underline{e}$ are perpendicular to each other.

As for three-dimensional lattices, translational and point symmetry elements of a netplane are subject to **compatibility constraints**, which limits the number of possible point symmetry operations as well as their relationship with translations. This subject is treated in full mathematical detail in this section, going beyond the analogous discussion of symmetry for three-dimensional lattices in this book. First, the different symmetry operations, mentioned earlier, are defined and their interplay with translational symmetry is discussed in Sections 3.6.2–3.6.5. Then, Section 3.6.6 combines all symmetry elements to **symmetry groups**, which can be used to classify the different types of netplanes according to their symmetry. These sections are rather formal and filled with mathematical details. Thus, readers, who are less interested in mathematics, may inspect the conclusions provided at the end of each of the Sections 3.6.2–3.6.5 or may skip these sections altogether and move to Section 3.7.

3.6.1
Centered Netplanes

Centering of three-dimensional lattices was shown in Section 2.2.3 to allow primitive lattice descriptions in cases where the morphological unit cells of the initial lattice are not primitive. The same considerations can also be applied to two-dimensional netplanes where the reasoning is identical to that for cell faces of three-dimensional unit cells. Consider a netplane defined by lattice vectors $\underline{R}_1$ and $\underline{R}_2$, where both vectors are assumed to be of smallest length along their direction. Then the morphological unit cell spanned by $\underline{R}_1$ and $\underline{R}_2$ may be **primitive**, that is, of smallest area compared with all possible unit cells in the netplane. If the cell is **non-primitive** then its area is not the smallest and, in analogy to the discussion for cell faces in Section 2.2.3, there is at least one **additional lattice point** inside the morphological unit cell, described by vector $\underline{R}'$ with

$$\underline{R}' = \kappa_1 \underline{R}_1 + \kappa_2 \underline{R}_2, \quad 0 < \kappa_1, \kappa_2 < 1 \tag{3.60}$$

and for general values κ_1, κ_2 there is always a second lattice point inside the cell with vector $\underline{R}''$ given by

$$\underline{R}'' = \underline{R}_1 + \underline{R}_2 - \underline{R}' = (1 - \kappa_1)\underline{R}_1 + (1 - \kappa_2)\underline{R}_2 \tag{3.61}$$

If, however, the cell is assumed to contain **only one** additional lattice point then the vectors $\underline{R}'$ and $\underline{R}''$ must coincide, that is,

$$\underline{R}'' - \underline{R}' = (1 - 2\kappa_1)\underline{R}_1 + (1 - 2\kappa_2)\underline{R}_2 = 0 \tag{3.62}$$

Here the expressions in brackets must all be zero since the vectors $\underline{R}_1$ and $\underline{R}_2$ are linearly independent. This leads to

$$\kappa_1 = \kappa_2 = 1/2, \quad \underline{R}' = 1/2\,(\underline{R}_1 + \underline{R}_2) \tag{3.63}$$

yielding a lattice vector $\underline{R}'$ in the **center** of the **cell** spanned by $\underline{R}_1$ and $\underline{R}_2$ describing a **centered** netplane. In the symmetry classification of netplanes, discussed in Section 3.8, it will be shown that centering of a netplane of given symmetry type results in a new type of netplane only if the initial (non-primitive) netplane is rectangular, leading to a centered rectangular or c-rectangular netplane. As an illustration, Figure 3.5 shows morphological unit cells of different primitive and non-primitive lattice descriptions of a centered rectangular netplane. This figure also illustrates the general result that centered netplanes imply glide reflection symmetry as discussed in detail in Section 3.6.5.

3.6.2
Inversion

Inversion operations $i(\underline{r}_o)$ convert any point $\underline{r}$ on the netplane into its image $\underline{r}'$ such that the **inversion center** $\underline{r}_o$ cuts the connecting line between $\underline{r}$ and $\underline{r}'$ into

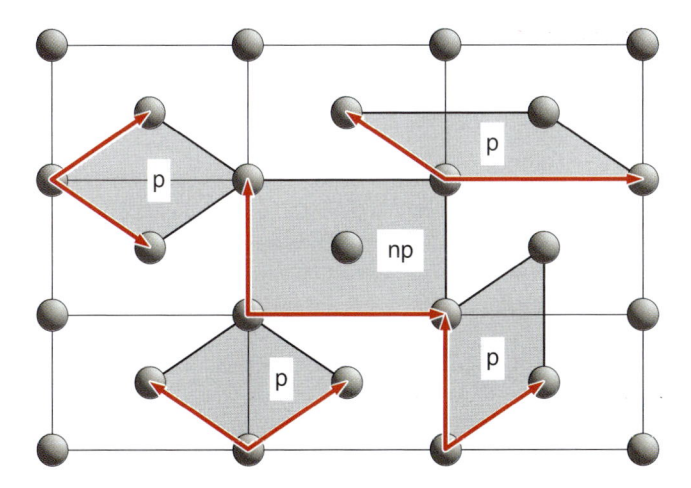

Figure 3.5 Morphological unit cells of different primitive and non-primitive lattice descriptions of a centered rectangular netplane. The cells are shaded in gray with primitive and non-primitive cells labeled "p" and "np," respectively. Corresponding lattice vectors are shown as red arrows.

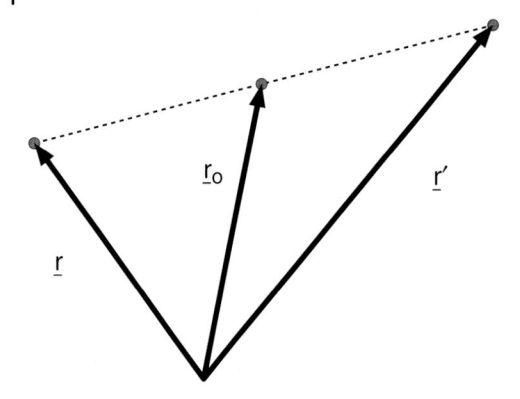

Figure 3.6 Sketch of an inversion operation applied to vector $\underline{r}$ to yield $\underline{r}'$ with the inversion center at $\underline{r}_o$.

half, see Figure 3.6. This can be expressed **mathematically** by a coordinate transformation of points on the netplane

$$\underline{r} = \underline{r}_o + (\underline{r} - \underline{r}_o) \rightarrow \underline{r}' = \underline{r}_o - (\underline{r} - \underline{r}_o) = 2\underline{r}_o - \underline{r} = \underline{r}_o + i(\underline{r} - \underline{r}_o) \quad (3.64)$$

where i is formally defined as the inversion operator. This is connected with a two-dimensional Cartesian **coordinate transformation** with respect to $\underline{r}_o$, applying a 2×2 matrix $\underline{i}$, where

$$\begin{pmatrix} x' \\ y' \end{pmatrix} = \underline{i} \cdot \begin{pmatrix} x \\ y \end{pmatrix}, \quad \underline{i} = \begin{pmatrix} -1 & 0 \\ 0 & -1 \end{pmatrix} = -\underline{1} \quad (3.65)$$

The definition of general lattice points $\underline{R}$ on a netplane with primitive lattice vectors $\underline{R}_1$ and $\underline{R}_2$ by

$$\underline{R} = n_1 \underline{R}_1 + n_2 \underline{R}_2, \quad n_1, n_2 \text{ integer} \quad (3.66)$$

implies that the **origin** of a netplane is also an **inversion center**, since for any combination of integers (n_1, n_2) there is the negative counterpart $(-n_1, -n_2)$ converting $\underline{R}$ into $-\underline{R}$. In addition, the translational symmetry of the netplane yields inversion centers at all lattice points (Eq. (3.66)) of the netplane.

There are also **other** inversion centers $\underline{r}_o$ inside the morphological unit cell of the netplane. The inversion operation (Eq. (3.64)) at a lattice point $\underline{R}$ given by Eq. (3.66) can be rewritten as

$$\underline{r}' = \underline{R} - (\underline{r} - \underline{R}) = 2\underline{R} - \underline{r} = [2(\underline{R} - \underline{r}_o) - \underline{r}] + 2\underline{r}_o$$
$$= (\underline{R} - \underline{r}_o) - (\underline{r} - (\underline{R} - \underline{r}_o)) + 2\underline{r}_o \quad (3.67)$$

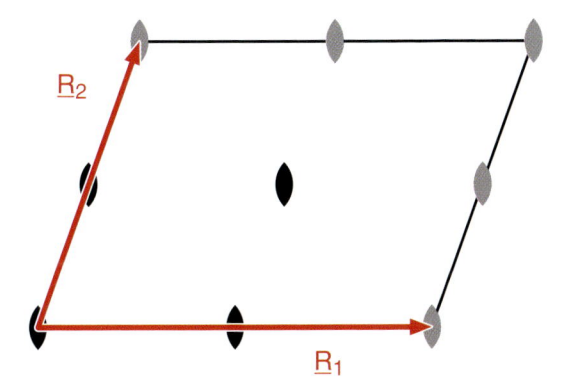

Figure 3.7 Morphological unit cell of a netplane with lattice vectors $\underline{R}_1$, $\underline{R}_2$. Non-equivalent inversion centers inside the cell are shown as black ellipses while translationally equivalent inversion centers are given by gray ellipses.

where $\underline{r}_o$ may be a point inside the morphological unit cell, given by

$$\underline{r}_o = \gamma_1 \underline{R}_1 + \gamma_2 \underline{R}_2, \quad 0 \leq \gamma_i < 1, \quad i = 1, 2 \tag{3.68}$$

Thus, Eq. (3.67) can be interpreted as an inversion at lattice point $(\underline{R} - \underline{r}_o)$ followed by a shift by vector $(2\,\underline{r}_o)$. As a consequence, if a netplane reproduces itself for inversions at $\underline{R}$ according to Eq. (3.66), its translational symmetry also yields inversion symmetry with respect to $(\underline{R} - \underline{r}_o)$ as long as $(2\,\underline{r}_o)$ is a general lattice vector. The latter allows, together with the constraints in Eq. (3.68), only parameter values $\gamma_i = 0$ or $\gamma_i = 1/2$. This results, altogether, in **four** different possible inversion centers $\underline{r}_o$ inside the unit cell at

$$\underline{r}_o^{(1)} = \underline{0}, \quad \underline{r}_o^{(2)} = 1/2\,\underline{R}_1, \quad \underline{r}_o^{(3)} = 1/2\,\underline{R}_2, \quad \underline{r}_o^{(4)} = 1/2\,(\underline{R}_1 + \underline{R}_2) \tag{3.69}$$

Thus, the primitive morphological unit cell of a netplane with inversion symmetry can contain inversion centers only at its **origin**, one at the cell **center**, and two at the **midpoints** of the cell edges, see Figure 3.7.

This means, in particular, that, if a netplane, given by $\underline{R}_1$ and $\underline{R}_2$, possesses inversion symmetry, then it can always be arranged such that there are inversion centers at positions

$$\underline{R} = n_1 \underline{R}_1 + n_2 \underline{R}_2 + \underline{r}_o^{(i)}, \quad i = 1-4, \quad n_1, n_2 \text{ integer} \tag{3.70}$$

which, together with Eq. (3.69), can also be written as

$$\underline{R} = n_1 (\underline{R}_1/2) + n_2 (\underline{R}_2/2), \quad n_1, n_2 \text{ integer} \tag{3.71}$$

Equation (3.71) describes general lattice points on a netplane with lattice vectors $\underline{R}_1/2$, $\underline{R}_2/2$.

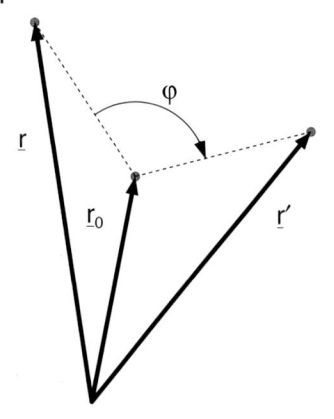

Figure 3.8 Sketch of a rotation operation applied to vector $\underline{r}$ to yield $\underline{r}'$ with the rotation center at $\underline{r}_o$ and rotation angle φ labeled accordingly.

In conclusion, inversion centers existing as symmetry elements of a netplane, defined by primitive lattice vectors $\underline{R}_1$ and $\underline{R}_2$, are subject to the following constraint:

- All inversion centers form a netplane with lattice vectors $\underline{R}_1/2$, $\underline{R}_2/2$ (3.72)

3.6.3 Rotation

Rotation operations $C_\varphi(\underline{r}_o)$ rotate any point $\underline{r}$ on the netplane by an angle φ (**rotation angle**) about a center $\underline{r}_o$ (**rotation center**) to yield an image point $\underline{r}'$, see Figure 3.8. This can be expressed **mathematically** by a transformation of points on the netplane

$$\underline{r} = \underline{r}_o + (\underline{r} - \underline{r}_o) \rightarrow \underline{r}' = \underline{r}_o + C_\varphi(\underline{r} - \underline{r}_o) \quad (3.73)$$

where C_φ is formally defined as the rotation operator. This is connected with a two-dimensional Cartesian **coordinate transformation** with respect to the rotation center $\underline{r}_o$, applying a 2×2 matrix $\underline{\underline{C}}_\varphi$, where

$$\begin{pmatrix} x' \\ y' \end{pmatrix} = \underline{\underline{C}}_\varphi \cdot \begin{pmatrix} x \\ y \end{pmatrix}, \quad \underline{\underline{C}}_\varphi = \begin{pmatrix} \cos\varphi & -\sin\varphi \\ \sin\varphi & \cos\varphi \end{pmatrix} \quad (3.74)$$

Possible rotation **angles** φ that transform netplanes into themselves, are subject to **constraints**. In particular, if a rotation by angle φ reproduces a netplane, then a rotation by 2φ must also reproduce the netplane. A repeated rotation by φ will eventually lead to a full circular movement of the netplane after a finite number of steps. Thus, possible rotation angles φ can only be fractions of $360°$, that is,

$$\varphi = p\,(360°/n), \quad p = 1, \ldots, n, \quad n \text{ integer} \tag{3.75}$$

If a netplane reproduces itself, when rotated about a center $\underline{r}_o$ by all angles φ with $p = 1, \ldots, n$ in Eq. (3.75), it is said to possess an **n-fold rotation axis** **n (C_n)** at $\underline{r}_o$. Here, the symmetry symbols **n**, **C_n** refer to the Hermann–Mauguin and Schönflies notation for rotation axes, discussed in Section 2.4. Both notations will be used in the following with the Schönflies notation put in parentheses.

The angle $\varphi = 180°$, corresponding to a **twofold rotation axis**, is special. According to Eq. (3.74), the transformation matrix $\underline{\underline{C}}_\varphi$ for $\varphi = 180°$ is identical to the **inversion** matrix $\underline{\underline{i}} = -\underline{\underline{1}}$ in Eq. (3.65). Thus, in two dimensions a twofold rotation is equivalent to an inversion operation. As a consequence, the four possible inversion centers $\underline{r}_o$, given by Eq. (3.69) are the only centers inside the unit cell, where twofold rotation axes can exist.

The **compatibility** of rotational with translational symmetry of a netplane imposes **constraints** on possible **rotation angles** φ. Let us assume a netplane to possess a rotation center at $\underline{r}_o$, where the netplane can always be shifted such that $\underline{r}_o$ coincides with the netplane origin. Then, as a result of translational symmetry, there are infinitely many rotation centers at all lattice points (Eq. (3.66)) of the netplane. Thus, each rotation center A has an equivalent center B separated by lattice vector $\underline{R}$, which can be assumed to be the smallest lattice vector of the netplane, see Figure 3.9. Rotating the netplane anticlockwise about center A by angle φ transforms center B to B′ while the clockwise rotation about center B transforms A to A′.

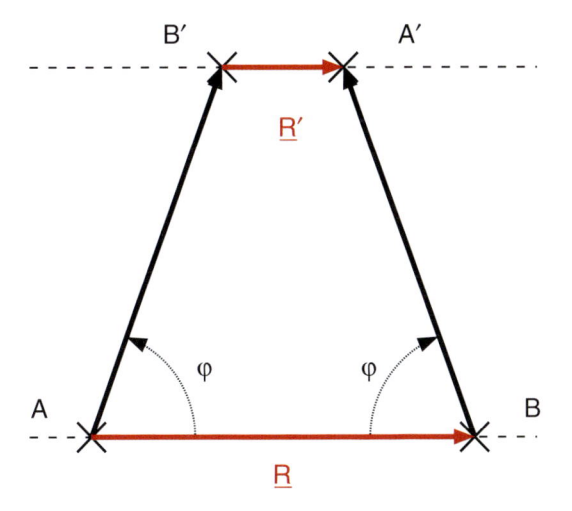

Figure 3.9 Rotations by angles $\pm \varphi$ applied to translationally equivalent rotation centers A and B (indicated by crosses). Lattice vector $\underline{R}$ connects center A with B while $\underline{R}'$ connects the image centers B′ and A′.

Table 3.8 List of possible rotation angles φ and corresponding n-fold rotation axes allowed for netplanes.

$p = \|\underline{R}'\|/\|\underline{R}\|$	φ (°)	Rotation axis
0	60	Sixfold
1	90	Fourfold
2	120	Threefold
3	180	Twofold

Vector $\underline{R}'$, connecting B' with A' must be parallel to vector $\underline{R}$ and simple algebra yields

$$\underline{R}' = \underline{R} - 2\underline{R}\cos\varphi = \underline{R}(1 - 2\cos\varphi) \tag{3.76}$$

Since the two rotations are assumed to transform the netplane into itself, vector $\underline{R}'$ must be an integer multiple of $\underline{R}$, which means that

$$(1 - 2\cos\varphi) = p \quad \text{or} \quad \cos\varphi = (1-p)/2, \quad p \text{ integer} \tag{3.77}$$

where the range of the cosine function, $|\cos\varphi| \leq 1$, limits the integer values p to $-1 \leq p \leq 3$. Here $p = -1$, corresponding to $\varphi = 0°$, can be ignored, which leaves four possible angles φ and the corresponding n-fold rotation axes listed in Table 3.8.

Altogether, the translational symmetry of netplanes allows only **two-, three-, four-, and sixfold** rotation axes. Relation (3.75) shows, in addition, that a fourfold rotation axis implies always a twofold axis at the same rotation center $\underline{r}_o$. Further, a sixfold rotation axis includes two- and threefold axes. Since twofold rotations are equivalent to inversion operations, possible rotation centers $\underline{r}_o$ for two-, four-, or sixfold rotation axes must coincide with inversion centers given by Eq. (3.69). In contrast, centers of true threefold rotation axes (i.e., excluding sixfold rotation) can never coincide with inversion centers.

Consider a netplane with **fourfold rotation symmetry** at $\underline{r}_o$, assumed to coincide with the netplane origin. Then its periodicity can be described by lattice vectors $\underline{R}_1$ and $\underline{R}_2$, where $|\underline{R}_1|$ denotes the smallest distance between lattice points of the netplane. Further, $\underline{R}_2$ can be constructed to be the image of $\underline{R}_1$, rotated by 90°, yielding the same length as vector $\underline{R}_1$. Thus, the corresponding morphological unit cell is of square shape, see Figure 3.10. A rotation of the lattice vectors $\underline{R}_1$ and $\underline{R}_2$ by 90° with respect to the netplane origin yields vectors $\underline{R}'_1$, $\underline{R}'_2$ with

$$\underline{R}'_1 = C_{90}(\underline{R}_1) = \underline{R}_2, \quad \underline{R}'_2 = C_{90}(\underline{R}_2) = -\underline{R}_1 \tag{3.78}$$

Therefore, a 90° rotation about a center $\underline{r}_o$ inside the unit cell, given by

$$\underline{r}_o = \gamma_1 \underline{R}_1 + \gamma_2 \underline{R}_2, \quad 0 \leq \gamma_i < 1, \quad i = 1, 2 \tag{3.79}$$

transforms general lattice vectors $\underline{R}$ of Eq. (3.66) to $\underline{R}'$ according to

$$\begin{aligned}\underline{R}' &= n_1' \underline{R}_1 + n_2' \underline{R}_2 \\ &= \underline{r}_o + C_{90}(\underline{R} - \underline{r}_o) = \underline{r}_o + C_{90}(\underline{R}) - C_{90}(\underline{r}_o) \\ &= \gamma_1 \underline{R}_1 + \gamma_2 \underline{R}_2 + n_1 \underline{R}_2 - n_2 \underline{R}_1 - \gamma_1 \underline{R}_2 + \gamma_2 \underline{R}_1 \\ &= (\gamma_1 + \gamma_2 - n_2) \underline{R}_1 + (\gamma_2 - \gamma_1 + n_1) \underline{R}_2 = \beta_1 \underline{R}_1 + \beta_2 \underline{R}_2 \end{aligned} \qquad (3.80)$$

where

$$\beta_1 = \gamma_1 + \gamma_2 - n_2, \quad \beta_2 = \gamma_2 - \gamma_1 + n_1$$
$$\gamma_1 = 1/2\,(\beta_1 - \beta_2 + n_1 + n_2), \quad \gamma_2 = 1/2\,(\beta_1 + \beta_2 - n_1 + n_2) \qquad (3.81)$$

If $\underline{r}_o$ is the center of a fourfold rotation axis of the netplane, then all transformed centers $\underline{R}'$ in Eq. (3.80) must coincide with the initial centers $\underline{R}$ given by Eq. (3.66). This requires that both β_1 and β_2 are integer-valued and, hence, the corresponding parameters γ_i in Eq. (3.81) must be integer multiples of 1/2 where the multiples are both even or both odd. Together with relation (3.79), only two parameter choices $(\gamma_1 = \gamma_2 = 0)$ and $(\gamma_1 = \gamma_2 = 1/2)$ are possible. This results in **two** different possible centers of **fourfold rotation** axes inside the unit cell at

$$\underline{r}_o^{(1)} = \underline{0}, \quad \underline{r}_o^{(2)} = 1/2\,(\underline{R}_1 + \underline{R}_2) \qquad (3.82)$$

which covers only half of the twofold rotation centers given by Eq. (3.69), as shown in Figure 3.10.

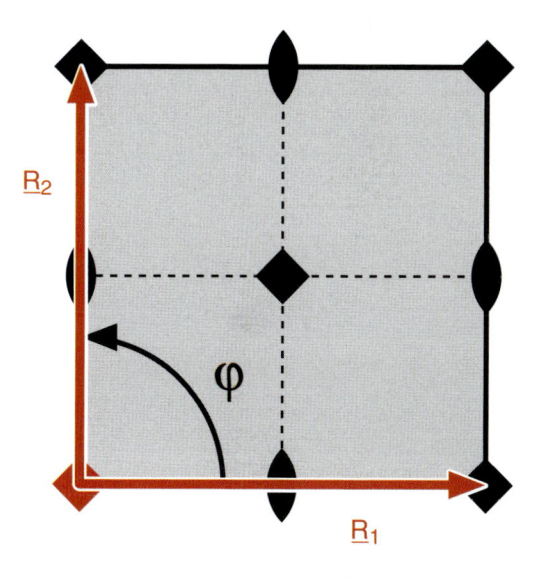

Figure 3.10 Rotation by $\varphi = 90°$ applied to lattice vector $\underline{R}_1$ to yield $\underline{R}_2$. The initial rotation center is indicated by a red square. The unit cell is emphasized in gray with its other two- and fourfold rotation centers shown by black ellipses and squares, respectively.

Next, consider a netplane with **true threefold rotation symmetry** at $\underline{r}_o$ (i.e., three- but not sixfold rotation), assumed to coincide with the netplane origin. Then its periodicity can be described by lattice vectors $\underline{R}_1$ and $\underline{R}_2$, where $|\underline{R}_1|$, as before, denotes the smallest distance between lattice points of the netplane. Further, $\underline{R}_2$ can be constructed to be the image of $\underline{R}_1$, rotated by 120°, yielding the same length as vector $\underline{R}_1$. Thus, the corresponding morphological unit cell takes the shape of a highly symmetric **rhombus**, see Figure 3.11. A rotation of the lattice vectors $\underline{R}_1$ and $\underline{R}_2$ by 120° with respect to the netplane origin yields vectors $\underline{R}'_1$, $\underline{R}'_2$ with

$$\underline{R}'_1 = C_{120}(\underline{R}_1) = \underline{R}_2, \quad \underline{R}'_2 = C_{120}(\underline{R}_2) = -(\underline{R}_1 + \underline{R}_2) \qquad (3.83)$$

Therefore, a 120° rotation about a center $\underline{r}_o$ inside the unit cell, given by

$$\underline{r}_o = \gamma_1 \underline{R}_1 + \gamma_2 \underline{R}_2, \quad 0 \leq \gamma_i < 1, \quad i = 1, 2 \qquad (3.84)$$

transforms general lattice vectors $\underline{R}$ of Eq. (3.66) to $\underline{R}'$ according to

$$\underline{R}' = \underline{r}_o + C_{120}(\underline{R} - \underline{r}_o) = \underline{r}_o + C_{120}(R) - C_{120}(\underline{r}_o)$$
$$= \gamma_1 \underline{R}_1 + \gamma_2 \underline{R}_2 + n_1 \underline{R}_2 - n_2(\underline{R}_1 + \underline{R}_2) - \gamma_1 \underline{R}_2 + \gamma_2(\underline{R}_1 + \underline{R}_2)$$
$$= (\gamma_1 + \gamma_2 - n_2)\underline{R}_1 + (2\gamma_2 - \gamma_1 + n_1 - n_2)\underline{R}_2 = \beta_1 \underline{R}_1 + \beta_2 \underline{R}_2 \qquad (3.85)$$

where

$$\beta_1 = \gamma_1 + \gamma_2 - n_2, \quad \beta_2 = 2\gamma_2 - \gamma_1 + n_1 - n_2$$
$$\gamma_1 = 1/3(2\beta_1 - \beta_2 + n_1 + n_2), \quad \gamma_2 = 1/3(\beta_1 + \beta_2 - n_1 + 2n_2) \qquad (3.86)$$

If $\underline{r}_o$ is the center of a threefold rotation axis of the netplane, then all transformed centers $\underline{R}'$ in Eq. (3.85) must reflect the initial centers $\underline{R}$ given by Eq. (3.66). This requires that both β_1 and β_2 are integer-valued and, therefore, the corresponding parameters γ_i must be integer multiples of 1/3. Together with relation (3.84), only

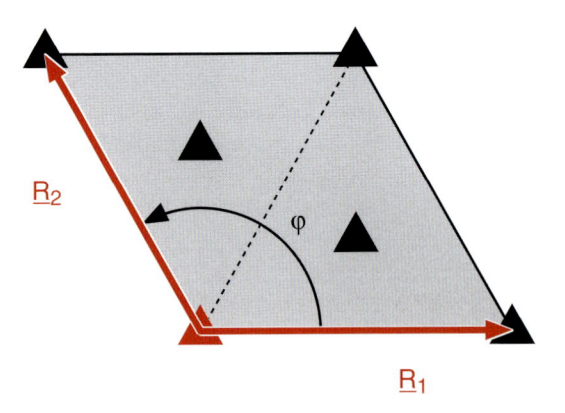

Figure 3.11 Rotation by $\varphi = 120°$ applied to lattice vector $\underline{R}_1$ to yield $\underline{R}_2$. The initial rotation center is indicated by a red triangle. The unit cell is emphasized in gray with its other threefold rotation centers shown by black triangles.

three parameter choices ($\gamma_1 = \gamma_2 = 0$), ($\gamma_1 = 2/3$, $\gamma_2 = 1/3$), and ($\gamma_1 = 1/3$, $\gamma_2 = 2/3$) are possible. This results in **three** different possible centers of true **threefold rotation** axes at

$$\underline{r}_o^{(1)} = \underline{0}, \quad \underline{r}_o^{(2)} = 1/3\,(2\underline{R}_1 + \underline{R}_2), \quad \underline{r}_o^{(3)} = 1/3\,(\underline{R}_1 + 2\underline{R}_2) \tag{3.87}$$

If a netplane possesses **sixfold rotation symmetry** at $\underline{r}_o$ this center must also serve for a two- and threefold rotation axis. Here, the former offers four distinct centers $\underline{r}_o$ inside the unit cell according to Eq. (3.69), while the latter allows three different centers, given by Eq. (3.87). The two sets of centers overlap only at $\underline{r}_o = \underline{0}$, the origin of the unit cell. Therefore, a netplane allows **sixfold rotation** symmetry only at centers coinciding with lattice points.

In conclusion, rotation axes existing as symmetry elements of a netplane, defined by primitive lattice vectors $\underline{R}_1$ and $\underline{R}_2$, are subject to the following constraints:

- Netplanes allow only two-, three-, four-, and sixfold rotation axes. (3.88a)
- Centers of twofold rotation axes form netplanes with lattice vectors $1/2\,\underline{R}_1$, $1/2\,\underline{R}_2$. (3.88b)
- Centers of twofold rotations are identical with inversion centers. (3.88c)
- Centers of true threefold rotations are restricted to hexagonal netplanes. They form netplanes with lattice vectors

$$1/3\,(2\underline{R}_1 + \underline{R}_2), \quad 1/3\,(\underline{R}_1 + 2\underline{R}_2) \quad \text{(obtuse representation)}$$
$$1/3\,(\underline{R}_1 + \underline{R}_2), \quad 1/3\,(-\underline{R}_1 + 2\underline{R}_2) \quad \text{(acute representation)}. \tag{3.88d}$$

- Centers of fourfold rotation axes are restricted to square netplanes. They form netplanes with lattice vectors $1/2\,(\underline{R}_1 + \underline{R}_2)$, $1/2\,(\underline{R}_1 - \underline{R}_2)$. Thus, fourfold rotation centers coincide with half of the twofold rotation centers, given in Eq. (3.88b). (3.88e)
- Centers of sixfold rotation axes combine two- with threefold rotation axes. They form netplanes with lattice vectors $\underline{R}_1$, $\underline{R}_2$, reflecting the initial netplane. (3.88f)

3.6.4
Mirror Operation

Mirror operations $\sigma\,(\underline{r}_o, \underline{e})$ with respect to a mirror line along vector $\underline{e}$ on the netplane create, for any point $\underline{r}$ on one side of the line, an image point $\underline{r}'$ on the other side, such that the connecting line between the two points is perpendicular to the mirror line and their distances from the mirror line are the same, see Figure 3.12. This can be expressed **mathematically** by a transformation of points on the netplane

$$\underline{r} \rightarrow \underline{r}' = \underline{r} - 2\,[(\underline{r} - \underline{r}_o)\,\underline{m}]\,\underline{m} = \underline{r}_o + \sigma_m(\underline{r} - \underline{r}_o) \tag{3.89}$$

where the **mirror line** is defined by its origin $\underline{r}_o$ (**mirror center**), a direction vector $\underline{e}$ along the line (**mirror line vector**), and a normal vector $\underline{m}$ (**mirror line normal**

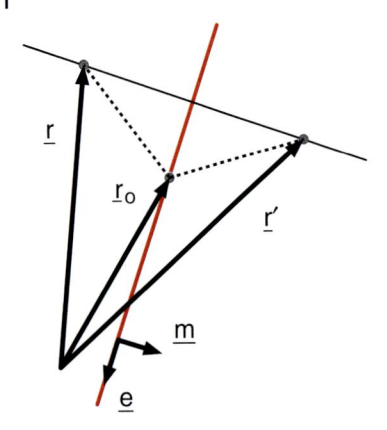

Figure 3.12 Sketch of a mirror operation applied to vector $\underline{r}$ to yield $\underline{r}'$. The mirror center $\underline{r}_o$, mirror line vector $\underline{e}$, and mirror line normal vector $\underline{m}$ are labeled accordingly.

vector) of unit length perpendicular to vector $\underline{e}$, sketched in Figure 3.12. The mirror operation can also be connected with a two-dimensional Cartesian **coordinate transformation** with respect to the mirror center $\underline{r}_o$ applying a 2×2 matrix $\underline{\sigma}_m$ where

$$\begin{pmatrix} x' \\ y' \end{pmatrix} = \underline{\sigma}_m \cdot \begin{pmatrix} x \\ y \end{pmatrix}, \quad \underline{\sigma}_m = \begin{pmatrix} 1 - 2m_x^2 & -2m_x m_y \\ -2m_x m_y & 1 - 2m_y^2 \end{pmatrix} \quad (3.90)$$

with

$$\underline{m} = (m_x, m_y) = (-e_y, e_x) \quad \text{and} \quad \underline{e} = (e_x, e_y) \text{ with } e_x^2 + e_y^2 = 1 \quad (3.91)$$

The mirror line vector $\underline{e}$ may also be written in Cartesian coordinates as

$$\underline{e} = (\cos \Phi, \; \sin \Phi) \quad (3.92)$$

where Φ denotes the angle of the mirror line with respect to an x axis. Hence mirror line normal vector $\underline{m}$ is given by

$$\underline{m} = (-\sin \Phi, \; \cos \Phi) \quad (3.93)$$

which, according to Eq. (3.90), leads to

$$\underline{\sigma}_m = \underline{\sigma}_\Phi = \begin{pmatrix} 1 - 2(\sin \Phi)^2 & 2 \sin \Phi \cos \Phi \\ 2 \sin \Phi \cos \Phi & 1 - 2(\cos \Phi)^2 \end{pmatrix} = \begin{pmatrix} \cos 2\Phi & \sin 2\Phi \\ \sin 2\Phi & -\cos 2\Phi \end{pmatrix} \quad (3.94)$$

Here the **mirror center** $\underline{r}_o$ can be chosen **arbitrarily** along the mirror line, since due to the orthogonality of $\underline{e}$ and $\underline{m}$ a shift

$$\underline{r}_o \to \underline{r}'_o = \underline{r}_o + \chi \underline{e} \quad (3.95)$$

results, according to Eq. (3.89) with $(\underline{e}\,\underline{m}) = 0$, in a transformation

$$\underline{r}'' = \underline{r} - 2[(\underline{r} - \underline{r}'_o)\,\underline{m}]\,\underline{m} = \underline{r} - 2[(\underline{r} - \underline{r}_o)\,\underline{m}]\,\underline{m} + 2\chi\,(\underline{e}\,\underline{m})\,\underline{m} = \underline{r}' \quad (3.96)$$

Parallel mirror lines given by mirror centers $\underline{r}_{1o}, \underline{r}_{2o}$ with identical normal vectors $\underline{m}$ are separated by a distance

$$d_m = |(\underline{r}_{2o} - \underline{r}_{1o})\,\underline{m}| \quad (3.97)$$

Here a combined mirror operation with respect to the two lines can be written as

$$\underline{r} \to \underline{r}' : \quad \underline{r}' = \underline{r} - 2[(\underline{r} - \underline{r}_{1o})\,\underline{m}]\,\underline{m} \quad \text{(mirror line 1)} \quad (3.98a)$$

$$\underline{r}' \to \underline{r}'' : \quad \underline{r}'' = \underline{r}' - 2[(\underline{r}' - \underline{r}_{2o})\,\underline{m}]\,\underline{m} \quad \text{(mirror line 2)} \quad (3.98b)$$

yielding, after some calculus

$$\underline{r}'' = \underline{r} + 2[(\underline{r}_{2o} - \underline{r}_{1o})\,\underline{m}]\,\underline{m} \quad (3.99)$$

This corresponds to a simple **shift operation** by a shift vector perpendicular to the mirror lines with, according to Eq. (3.97), a length equal to twice the distance between the two mirror lines.

Centers $\underline{r}_o$ of two **mirror lines** that **cross** each other can always be chosen so as to coincide with the crossing point. Thus, the mirror lines differ only by their transformation matrices $\underline{\underline{\sigma}}_m$, where, assuming a crossing angle Θ between the mirror lines leads, according to Eq. (3.94), to matrices $\underline{\underline{\sigma}}_m(\Phi)$ and $\underline{\underline{\sigma}}_m(\Phi + \Theta)$ with

$$\underline{\underline{\sigma}}_m(\Phi) = \begin{pmatrix} \cos 2\Phi & \sin 2\Phi \\ \sin 2\Phi & -\cos 2\Phi \end{pmatrix} \quad (3.100)$$

and

$$\underline{\underline{\sigma}}_m(\Phi + \Theta) = \begin{pmatrix} \cos 2(\Phi + \Theta) & \sin 2(\Phi + \Theta) \\ \sin 2(\Phi + \Theta) & -\cos 2(\Phi + \Theta) \end{pmatrix} \quad (3.101)$$

Therefore, a combination of the two mirror operations (in the two different sequences) results in symmetry operations with their centers at $\underline{r}_o$ and transformation matrices $\underline{\underline{T}}, \underline{\underline{T}}'$ given by the different products of $\underline{\underline{\sigma}}_m(\Phi)$ and $\underline{\underline{\sigma}}_m(\Phi + \Theta)$, that is, by

$$\underline{\underline{T}} = \underline{\underline{\sigma}}_m(\Phi + \Theta) \cdot \underline{\underline{\sigma}}_m(\Phi) = \begin{pmatrix} \cos 2(\Phi + \Theta) & \sin 2(\Phi + \Theta) \\ \sin 2(\Phi + \Theta) & -\cos 2(\Phi + \Theta) \end{pmatrix}$$

$$\cdot \begin{pmatrix} \cos 2\Phi & \sin 2\Phi \\ \sin 2\Phi & -\cos 2\Phi \end{pmatrix}$$

$$= \begin{pmatrix} \cos 2\Theta & -\sin 2\Theta \\ \sin 2\Theta & \cos 2\Theta \end{pmatrix} = \underline{\underline{C}}_{2\Theta} \quad (3.102)$$

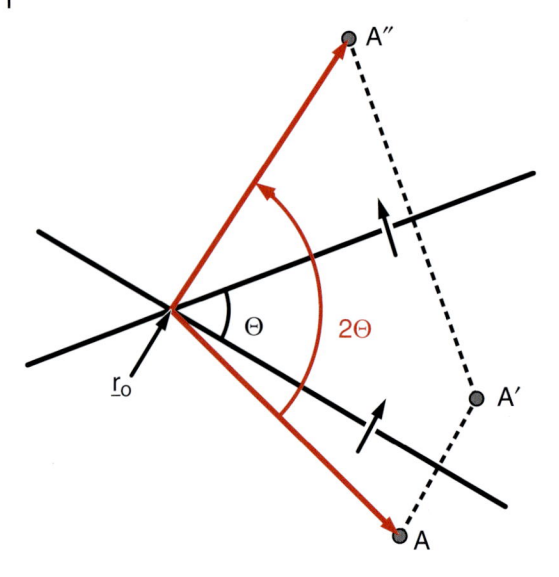

Figure 3.13 Two subsequent mirror line operations with mirror lines crossing at $\underline{r}_o$ and forming an angle Θ. Point A is transformed to A′, then to A″. The mirror lines are indicated by black full lines with small mirror normal arrows. Angles Θ and 2Θ are labeled accordingly.

and

$$\underline{\underline{T}}' = \underline{\underline{\sigma}}_m(\Phi) \cdot \underline{\underline{\sigma}}_m(\Phi + \Theta) = \begin{pmatrix} \cos 2\Phi & \sin 2\Phi \\ \sin 2\Phi & -\cos 2\Phi \end{pmatrix}$$
$$\cdot \begin{pmatrix} \cos 2(\Phi + \Theta) & \sin 2(\Phi + \Theta) \\ \sin 2(\Phi + \Theta) & -\cos 2(\Phi + \Theta) \end{pmatrix}$$
$$= \begin{pmatrix} \cos 2\Theta & \sin 2\Theta \\ -\sin 2\Theta & \cos 2\Theta \end{pmatrix} = \underline{\underline{C}}_{-2\Theta} \qquad (3.103)$$

Comparing with Eq. (3.74), matrices $\underline{\underline{C}}_{2\Theta}$ and $\underline{\underline{C}}_{-2\Theta}$ represent clockwise and anticlockwise rotations by an angle 2Θ, as sketched in Figure 3.13 for the anticlockwise case. Therefore, two mirror symmetry lines of a netplane which cross at $\underline{r}_o$ with a finite angle Θ are always connected with (clockwise and anticlockwise) **rotations** by 2Θ. If the two mirror lines are symmetry elements of a netplane the resulting rotation by $\pm 2\Theta$ must also be a symmetry element. Thus, it must belong to an n-fold rotation axis of the netplane. Since these axes can only be two-, three-, four-, or sixfold, possible angles Θ with $0 < |\Theta| \leq 90$ amount to

$$|\Theta| = 30°, 45°, 60°, 90° \qquad (3.104)$$

In particular, matrix $\underline{\underline{T}}$ of Eq. (3.102) for $\Theta = \pm 90°$ reflects a 180° rotation, corresponding to a twofold rotation axis, which was shown to be also equivalent to an inversion operation. Thus, the **crossing** point of two **orthogonal mirror** lines is always the center of a **twofold rotation** or **inversion**.

Relation (3.102) can be modified to read

$$\underline{C}_\Theta \cdot \underline{\sigma}_m(\Phi) = \underline{\sigma}_m(\Phi + \Theta/2) \cdot \underline{\sigma}_m(\Phi) \cdot \underline{\sigma}_m(\Phi) = \underline{\sigma}_m(\Phi + \Theta/2) \quad (3.105)$$

This shows that a **mirror** operation followed by a **rotation** by an angle Θ with respect to a center $\underline{r}_o$ on the mirror line is equivalent to another **mirror** operation. The latter mirror line crosses the initial line at $\underline{r}_o$ and forms an angle $\Theta/2$ with it. This result becomes important for the symmetry analysis of netplanes discussed in the following.

The **compatibility** between translational and mirror symmetry imposes **constraints** on possible positions and directions of mirror lines in a netplane. Let us assume that a netplane with lattice vectors $\underline{R}_1$ and $\underline{R}_2$ possesses mirror symmetry with a mirror line through $\underline{r}_o$. Then the netplane can always be shifted such that its origin coincides with $\underline{r}_o$ (setting $\underline{r}_o = \underline{0}$). According to Eq. (3.89), a general lattice vector $\underline{R}$, given by Eq. (3.66) and not located at a mirror line, will have a mirror image $\underline{R}'$. Then the difference vector $\underline{R}_{r2}$, given by

$$\underline{R}_{r2} = \underline{R} - \underline{R}' = 2\,(\underline{R}\,\underline{m})\,\underline{m} = k_1\,\underline{R}_1 + k_2\,\underline{R}_2, \quad k_1,\ k_2 \text{ integer} \quad (3.106)$$

must also be a general lattice vector. This shows, in particular, that normal vectors $\underline{m}$ of mirror lines must always point along general lattice vectors of the netplane. Further, the vector sum $\underline{R}_{r1}$ is given by

$$\underline{R}_{r1} = \underline{R} + \underline{R}' = 2\,[\underline{R} - (\underline{R}\,\underline{m})\,\underline{m}] = (2n_1 - k_1)\,\underline{R}_1 + (2n_2 - k_2)\,\underline{R}_2 \quad (3.107)$$

with

$$(\underline{R}_{r1}\,\underline{R}_{r2}) = 4\,[\underline{R} - (\underline{R}\,\underline{m})\,\underline{m}]\,[(\underline{R}\,\underline{m})\,\underline{m}] = 0 \quad (3.108)$$

Thus, the general lattice vector $\underline{R}_{r1}$ is perpendicular to $\underline{R}_{r2}$ and, hence, points along the mirror line.

Altogether, the existence of a **mirror plane** implies two **orthogonal** general lattice vectors $\underline{R}_{r1}$ and $\underline{R}_{r2}$ in the netplane. These may be integer multiples of smaller lattice vectors along their directions, where we consider, in each case, the smallest possible vector, calling it $\underline{R}_1$ and $\underline{R}_2$. Vectors $\underline{R}_1$ and $\underline{R}_2$ can then be used as lattice vectors to describe the netplane periodicity and, since the vectors are perpendicular to each other the **netplane** must be **rectangular.** Therefore, in the following we will use mutually orthogonal lattice vectors $\underline{R}_1$ and $\underline{R}_2$ to describe netplanes with mirror symmetry, where primitive and centered rectangular netplanes, see Section 3.6.1, can be treated on the same footing.

Consider a **primitive rectangular** netplane defined by orthogonal lattice vectors $\underline{R}_1$ and $\underline{R}_2$ and a mirror operation with its **mirror line** through the netplane origin and **parallel to** $\underline{R}_1$, that is,

$$\underline{m} = \underline{R}_2/|\underline{R}_2|, \quad \underline{r}_o = \underline{0} \quad (3.109)$$

Then, as a result of translational symmetry, there are **infinitely** many parallel **mirror lines** through mirror centers at $\underline{r}_{on}$ with

$$\underline{r}_{on} = n\,\underline{R}_2, \quad n \text{ integer} \quad (3.110)$$

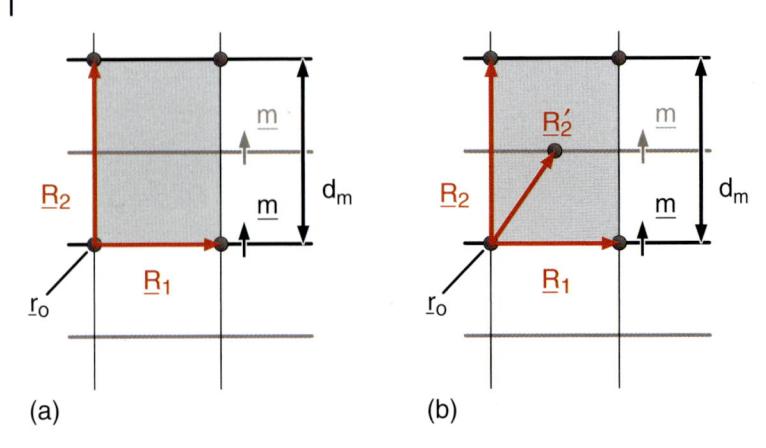

Figure 3.14 Periodic sets of horizontal mirror lines for (a) primitive and (b) centered rectangular netplanes. The mirror lines are shown by black and gray horizontal lines parallel to lattice vector R_1 (some with small mirror normal arrows). The unit cell is emphasized in gray.

indicated by black horizontal lines in Figure 3.14a, where adjacent lines are separated by a distance d_m with

$$d_m = |\underline{R}_2 \underline{m}| = R_2 \tag{3.111}$$

The translational symmetry of the netplane yields **additional** parallel mirror lines beyond those given by Eq. (3.110). The mirror operation Eq. (3.89), together with Eqs. (3.109) and (3.110), can be written as

$$\underline{r}' = \underline{r} - 2\,[(\underline{r} - \underline{r}_{on})\,\underline{m}]\,\underline{m} = \underline{r} - 2\,[(\underline{r} - (\underline{r}_{on} + \underline{R}_2/2))\,\underline{m}]\,\underline{m} - \underline{R}_2 \tag{3.112}$$

which can be interpreted as a parallel mirror operation with respect to a mirror line at origin

$$\underline{r}'_{on} = (n + 1/2)\,\underline{R}_2, \quad \text{n integer} \tag{3.113}$$

followed by a (backwards) shift by lattice vector $\underline{R}_2$. Thus, if a netplane reproduces itself for mirror operations with mirror lines at origins $\underline{r}_{on}$, according to Eq. (3.110), its translational symmetry also yields mirror symmetry with respect to additional parallel lines. The latter originate at $\underline{r}'_{on}$, given by Eq. (3.113), and, thus, are located in the **middle** between the adjacent mirror lines of the initial set Eq. (3.110), as indicated by gray horizontal lines in Figure 3.14a. The two sets, defined by origins of Eqs. (3.110) and (3.113), can be combined to one set of parallel mirror lines with $\underline{m} = \underline{R}_2/|\underline{R}_2|$ and originating at mirror centers

$$\underline{r}_{on} = n\,\underline{R}_2/2, \quad \text{n integer} \tag{3.114}$$

where even values n refer to Eq. (3.110) and odd ones to Eq. (3.113). Mirror symmetry of the netplane for additional lines parallel to those given by Eq. (3.114) cannot occur since this would violate the translational symmetry of the netplane.

The results for sets of mirror lines **parallel to** $\underline{R}_2$ are completely analogous to those when mirror lines parallel to $\underline{R}_1$ are considered. Interchanging $\underline{R}_1$ with $\underline{R}_2$ in

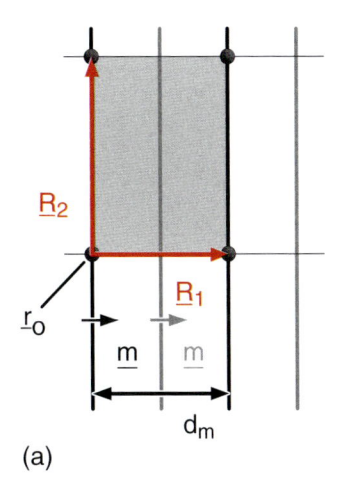

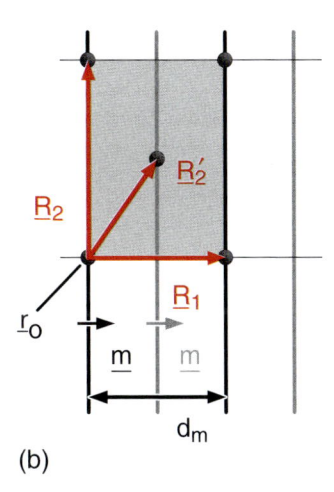

Figure 3.15 Periodic sets of vertical mirror lines for (a) primitive and (b) centered rectangular netplanes. The mirror lines are shown by black and gray vertical lines parallel to lattice vector R_2 (some with small mirror normal arrows). The unit cell is emphasized in gray.

the earlier discussion shows that, as a result of translational symmetry, there are **infinitely** many parallel **mirror lines** through mirror centers at $\underline{r}_{on}$ with

$$\underline{m} = \underline{R}_1/|\underline{R}_1|, \quad \underline{r}_{on} = n\,\underline{R}_1/2, \quad n \text{ integer} \tag{3.115}$$

as sketched in Figure 3.15a. The sets of mirror lines parallel to lattice vectors $\underline{R}_1$ and $\underline{R}_2$ of a primitive rectangular netplane, given by origins Eqs. (3.114) and (3.115), must also exist when the orthogonal lattice vectors $\underline{R}_1$ and $\underline{R}_2$ refer to a non-primitive representation of a **centered rectangular netplane**, as illustrated in Figures 3.14b and 3.15b. The additional netplane point in the center of the morphological unit cell of the centered rectangular netplane is always positioned on a mirror line, shown in gray in the two figures. Thus, it cannot give rise to additional mirror lines and the set of parallel mirror lines at origins Eqs. (3.114) and (3.115) is also **complete** for centered rectangular lattices.

Netplanes with **mirror symmetry** may also include **inversion** centers as symmetry elements. Let us assume a **primitive rectangular** netplane with orthogonal lattice vectors $\underline{R}_1$ and $\underline{R}_2$ to possess mirror lines **parallel to** $\underline{R}_1$ through centers $\underline{r}_{on}$, given by Eq. (3.114), as symmetry elements. If the netplane also possesses inversion symmetry, then there is, according to Section 3.6.2, an infinite number of inversion centers that form a netplane on their own (inversion netplane), with lattice vectors $1/2\,\underline{R}_1$ and $1/2\,\underline{R}_2$, see Eq. (3.72). The **origin** of the inversion netplane may be shifted with respect to that of the initial netplane defining the mirror lines, but the two have to be **compatible.** Thus, mirror operations with respect to any mirror line through centers $\underline{r}_{on}$, given by Eq. (3.114) must reproduce all inversion centers, and inversions with respect to any inversion center must reproduce all mirror lines. This is possible only if, either, the origins of the initial and its inversion netplane coincide, or the origins are shifted with respect to each other by

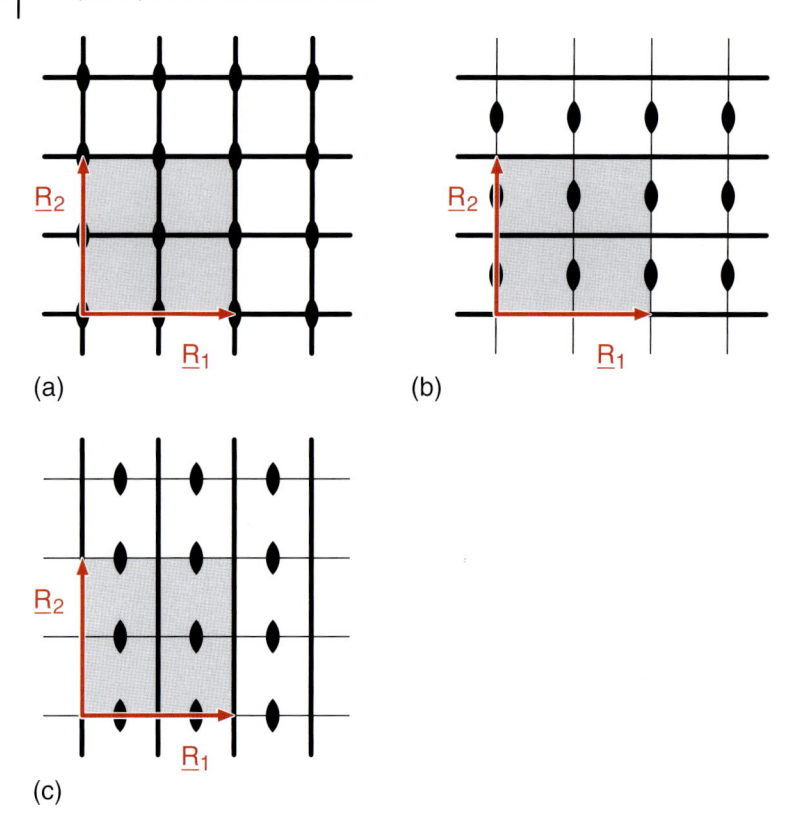

Figure 3.16 Primitive rectangular netplanes with coexisting inversion centers (twofold rotation axes) and mirror lines. (a) Mirror lines parallel to $\underline{R}_1$ and $\underline{R}_2$ with inversion centers on mirror lines. (b, c) Mirror lines with inversion centers between mirror lines parallel to $\underline{R}_1$ and parallel to $\underline{R}_2$, respectively. Corresponding unit cells are emphasized in gray with lattice vectors $\underline{R}_1$, $\underline{R}_2$ indicated accordingly. Mirror lines are shown by thick lines and twofold rotation centers by black ellipses.

$\underline{R}_2/4$. In the former case all inversion centers lie on mirror lines, while, in the latter, inversion centers lie only in the middle between mirror lines, as illustrated in Figure 3.16.

According to Eq. (3.105) an inversion center $\underline{r}_o$ on a mirror line parallel to $\underline{R}_1$ and through $\underline{r}_o$ implies another mirror line through $\underline{r}_o$ but parallel to $\underline{R}_2$, that is, perpendicular to $\underline{R}_1$. This shows that, when all inversion centers lie on mirror lines parallel to $\underline{R}_1$, the netplane also includes the set of mirror lines parallel to $\underline{R}_2$ (Eq. 3.115). This is illustrated in Figure 3.16a, where mirror lines of both sets are sketched.

The above results apply analogously to **primitive rectangular** netplanes with orthogonal lattice vectors $\underline{R}_1$ and $\underline{R}_2$ including mirror lines **parallel to $\underline{R}_2$** (through centers $\underline{r}_{on}$ given by Eq. (3.115)) as symmetry elements. Interchanging

$\underline{R}_1$ with $\underline{R}_2$ in the previous discussion shows that inversion symmetry of the netplane is only possible if the origins of the initial and its inversion netplane coincide or, if the origins are shifted with respect to each other by $\underline{R}_1/4$. This means that, either, all inversion centers lie on mirror lines, which implies two orthogonal sets of mirror lines, as discussedearlier, see Figure 3.16a, or that inversion centers will lie only in the middle between mirror lines, as illustrated in Figure 3.16c.

The same arguments concerning compatibility of **mirror** and **inversion** symmetry for primitive rectangular netplanes can also be used for **centered rectangular** netplanes. Starting from a non-primitive description of a centered rectangular netplane by orthogonal lattice vectors $\underline{R}_1$ and $\underline{R}_2$, a primitive description can be obtained by lattice vectors $\underline{R}_1$ and $\underline{R}'_2$ with

$$\underline{R}'_2 = 1/2\,(\underline{R}_1 + \underline{R}_2) \tag{3.116}$$

Thus, a centered rectangular netplane with inversion symmetry possesses, according to Section 3.6.2, an infinite number of inversion centers forming an inversion netplane with lattice vectors $(\underline{R}_1/2)$, $(\underline{R}'_2/2) = 1/4\,(\underline{R}_1 + \underline{R}_2)$. On the other hand, it was shown earlier that the set of mirror lines parallel to $\underline{R}_1$ is identical for primitive and centered rectangular netplanes. Therefore, **compatibility** of mirror and inversion symmetry requires that the origins of the initial and its inversion netplane coincide. This is the only choice and results in **half** of the inversion centers lying **at mirror lines** and the other half in the **middle** between mirror lines, as illustrated in Figure 3.17. Further, the inversion centers on the mirror lines lead

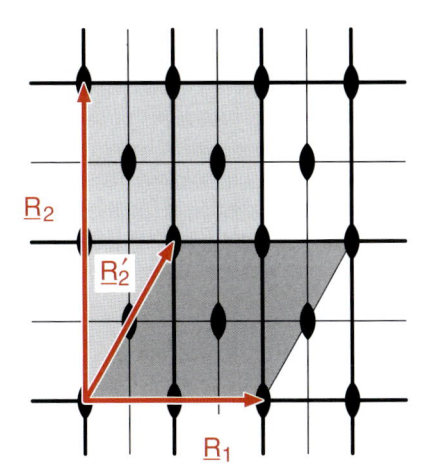

Figure 3.17 Centered rectangular netplane with coexisting inversion centers (twofold rotation axes) and mirror lines parallel to $\underline{R}_1$, $\underline{R}_2$. Corresponding non-primitive and primitive unit cells are emphasized in light and dark gray, respectively, with lattice vectors $\underline{R}_1$, $\underline{R}_2$, $\underline{R}'_2$ indicated accordingly. Mirror lines are shown by thick lines and twofold rotation centers by black ellipses.

to two orthogonal sets of mirror lines, as discussed for primitive rectangular netplanes and sketched in Figure 3.17.

Since in two dimensions inversion centers and twofold rotation axes are equivalent, the preceding discussion also applies to coexisting **mirror** and **twofold rotation symmetry** in netplanes.

Netplanes with **mirror symmetry** may also possess **fourfold rotation** axes as symmetry elements. According to the discussion in Section 3.6.3, fourfold rotation symmetry in a netplane results in symmetry-adapted lattice vectors $\underline{R}_1$ and $\underline{R}_2$ which are mutually orthogonal and both of the same length. This yields a primitive morphological unit cell of square shape.

Let us assume a **square** netplane with orthogonal lattice vectors $\underline{R}_1$ and $\underline{R}_2$ of equal length to possess **mirror lines** parallel to $\underline{R}_1$ through centers $\underline{r}_{on}$, given by Eq. (3.114), as symmetry elements. If the netplane is also symmetric with respect to **fourfold rotation**, the corresponding rotation centers must coincide with positions of twofold rotation centers on the netplane, since fourfold rotation implies twofold rotation. As shown earlier, the complete set of twofold rotation centers (= inversion centers) forms an inversion netplane with lattice vectors $1/2\,\underline{R}_1$ and $1/2\,\underline{R}_2$. Adding fourfold rotation centers according to Eq. (3.88e) covers half of the inversion centers in a checkerboard type arrangement, see Figure 3.18a. The origin of this modified inversion netplane can only coincide with that of the initial square netplane or its origin is shifted, by $\underline{R}_1/4$, $\underline{R}_2/4$, or $(\underline{R}_1+\underline{R}_2)/4$, see preceding discussion. Here, the presence of fourfold rotation centers is found to yield the same structures for all cases.

Thus, only the coincidence geometry needs to be considered, where it can be assumed that the netplane origin $\underline{r}_o$ also coincides with a fourfold rotation center, as indicated in Figure 3.18a. As a consequence, this fourfold rotation center $\underline{r}_o$

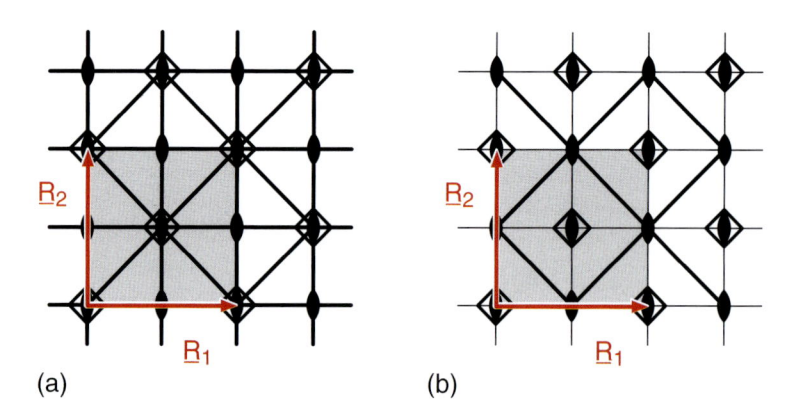

Figure 3.18 Square netplane with coexisting two-, fourfold rotation axes and mirror lines. (a) Mirror lines parallel to $\underline{R}_1$, $\underline{R}_2$, and to diagonals. (b) Mirror lines parallel to diagonals between true twofold rotation axes only. Corresponding unit cells are emphasized in gray. Mirror lines are shown by thick lines and two-, fourfold rotation centers by black ellipses and open squares, respectively.

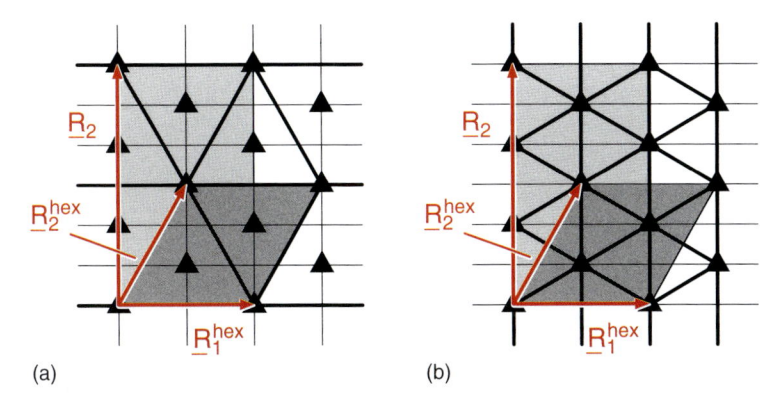

Figure 3.19 Hexagonal netplane with coexisting true threefold rotation axes and mirror lines. (a) Mirror lines parallel to $\underline{R}_1^{hex}$ and corresponding rotated mirror lines. (b) Mirror lines parallel to $\underline{R}_2$ and corresponding rotated mirror lines. Non-primitive and primitive unit cells are emphasized in light and dark gray, respectively, with lattice vectors $\underline{R}_1^{hex}$, $\underline{R}_2^{hex}$, $\underline{R}_2$ indicated accordingly. Mirror lines are shown by thick lines and threefold rotation centers by black triangles.

implies other mirror lines crossing $\underline{r}_o$, where, according to Eq. (3.105), mirror lines parallel to $\underline{R}_1$ and $\underline{R}_2$, as well as parallel to the two diagonals $(\underline{R}_1 \pm \underline{R}_2)$ cover all cases, illustrated by Figure 3.18a. This combination of mirror and fourfold rotation symmetry does not include mirror lines parallel to the two diagonals $(\underline{R}_1 \pm \underline{R}_2)$ and connecting true twofold rotation centers. The existence of such mirror lines can be shown to exclude all previous mirror lines shown in Figure 3.18a, yielding an alternative geometry for coexisting mirror and fourfold rotation symmetry, as shown in Figure 3.18b.

The coexistence of **mirror lines** and **true threefold** (i.e., three- but not sixfold) **rotation** axes as symmetry elements of a netplane is slightly more involved. First, we note that, as discussed in Section 3.6.3, threefold rotations are restricted to **hexagonal** netplanes, described by two primitive lattice vectors $\underline{R}_1^{hex}$, $\underline{R}_2^{hex}$ of equal length, forming an angle of 120° or 60°. Here we use the acute representation (i.e., 60°) which yields

$$|\underline{R}_1^{hex}| = |\underline{R}_2^{hex}| = a, \quad (\underline{R}_1^{hex} \underline{R}_2^{hex}) = a^2/2 \tag{3.117}$$

where a is the lattice constant of the netplane. This allows the definition of a vector $\underline{R}_2$

$$\underline{R}_2 = 2\underline{R}_2^{hex} - \underline{R}_1^{hex} \tag{3.118}$$

which is orthogonal to $\underline{R}_1^{hex}$. Therefore, vectors $\underline{R}_1^{hex}$, $\underline{R}_2$, together with $\underline{R}_2^{hex}$ yield a non-primitive lattice description of the hexagonal netplane by a **centered rectangular** netplane, which will be used in the following.

Consider a **hexagonal** netplane, described as centered rectangular by orthogonal non-primitive lattice vectors $\underline{R}_1^{hex}$, $\underline{R}_2$ ($|\underline{R}_2| = \sqrt{3}\,|\underline{R}_1^{hex}|$) and a centering lattice vector $\underline{R}_2^{hex}$. This netplane is further assumed to possess **mirror lines** parallel to $\underline{R}_1^{hex}$ through centers $\underline{r}_{on}$, given by Eq. (3.114), as symmetry elements. If the netplane also includes symmetry with respect to **true threefold rotation**, then corresponding rotation centers form, according to Eq. (**3.88d**) and after some calculus, a separate rotation netplane with lattice vectors 1/3 ($\underline{R}_1^{hex} + \underline{R}_2^{hex}$) and 1/3 $\underline{R}_2$. The origin $\underline{r}_o$ of this rotation netplane, which can be assumed to be a threefold rotation center, must coincide with the origin of the initial hexagonal netplane, as indicated in Figure 3.19a. The threefold rotation center at $\underline{r}_o$, positioned on a mirror line parallel to $\underline{R}_1^{hex}$, implies, according to Eq. (3.105), two additional mirror lines crossing $\underline{r}_o$, which are rotated by 60° and 120°. This yields the netplane geometry illustrated by Figure 3.19a. The discussion for a **hexagonal** netplane assumed to possess **mirror symmetry** parallel to $\underline{R}_2$ together with **threefold rotation** symmetry is completely analogous and leads to the netplane geometry illustrated by Figure 3.19b.

The coexistence of **mirror lines** and **sixfold rotation** axes as symmetry elements of a netplane is, analogous to the case of threefold rotation symmetry, also restricted to **hexagonal** netplanes. As before, we describe the hexagonal netplane as centered rectangular by orthogonal non-primitive lattice vectors $\underline{R}_1^{hex}$, $\underline{R}_2$ ($|\underline{R}_2| = \sqrt{3}\ |\underline{R}_1^{hex}|$) and a centering lattice vector $\underline{R}_2^{hex}$. The netplane is further assumed to possess **mirror lines** parallel to $\underline{R}_1^{hex}$ through centers $\underline{r}_{on}$, given by Eq. (3.114), as symmetry elements. Allowing, in addition, for **sixfold rotational** symmetry leads to corresponding rotation centers, which form, according to Eq. (3.88f) a rotation netplane with lattice vectors $\underline{R}_1^{hex}$, $\underline{R}_2^{hex}$, that is, the rotation netplane is of the same periodicity than the initial hexagonal netplane. The origin $\underline{r}_o$ of this rotation netplane, which can be assumed to be a sixfold rotation center, must coincide with the origin of the initial hexagonal netplane, as indicated in Figure 3.20. The sixfold rotation center at $\underline{r}_o$, positioned on a mirror line parallel to $\underline{R}_1^{hex}$, implies, according to Eq. (3.105), five additional mirror lines crossing $\underline{r}_o$, which are rotated by 30°, 60°, 90°, 120°, and 150°. This yields the netplane geometry illustrated by Figure 3.20.

In conclusion, mirror lines existing as symmetry elements of a netplane, defined by lattice vectors $\underline{R}_1$ and $\underline{R}_2$, are subject to the following constraints:

- Netplanes with mirror symmetry must be either primitive or centered rectangular (including square and hexagonal netplanes as special cases). (**3.119a**)
- Netplanes with mirror symmetry include infinite sets of parallel mirror lines. If $\underline{R}_1$ and $\underline{R}_2$, are orthogonal lattice vectors describing the primitive or non-primitive morphological unit cell, then mirror line sets can be given by normal vectors $\underline{m} = \underline{R}_1/|\underline{R}_1|$ and centers $\underline{r}_{on} = n\ \underline{R}_1/2$, or by normal vectors $\underline{m} = \underline{R}_2/|\underline{R}_2|$ and centers $\underline{r}_{on} = n\ \underline{R}_2/2$. (**3.119b**)
- In primitive rectangular netplanes with mirror symmetry inversion (twofold rotation) centers lie either all on mirror lines or all in the middle between adjacent parallel mirror lines. (**3.119c**)
- In centered rectangular netplanes with mirror symmetry inversion (twofold rotation) centers induce two orthogonal sets of parallel mirror lines. The

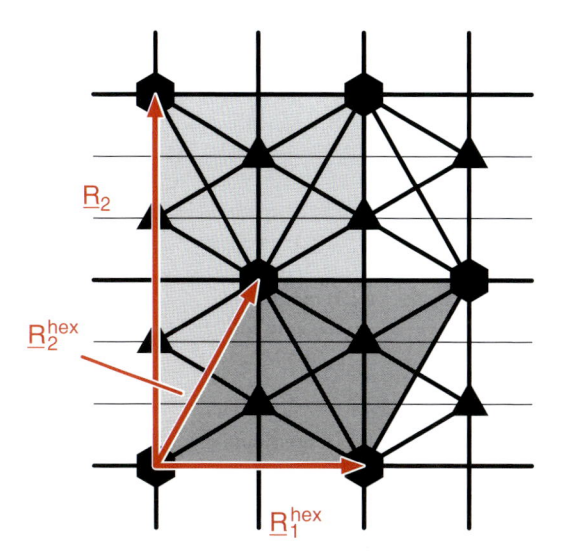

Figure 3.20 Hexagonal netplane with coexisting three-, sixfold rotation axes and mirror lines parallel to $\underline{R}_1^{hex}$, $\underline{R}_2^{hex}$ and corresponding rotated mirror lines. Non-primitive and primitive unit cells are emphasized in light and dark gray, respectively, with lattice vectors $\underline{R}_1^{hex}$, $\underline{R}_2^{hex}$, $\underline{R}_2$ indicated accordingly. Mirror lines are shown by thick lines and three-, sixfold rotation centers by black triangles and hexagons, respectively.

inversion centers lie both on mirror lines and in the middle between adjacent parallel mirror lines. (**3.119d**)
- In square netplanes with mirror symmetry fourfold rotation centers induce two or four orthogonal sets of parallel mirror lines. The mirror lines will either cross two- and fourfold rotation centers (four sets) or will only connect twofold centers between fourfold rotation centers (two sets). (**3.119e**)
- In hexagonal netplanes with mirror symmetry true threefold rotation centers induce sets of parallel mirror lines that cut each other at angles of 60°. (**3.119f**)
- In hexagonal netplanes with mirror symmetry sixfold rotation centers induce sets of parallel mirror lines that cut each other at angles of 30°, 60°, and 90°, respectively. (**3.119g**)

3.6.5
Glide Reflection

Glide reflections $g(\underline{r}_o, \underline{g})$ combine mirroring with translation and are, therefore, not point symmetry operations in the strict sense. They create, for any point $\underline{r}$ on one side of a glide line along $\underline{g}$, a mirror point on the other side (the glide line acting as a mirror line), which is then shifted by a vector $\underline{g}$ parallel to the glide line

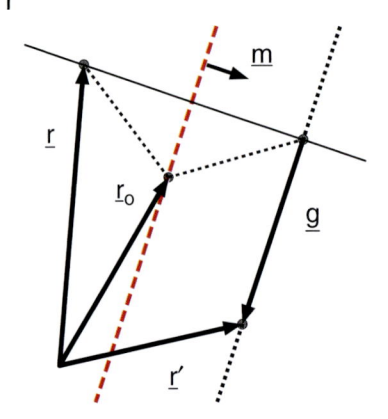

Figure 3.21 Sketch of a glide reflection operation applied to vector $\underline{r}$ and yielding $\underline{r}'$. The glide line center $\underline{r}_o$, glide line vector $\underline{g}$, and glide line normal vector $\underline{m}$ are labeled accordingly.

to yield the image point $\underline{r}'$, see Figure 3.21. This can be expressed **mathematically** by a transformation of points on the netplane

$$\underline{r} \to \underline{r}' = \underline{r} - 2\,[(\underline{r} - \underline{r}_o)\,\underline{m}]\,\underline{m} + \underline{g} = \underline{r}_o + \underline{\sigma}_m(\underline{r} - \underline{r}_o) + \underline{g} \tag{3.120}$$

where the **glide line** is defined by its origin $\underline{r}_o$ (**glide line center**), a shift vector $\underline{g}$ along the line (**glide line vector**), and a normal vector $\underline{m}$ (**glide line normal vector**) of unit length perpendicular to vector $\underline{g}$, sketched in Figure 3.21. The glide reflection can also be connected with a two-dimensional Cartesian **coordinate transformation** with respect to the mirror center $\underline{r}_o$ applying a 2×2 matrix $\underline{\sigma}_m$ where

$$\begin{pmatrix} x' \\ y' \end{pmatrix} = \underline{\sigma}_m \cdot \begin{pmatrix} x \\ y \end{pmatrix} + \begin{pmatrix} g_x \\ g_y \end{pmatrix}, \quad \underline{\sigma}_m = \begin{pmatrix} 1 - 2m_x^2 & -2m_x m_y \\ -2m_x m_y & 1 - 2m_y^2 \end{pmatrix} \tag{3.121}$$

where matrix $\underline{\sigma}_m$ is identical to that for mirroring, see Eq. (3.90), and

$$\underline{m} = (m_x, m_y),\ \underline{g} = (g_x, g_y),\ (\underline{m}\,\underline{g}) = 0 \tag{3.122}$$

As for mirroring the **glide line center** $\underline{r}_o$ can be chosen **arbitrarily** along the glide line. Further, repeating a glide reflection Eq. (3.120) with the same glide line vector $\underline{g}$ results in an operation given by

$$\underline{r} \to \underline{r}' \to \underline{r}'' = \underline{r}' - 2\,[(\underline{r}' - \underline{r}_o)\,\underline{m}]\,\underline{m} + \underline{g}$$

$$= \underline{r} - 2\,[(\underline{r} - \underline{r}_o)\,\underline{m}]\,\underline{m} + 2\underline{g} + 2\,[(\underline{r} - \underline{r}_o)\,\underline{m}]\,\underline{m}$$

$$= \underline{r} + 2\underline{g} \tag{3.123}$$

Thus, two subsequent glide reflections with identical $\underline{g}$ are equivalent to a **translation** by vector $2\underline{g}$. In a netplane possessing glide reflection symmetry this vector must be a general lattice vector, that is,

$$2\underline{g} = \underline{R} = n_1\,\underline{R}_1 + n_2\,\underline{R}_2, \quad n_1, n_2 \text{ integer}$$

or

$$\underline{g} = 1/2\,\underline{R} = 1/2\,(n_1\,\underline{R}_1 + n_2\,\underline{R}_2) \tag{3.124}$$

This restricts possible translation vectors $\underline{g}$ of glide lines and is the main condition for compatibility between translational and glide line symmetry in netplanes.

A **glide line** $g(\underline{r}_o, \underline{g})$ as a symmetry element of a netplane was shown to point always along one of its **general lattice vectors** $\underline{R}$. Of these, the smallest vector along $\underline{R}$, denoted $\underline{R}_o$, is given, according to Eq. (3.124) by mixing factors n_1, n_2, whose common divisor is not greater than 1. This yields glide reflections $g(\underline{r}_o, \underline{g} = \underline{R}_o/2)$ with $\underline{R}_o$ determining the translational periodicity of the netplane along the glide line. Relation (3.124) also allows **multiples** of $\underline{R}_o/2$ as possible glide line vectors $\underline{g}$, that is, in general

$$\underline{g} = p\,\underline{R}_o/2, \quad p \text{ integer} \tag{3.125}$$

Here, **even** p values yield vectors $\underline{g}$ that are general lattice vectors themselves. Thus, the definition (3.120) of a glide reflection together with translational symmetry of the netplane leads to a **mirror operation.** On the other hand, **odd** p values correspond to the **glide reflection** $g(\underline{r}_o, \underline{g} = \underline{R}_o/2)$ up to a shift by a general lattice vector, which can be ignored due to the translational symmetry of the netplane. Thus, in the following true glide reflection symmetry will always be connected with symmetry operations $g(\underline{r}_o, \underline{R}_o/2)$, where $\underline{R}_o$ denotes the smallest vector along its direction.

The **compatibility** between translational and glide reflection symmetry imposes constraints on the possible positions and directions of glide lines in a netplane. Let us assume a netplane with lattice vectors $\underline{R}_1$ and $\underline{R}_2$ to possess glide reflection symmetry with respect to a glide line along $\underline{R}_o$ and through $\underline{r}_o$. Then the netplane can always be shifted such that its origin coincides with $\underline{r}_o$ (setting $\underline{r}_o = 0$). Further, since $\underline{R}_o$ was shown to be a general lattice vector of smallest length along its direction, it can be used to define one of the lattice vectors of the netplane, for example, setting $\underline{R}_1 = \underline{R}_o$. According to Eq. (3.120), a general lattice vector $\underline{R}$, given by Eq. (3.66) and not positioned on a glide line, will have a glide reflection image $\underline{R}'$, where

$$\underline{R}' = \underline{R} - 2\,(\underline{R}\,\underline{m})\,\underline{m} + \underline{R}_o/2 \tag{3.126}$$

and applying a glide reflection to $\underline{R}'$ creates a second image $\underline{R}''$ with

$$\underline{R}'' = \underline{R}' - 2\,(\underline{R}'\underline{m})\,\underline{m} + \underline{R}_o/2 \tag{3.127}$$

where both $\underline{R}'$ and $\underline{R}''$ are general lattice vectors. Therefore, vector $\underline{R}_{g2}$, given by

$$\underline{R}_{g2} = \underline{R} + \underline{R}'' - 2\underline{R}' = 2\,[(\underline{R} - \underline{R}')\,\underline{m}]\,\underline{m} \tag{3.128}$$

must also be a general lattice vector, which is orthogonal to $\underline{R}_1 = \underline{R}_o$ since $(\underline{m}\,\underline{R}_o) = 0$ according to Eq. (3.122). Thus, the lattice vector of smallest length along $\underline{R}_{g2}$, together with $\underline{R}_1$ provides an **orthogonal** set of (primitive or non-primitive) lattice vectors describing the netplane periodicity. This proves that the existence of **glide reflection** symmetry is always connected with primitive or non-primitive **rectangular** netplanes. Therefore, in the following we will use mutually orthogonal lattice vectors $\underline{R}_1$ and $\underline{R}_2$ to describe netplanes with glide reflection symmetry.

First, consider a **rectangular** netplane defined by orthogonal lattice vectors $\underline{R}_1$ and $\underline{R}_2$ and a glide reflection with its **glide line** through the netplane origin and **parallel to** $\underline{R}_1$, that is,

$$\underline{m} = \underline{R}_2/|\underline{R}_2|, \quad \underline{r}_o = \underline{0} \tag{3.129}$$

Then, in complete analogy with the discussion of mirror operations in Section 3.6.4, resulting in Eq. (3.113), there are **infinitely** many parallel **glide lines** through glide line centers at $\underline{r}_{on}$ with

$$\underline{r}_{on} = n\,\underline{R}_2/2, \quad n \text{ integer} \tag{3.130}$$

which applies to both primitive and centered rectangular netplanes, as illustrated in Figure 3.22a for the primitive rectangular case.

Analogously, glide reflections with their glide lines **parallel to** $\underline{R}_2$ form an infinite set with glide line centers at

$$\underline{r}_{on} = n\,\underline{R}_1/2, \quad n \text{ integer} \tag{3.131}$$

for primitive and centered rectangular netplanes, as shown in Figure 3.22b.

A **glide line** $g(\underline{r}_o, \underline{g} = \underline{R}_o/2)$ can **never** coincide with a **mirror line** $\sigma(\underline{r}_o, \underline{e} = \underline{R}_o/|\underline{R}_o|)$ along the same direction in a netplane, since this would result in lattice vectors $\underline{R}'_o = \underline{R}_o/2$ contradicting $\underline{R}_o$ to be of smallest length along

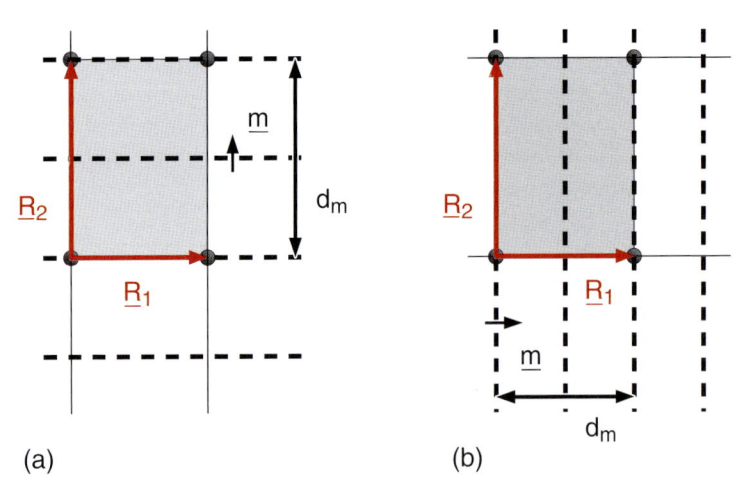

Figure 3.22 Periodic sets of glide lines (a) parallel to lattice vector $\underline{R}_1$, (b) parallel to lattice vector $\underline{R}_2$ for primitive rectangular netplanes. The glide lines are shown by black horizontal and vertical dashed lines. The unit cell is emphasized in gray.

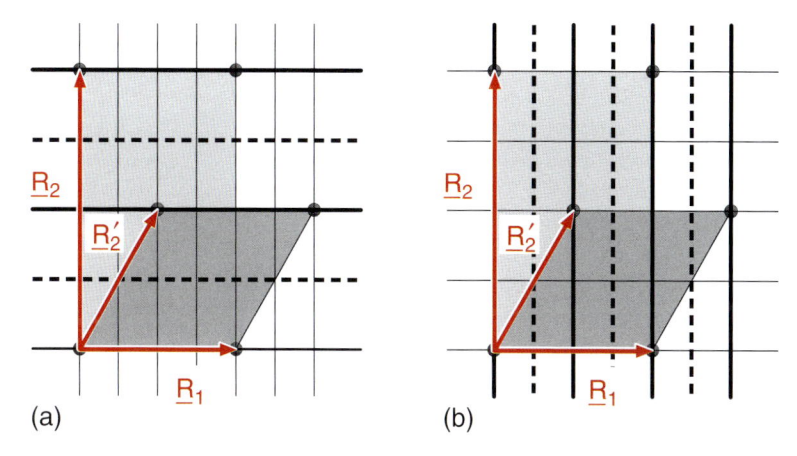

Figure 3.23 Morphological unit cells of centered rectangular netplanes with coexisting mirror and glide line symmetry, glide and mirror lines (a) parallel to $\underline{R}_1$, (b) parallel to $\underline{R}_2$. The non-primitive lattice vectors $\underline{R}_1$, $\underline{R}_2$, as well as the centering vector $\underline{R}_2'$ are labeled accordingly. The unit cells are emphasized in light and dark gray (non-primitive and primitive cells, respectively) with mirror lines indicated by thick and glide lines by dashed lines.

its direction. However, **glide lines** may exist **between** adjacent parallel **mirror lines.** Let us assume a netplane with orthogonal lattice vectors $\underline{R}_1$ and $\underline{R}_2$ to possess mirror symmetry with respect to a mirror line parallel to $\underline{R}_1$ through $\underline{r}_o$, chosen as the origin of the netplane. Then, according to the discussion in Section 3.6.4, the netplane includes an infinite set of parallel mirror lines along $\underline{R}_1$ through centers $\underline{r}_{on} = n\,\underline{R}_2/2$, n integer, as symmetry elements. Thus, mirror lines with n = 0, 1, 2 confine or cut the morphological unit cell, see Figure 3.23a. Let us assume that the netplane has also glide reflection symmetry with respect to a glide line parallel to $\underline{R}_1$ and cutting the morphological unit cell, which can be achieved by setting the glide line center $\underline{r}_{go}$ at

$$\underline{r}_{go} = \gamma\,\underline{R}_2, \quad 0 < \gamma < 1 \tag{3.132}$$

Then a general lattice vector $\underline{R}$ according to Eq. (3.66) will be transformed by a glide reflection Eq. (3.120) to yield another general lattice vector $\underline{R}'$, where

$$\underline{R}' = \underline{R} - 2\,[(\underline{R} - \underline{r}_{go})\,\underline{m}]\,\underline{m} + \underline{R}_1/2, \quad \underline{m} = \underline{R}_2/|\underline{R}_2| \tag{3.133}$$

or, using representation Eq. (3.66) together with the orthogonality of $\underline{R}_1$ and $\underline{R}_2$,

$$\underline{R} = n_1\,\underline{R}_1 + n_2\,\underline{R}_2, \quad n_1,\ n_2\ \text{integer}$$
$$\underline{R}' = (n_1 + 1/2)\,\underline{R}_1 + (2\,\gamma - n_2)\,\underline{R}_2, \quad n_1,\ n_2\ \text{integer} \tag{3.134}$$

which can be written as

$$\underline{R}' = 1/2\,(\underline{R}_1 + \underline{R}_2) + n_1\,\underline{R}_1 + [2(\gamma - 1/4) - n_2]\,\underline{R}_2 \tag{3.135}$$

This shows that the transformed general lattice vector $\underline{R}'$ must belong to a **centered rectangular** lattice. Further, the prefactor $[2\,(\gamma - 1/4) - n_2]$ in front of $\underline{R}_2$

must be an integer, which constrains values of γ to

$$2(\gamma - 1/4) = p, \quad \gamma = (2p + 1)/4, \quad p \text{ integer} \tag{3.136}$$

Thus, γ must be an odd-valued multiple of 1/4, which, together with constraints Eq. (3.132), allows only values $\gamma = 1/4$ or $\gamma = 3/4$, yielding glide lines in the middle between mirror lines through $\underline{r}_o = \underline{0}$ and $\underline{r}_o = \underline{R}_2/2$, as well as through $\underline{r}_o = \underline{R}_2/2$ and $\underline{r}_o = \underline{R}_2$. Thus, a combination of glide reflection and mirror symmetry parallel to $\underline{R}_1$ is possible only if the glide lines exist **in the middle** between adjacent mirror lines. In addition, the netplane must allow a **centered rectangular** representation by non-primitive orthogonal lattice vectors $\underline{R}_1$ and $\underline{R}_2$, as illustrated in Figure 3.23a. The corresponding result for netplane symmetry with glide and mirror lines parallel to $\underline{R}_2$ is analogous and illustrated in Figure 3.23b.

The result of the previous discussion also has consequences for **primitive rectangular** netplanes described by orthogonal lattice vectors $\underline{R}_1$ and $\underline{R}_2$. There, symmetry with respect to a glide line parallel to $\underline{R}_1$ (or $\underline{R}_2$) excludes mirror symmetry with lines parallel to the glide line. Primitive rectangular netplanes do **not** allow glide lines **parallel** to mirror lines.

Netplanes with **glide reflection** symmetry may also possess **inversion** symmetry. First, consider **primitive rectangular** netplanes with primitive lattice vectors $\underline{R}_1$ and $\underline{R}_2$. According to Eq. (3.130) the complete set of glide lines parallel to $\underline{R}_1$ can be positioned at centers $\underline{r}_{on} = n\,\underline{R}_2/2$. Then inversion centers, forming a netplane with lattice vectors $1/2\,\underline{R}_1$ and $1/2\,\underline{R}_2$, see Section 3.6.2, can only either all lie on glide lines or all in the middle between adjacent glide lines, as illustrated in Figure 3.24. (The proof is completely analogous to that for mirror lines given in Section 3.6.4.) In the former case inversion centers $\underline{r}_{io}$ are given by

$$\underline{r}_{io} = p\,\underline{R}_1/2 + q\,\underline{R}_2/2, \quad p, q \text{ integer} \tag{3.137}$$

while in the latter $\underline{r}_{io}$ can be defined by

$$\underline{r}_{io} = p\,\underline{R}_1/2 + q\,\underline{R}_2/2 + \underline{R}_2/4, \quad p, q \text{ integer} \tag{3.138}$$

with glide line centers $\underline{r}_{go}$ given by

$$\underline{r}_{go} = n_2\,\underline{R}_2/2, \quad n_2 \text{ integer} \tag{3.139}$$

Then a glide reflection $g(\underline{r}_{go}, \underline{R}_1/2)$ followed by an inversion $i(\underline{r}_{io})$ according to Eq. (3.64), Eq. (3.120) is given by a transformation $\underline{r} \to \underline{r}' \to \underline{r}''$, where

$$\underline{r}' = \underline{r} - 2[(\underline{r} - \underline{r}_{go})\,\underline{m}]\,\underline{m} + \underline{R}_1/2, \quad \underline{m} = \underline{R}_2/|\underline{R}_2|$$

$$\underline{r}'' = 2\underline{r}_{io} - \underline{r}' = -\underline{r} + 2[(\underline{r} - \underline{r}_{go})\,\underline{m}]\,\underline{m} + 2\underline{r}_{io} - \underline{R}_1/2$$

$$= -\underline{r} + 2(\underline{r}\,\underline{m})\,\underline{m} + 2(\underline{r}_{io} - \underline{r}_{go}) - \underline{R}_1/2 \tag{3.140}$$

and with

$$\underline{m}' = \underline{R}_1/|\underline{R}_1|, \quad (\underline{m}\,\underline{m}') = 0, \quad \underline{r} = (\underline{r}\,\underline{m})\,\underline{m} + (\underline{r}\,\underline{m}')\,\underline{m}' \tag{3.141}$$

we obtain

$$\underline{r}'' = \underline{r} - 2(\underline{r}\,\underline{m}')\,\underline{m}' + 2(\underline{r}_{io} - \underline{r}_{go}) - \underline{R}_1/2 \tag{3.142}$$

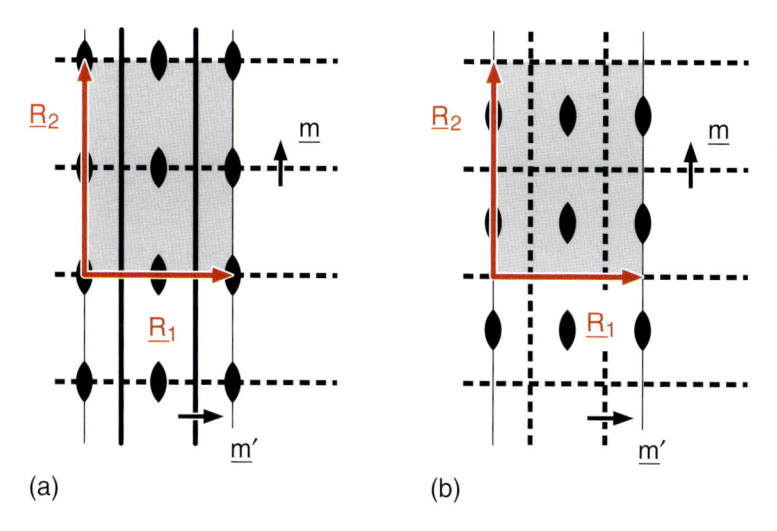

Figure 3.24 Periodic sets of glide lines (dashed lines), parallel to $\underline{R}_1$, combined with corresponding inversion centers (ellipses) for a primitive rectangular netplane. (a) Inversion centers positioned on glide lines and perpendicular mirror lines, shown by thick lines. (b) Inversion centers positioned in the middle between glide lines and perpendicular glide lines. The unit cells are emphasized in gray.

If all inversion centers are positioned at glide lines, that is, for Eq. (3.137), relation (3.142) reads

$$\underline{r}'' = \underline{r} - 2\,(\underline{r}\,\underline{m}')\,\underline{m}' + p\,\underline{R}_1 + (q - n_2)\,\underline{R}_2 - \underline{R}_1/2$$
$$= \underline{r} - 2\,[(\underline{r} - (2p-1)\,\underline{R}_1/4)\,\underline{m}']\,\underline{m}' + (q - n_2)\,\underline{R}_2 \qquad (3.143)$$

This corresponds to mirror operations $\sigma((2p-1)\,\underline{R}_1/4,\,\underline{R}_1/|\underline{R}_1|)$ with mirror lines perpendicular to the initial glide lines, followed by translations by general lattice vectors. Thus, a netplane, which is symmetric with respect to the initial glide lines parallel to $\underline{R}_1$ and contains inversion centers on the glide lines, will also include a set of **perpendicular mirror lines** as symmetry elements, as illustrated in Figure 3.24a.

If, on the other hand, all inversion centers are positioned in the middle between adjacent glide lines, that is, for Eq. (3.138), then relation (3.142) reads

$$\underline{r}'' = \underline{r} - 2\,(\underline{r}\,\underline{m}')\,\underline{m}' + p\,\underline{R}_1 + (q - n_2)\,\underline{R}_2 - \underline{R}_1/2 + \underline{R}_2/2$$
$$= \underline{r} - 2\,[(\underline{r} - (2p-1)\,\underline{R}_1/4)\,\underline{m}']\,\underline{m}' + \underline{R}_2/2 + (q - n_2)\,\underline{R}_2 \qquad (3.144)$$

This corresponds to glide reflections $g((2p-1)\,\underline{R}_1/4,\,\underline{R}_2/2)$ with glide lines perpendicular to the initial glide lines, followed by translations by general lattice vectors. Thus, a netplane, which is symmetric with respect to the initial glide lines parallel to $\underline{R}_1$ and contains inversion centers between the glide lines, will also include a set of **perpendicular glide lines** as symmetry elements, as illustrated in Figure 3.24b.

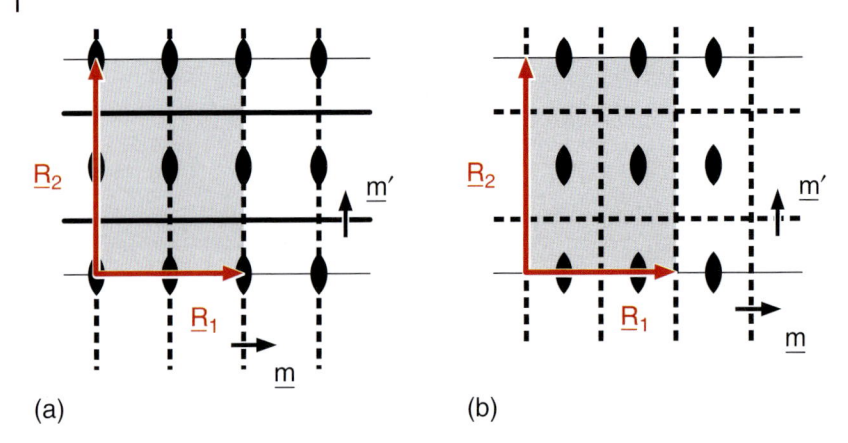

Figure 3.25 Periodic sets of glide lines (dashed lines), parallel to $\underline{R}_2$, combined with corresponding inversion centers (ellipses) for a primitive rectangular netplane. (a) Inversion centers positioned on glide lines and perpendicular mirror lines, shown by thick lines. (b) Inversion centers positioned in the middle between glide lines and perpendicular glide lines. The unit cells are emphasized in gray.

Primitive rectangular netplanes with **inversion** and **glide reflection** symmetry, where glide lines are parallel to $\underline{R}_2$, yield structures, which are analogous to those obtained for glide lines parallel to $\underline{R}_1$. Here the two cases, inversion centers on glide lines and between glide lines, lead to additional orthogonal mirror lines and glide lines, respectively, as shown in Figure 3.25, which is quite similar to Figure 3.24. In fact, the structures of Figures 3.24b and 3.25b differ only by a shift of the netplane origins.

Centered rectangular netplanes with **inversion** and **glide reflection** symmetry, where glide lines are parallel to $\underline{R}_1$ or $\underline{R}_2$, always yield inversion centers positioned on glide lines as well as between. According to the previous discussion, this results in additional perpendicular glide and mirror lines as symmetry elements and leads to the geometry shown in Figure 3.26.

In conclusion, glide lines existing as symmetry elements of a netplane, defined by lattice vectors $\underline{R}_1$ and $\underline{R}_2$, are subject to the following constraints:

- Netplanes with glide reflection symmetry must be either primitive or centered rectangular (including square and hexagonal netplanes as special cases). \hfill (**3.145a**)
- Netplanes with glide reflection symmetry include infinite sets of parallel glide lines. If $\underline{R}_1$ and $\underline{R}_2$, are orthogonal lattice vectors describing the morphological unit cell, then glide line sets can be given by normal vectors $\underline{m} = \underline{R}_1/|\underline{R}_1|$ and centers $\underline{r}_{on} = n\underline{R}_1/2$, or by $\underline{m} = \underline{R}_2/|\underline{R}_2|$ and centers $\underline{r}_{on} = n\underline{R}_2/2$. \hfill (**3.145b**)
- Glide lines of netplanes with glide reflection and mirror symmetry can never coincide with mirror lines. Glide lines parallel to mirror lines exist only in centered rectangular netplanes and appear in alternating equidistant sequences.

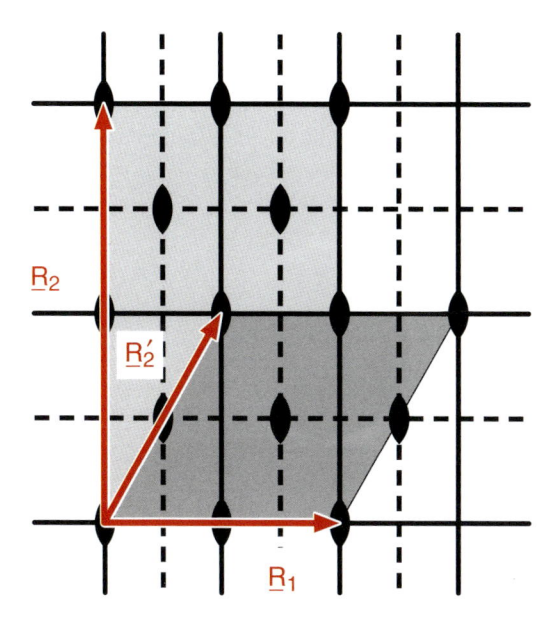

Figure 3.26 Periodic sets of glide lines (dashed lines), parallel to $\underline{R}_1$, $\underline{R}_2$, combined with corresponding inversion centers (ellipses) for a centered rectangular netplane. The non-primitive lattice vectors $\underline{R}_1$, $\underline{R}_2$, as well as the centering vector $\underline{R}'_2$ are shown by red arrows and labeled accordingly. The unit cells are emphasized in light and dark gray (non-primitive and primitive cells, respectively). The resulting mirror lines are indicated by thick lines.

- Primitive netplanes do not allow parallel glide and mirror lines as symmetry elements. (**3.145c**)
- In primitive rectangular netplanes with glide reflection and inversion symmetry inversion (twofold rotation) centers lie either all on glide lines or all in the middle between adjacent parallel glide lines. (**3.145d**)
- Primitive rectangular netplanes with glide reflection and inversion symmetry include an additional set of mirror lines perpendicular to the glide lines, if the inversion centers lie on the glide lines, and an additional set of glide lines perpendicular to the initial glide lines, if the inversion centers lie between glide lines. (**3.145e**)
- Centered rectangular netplanes with glide reflection and inversion symmetry include orthogonal sets of parallel glide lines as well as of mirror lines as symmetry elements. (**3.145f**)

3.6.6
Symmetry Groups

The **point symmetry** operations, that is, inversion, n-fold rotation, and mirroring, discussed in the previous sections together with corresponding translation operations, form a **complete** set that can be used to **classify** the different types of netplanes according to their point symmetry behavior. In the following point symmetry operations are denoted by **Schönflies** symbols and their origins r_o are assumed to coincide with the origin of a two-dimensional Cartesian coordinate system (x, y). Further, rotation and mirror line angles are defined with respect to the x axis.

Overall, there are **16 true point symmetry operations** that can be applied and may reproduce a netplane. These are

- **eight rotation** operations $C_\varphi(r_o)$, see Section 3.6.3, where, due to compatibility with translational symmetry, rotation angles φ are restricted to finite values $\varphi = \pm 60°, \pm 90°, \pm 120°, 180°$, with the latter also shown to reflect an **inversion** operation. Further, a rotation with $\varphi = 0°$ describes the **identity** operation, which leaves a netplane unchanged and is introduced for mathematical completeness only. The corresponding 2×2 coordinate transformation matrices $\underline{\underline{C}}_\varphi$ are given by Eq. (3.74), that is, by

$$\underline{\underline{C}}_\varphi = \begin{pmatrix} \cos\varphi & -\sin\varphi \\ \sin\varphi & \cos\varphi \end{pmatrix} \tag{3.146}$$

- **eight mirror** operations $\sigma_\Phi(r_o)$, see Section 3.6.4, where, due to compatibility with translational symmetry, mirror line angles Φ are restricted to values $\Phi = 0°$ (mirror line along x axis), $\pm 30°, \pm 45°$ (mirror line along $\pm$x/y diagonal), $\pm 60°$, and $90°$ (mirror line along y axis). The corresponding 2×2 coordinate transformation matrices $\underline{\underline{\sigma}}_\Phi$ are given by Eq. (3.94), that is, by

$$\underline{\underline{\sigma}}_\Phi = \begin{pmatrix} \cos(2\Phi) & \sin(2\Phi) \\ \sin(2\Phi) & -\cos(2\Phi) \end{pmatrix} \tag{3.147}$$

Usually, a netplane includes several of these 16 operations as elements describing its full point symmetry. If a netplane transforms into itself upon applying two point symmetry operations A, B separately, then it must also be symmetric with respect to a subsequent application of the two, formally described as a **product** operation $(A\,B)$ whose 2×2 transformation matrix is given by the product of the two matrices $\underline{\underline{A}}, \underline{\underline{B}}$ characterizing the operations. Thus,

- the **product** of two **rotation** operations C_φ, C_χ (the explicit mention of the common rotation center r_o is omitted in the following) is characterized according to Eq. (3.74) by a 2×2 matrix $\underline{\underline{T}}$ with

$$\underline{\underline{T}} = \underline{\underline{C}}_\varphi \cdot \underline{\underline{C}}_\chi = \begin{pmatrix} \cos\varphi & -\sin\varphi \\ \sin\varphi & \cos\varphi \end{pmatrix} \cdot \begin{pmatrix} \cos\chi & -\sin\chi \\ \sin\chi & \cos\chi \end{pmatrix}$$

$$= \begin{pmatrix} \cos(\varphi+\chi) & -\sin(\varphi+\chi) \\ \sin(\varphi+\chi) & \cos(\varphi+\chi) \end{pmatrix} = \underline{\underline{C}}_\chi \cdot \underline{\underline{C}}_\varphi = \underline{\underline{C}}_{\varphi+\chi} \tag{3.148}$$

which describes $C_{\varphi+\chi}$, a rotation by an angle equal to the sum of the two initial angles.

- the **product** of two **mirror** operations $\underline{\sigma}_\Phi$, $\underline{\sigma}_\Theta$ is characterized according to Eq. (3.94) by a 2×2 matrix $\underline{\underline{T}}$ with

$$\underline{\underline{T}} = \underline{\underline{\sigma}}_\Phi \cdot \underline{\underline{\sigma}}_\Theta = \begin{pmatrix} \cos(2\Phi) & \sin(2\Phi) \\ \sin(2\Phi) & -\cos(2\Phi) \end{pmatrix} \cdot \begin{pmatrix} \cos(2\Theta) & \sin(2\Theta) \\ \sin(2\Theta) & -\cos(2\Theta) \end{pmatrix}$$

$$= \begin{pmatrix} \cos 2(\Phi - \Theta) & -\sin 2(\Phi - \Theta) \\ \sin 2(\Phi - \Theta) & \cos 2(\Phi - \Theta) \end{pmatrix} = \underline{\underline{C}}_{2(\Phi-\Theta)} \quad (3.149)$$

which describes a rotation $C_{2(\Phi-\Theta)}$ by an angle equal to twice that between the two mirror lines.

- the **mixed products** of a **rotation** and a **mirror** operation C_φ, σ_Φ are characterized according to Eqs. (3.74) and (3.94) by 2×2 matrices $\underline{\underline{T}}$ with

$$\underline{\underline{T}} = \underline{\underline{C}}_\varphi \cdot \underline{\underline{\sigma}}_\Phi = \begin{pmatrix} \cos\varphi & -\sin\varphi \\ \sin\varphi & \cos\varphi \end{pmatrix} \cdot \begin{pmatrix} \cos(2\Phi) & \sin(2\Phi) \\ \sin(2\Phi) & -\cos(2\Phi) \end{pmatrix}$$

$$= \begin{pmatrix} \cos(2\Phi + \varphi) & \sin(2\Phi + \varphi) \\ \sin(2\Phi + \varphi) & -\cos(2\Phi + \varphi) \end{pmatrix} = \underline{\underline{\sigma}}_{(\Phi+\varphi/2)} \quad (3.150)$$

$$\underline{\underline{T}} = \underline{\underline{\sigma}}_\Phi \cdot \underline{\underline{C}}_\varphi = \begin{pmatrix} \cos(2\Phi) & \sin(2\Phi) \\ \sin(2\Phi) & -\cos(2\Phi) \end{pmatrix} \cdot \begin{pmatrix} \cos\varphi & -\sin\varphi \\ \sin\varphi & \cos\varphi \end{pmatrix}$$

$$= \begin{pmatrix} \cos(2\Phi - \varphi) & \sin(2\Phi - \varphi) \\ \sin(2\Phi - \varphi) & -\cos(2\Phi - \varphi) \end{pmatrix} = \underline{\underline{\sigma}}_{(\Phi-\varphi/2)} \quad (3.151)$$

which describe mirror operations $\sigma_{(\Phi\pm\varphi/2)}$ with mirror lines rotated by $\pm\varphi/2$ with respect to the initial mirror line referring to σ_Φ, where the two results depend on the sequence of the product formation.

The rotation by $0°$ (**identity** operation), described as C_0 and represented, according to Eq. (3.74), by the 2×2 unit matrix $\underline{\underline{C}}_0$ with

$$\underline{\underline{C}}_0 = \begin{pmatrix} \cos\varphi & -\sin\varphi \\ \sin\varphi & \cos\varphi \end{pmatrix}\bigg|_{\varphi=0°} = \begin{pmatrix} 1 & 0 \\ 0 & 1 \end{pmatrix} \quad (3.152)$$

can be considered a **unit operation** whose products with rotations and mirror operations yield

$$C_\varphi C_0 = C_0 C_\varphi = C_\varphi, \quad \sigma_\Phi C_0 = C_0 \sigma_\Phi = \sigma_\Phi \quad (3.153)$$

On the other hand, relation (3.148) shows together with Eq. (3.152) that any rotation by angle φ can be undone by an inverted rotation with angle $\varphi' = -\varphi$. Further, applying $\Theta = \Phi$ in Eq. (3.149) shows that any mirror operation can be undone by its own mirror operation. These two results, which are intuitively clear, can be formally expressed by the statement that each symmetry operation A of the above set possesses an **inverted operation** A^{-1}, where

$$A A^{-1} = A^{-1} A = C_0 \quad (3.154)$$

As a last result we mention the **associativity** of subsequent applications of symmetry operations A, B, C, that is,

$$A(BC) = (AB)C \quad (3.155)$$

which is clear from the associativity of the matrix multiplication.

Any subset of the 16 point symmetry operations listed above forms a **point symmetry group** if all its product operations belong to the same subset. This is an example of a mathematical group $\mathbf{G} = \{g_1, \ldots, g_p\}$ with a finite number p of elements g_i, which is defined formally [40] by the following four properties:

1) A **product** of two group elements $g_i g_j$ is defined and always yields an element g_k of the group (closure condition). (3.156a)
2) The group contains a **unit element** e with $g_i e = e g_i = g_i$ for all elements g_i of the group, analogous to Eq. (3.153). (3.156b)
3) The group contains an **inverse element** g_i^{-1} with $g_i g_i^{-1} = g_i^{-1} g_i = e$ for each element g_i of the group, analogous to Eq. (3.154). (3.156c)
4) Group products are **associative**, that is, $g_i(g_j g_k) = (g_i g_j)g_k$ for all elements g_i of the group, analogous to Eq. (3.155). (3.156d)

As an example, the three rotation operations defining a threefold rotation axis form a set $\{C_0, C_{120}, C_{-120}\}$, which defines a point symmetry group (denoted C_3 or **3**).

The complete collection of two-dimensional point symmetry groups allowed for netplanes can be determined by considering the above 16 point symmetry operations together with their products given by Eqs. (3.148)–(3.151). Then, collecting the sets of operations such that they satisfy the four group properties yields, altogether, **10** different **point symmetry groups.** These are listed by their operations as well as their symmetries in Table 3.9, where symmetry operation C_0 defines the identity operation while C_{180} (twofold rotation) is equivalent to inversion.

Table 3.9 List of all point symmetry groups allowed for netplanes.

Group	Symmetries	Members	Group	Symmetries	Members
1 (C_1)	Identity	C_0	m (C_S)	One mirror	C_0, σ_0
2 (C_2)	Twofold rotation	C_0, C_{180}	2mm (C_{2v})	Twofold rotation, two mirrors	$C_0, C_{180}, \sigma_0, \sigma_{90}$
3 (C_3)	Threefold rotation	C_0, C_{120}, C_{-120}	3m1, 31m (C_{3v})	Threefold rotation, three mirrors	$C_0, C_{120}, C_{-120}, \sigma_{60}, \sigma_{-60}, \sigma_0$
4 (C_4)	Two-, fourfold rotation	$C_0, C_{90}, C_{180}, C_{-90}$	4mm (C_{4v})	Two-, fourfold rotation, four mirrors	$C_0, C_{90}, C_{180}, C_{-90}, \sigma_0, \sigma_{45}, \sigma_{90}, \sigma_{-45}$
6 (C_6)	Two-, three-, sixfold rotation	$C_0, C_{60}, C_{120}, C_{180}, C_{-120}, C_{-60}$	6mm (C_{6v})	Two-, three-, sixfold rotation, six mirrors	$C_0, C_{60}, C_{120}, C_{180}, C_{-120}, C_{-60}, \sigma_0, \sigma_{30}, \sigma_{60}, \sigma_{90}, \sigma_{-60}, \sigma_{-30}$

The list includes group members, rotations C_φ with φ denoting the rotation angle, and mirror operations σ_Φ, where Φ denotes the angle of the mirror line with respect to the x axis. The groups are labeled according to both the Hermann–Mauguin and the Schönflies notation, where the latter is put in parentheses.

Table 3.10 List of all point symmetry groups allowed for netplanes together with corresponding subgroups.

Group	Subgroups	Group	Subgroups
1 (C_1)	—	m (C_S)	1
2 (C_2)	1	2mm (C_{2v})	1, 2, m
3 (C_3)	1	3m1, 31m (C_{3v})	1, 3, m
4 (C_4)	1, 2	4mm (C_{4v})	1, 2, 4, m, 2mm
6 (C_6)	1, 2, 3	6mm (C_{6v})	1, 2, 3, 6, m, 2mm, 3m1, 31m

The groups are labeled according to both the Hermann–Mauguin and the Schönflies notation, where the latter is put in parentheses. Subgroups are given in Hermann–Mauguin notation only.

Except for the identity group C_1, each point symmetry group in Table 3.9 includes subsets of symmetry elements, which themselves form groups, also called **subgroups** of the initial point symmetry group. This is shown in Table 3.10 where all subgroups of the point symmetry groups of Table 3.9 have been listed.

The formal group definition (3.156) can also be applied to describe the translational symmetry of a netplane by its **translation group**. The characterization of a periodic netplane by its lattice vectors $\underline{R}_1$ and $\underline{R}_2$ means in particular that the netplane is symmetric with respect to translations by general lattice vectors $\underline{R}$ with $\underline{R} = n_1 \underline{R}_1 + n_2 \underline{R}_2$. This can be phrased by a **translation symmetry** operation

$$\underline{r} \rightarrow \underline{r}' = T_{(n1, n2)}\underline{r} = \underline{r} + (n_1 \underline{R}_1 + n_2 \underline{R}_2), \quad n_1, n_2 \text{ integer} \quad (3.157)$$

where the infinite set of operations $T_{(n1,n2)}$, n_1, n_2 integer, together with a **product** definition,

$$T_{(n1, n2)} T_{(n1', n2')} = T_{(n1+n1', n2+n2')} \quad (3.158)$$

a **unit** element $T_{(0,0)}$, and **inverse** elements,

$$T^{-1}_{(n1, n2)} = T_{(-n1, -n2)} \quad (3.159)$$

defines the (infinite) translation group of a netplane. This group can be combined with a corresponding point symmetry group of a netplane to yield a **two-dimensional space group**, sometimes also called **plane group**. If this space group describes all the symmetry properties of a netplane, the netplane will be said to be characterized by a **simple** or **symmorphic** space group. This includes groups, where, due to translational symmetry, the initial point symmetry operations also allow glide reflections, which combine mirroring with translation, see Section 3.6.5. (As discussed in Section 3.6.5, glide lines will appear if a netplane with mirror lines includes a primitive lattice vector that is inclined with respect to a mirror line.) In addition, there are cases where the full symmetry of a netplane requires, in addition to true point symmetry operations, glide reflections as initial generating symmetry elements. In this case, the netplane will be said to be characterized by a **non-symmorphic** space group. The distinction between the different types of space groups becomes clear in Section 3.8, where space groups are used to classify all netplanes with respect to their possible symmetry properties.

3.7
Crystal Systems and Bravais Lattices in Two Dimensions

Properties of the different point symmetry operations, discussed in Section 3.6, together with translational symmetry allow a symmetry **classification** of all netplanes by their corresponding **space groups.** This is analogous to the classification of three-dimensional lattices discussed in Section 2.4. As in the three-dimensional case, we can also use the four different rotation axes, two-, three-, four-, and sixfold, to distinguish between the different types of netplanes. Here the netplane origin is assumed to coincide with a rotation center. Further, the two lattice vectors $\underline{R}_1$ and $\underline{R}_2$ that describe the netplane periodicity will be described by their lengths a, b and the angle γ between them, that is,

$$|\underline{R}_1| = a, \quad |\underline{R}_2| = b, \quad \angle(\underline{R}_1, \underline{R}_2) = \gamma, \quad (\underline{R}_1 \underline{R}_2) = a b \cos(\gamma) \qquad (3.160)$$

First, we note that, as a result of their periodicity, netplanes without further symmetry constraints always include an inversion center as a symmetry element at their origin, which is equivalent to **twofold rotation symmetry.** Then the most general type of netplanes is described by primitive lattice vectors $\underline{R}_1$ and $\underline{R}_2$ of different length, $a \neq b$, forming an angle γ different from 60°, 90°, and 120°. This corresponds to a morphological unit cell of parallelogram shape as shown in Figure 3.27a, and the respective netplanes are called **oblique**, forming the oblique **Bravais lattice**.

A second type of netplanes is described by primitive lattice vectors $\underline{R}_1$ and $\underline{R}_2$ of different length that are orthogonal to each other, that is, $a \neq b$, and $\gamma = 90°$. Here the morphological unit cell is of rectangular shape, as shown in Figure 3.27b. These netplanes are called **primitive rectangular** or **p-rectangular**, referring to the p-rectangular **Bravais lattice**.

A third type of netplane, which includes **fourfold rotation symmetry**, is described by primitive lattice vectors $\underline{R}_1$ and $\underline{R}_2$ of the same length that are orthogonal to each other, that is, $a = b$ and $\gamma = 90°$. Here the morphological unit cell is of square shape, as shown in Figure 3.27d. These netplanes are called **square**, referring to the square **Bravais lattice**.

A forth type of netplanes, which includes **three and sixfold rotation symmetry**, is described by primitive lattice vectors $\underline{R}_1$ and $\underline{R}_2$ of identical length, $a = b$, and $\gamma = 60°$ (acute representation) or 120° (obtuse representation). Here the morphological unit cell has the shape of two equilateral triangles joining on one side, as shown in Figure 3.27e. These netplanes are called **hexagonal**, referring to the hexagonal **Bravais lattice**.

So far, the discussion was restricted to Bravais netplanes whose periodicity was described by primitive lattice vectors $\underline{R}_1$ and $\underline{R}_2$. Assuming non-primitive lattice vectors $\underline{R}_1$ and $\underline{R}_2$, the corresponding morphological unit cell will contain additional lattice points. In Section 3.6.1 it was shown that, if vectors $\underline{R}_1$ and $\underline{R}_2$ are of

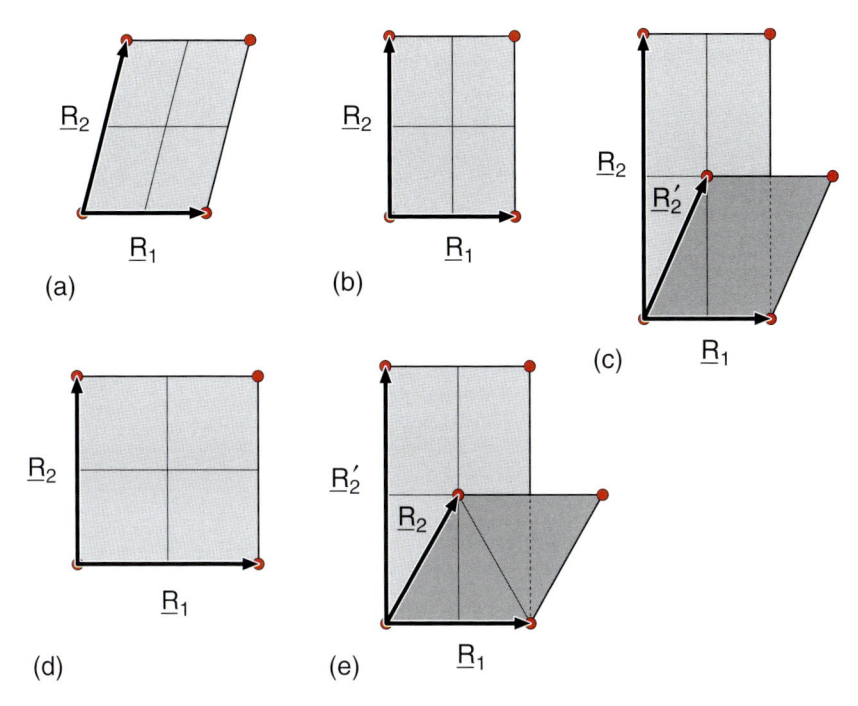

Figure 3.27 Morphological unit cells of netplanes referring to the five Bravais lattices (a) oblique, (b) primitive rectangular, (c) centered rectangular, (d) square, and (e) hexagonal. The cells are emphasized in gray with lattice vectors $\underline{R}_1$, $\underline{R}_2$, ($\underline{R}'_2$) sketched accordingly.

smallest length along their direction, there can only be one lattice point located in the center of the unit cell, given by a vector $\underline{R}' = 1/2\,(\underline{R}_1 + \underline{R}_2)$, see Eq. (3.63). This results in a **centered netplane**, which may be defined by primitive lattice vectors $\underline{R}'_1$ and $\underline{R}'_2$ with

$$\underline{R}'_1 = 1/2\,(\underline{R}_1 - \underline{R}_2), \quad \underline{R}'_2 = 1/2\,(\underline{R}_1 + \underline{R}_2) \tag{3.161}$$

Centering an **oblique** netplane, sketched in Figure 3.28a, cannot lead to a different symmetry that would change the type of Bravais lattice. Therefore, a centered oblique Bravais lattice is, in its type, identical to its primitive oblique counterpart. In contrast, centering a **primitive rectangular** netplane, sketched in Figure 3.28b, creates a new type of netplane, defining the **centered rectangular** or **c-rectangular** Bravais lattice. This lattice is not generally available by any of the other Bravais lattices. Due to the orthogonality of the initial non-primitive lattice vectors $\underline{R}_1$ and $\underline{R}_2$, the corresponding primitive vectors $\underline{R}'_1$, $\underline{R}'_2$, according to Eq. (3.161), are of equal length and can form any angle γ between them. This can also be used as a **definition** of the **c-rectangular** Bravais lattice being described by primitive lattice vectors $\underline{R}'_1$, $\underline{R}'_2$ of identical length, $a = b$, and any angle γ ($\neq 60°$, 90°, 120°, see below). With this vector definition, centering the c-rectangular Bravais lattice according to Eq. (3.161) yields a primitive rectangular netplane as sketched in Figure 3.28d. Alternatively, one could use vectors $\underline{R}_1$ and $\underline{R}'_2$ as a

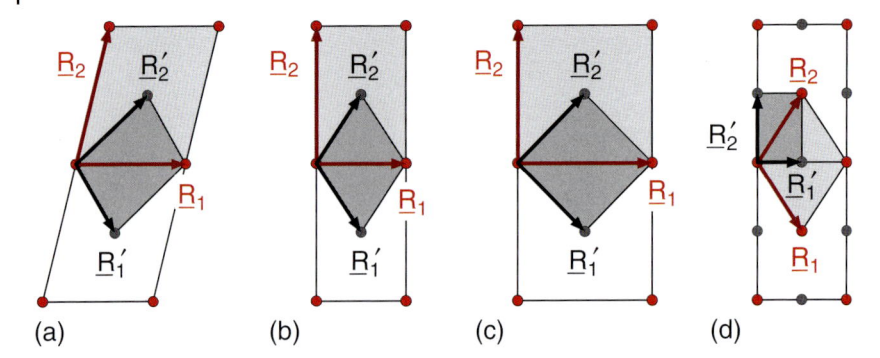

Figure 3.28 Morphological unit cells of centered Bravais lattices, (a) oblique, (b) p-rectangular, (c) square, and (d) c-rectangular. Lattice vectors $\underline{R}_1$, $\underline{R}_2$ refer to non-primitive and $\underline{R}'_1$, $\underline{R}'_2$ to primitive lattice descriptions, respectively. The cells are emphasized in light (non-primitive) and dark (primitive) gray. Lattice points of the initial lattice are painted red while those added by centering are painted gray.

primitive description, which is shown in Figure 3.27c where centering results in an oblique netplane unless $|\underline{R}_1| = |\underline{R}'_2|$, corresponding to a hexagonal netplane, where centering leads to a rectangular netplane.

A centered **square** netplane is described by non-primitive lattice vectors $\underline{R}_1$ and $\underline{R}_2$ of equal length and $\gamma = 90°$. This results in primitive vectors $\underline{R}'_1$, $\underline{R}'_2$ according to Eq. (3.161), which are also of the same length and orthogonal, as sketched in Figure 3.28c. Thus, the centered square netplane is identical in its type with its primitive counterpart. **Hexagonal** netplanes are described by primitive lattice vectors $\underline{R}_1$ and $\underline{R}_2$ of identical length, $a = b$, and $\gamma = 60°$, $120°$. Thus, they can be considered special cases of centered rectangular netplanes according to the definition given earlier. This is also shown in Figure 3.27e, where the non-primitive rectangular unit cell of a hexagonal netplane with its orthogonal lattice vectors $\underline{R}_1$ and $\underline{R}'_2$ is included together with the primitive unit cell, given by $\underline{R}_1$ and $\underline{R}_2$. As a consequence of this close relationship between hexagonal and centered rectangular netplanes, centering hexagonal netplanes yields a primitive rectangular netplane analogous to that shown in Figure 3.28d.

This concludes the **symmetry classification** of netplanes and yields, altogether, **five** different types, oblique, primitive and centered rectangular, square, and hexagonal. The classification describes the **two-dimensional crystal system** given by five different **Bravais lattices** in two dimensions with definitions collected in Table 3.11. In this table all lattice vectors are assumed to be primitive and defined according to Eq. (3.160).

The listing in Table 3.11 shows a **hierarchy** of the different Bravais lattices, where the oblique lattice exhibits the lowest symmetry. Thus, primitive and centered rectangular Bravais lattices can always be considered as special cases of oblique lattices. Further, the square Bravais lattice can be thought of as a special case of both primitive and centered rectangular lattices, while the hexagonal Bravais lattice represents a special case of a centered rectangular lattice. The

3.7 Crystal Systems and Bravais Lattices in Two Dimensions

Table 3.11 Bravais lattice members described by lattice constants a, b, and angle γ.

Crystal system, Bravais lattice	Definitions
Oblique	$a \neq b$, any $\gamma \neq 60°, 90°, 120°$
Primitive rectangular	$a \neq b$, $\gamma = 90°$
Centered rectangular	$a \neq b$, $(\underline{R}_1 \underline{R}_2) = \pm 1/2\, a^2$
	or
	$a \neq b$, $(\underline{R}_1 \underline{R}_2) = \pm 1/2\, b^2$
	or
	$a = b$, any $\gamma \neq 60°, 90°, 120°$
Square	$a = b$, $\gamma = 90°$
Hexagonal	$a = b$, $\gamma = 60°, 120°$

distinction between the different lattice types becomes important when all symmetry elements of the netplanes are considered.

The **relationship** between the different **Bravais lattices** is further illustrated in Figure 3.29, which shows all possible types of two-dimensional Bravais lattices described by Minkowski-reduced lattice vectors $\underline{R}_1$ and $\underline{R}_2$, see Section 3.3. Here, lattice vector $\underline{R}_1$ is kept fixed and defines the x axis of the graph while all possible vectors $\underline{R}_2$ (constrained to be Minkowski-reduced) are obtained from points inside the gray shaded area: dark gray for obtuse and light gray for acute lattice representations. The figure shows very clearly the network of lines referring to Bravais lattices of different types, where transitions between the types occur at line intersections. Further, the distinction between acute, $60° \leq \gamma \leq 90°$, and obtuse lattice vector sets, $90° < \gamma \leq 120°$, becomes clear. Centered rectangular and hexagonal lattices yield equivalent lattice vector sets for $\cos(\gamma) = A/2$ and $\cos(\gamma) = -A/2$ (with $A = \min(R_1/R_2, R_2/R_1)$ according to Eq. (3.17)) corresponding to acute and obtuse vector sets, which can both be considered to be Minkowski-reduced. Here, crystallographers prefer the obtuse representation referring to the strict definition given in Eq. (3.16).

The **symmetry classification** of netplanes discussed earlier also has consequences for the possible shape of compact two-dimensional unit cells, the **Wigner–Seitz cells**, discussed in Section 2.3. These are of interest, in particular, for theoretical studies on electronic or vibronic properties of two-dimensionally periodic systems, such as single crystal surfaces, discussed below. The basic definition of a Wigner–Seitz cell in two dimensions starts from lattice vectors $\underline{R}_1$ and $\underline{R}_2$ of a netplane and subdivides the netplane area into identical cells enclosing general lattice points $\underline{R} = n_1 \underline{R}_2 + n_2 \underline{R}_2$. Each cell is defined by including all points $\underline{r}$ that are the nearest to $\underline{R}$ compared with all other lattice points. Then, adjacent cells about $\underline{R}$ and $\underline{R}'$ must be separated by sections of straight lines defined by $\underline{r}$ assuming equal distance with respect to $\underline{R}$ and $\underline{R}'$. Hence, two-dimensional Wigner–Seitz cells must be **polygonal** in shape. The edge points of these cells can be constructed for the five Bravais lattices by simple geometry considerations where we mention only the final result derived in [84].

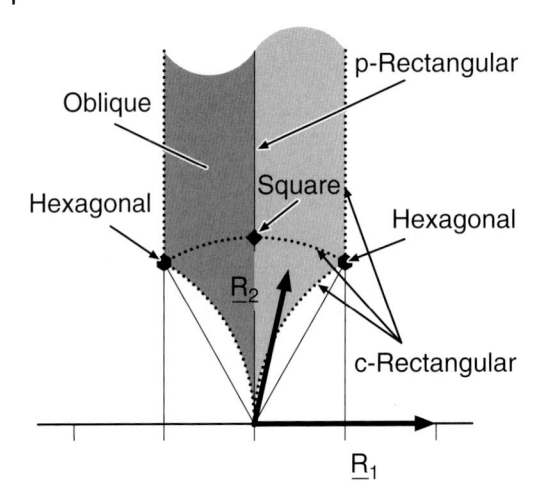

Figure 3.29 Minkowski-reduced lattice vectors of general netplanes. Lines of Bravais lattices of different types are labeled accordingly. The filled square and the two filled hexagons indicate square and hexagonal Bravais lattices, respectively.

Table 3.12 Edge vectors of a general two-dimensional Wigner–Seitz cell based on Minkowski-reduced lattice vectors $\underline{R}_1$, $\underline{R}_2$.

Edge vector $\underline{r}_i$	κ_{i1}	κ_{i2}	
$\underline{r}_1$	η_{21}	η_{12}	$\eta_{12} = \lambda\,[|\underline{R}_1|^2\,(|\underline{R}_2|^2 - \underline{R}_1\underline{R}_2)]$
$\underline{r}_2$	$-\eta_{21}$	$1 - \eta_{12}$	$\eta_{21} = \lambda\,[|\underline{R}_2|^2\,(|\underline{R}_1|^2 - \underline{R}_1\underline{R}_2)]$
$\underline{r}_3$	$\eta_{21} - 1$	η_{12}	$\lambda = 1/2\,[|\underline{R}_1|^2\,|\underline{R}_2|^2 - (\underline{R}_1\underline{R}_2)^2]^{-1}$
$\underline{r}_4\,(\equiv -\underline{r}_1)$	$-\eta_{21}$	$-\eta_{12}$	
$\underline{r}_5\,(\equiv -\underline{r}_2)$	η_{21}	$\eta_{12} - 1$	
$\underline{r}_6\,(\equiv -\underline{r}_3)$	$1 - \eta_{21}$	$-\eta_{12}$	

Let us consider a netplane with its periodicity described by Minkowski-reduced lattice vectors $\underline{R}_1$ and $\underline{R}_2$. Then its (two-dimensional) Wigner–Seitz cell about the netplane origin forms, in general, an irregular hexagon, see Figure 3.30, given by edge vectors

$$\underline{r}_i = \kappa_{i1}\,\underline{R}_1 + \kappa_{i2}\,\underline{R}_2, \quad i = 1, \ldots, 6 \qquad (3.162)$$

where the mixing coefficients κ_{ij} are listed in Table 3.12.

As a result, the edge points of Table 3.12 yield Wigner–Seitz cells shaped as

1) **irregular hexagons** for **oblique** netplanes, see Figure 3.31a;
2) **rectangles** (coinciding edges $\underline{r}_2$, $\underline{r}_3$ and $\underline{r}_5$, $\underline{r}_6$) for **primitive rectangular** netplanes, see Figure 3.31b;

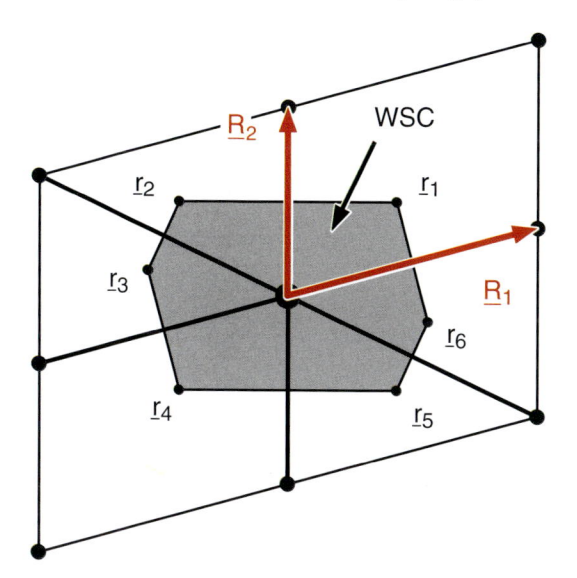

Figure 3.30 Wigner–Seitz cell (WSC) for a general netplane with lattice vectors $\underline{R}_1$, $\underline{R}_2$. The edges $\underline{r}_i$, i = 1, ..., 6 of the polygonal cell are labeled accordingly.

3) symmetrically **stretched hexagons** (two perpendicular mirror lines, one through two opposing edges) for **centered rectangular** netplanes, see Figure 3.31c;
4) **squares** (coinciding edges $\underline{r}_2$, $\underline{r}_3$ and $\underline{r}_5$, $\underline{r}_6$) for **square** netplanes, see Figure 3.31d;
5) **regular hexagons** for **hexagonal** netplanes, see Figure 3.31e.

3.8
Crystallographic Classification of Netplanes and Monolayers

The symmetry classification discussed in Section 3.7 gave a first overview of the possible types of netplanes, yielding the two-dimensional crystal systems with their five Bravais lattices. Here, the discrimination between the lattice types was based only on the interplay of the translational and rotational symmetry of netplanes. Monolayers are described in their translational symmetry by corresponding netplanes and can, therefore, also be classified according to the five Bravais lattices. However, a **complete crystallographic classification** of all possible monolayer types must take into account the full set of point symmetry elements, including glide reflections, together with translational symmetry. This can be achieved formally by considering, for each of the five Bravais lattices, the appropriate symmetry operations of all 10 point symmetry groups, listed in Tables 3.9 and 3.10, as well as glide reflections. Here, the symmetry of a monolayer may turn out to be lower than that of its corresponding netplane due to the appearance

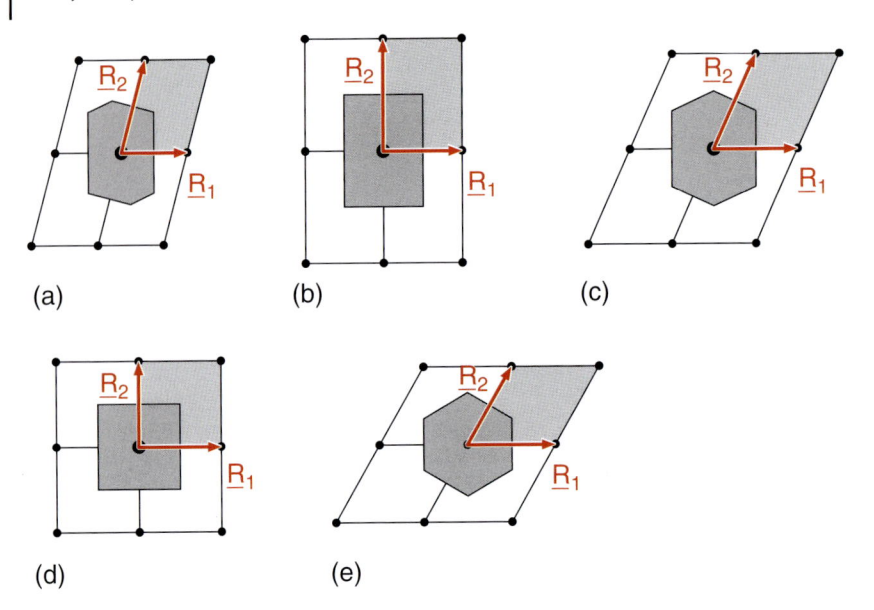

Figure 3.31 Wigner–Seitz cells (WSCs) of the five Bravais lattices in two dimensions, (a) oblique, (b) p-rectangular, (c) c-rectangular, (d) square, and (e) hexagonal. The WSCs, emphasized in dark gray, are compared with morphological unit cells spanned by lattice vectors $\underline{R}_1$, $\underline{R}_2$, shown in lighter gray.

of additional monolayer atoms at off-symmetry sites of the elementary unit cell given by the netplane. Altogether, the analysis leads to **17** different symmetry types describing netplanes and monolayers, the so-called **two-dimensional space groups**, which will be discussed in Sections 3.8.1–3.8.5. These sections are rather formal and filled with geometric and group theoretical details. Thus, readers, who are less interested in the mathematics behind two-dimensional space groups, may skip these sections and move to the conclusion of the analysis, given in Section 3.8.6.

The classification distinguishes between symmorphic and non-symmorphic space groups. **Symmorphic space groups** result from considering any of the 10 point symmetry groups of Table 3.9 (i.e., excluding glide reflections as initial symmetry elements) and adding the translation symmetry group of the Bravais lattice defined by lattice vectors $\underline{R}_1$ and $\underline{R}_2$. As a consequence of combining point symmetry elements, centered at different points of the Bravais lattice, glide reflections may also appear as symmetry elements in symmorphic space groups. However, they do not act as symmetry generating elements. In contrast, **non-symmorphic space groups** arise from applying glide reflections as initial symmetry elements of a netplane or monolayer and combining them with selected point symmetry operations as well as with translational symmetry. Here, only **rectangular** and **square** Bravais lattices need to be considered since glide reflection symmetry is connected with rectangular symmetry, as discussed in Section 3.6.5.

In the following sections, we consider the center $\underline{r}_o$ of the **highest** point symmetry group **G** of a netplane to coincide with the netplane origin. (The highest point symmetry group is defined as the group with the largest number of point symmetry operations that transform the netplane into itself.) Further, the symmetry elements of each space group will be sketched in **symbolic** form, according to the conventional crystallographic notation used in the International Tables for Crystallography [33] (denoted as **ITC**). With this notation, mirror lines are indicated by thick lines and glide lines by dashed lines while twofold rotation (and inversion) centers are shown by ellipses, and three-, four-, and sixfold rotation centers by filled triangles, squares, and hexagons, respectively. In addition, point symmetry and space groups will be denoted by both the international notation according to **Hermann–Mauguin** and by the **Schönflies** notation, see Section 2.4, where the latter will be put in parentheses. As examples, we mention the point symmetry group **4** (C_4) and the space group **p31m** (C_{3v}^2).

3.8.1
Oblique Netplanes

The oblique Bravais lattice with the **largest** number of symmetry elements is described by point symmetry group **2** (C_2). The corresponding **symmorphic** space group is denoted as **p2** (C_2) and is included as No. 2 in the ITC. Figure 3.32a shows a morphological unit cell, where the circular patterns in Figure 3.32a represent example atoms or groups of atoms (**motifs**). The motifs are placed at **general positions** of the morphological unit cell, that is, not located at symmetry points of the cell. As a consequence, each operation of the point symmetry group **2** (C_2) generates a new motif at a different location, where the transformation is illustrated by the orientation of the black wedges inside the circles of the patterns. Motifs illustrating symmetry elements of the unit cell will be used throughout the following discussion. Figure 3.32b sketches all symmetry elements of the unit cell in a symbolic form, where centers of twofold rotation axes (inversion centers) are indicated by ellipses, according to the conventional crystallographic notation [33].

According to Table 3.10 of Section 3.6.6, point symmetry group **2** (C_2) contains only one subgroup, the identity group **1** (C_1), which can be combined with translational symmetry to yield another space group, where corresponding monolayers of the oblique Bravais lattice exhibit only translation symmetry. This **symmorphic** space group is denoted as **p1** (C_1) and is included as No. 1 in the ITC. Figure 3.33a shows a morphological unit cell of the space group with the circular patterns illustrating its (missing) symmetry while Figure 3.33b sketches all (i.e., none) symmetry elements of the unit cell and is added only for completeness.

3.8.2
Primitive Rectangular Netplanes

The primitive rectangular Bravais lattice with the **largest** number of symmetries is described by point symmetry group **2mm** (C_{2v}). The corresponding **symmorphic** space group is denoted as **p2mm** (C_{2v}^1), or in short as **pmm**, and is included as

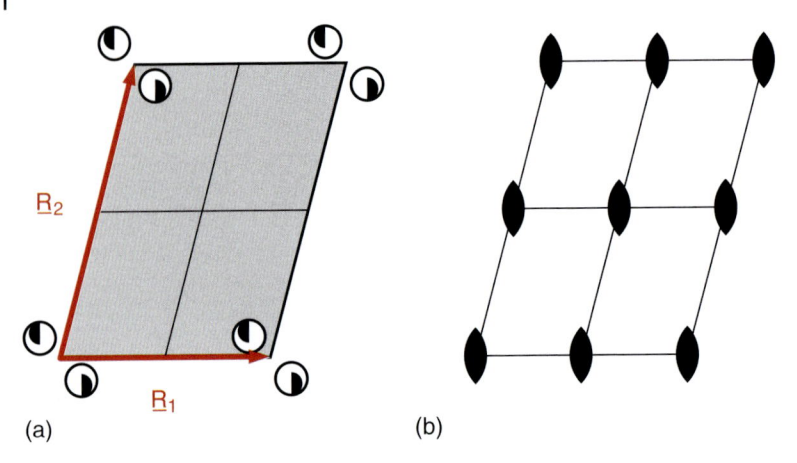

Figure 3.32 Symmorphic oblique space group **p2**, No. 2 in the ITC. (a) Morphological unit cell (in gray) spanned by lattice vectors $\underline{R}_1$, $\underline{R}_2$ with example motifs shown by circular patterns. (b) Sketch of all symmetry elements of the unit cell according to ITC.

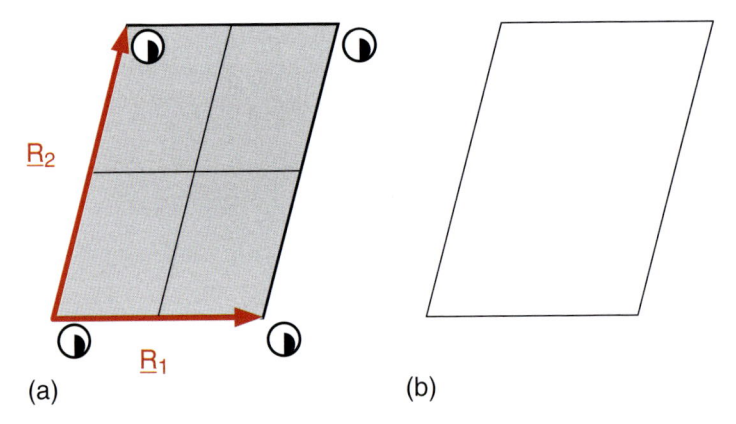

Figure 3.33 Symmorphic oblique space group **p1**, No. 1 in the ITC. (a) Morphological unit cell (in gray) spanned by lattice vectors $\underline{R}_1$, $\underline{R}_2$ with example motifs shown by circular patterns. (b) Sketch of all (= no) symmetry elements of the unit cell.

No. 6 in the ITC. Figure 3.34a shows a morphological unit cell of the space group with the circular patterns illustrating its symmetry. Figure 3.34b sketches all symmetry elements of the unit cell following the ITC notation given in Section 3.8.

According to Table 3.10, point symmetry group **2mm** ($\mathbf{C}_{2v}$) contains three subgroups, **1** ($\mathbf{C}_1$), **2** ($\mathbf{C}_2$), and **m** ($\mathbf{C}_s$), which can be combined with translational symmetry of the primitive rectangular Bravais lattice to yield three other space groups with less symmetry elements than the initial space group **p2mm** ($\mathbf{C}_{2v}^1$). Here the cases of point symmetry groups **1** ($\mathbf{C}_1$) and **2** ($\mathbf{C}_2$), resulting in space groups **p1** ($\mathbf{C}_1$) and **p2** ($\mathbf{C}_2$), respectively, have already been discussed in Section 3.8.1.

Point symmetry group **m** ($\mathbf{C}_s$) combined with a primitive rectangular Bravais lattice results in the **symmorphic** space group denoted as **p1m1** ($\mathbf{C}_s^1$), or in short

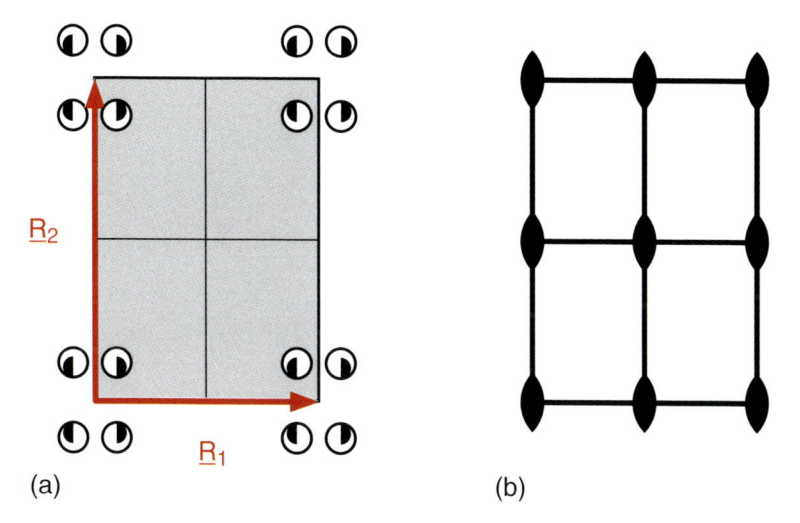

Figure 3.34 Symmorphic primitive rectangular space group **p2mm**, No. 6 in the ITC. (a) Morphological unit cell (in gray) spanned by lattice vectors $\underline{R}_1$, $\underline{R}_2$ with example motifs shown by circular patterns. (b) Sketch of all symmetry elements of the unit cell according to ITC.

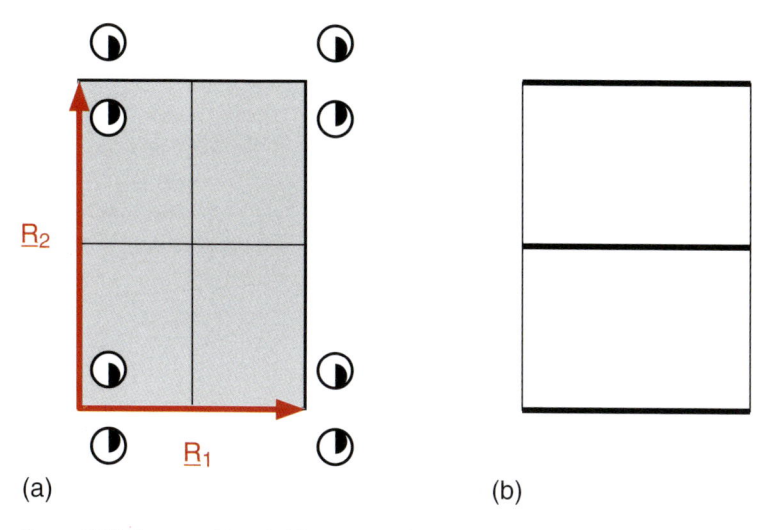

Figure 3.35 Symmorphic primitive rectangular space group **p1m1**, No. 3 in the ITC. (a) Morphological unit cell (in gray) spanned by lattice vectors $\underline{R}_1$, $\underline{R}_2$ with example motifs shown by circular patterns. (b) Sketch of all symmetry elements of the unit cell according to ITC.

as **pm**, and is included as No. 3 in the ITC. Figure 3.35a shows a morphological unit cell of the space group with the circular patterns illustrating its symmetry. Figure 3.35b sketches all symmetry elements of the unit cell following the ITC notation given in Section 3.8.

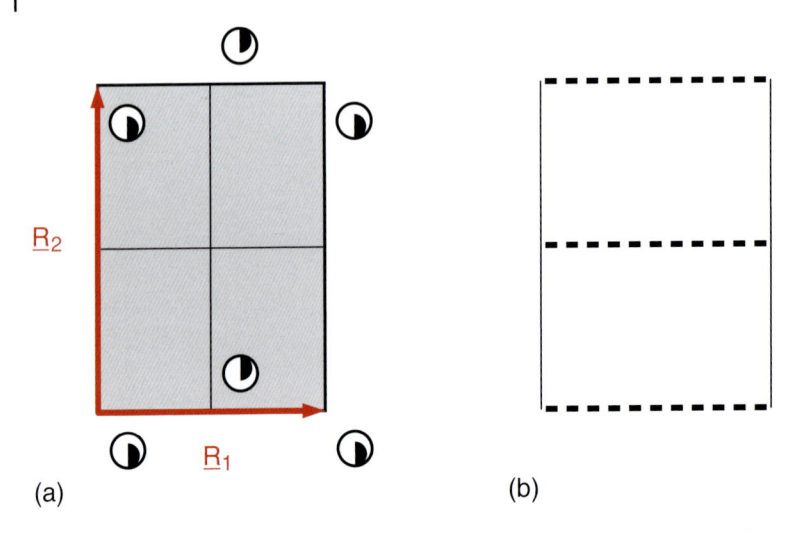

Figure 3.36 Non-symmorphic primitive rectangular space group **p1g1**, No. 4 in the ITC. (a) Morphological unit cell (in gray) spanned by lattice vectors $\underline{R}_1$, $\underline{R}_2$ with example motifs shown by circular patterns. (b) Sketch of all symmetry elements of the unit cell according to ITC.

In Section 3.6.5 it was shown that the existence of glide line symmetry is always connected with rectangular Bravais lattices. In particular, glide reflections as generating symmetry elements can be combined with primitive rectangular Bravais lattices to yield the corresponding **non-symmorphic** space groups.

Combining the infinite set of parallel glide lines, shown to exist in primitive rectangular Bravais lattices, see Section 3.6.5 and Figure 3.22, with translational symmetry leads to the **non-symmorphic** space group denoted as **p1g1** ($\mathbf{C}_S^2$), or in short as **pg**, and is included as No. 4 in the ITC. Figure 3.36a shows a morphological unit cell of the space group with the circular patterns illustrating its symmetry. Figure 3.36b sketches all symmetry elements of the unit cell following the ITC notation given in Section 3.8.

Infinite sets of parallel glide lines in primitive rectangular Bravais lattices can also be combined with symmetry described by point symmetry group **2** ($\mathbf{C}_2$), which includes twofold rotation (inversion) symmetry. As discussed in Section 3.6.5, there are two possible arrangements of inversion centers with respect to parallel glide lines,

1) All inversion centers lie on glide lines, see Figure 3.24a. This was shown to yield additional mirror lines perpendicular to the glide lines, resulting in a **non-symmorphic** space group, denoted as **p2mg** ($\mathbf{C}_{2v}^2$), or in short as **pmg**, and is included as No. 7 in the ITC. Figure 3.37a shows a morphological unit cell of the space group with the circular patterns illustrating its symmetry. Figure 3.37b sketches all symmetry elements of the unit cell following the ITC notation given in Section 3.8. Note that the unit cells of Figure 3.37 are

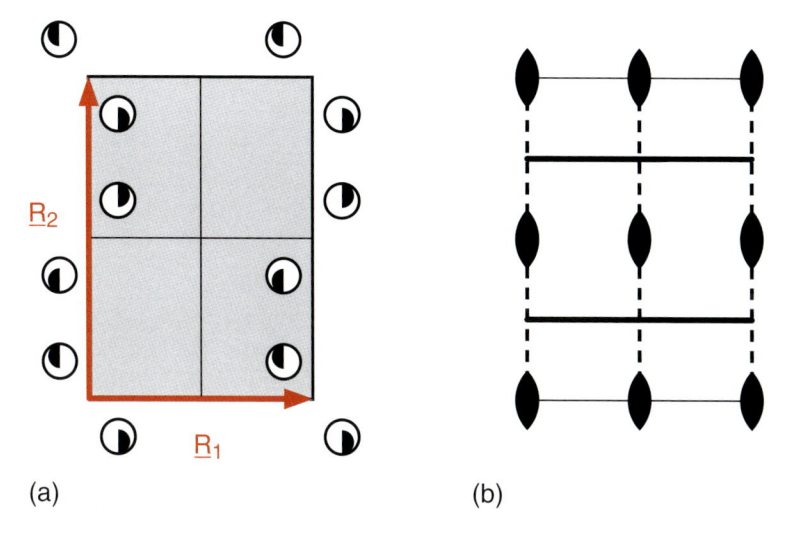

3.8 Crystallographic Classification of Netplanes and Monolayers | 155

Figure 3.37 Non-symmorphic primitive rectangular space group **p2mg**, No. 7 in the ITC. (a) Morphological unit cell (in gray) spanned by lattice vectors $\underline{R}_1$, $\underline{R}_2$ with example motifs shown by circular patterns. (b) Sketch of all symmetry elements of the unit cell according to ITC.

rotated by 90° with respect to that of Figure 3.24a in order to comply with the conventional notation used in the ITC.

2) All inversion centers lie in the middle between glide lines, see Figure 3.24b. This was shown to yield additional glide lines perpendicular to the initial glide lines resulting in an orthogonal set of glide lines. This glide line network combined with a primitive rectangular Bravais lattice leads to another **non-symmorphic** space group, denoted as **p2gg** (C_{2v}^3), or in short as **pgg**, and is included as No. 8 in the ITC. Figure 3.38a shows a morphological unit cell of the space group with the circular patterns illustrating its symmetry. Figure 3.38b sketches all symmetry elements of the unit cell following the ITC notation given in Section 3.8. Note that the unit cells of Figure 3.38 are shifted by 1/4 $\underline{R}_2$ with respect to that of Figure 3.24b in order to comply with the conventional notation used in the ITC.

3.8.3
Centered Rectangular Netplanes

The centered rectangular Bravais lattice with the **largest** number of symmetries is, as the primitive rectangular lattice, described by point symmetry group **2mm** (C_{2v}). The corresponding **symmorphic** space group is denoted as **c2mm** (C_{2v}^4), or in short as **cmm**, and is included as No. 9 in the ITC. Figure 3.39a shows two morphological unit cells of the space group, with the circular patterns illustrating its symmetry. The primitive skewed cell,

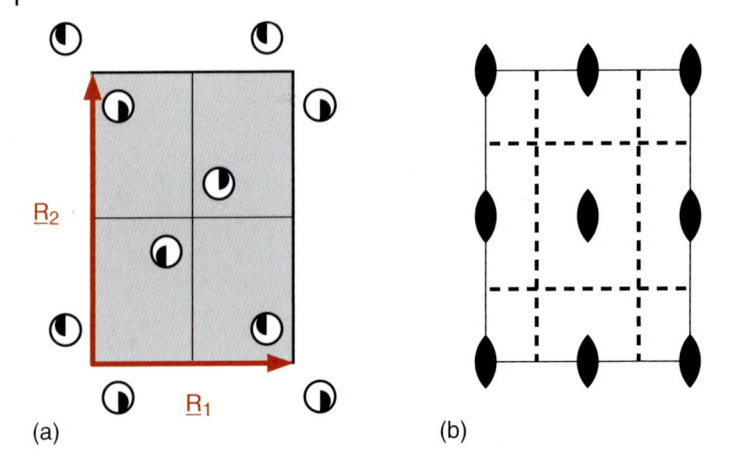

Figure 3.38 Non-symmorphic primitive rectangular space group **p2gg**, No. 8 in the ITC. (a) Morphological unit cell (in gray) spanned by lattice vectors $\underline{R}_1$, $\underline{R}_2$ with example motifs shown by circular patterns. (b) Sketch of all symmetry elements of the unit cell according to ITC.

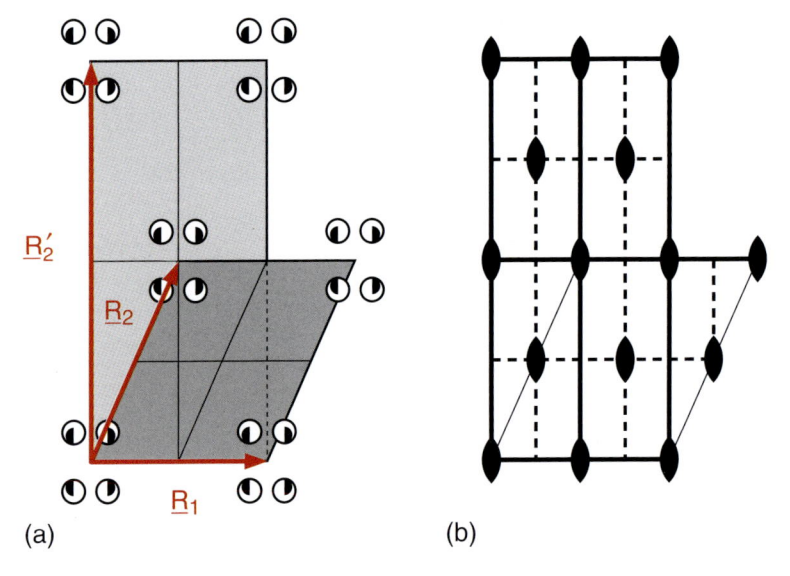

Figure 3.39 Symmorphic centered rectangular space group **c2mm**, No. 9 in the ITC. (a) Morphological unit cells (in gray) spanned by lattice vectors $\underline{R}_1$, $\underline{R}_2$ (skewed cell) and $\underline{R}_1$, $\underline{R}'_2$ (rectangular cell) with example motifs shown by circular patterns. (b) Sketch of all symmetry elements of the unit cell according to ITC.

defined by lattice vectors $\underline{R}_1$ and $\underline{R}_2$, is emphasized by dark gray painting, while the non-primitive rectangular cell, given by vectors $\underline{R}_1$ and $\underline{R}'_2$ and preferred by crystallographers, is painted light gray. These two cells will be shown for all centered lattices discussed in the following. Figure 3.39b sketches

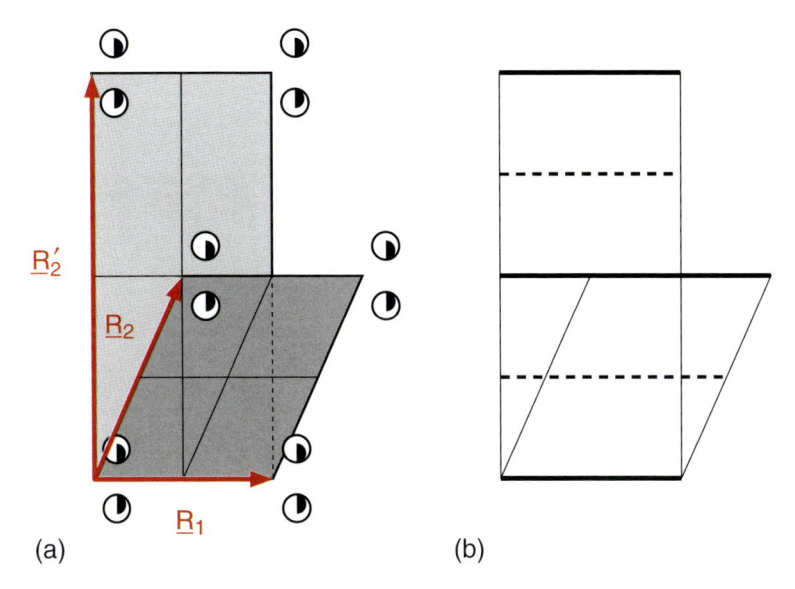

Figure 3.40 Symmorphic centered rectangular space group **c1m1**, No. 5 in the ITC. (a) Morphological unit cells (in gray) spanned by lattice vectors $\underline{R}_1$, $\underline{R}_2$ (skewed cell) and $\underline{R}_1$, $\underline{R}_2'$ (rectangular cell) with example motifs shown by circular patterns. (b) Sketch of all symmetry elements of the unit cell according to ITC.

all symmetry elements of the unit cell following the ITC notation given in Section 3.8.

According to Table 3.10, point symmetry group **2mm** (**C$_{2v}$**) contains three subgroups, **1** (**C$_1$**), **2** (**C$_2$**), and **m** (**C$_s$**), which can be combined with translational symmetry of the centered rectangular Bravais lattice to yield three other space groups. Analogous to the discussion for the primitive rectangular case, see Section 3.8.2, point symmetry groups **1** (**C$_1$**) and **2** (**C$_2$**) can be ignored since they have been treated in Section 3.8.1.

Point symmetry group **m** (**C$_s$**) combined with a centered rectangular Bravais lattice results in the **symmorphic** space group denoted as **c1m1** (**C$_s^2$**), or in short as **cm**, and is included as No. 5 in the ITC. Figure 3.40a shows two morphological unit cells (primitive skewed and non-primitive rectangular) of the space group with the circular patterns illustrating its symmetry. Figure 3.40b sketches all symmetry elements of the unit cell following the ITC notation given in Section 3.8.

3.8.4
Square Netplanes

The square Bravais lattice with the **largest** number of symmetries is described by point symmetry group **4mm** (**C$_{4v}$**). The corresponding **symmorphic** space group

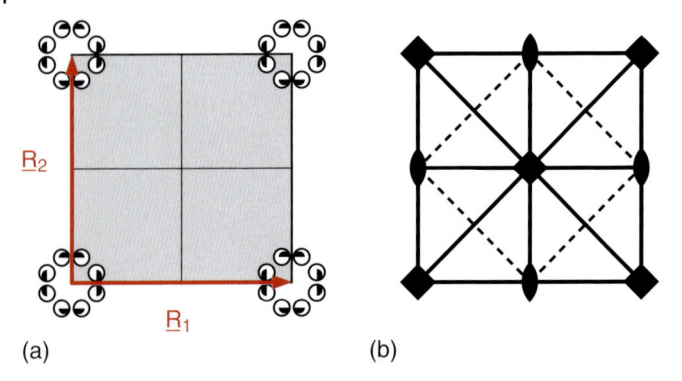

Figure 3.41 Symmorphic square space group **p4mm**, No. 11 in the ITC. (a) Morphological unit cell (in gray) spanned by lattice vectors $\underline{R}_1$, $\underline{R}_2$ with example motifs shown by circular patterns. (b) Sketch of all symmetry elements of the unit cell according to ITC.

is denoted as **p4mm** ($\mathbf{C}_{4v}^1$), or in short as **p4m**, and is included as No. 11 in the ITC. Figure 3.41a shows a morphological unit cell of the space group with the circular patterns illustrating its symmetry. Figure 3.41b sketches all symmetry elements of the unit cell following the ITC notation given in Section 3.8.

According to Table 3.10, point symmetry group **4mm** ($\mathbf{C}_{4v}$) contains five subgroups, **1** ($\mathbf{C}_1$), **2** ($\mathbf{C}_2$), **4** ($\mathbf{C}_4$), **m** ($\mathbf{C}_s$), and **2mm** ($\mathbf{C}_{2v}$), which can be combined with translational symmetry of the square Bravais lattice to yield five other space groups. Here, only point symmetry groups **4** ($\mathbf{C}_4$) need to be considered since the other groups have been dealt with earlier, see Sections 3.8.1–3.8.3.

Point symmetry group **4** ($\mathbf{C}_4$) combined with a square Bravais lattice results in the **symmorphic** space group denoted as **p4** ($\mathbf{C}_4$) and is included as No. 10 in the ITC. Figure 3.42a shows a morphological unit cell of the space group with the circular patterns illustrating its symmetry. Figure 3.42b sketches all symmetry elements of the unit cell following the ITC notation given in Section 3.8.

Infinite sets of parallel glide lines in square Bravais lattices can also be combined with symmetry described by point symmetry group **4** ($\mathbf{C}_4$), which includes two- and fourfold rotation symmetry. We mention without further proof that there is only one choice, where the corresponding **non-symmorphic** space group is denoted as **p4gm** ($\mathbf{C}_{4v}^2$), or in short as **p4g**, and is included as No. 12 in the ITC. Figure 3.43a shows a morphological unit cell of the space group with the circular patterns illustrating its symmetry. Figure 3.43b sketches all symmetry elements of the unit cell following the ITC notation given in Section 3.8.

3.8.5
Hexagonal Netplanes

In Section 3.7 it was shown that a **hexagonal** lattice, defined by primitive lattice vectors $\underline{R}_1$ and $\underline{R}_2$ in acute representation ($\gamma = 60°$), can be described

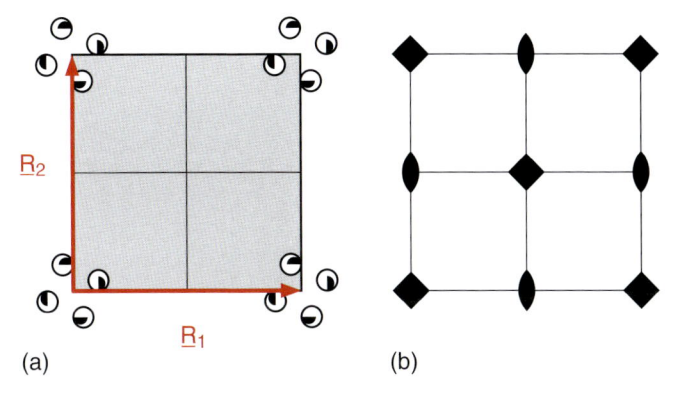

3.8 Crystallographic Classification of Netplanes and Monolayers | 159

Figure 3.42 Symmorphic square space group **p4**, No. 10 in the ITC. (a) Morphological unit cell (in gray) spanned by lattice vectors $\underline{R}_1$, $\underline{R}_2$ with example motifs shown by circular patterns. (b) Sketch of all symmetry elements of the unit cell according to ITC.

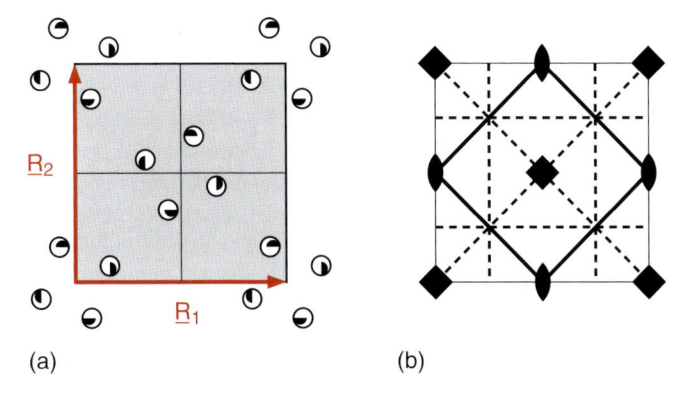

Figure 3.43 Non-symmorphic square space group **p4gm**, No. 12 in the ITC. (a) Morphological unit cell (in gray) spanned by lattice vectors $\underline{R}_1$, $\underline{R}_2$ with example motifs shown by circular patterns. (b) Sketch of all symmetry elements of the unit cell according to ITC.

alternatively as a **centered rectangular** lattice with non-primitive lattice vectors $\underline{R}_1$ and $\underline{R}'_2 = 2\,\underline{R}_2 - \underline{R}_1$, which are orthogonal. Therefore, all symmetry diagrams for hexagonal space groups, discussed in the following, will be shown with **two** morphological unit cells of the space group, analogous to the centered rectangular case in Section 3.8.3. The primitive rhombic cell, defined by lattice vectors $\underline{R}_1$ and $\underline{R}_2$, is emphasized by dark gray, and the non-primitive rectangular cell, given by vectors $\underline{R}_1$ and $\underline{R}'_2$ and preferred by crystallographers, is painted light gray.

The hexagonal Bravais lattice with the **largest** number of symmetries is described by point symmetry group **6mm** (C_{6v}). The corresponding **symmorphic** space group is denoted as **p6mm** (C_{6v}), or in short as **p6m**, and is included as No. 17 in the ITC. Figure 3.44a shows two morphological unit cells (primitive rhombic and non-primitive rectangular) of the space group with the circular

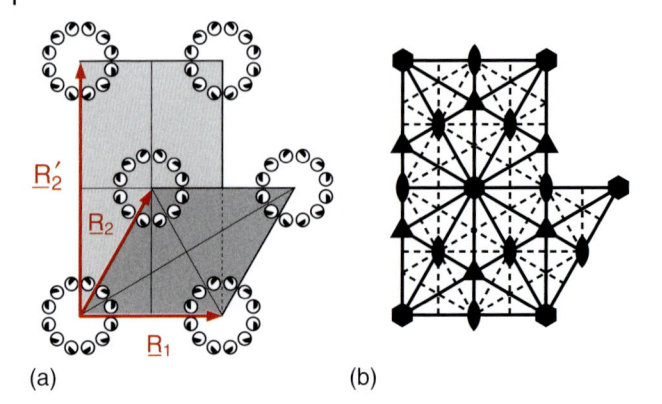

Figure 3.44 Symmorphic hexagonal space group **p6mm**, No. 17 in the ITC. (a) Morphological unit cells (in gray) spanned by lattice vectors $\underline{R}_1$, $\underline{R}_2$ (rhombohedral cell) and $\underline{R}_1$, $\underline{R}_2'$ (rectangular cell) with example motifs shown by circular patterns. (b) Sketch of all symmetry elements of the unit cell according to ITC.

patterns illustrating its symmetry. Figure 3.44b sketches all symmetry elements of the unit cell following the ITC notation given in Section 3.8.

According to Table 3.10, point symmetry group **6mm** (C_{6v}) contains seven subgroups, **1** (C_1), **2** (C_2), **3** (C_3), **6** (C_6), **m** (C_s), **2mm** (C_{2v}), and **3m1, 31m** (C_{3v}), which can be combined with translational symmetry of the hexagonal Bravais lattice to yield other space groups. Here, only point symmetry groups **3** (C_3), **6** (C_6), and **3m1, 31m** (C_{3v}) need to be considered since the other groups have been dealt with earlier, see Sections 3.8.1–3.8.3.

Point symmetry group **3** (C_3) combined with a hexagonal Bravais lattice results in the **symmorphic** space group denoted as **p3** (C_3) and is included as No. 13 in the ITC. Figure 3.45a shows two morphological unit cells (primitive and nonprimitive) of the space group with the circular patterns illustrating its symmetry. Figure 3.45b sketches all symmetry elements of the unit cell following the ITC notation given in Section 3.8.

Point symmetry group **6** (C_6) combined with a hexagonal Bravais lattice results in the **symmorphic** space group denoted as **p6** (C_6) and is included as No. 16 in the ITC. Figure 3.46a shows two morphological unit cells of the space group with the circular patterns illustrating its symmetry. Figure 3.46b sketches all symmetry elements of the unit cell following the ITC notation given in Section 3.8.

Point symmetry group **3m1, 31m** (C_{3v}) can also be combined with a hexagonal Bravais lattice. Here, the two variants depend on the orientation of the mirror lines with respect to the primitive lattice vectors $\underline{R}_1$ and $\underline{R}_2$ (assuming an acute representation, $\gamma = 60°$),

1) Mirror lines point between lattice vectors $\underline{R}_1$ and $\underline{R}_2$, resulting in angles of 30° between lattice vectors and mirror lines, described by point symmetry group variant **3m1** (C_{3v}). This results in the **symmorphic** space group denoted

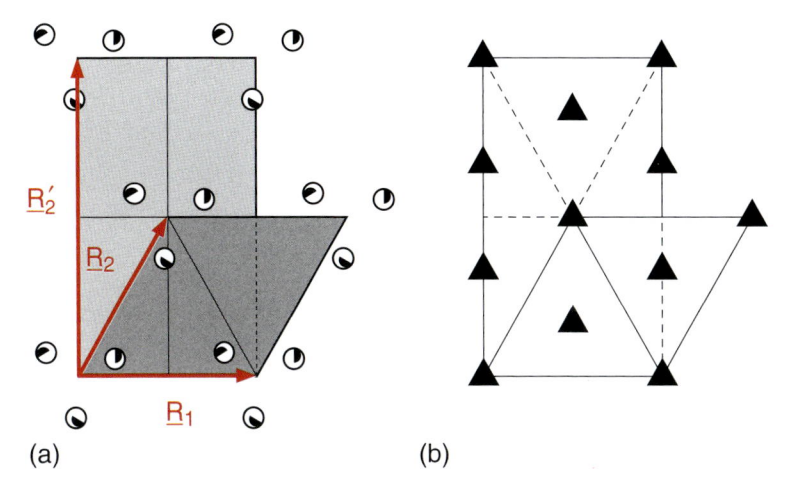

Figure 3.45 Symmorphic hexagonal space group **p3**, No. 13 in the ITC. (a) Morphological unit cells (in gray) spanned by lattice vectors $\underline{R}_1$, $\underline{R}_2$ (rhombohedral cell) and $\underline{R}_1$, $\underline{R}'_2$ (rectangular cell) with example motifs shown by circular patterns. (b) Sketch of all symmetry elements of the unit cell according to ITC.

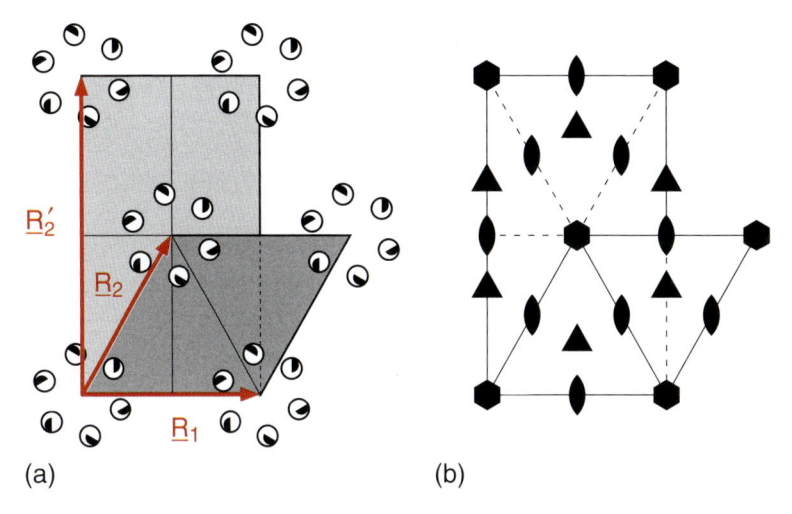

Figure 3.46 Symmorphic hexagonal space group **p6**, No. 16 in the ITC. (a) Morphological unit cells (in gray) spanned by lattice vectors $\underline{R}_1$, $\underline{R}_2$ (rhombohedral cell) and $\underline{R}_1$, $\underline{R}'_2$ (rectangular cell) with example motifs shown by circular patterns. (b) Sketch of all symmetry elements of the unit cell according to ITC.

as **p3m1** ($\mathbf{C}^1_{3v}$), which is included as No. 14 in the ITC. Figure 3.47a shows two morphological unit cells of the space group with the circular patterns illustrating its symmetry. Figure 3.47b sketches all symmetry elements of the unit cell following the ITC notation given in Section 3.8.

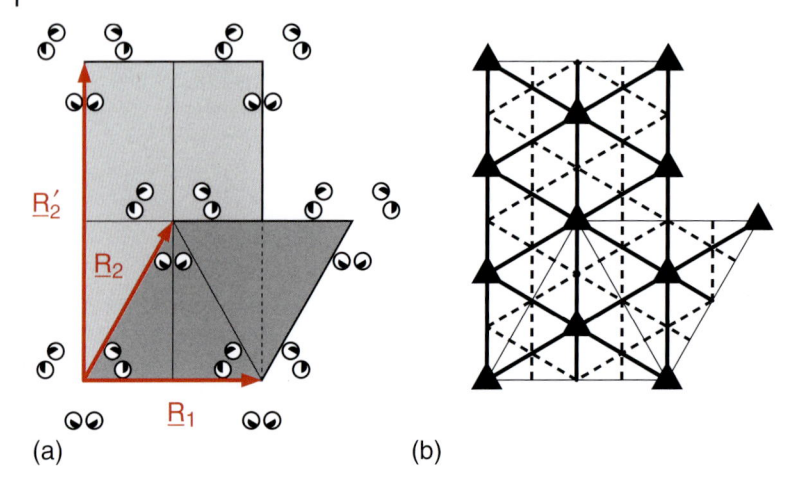

Figure 3.47 Symmorphic hexagonal space group **p3m1**, No. 14 in the ITC. (a) Morphological unit cells (in gray) spanned by lattice vectors $\underline{R}_1$, $\underline{R}_2$ (rhombohedral cell) and $\underline{R}_1$, $\underline{R}_2'$ (rectangular cell) with example motifs shown by circular patterns. (b) Sketch of all symmetry elements of the unit cell according to ITC.

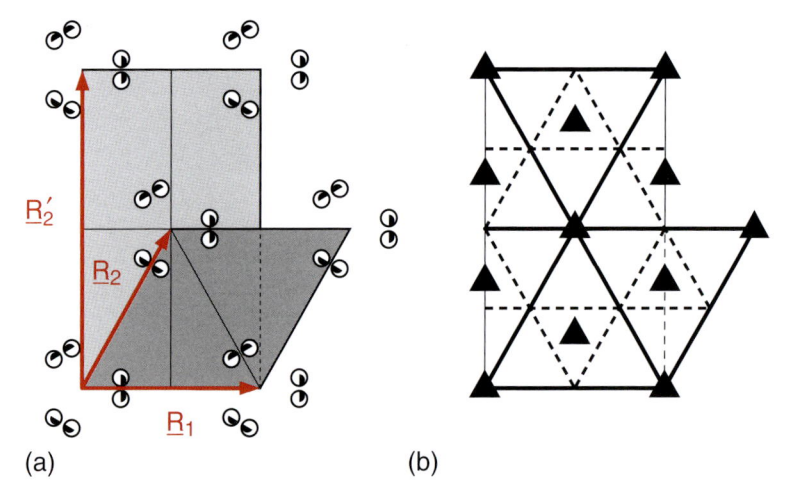

Figure 3.48 Symmorphic hexagonal space group **p31m**, No. 15 in the ITC. (a) Morphological unit cells (in gray) spanned by lattice vectors $\underline{R}_1$, $\underline{R}_2$ (rhombohedral cell) and $\underline{R}_1$, $\underline{R}_2'$ (rectangular cell) with example motifs shown by circular patterns. (b) Sketch of all symmetry elements of the unit cell according to ITC.

2) Mirror lines coincide with lattice vectors $\underline{R}_1$ and $\underline{R}_2$, described by point symmetry group variant **31m** ($\mathbf{C}_{3v}$). This results in the **symmorphic** space group denoted as **p31m** ($\mathbf{C}_{3v}^2$), which is included as No. 15 in the ITC. Figure 3.48a shows two morphological unit cells of the space group with the circular patterns illustrating its symmetry. Figure 3.48b sketches all symmetry elements of the unit cell following the ITC notation given in Section 3.8.

3.8 Crystallographic Classification of Netplanes and Monolayers | 163

Table 3.13 Properties of the 17 space groups in two dimensions.

ITC no.	Space group	Point group	Bravais lattice
1	p1 (C_1)	1 (C_1)	Oblique
2+	p2 (C_2)	2 (C_2)	Oblique
3	p1m1/pm (C_S^1)	m (C_s)	p-Rectangular
4*	p1g1/pg (C_S^2)	1 (C_1)	p-Rectangular
5	c1m1/cm (C_S^2)	m (C_s)	c-Rectangular
6+	p2mm/pmm (C_{2v}^1)	2mm (C_{2v})	p-Rectangular
7*	p2mg/pmg (C_{2v}^2)	2 (C_2)	p-Rectangular
8*	p2gg/pgg (C_{2v}^3)	2 (C_2)	p-Rectangular
9+	c2mm/cmm (C_{2v}^4)	2mm (C_{2v})	c-Rectangular
10	p4 (C_4)	4 (C_4)	Square
11+	p4mm/p4m (C_{4v}^1)	4mm (C_{4v})	Square
12*	p4gm/p4g (C_{4v}^2)	4 (C_4)	Square
13	p3 (C_3)	3 (C_3)	Hexagonal
14	p3m1 (C_{3v}^1)	3m1 (C_{3v})	Hexagonal
15	p31m (C_{3v}^2)	31m (C_{3v})	Hexagonal
16	p6 (C_6)	6 (C_6)	Hexagonal
17+	p6mm/p6m (C_{6v})	6mm (C_{6v})	Hexagonal

The ITC number refers to the International Tables for Crystallography [33], where non-symmorphic space groups are labeled by asterisks (*) and space groups with highest point symmetry of a given Bravais lattice by (+). Space and point group names are listed in Hermann–Mauguin and Schönflies notation, the latter in parentheses. The Hermann–Mauguin notation of the space groups includes also short names separated by slashes.

3.8.6
Classification Overview

Sections 3.8.1–3.8.5 have covered all possible symmorphic and non-symmorphic space groups that are available for a classification of netplanes and monolayers by their symmetry behavior. Altogether, there are **17 different space groups** (13 symmorphic, 4 non-symmorphic) in **two dimensions**, where two space groups refer to the oblique, five to the primitive rectangular, two to the centered rectangular, three to the square, and five to the hexagonal Bravais lattice. Table 3.13 collects all two-dimensional space groups and their properties with the numbering scheme following the sequence in the ITC [33].

Figure 3.49 gives an overview over all **symmetry elements** inside morphological unit cells of netplanes and monolayers described by the 17 two-dimensional **space groups** where the sequence follows the numbering scheme and the symbolic notation used in the ITC [33]. This collects the sketches shown already in the extended discussion of space groups in Sections 3.8.1–3.8.5. Further details concerning symmetry properties of two-dimensional space groups can be found in Refs. [23, 33, 34, 40, 86, 87].

164 | *3 Crystal Layers: Two-Dimensional Lattices*

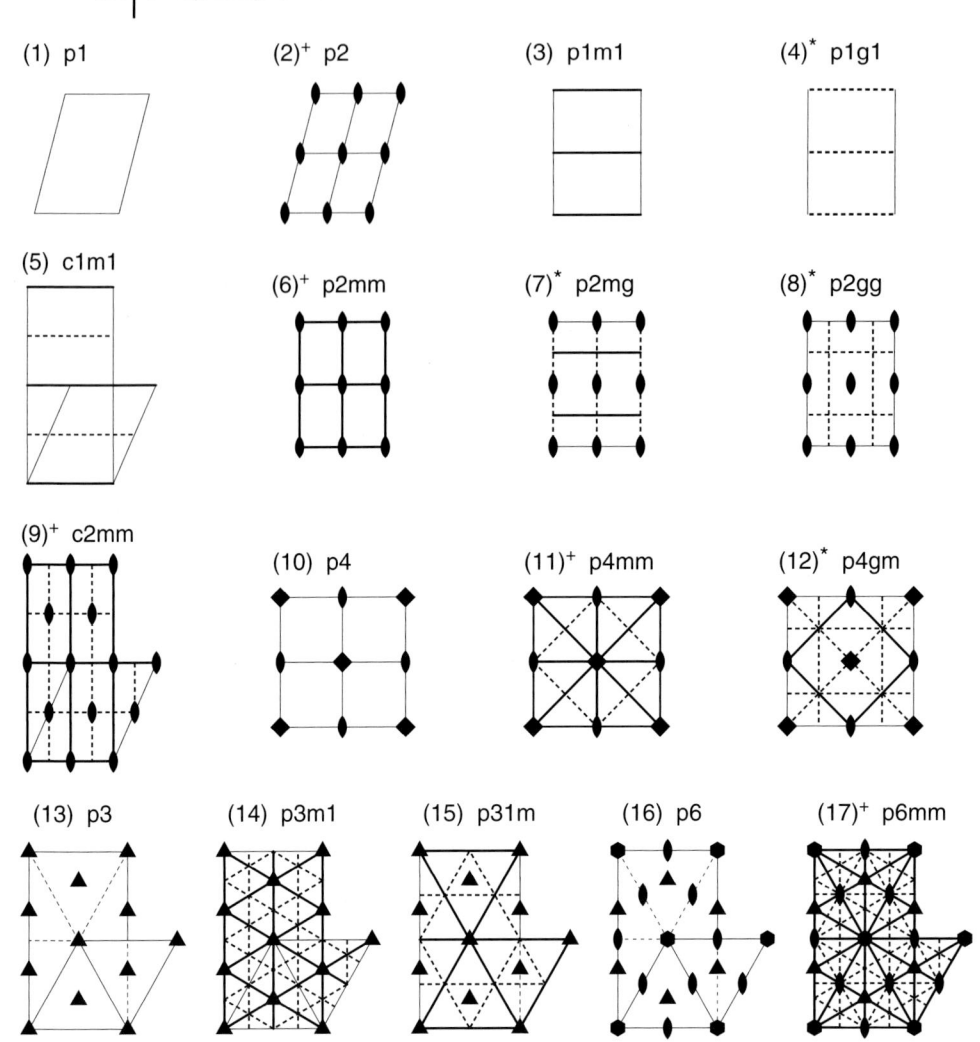

(1) p1   (2)⁺ p2   (3) p1m1   (4)* p1g1

(5) c1m1   (6)⁺ p2mm   (7)* p2mg   (8)* p2gg

(9)⁺ c2mm   (10) p4   (11)⁺ p4mm   (12)* p4gm

(13) p3   (14) p3m1   (15) p31m   (16) p6   (17)⁺ p6mm

Figure 3.49 Symmetry elements inside morphological unit cells of monolayers described by all 17 two-dimensional space groups. Non-symmorphic space groups are labeled by asterisks (*) and space groups of highest point symmetry for a given Bravais lattice by (⁺). Mirror lines are indicated by thick and glide lines by dashed lines. Two-, three-, four-, and sixfold rotation centers are shown by ellipses, triangles, squares, and hexagons, respectively, following the ITC notation.

3.9
Exercises

3.1 Show that the (a) sc, (b) fcc, (c) bcc, and (d) hcp lattices can be built by stacking hexagonal netplanes. For each lattice, determine corresponding Miller indices $(h\,k\,l)$ of the netplanes and relationships between the intrinsic bulk lattice constant a, the netplane lattice constant $a_{(h\,k\,l)}$, and the distance between adjacent netplanes $d_{(h\,k\,l)}$.

3.2 Calculate netplane-adapted lattice vectors for $(h\,k\,l)$ monolayers described as
(a) fcc Ni (1 1 1)
(b) bcc Fe (1 1 0)
(c) fcc Pd (1 1 3)
(d) hcp Co (1 1 2)
(e) diamond (1 2 3)
(f) fcc (m 1 1), fcc (m 0 1), m > 1
For cubic crystals $(h\,k\,l)$ refers to the sc notation. Determine the atom densities of the monolayers and compare with those of the densest monolayers.

3.3 Find the densest and second densest monolayers of the (a) sc, (b) fcc, (c) bcc, and (d) hcp, (e) NaCl crystals. For each lattice, determine how many and which directions yield densest monolayers.

3.4 Determine the netplane-adapted lattice vectors and atom densities of monolayers of the graphite crystal with (a) (0 0 0 1), (b) (1 1 2 1), (c) (1 −2 1 1), (d) (1 −1 0 1) orientation (Miller–Bravais indices, obtuse bulk lattice vectors). Characterize their geometric structures. What are the corresponding three-index Miller indices?

3.5 Consider a hexagonal lattice with (a) obtuse and (b) acute lattice vector representation and netplanes with orientations given in three-index notation $(h\,k\,l)$. Determine Miller–Bravais indices $(l\,m\,n\,q)$ of the (0 0 1), (1 0 1), (1 1 1), (1 0 −1), (1 2 1), and (1 −3 5) oriented netplanes.

3.6 The hcp crystal can be defined by lattice vectors $\underline{R}_1, \underline{R}_2, \underline{R}_3$ and a basis of two atoms where

$$\underline{R}_1 = a\,(1,\,0,\,0), \quad \underline{R}_2 = a\,(-1/2,\,(\sqrt{3})/2,\,0\,), \quad \underline{R}_3 = c\,(0,\,0,\,1)$$
$$\underline{r}_1 = (0,\,0,\,0), \quad \underline{r}_2 = a\,(\,1/2,\,1/(\sqrt{12}),\,\sqrt{(2/3)}) \quad c/a = \sqrt{(8/3)}$$

with $\underline{R}_i$, i = 1, 2, 3 and $\underline{r}_1, \underline{r}_2$ in Cartesian coordinates. Show that Miller indices $(h\,k\,l)$ of monolayers which contain both types of atoms fulfill the Diophantine equation $2\,h + 4\,k + 3\,l = 6\,N$.

3.7 Consider the (1 2 3) oriented netplane of the (a) fcc, (b) bcc, (c) diamond, (c) cubic zincblende, and (d) graphite crystal. How many and which netplanes of the crystal are symmetrically equivalent?

3.8 Determine the two-dimensional morphological unit cells and Wigner–Seitz cells of the monolayers described in Exercise 3.2.

3.9 Determine the Miller indices $(h\,k\,l)$ of all symmetry equivalent netplanes derived from a given netplane $(h_o\,k_o\,l_o)$ of a (a) fcc, (b) bcc, and (c) hcp lattice.

3.10 Consider a crystal lattice described by initial lattice vectors $\underline{R}_{o1}, \underline{R}_{o2}, \underline{R}_{o3}$ and by $(h\,k\,l)$ netplane-adapted lattice vectors $\underline{R}_1, \underline{R}_2, \underline{R}_3$ where the transformation is given by

$$\begin{pmatrix}\underline{R}_1\\ \underline{R}_2\\ \underline{R}_3\end{pmatrix}=\begin{pmatrix}t_{11}&t_{12}&t_{13}\\ t_{21}&t_{22}&t_{23}\\ t_{31}&t_{32}&t_{33}\end{pmatrix}\cdot\begin{pmatrix}\underline{R}_{o1}\\ \underline{R}_{o2}\\ \underline{R}_{o3}\end{pmatrix}=\underline{\underline{T}}^{(h\,k\,l)}\cdot\begin{pmatrix}\underline{R}_{o1}\\ \underline{R}_{o2}\\ \underline{R}_{o3}\end{pmatrix}$$

with integer-valued matrix elements t_{ij}. The corresponding reciprocal lattice vectors of the two lattice representations are $\underline{G}_{o1}, \underline{G}_{o2}, \underline{G}_{o3}$ and $\underline{G}_1, \underline{G}_2, \underline{G}_3$, respectively, with

$$\begin{pmatrix}\underline{G}_1\\ \underline{G}_2\\ \underline{G}_3\end{pmatrix}=\begin{pmatrix}q_{11}&q_{12}&q_{13}\\ q_{21}&q_{22}&q_{23}\\ q_{31}&q_{32}&q_{33}\end{pmatrix}\cdot\begin{pmatrix}\underline{G}_{o1}\\ \underline{G}_{o2}\\ \underline{G}_{o3}\end{pmatrix}=\underline{\underline{Q}}^{(h\,k\,l)}\cdot\begin{pmatrix}\underline{G}_{o1}\\ \underline{G}_{o2}\\ \underline{G}_{o3}\end{pmatrix}$$

Determine the relationship between the transformation matrices $\underline{\underline{T}}^{(h\,k\,l)}$ and $\underline{\underline{Q}}^{(h\,k\,l)}$.

3.11 Consider a crystal lattice described by initial lattice vectors $\underline{R}_{o1}, \underline{R}_{o2}, \underline{R}_{o3}$ and by $(h\,k\,l)$ netplane-adapted lattice vectors $\underline{R}_1, \underline{R}_2, \underline{R}_3$ with a vector transformation $\underline{\underline{T}}^{(h\,k\,l)}$ as in Exercise 3.10. Determine the Miller indices $(h'\,k'\,l')$ corresponding to a vector transformation $\underline{\underline{T}}^{(h'\,k'\,l')}=(\underline{\underline{T}}^{(h\,k\,l)})^{-1}$.

3.12 Consider a crystal lattice described by lattice vectors $\underline{R}_{o1}, \underline{R}_{o2}, \underline{R}_{o3}$ and Miller indices $(h_o\,k_o\,l_o)$ with respect to the corresponding reciprocal lattice vectors. Further, transformed lattice vectors $\underline{R}_1, \underline{R}_2, \underline{R}_3$ with a vector transformation $\underline{\underline{T}}^{(h\,k\,l)}$ are assumed to provide an equivalent lattice description, analogous to Exercise 3.10, and yield Miller indices $(h\,k\,l)$. Determine the Miller index transformation matrix $\underline{\underline{M}}$ where

$$\begin{pmatrix}h\\ k\\ l\end{pmatrix}=\begin{pmatrix}m_{11}&m_{12}&m_{13}\\ m_{21}&m_{22}&m_{23}\\ m_{31}&m_{32}&m_{33}\end{pmatrix}\cdot\begin{pmatrix}h_o\\ k_o\\ l_o\end{pmatrix}=\underline{\underline{M}}\cdot\begin{pmatrix}h_o\\ k_o\\ l_o\end{pmatrix}$$

3.13 A netplane is defined in a Cartesian coordinate system (x, y) by lattice vectors

$$\underline{R}_1=a\,(0,\sqrt{3}),\quad \underline{R}_2=a/2\,(1,\sqrt{27})$$

Show by Minkowski reduction that the netplane is hexagonal.

3.14 Determine the Miller indices of polar and non-polar monolayers of the NaCl and CsCl crystals. Hint: Polar monolayers are monoatomic.

3.15 Show that the atom density $\rho_{(h\,k\,l)}$ of $(h\,k\,l)$ monolayers of sc, fcc, and bcc crystals (Miller indices in sc notation) is determined by $\rho_{(h\,k\,l)}=p\,(h^2+k^2+l^2)^{-1/2}$ with $p = 1, 4, 2$ for sc, fcc, and bcc, respectively.

3.16 Find structurally different monolayers of equal atom density for crystals with (a) fcc and (b) bcc lattice.

3.17 Consider a netplane with two inversion centers at $\underline{r}_{o1}$ and $\underline{r}_{o2}$. Show that the combination of two inversion operations $i(\underline{r}_{o2})\,i(\underline{r}_{o1})$ corresponds to a shift by $\pm 2\,(\underline{r}_{o2} - \underline{r}_{o1})$.

3.18 The xy plane (in Cartesian coordinates) includes two fourfold rotation axes along z whose centers are separated by vector $\underline{R}_1 = a\,(1, 0)$. Consider subsequent rotation operations by 90°, 180°, 270° about the two centers generating new rotation centers in the xy plane. Show that infinitely many rotation operations create a periodic structure of rotation centers, corresponding to a square netplane with perpendicular lattice vectors $\underline{R}_1 = a\,(1, 0)$, $\underline{R}_2 = a\,(0, 1)$.

3.19 Consider a netplane with inversion centers at $\underline{r}^i = n_1\,\underline{r}_1 + n_2\,\underline{r}_2$, n_i integer. Determine the lattice vectors of the netplane.

3.20 Consider a netplane with n_1- and n_2-fold rotation axes at the same center $\underline{r}_{o1}$. Prove that the netplane has a p-fold rotation axis at $\underline{r}_{o1}$ with $p = \mathrm{lcm}(n_1, n_2)$.

3.21 Consider a rectangular netplane with orthogonal lattice vectors $\underline{R}_1$ and $\underline{R}_2$ and mirror symmetry. Show that if the netplane includes a mirror line, which does not point parallel or perpendicular to the lattice vectors, the netplane must also include glide lines.

3.22 Which of the point symmetry elements of the bulk lattice are conserved in the netplanes referring to the monolayers described in Exercise 3.2?

3.23 Which $(h\,k\,l)$ netplanes of an fcc lattice can be described by rectangular unit cells? Determine the general conditions for $(h\,k\,l)$.

3.24 The hcp crystal can be defined by lattice vectors $\underline{R}_1, \underline{R}_2, \underline{R}_3$ given in Exercise 3.6. Show that the atom density of primitive $(h\,k\,l)$ monolayers is given by

$$\rho_{(h\,k\,l)} = [8/3\,(h^2 + k^2 + h k) + 3/4\,l^2]^{-1/2}$$

3.25 Consider a primitive monolayer and its different netplane descriptions. Which alternative Bravais lattices can be used to describe a monolayer corresponding to a (a) square, (b) primitive rectangular, (c) centered rectangular, (d) hexagonal netplane?

3.26 Lattice vectors of trigonal lattices, $\underline{R}_1^{trg}, \underline{R}_2^{trg}, \underline{R}_3^{trg}$, can be represented by those of corresponding hexagonal lattices, $\underline{R}_1^{hex}, \underline{R}_2^{hex}, \underline{R}_3^{hex}$, where transformation (3.35)

$$\begin{pmatrix} \underline{R}_1^{hex} \\ \underline{R}_2^{hex} \\ \underline{R}_3^{hex} \end{pmatrix} = \begin{pmatrix} 1 & -1 & 0 \\ 0 & 1 & -1 \\ 1 & 1 & 1 \end{pmatrix} \cdot \begin{pmatrix} \underline{R}_1^{trg} \\ \underline{R}_2^{trg} \\ \underline{R}_3^{trg} \end{pmatrix}$$

is one example of an obtuse representation.

(a) Show that there are, altogether, six choices of obtuse and six of acute hexagonal representations and determine the corresponding transformations.

(b) Evaluate the Miller index transformation $(h^{trg}\, k^{trg}\, l^{trg}) \to (h^{hex}\, k^{hex}\, l^{hex})$ for each of the lattice transformations in (a).

(c) The Miller index transformations in (b) results in constraints for $h^{hex}, k^{hex}, l^{hex}$ reading

$$-h^{hex} + k^{hex} + l^{hex} = 3g, \quad h^{hex} - k^{hex} + l^{hex} = 3g,$$
$$h^{hex} + k^{hex} - l^{hex} = 3g, \quad h^{hex} + k^{hex} + l^{hex} = 3g$$

Which constraint belongs to which transformation?

4
Ideal Single Crystal Surfaces

Ideal single crystal surfaces, which result from **truncating** perfect three-dimensional bulk crystals, provide only an approximate description of structural properties of many crystal surfaces that appear in nature. However, these model surfaces can be treated in an **exact** way mathematically and show **general concepts** that can be easily applied to real crystal surfaces. Examples are the general classification of stepped and kinked surfaces and the treatment of chiral surfaces which will be discussed in detail in this chapter.

In the present chapter (also in Chapters 5 and 6) a number of examples deal with crystals of **cubic** symmetry, specifically face-centered cubic (fcc) and body-centered cubic (bcc). In these examples we will always use Miller indices which refer to **simple cubic** notation without further specification, see Section 3.4, since this notation is commonly used by surface scientists. Further, in all cases where Miller indices are used to denote only directions of netplane normal vectors, their values will be **normalized** such that the indices do not have a common divisor. As an example, Miller indices of (3 9 18) and (1 3 6) netplanes are equivalent, where the latter notation will be used. Further, **negative** Miller indices will be written with a **minus sign** in front, that is, $h = -2$, rather than given in crystallographic notation, that is, $h = \bar{2}$.

4.1
Basic Definition, Termination

The exact definition of an ideal single crystal surface starts from the **truncation** of a perfect three-dimensional **bulk** crystal parallel to one of its $(h\,k\,l)$ monolayers that acts as a top layer with the crystal substrate below and vacuum above. (By convention the corresponding reciprocal lattice vector $\underline{G}_{(h\,k\,l)}$ is defined such that it points from the substrate into vacuum.) As a result, the surface is **periodic** in two dimensions and its periodicity is determined by lattice vectors $\underline{R}_1$ and $\underline{R}_2$, which define the periodicity of the corresponding $(h\,k\,l)$ netplane. Therefore, one can use **Miller indices** $(h\,k\,l)$ to characterize the **surface orientation** and apply mathematical descriptions of monolayers and netplanes to ideal single crystal surfaces

Crystallography and Surface Structure: An Introduction for Surface Scientists and Nanoscientists,
Second Edition. Klaus Hermann.
© 2017 Wiley-VCH Verlag GmbH & Co. KGaA. Published 2017 by Wiley-VCH Verlag GmbH & Co. KGaA.

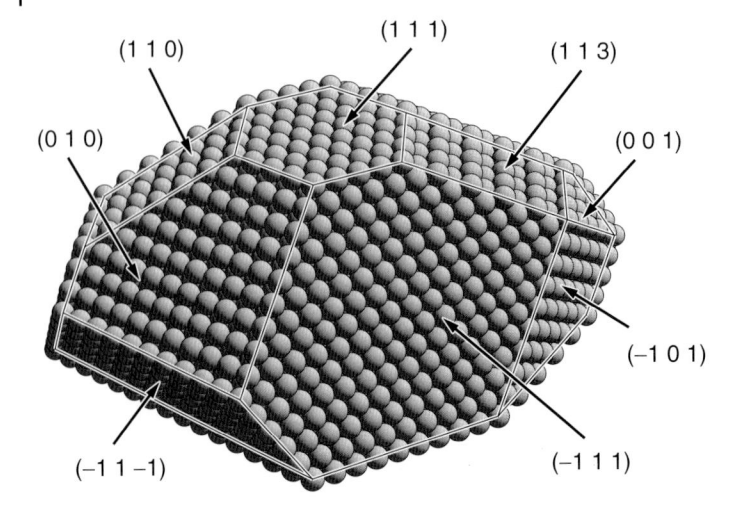

Figure 4.1 Different bulk truncated surface sections of an ideal fcc nickel single crystal with atoms shown as balls. The surfaces are indicated by their Miller indices.

as well. This is illustrated in Figure 4.1, which shows an irregular grain of an ideal fcc nickel single crystal exposing different bulk truncated surface sections that are labeled by their Miller indices $(h\,k\,l)$.

Single crystals with more than one atom type in the primitive unit cell can exhibit **differently terminated** surfaces for the same $(h\,k\,l)$ orientation. If the crystal contains $p > 1$ non-equivalent atoms in the primitive unit cell then there are p parallel primitive monolayers originating from each atom of the cell, which may describe the topmost layer of an ideal single crystal surface. Of these monolayers, some may fall on the **same spatial plane** depending on the actual $(h\,k\,l)$ direction. This yields monolayers with identical netplanes but a (planar) basis with more than one atom type. As a result, for polyatomic crystals there are $q \leq p$ different terminations of corresponding ideal single crystal surfaces described by $(h\,k\,l)$.

As an example, the **sodium chloride**, NaCl, crystal is described by an **fcc** lattice (as defined in Eq. (2.38)) of lattice constant a and **two** different elements, one sodium and one chlorine each, in its primitive unit cell, yielding $p = 2$. (The Na$^+$ and Cl$^-$ ions are positioned at $\underline{r}_1^{Na} = a\,(0, 0, 0)$ and $\underline{r}_2^{Cl} = a/2\,(1, 1, 1)$ inside the fcc unit cell.) For (1 0 0) monolayers one obtains $q = 1$ since Na$^+$ and Cl$^-$ ions fall on the same plane. This results in only one (1 0 0) surface termination, where Na$^+$ and Cl$^-$ ions exist in equal amounts giving rise to a nonpolar surface, see Figure 4.2a. On the other hand, (1 1 1) monolayers of Na$^+$ and Cl$^-$ ions are separated from each other, hence $q = 2$, which leads to two possible (1 1 1) surface terminations, one with Na$^+$ and one with Cl$^-$ ions at the top, see Figure 4.2b for the Na$^+$ termination. These surfaces are highly polar and quite difficult to prepare experimentally. The NaCl crystal structure also applies to **MgO**, resulting in the same surface terminations that were shown in Figure 1.1.

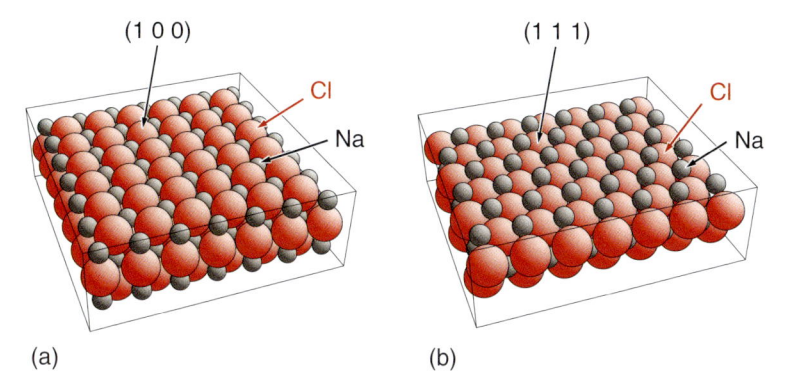

Figure 4.2 Different NaCl single crystal surfaces. (a) Nonpolar NaCl(1 0 0) surface, (b) polar NaCl(1 1 1) surface with Na termination.

A more complex example is given by the **vanadium sesquioxide**, V_2O_3, crystal whose structure is of corundum type with a **trigonal-R** (or equivalent hexagonal) lattice [31], see Section 2.4. The primitive unit cell of V_2O_3 contains **10** atoms, $4\times V$ and $6\times O$, yielding p = 10. Along the (1 1 1) direction (corresponding to (0 0 0 1) in the hexagonal four-index notation) there are two sets of three different hexagonal monolayers each, hence q = 6, where the two sets are connected by inversion symmetry. Each set contains two monolayers with V^{3+} ions (originating from two different V^{3+} ions in the unit cell) and one with O^{2-} ions of higher density (originating from three different O^{2-} ions in the unit cell). This allows three different (0 0 0 1) surface terminations as shown in Figure 4.3, the **full metal** termination $VV'O\ldots$, the **half metal** termination $V'OV\ldots$, and the **oxygen** termination $OVV'\ldots$. Experimental and theoretical studies on real $V_2O_3(0\,0\,0\,1)$ surfaces indicate that the half-metal $V'OV\ldots$ termination which is the least polar of the three terminations is energetically preferred [88].

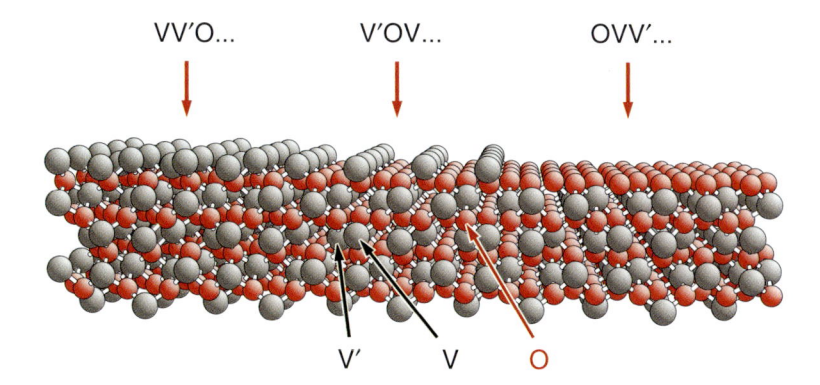

Figure 4.3 Structure of the V_2O_3 crystal (corundum lattice, trigonal-R) with three differently terminated (0 0 0 1) surfaces, denoted by $VV'O\ldots$, $V'OV\ldots$, $OVV'\ldots$ and indicated by red arrows.

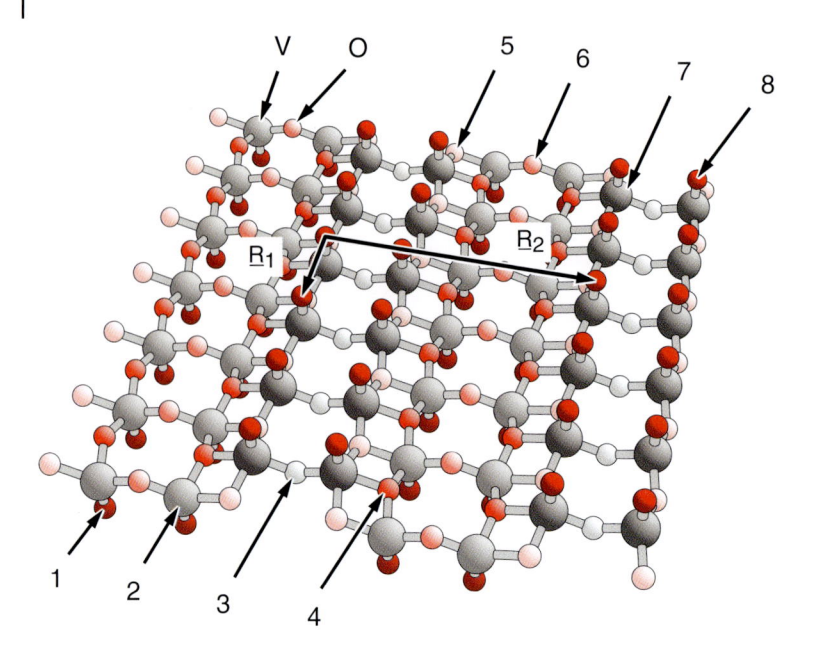

Figure 4.4 Structure of a physical layer of the (0 1 0) surface of a V_2O_5 crystal (orthorhombic lattice). Atoms (gray for V, red for O) are numbered with respect to their monolayer sequence inside the physical layer. The lattice vectors $\underline{R}_1$, $\underline{R}_2$ of the layer illustrate the surface periodicity.

An even more complex example is the **vanadium pentoxide**, V_2O_5, crystal with an **orthorhombic-P** lattice [31], see Section 2.4. The primitive unit cell of V_2O_5 contains **14** atoms, $4 \times V$, $10 \times O$, yielding p = 14. This crystal has a **layer structure** and can be described by a periodic arrangement of weakly binding physical layers along the (0 1 0) direction. (Note that, depending on the choice of the orthorhombic crystal axes, this termination may also be called (0 0 1) .) Each physical layer contains **eight** different monolayers (two with V^{5+} ions, six with O^{2-} ions) as indicated in Figure 4.4, hence q = 8. This results formally in **eight** different (0 1 0) surface terminations. However, for chemical reasons, that is, as a result of strong local binding between the atoms, the termination of the real surface is assumed to be described always as shown in Figure 4.4. Here, singly coordinated vanadyl oxygen (dark red balls), labeled "8" in the figure, forms the terminating monolayer [7].

All previous examples refer to single crystals that contain **inversion symmetry**. Therefore, $(h\ k\ l)$- and $(-h\ -k\ -l)$-oriented surfaces are equivalent and show the same termination schemes. This does not apply to single crystals without inversion centers, where $(h\ k\ l)$- and $(-h\ -k\ -l)$-oriented surfaces can be structurally different. This is also manifested in different physical and chemical properties. For example, **gallium arsenide**, GaAs, forms a crystal with a cubic **zincblende** lattice [31]. The primitive unit cell (fcc type) contains **two** atoms, Ga and As, yielding p = 2. This crystal has **no inversion symmetry** and, therefore, its monolayer stacking along the (1 1 1) direction differs from that along (−1 −1 −1). Figure 4.5 illustrates the structure of the two surfaces. The (1 1 1) surface allows two terminations, see Figure 4.5a. Termination 1 yields arsenic atoms sitting on top of gallium at a large perpendicular distance to form the topmost hexagonal surface layer. Termination 2 yields a hexagonal surface layer of gallium atoms at the top, where these atoms are threefold coordinated with respect to the underlying As at a small perpendicular distance. The (−1 −1 −1) surface, see Figure 4.5b, offers the same two types of terminations as (1 1 1) except that the gallium and arsenic atoms are interchanged. Experiments indicate that termination 2 is energetically preferred for both surfaces resulting in a gallium-terminated (1 1 1) and an arsenic-terminated (−1 −1 −1) surface.

The truncation of a perfect three-dimensional crystal by a (periodic) surface yields the crystal substrate with atoms below and vacuum above. This poses the question as to which of the substrate atoms are assigned to the surface and which need to be considered as bulk atoms. The definition of **surface atoms** is not unique and depends to some extent on the physical or chemical parameters

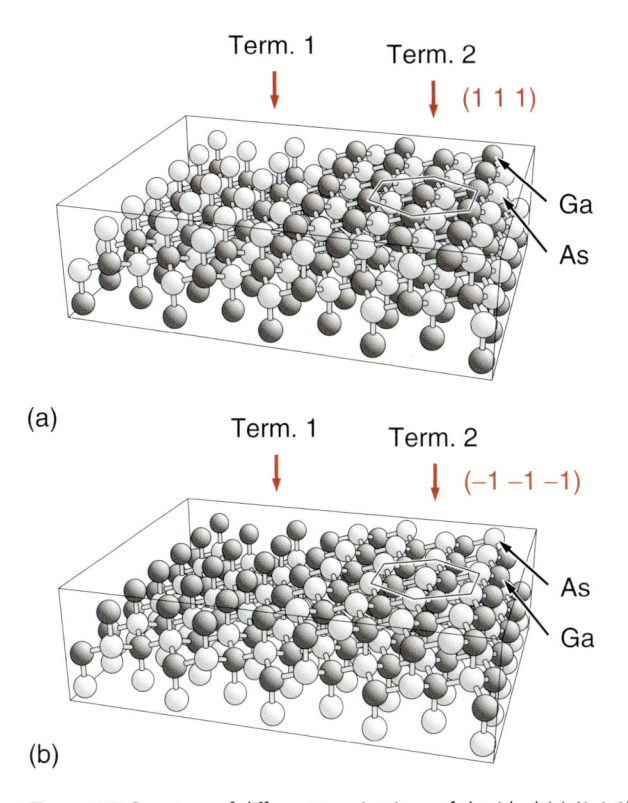

Figure 4.5 Structure of different terminations of the ideal (a) (1 1 1) and (b) (−1 −1 −1) oriented surface of GaAs. In both cases, the two terminations are indicated by "Term. 1" and "Term. 2" and atom balls are labeled accordingly. The hexagon to the right connects the neighboring atoms of the topmost surface layer.

that are to be described. Considering the structural properties, a reasonable choice is to define the surface atoms by their neighboring environment. As discussed in Section 2.6, each atom inside a bulk crystal is surrounded by atom neighbors which can be grouped into shells depending on their distance from the atom under consideration. The distances and numbers of atoms of the shells are determined only by the bulk crystal structure. When atoms from the substrate are close to the surface their neighbor shells become incomplete compared with the perfect bulk environment since atoms above the surface are missing. This shell behavior can be used to discriminate between bulk and surface atoms in a simple way: for a given number of shells, starting with the smallest, an atom will be considered to be a surface atom if any of these shells contains fewer atom members than those obtained for the perfect bulk. Otherwise, the atom will be denoted as a bulk atom. In practice, for many low-Miller-index surfaces, the first and second nearest neighbor shells are considered sufficient for a reasonable definition of surface atoms. The definition yields different results depending on the topmost layers of the $(h\,k\,l)$-oriented surfaces. In particular, the number of surface layers required to reach the bulk part of the substrate will vary with $(h\,k\,l)$. As an example, Table 4.1 lists the number of atoms in neighbor shells (first to third) surrounding atoms of the topmost four monolayers of the surface for selected $(h\,k\,l)$ surfaces of the fcc crystal. The different $(h\,k\,l)$ entries are grouped according to decreasing monolayer density, which shows that the number of topmost layers to reach the bulk from the surface increases as the monolayer density decreases. This is clear from the present definition of bulk atoms requiring a distance from the surface, which is determined by bulk crystal parameters irrespective of the surface orientation. In contrast, the $(h\,k\,l)$-oriented monolayers are separated by distances that decrease with decreasing monolayer density. This is illustrated for the vicinal stepped (9 9 7) surface with wide (1 1 1) terraces, as shown in Figure 4.6. Here the first and second atom neighbor shells are completed only

Table 4.1 Atom neighbor shell behavior of selected $(h\,k\,l)$ surfaces of the fcc crystal.

$(h\,k\,l)$	Layer 1	Layer 2	Layer 3	Layer 4	Bulk layer
(1 1 1)	9, 3, 15	**12, 6,** 21	**12, 6,** 24	**12, 6,** 24	3
(1 0 0)	8, 5, 12	**12,** 5, 20	**12, 6,** 24	**12, 6,** 24	3
(1 1 0)	7, 4, 14	11, 4, 18	**12, 6,** 20	**12, 6,** 24	4
(3 1 1)	7, 3, 14	10, 5, 16	**12,** 5, 19	**12, 6,** 23	5
(3 3 1)	7, 3, 12	9, 4, 16	11, 4, 19	**12, 6,** 19	6
(2 1 0)	6, 4, 14	9, 4, 16	11, 5, 16	**12,** 5, 20	7
(2 1 1)	7, 3, 12	9, 3, 16	10, 5, 17	**12,** 5, 19	7
(9 9 7)	7, 3, 12	9, 3, 14	9, 3, 15	9, 3, 15	18

Columns 2–5 give for each atom in monolayer one to four from the surface the number of atoms n_1, n_2, n_3 in its first to third neighbor shells. Numbers n_i are **boldfaced** if they reflect the bulk value. Column 6 lists smallest indices of layers with bulk atoms only, corresponding to $n_1, n_2, n_3 = 12, 6, 24$.

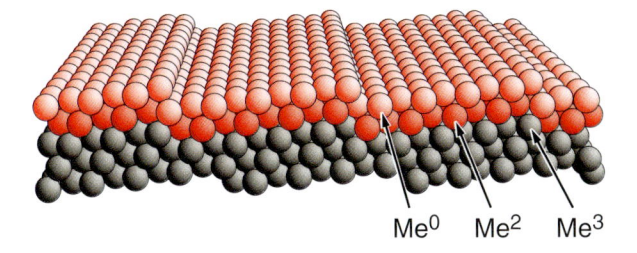

Figure 4.6 Structure of the (9 9 7) surface of an fcc crystal with atoms near the surface emphasized. Atom balls with all neighbor shells incomplete ("true" surface atoms, Me^0) are painted light red, those with first and second neighbor shells completed (Me^2) dark red, and those with the first to third neighbor shells completed ("true" bulk atoms, Me^3) gray.

after 10 monolayers, and the third atom neighbor shell is completed only after 18 monolayers. Thus, at stepped and (kinked) surfaces, all atoms on the terraces should be viewed as surface atoms.

4.2
Morphology of Surfaces, Stepped and Kinked Surfaces

The overall shapes (**morphology**) of $(h\ k\ l)$-oriented single crystal surfaces are only partly determined by the geometry of the corresponding $(h\ k\ l)$ monolayers. Local binding between atoms, which may involve several monolayers, are also important. This is particularly evident for oxide surfaces, where local binding dominates the detailed surface structure, as illustrated in Figure 4.4 for the V_2O_5 (0 1 0) surface. Here, atoms from eight different $(h\ k\ l)$ monolayers contribute to the shape of the surface. In more compact crystals, the morphology of $(h\ k\ l)$-oriented surfaces is characterized often by sections referring to **densest monolayers** of the crystal (microfacets) forming **terraces** and being separated by steps that may be straight **steps** or broken steps ("stepped steps," commonly called *kinked steps* or **kinks**). Since the $(h\ k\ l)$ orientation of these surfaces is often quite close to those of the densest monolayers, they are usually called **vicinal** surfaces. As an example, the (7 7 9) surface of an fcc crystal, see Figure 4.7, is described by (1 1 1) terraces (the (1 1 1) monolayers of the crystal are the densest) separated by steps originating from (0 0 1) monolayers. Thus, the (7 7 9) monolayer, which consists of a rather open set of parallel atom rows, shown in Figure 4.7 by light balls, does not characterize the surface morphology.

Further, the (5 6 8) surface of an fcc crystal, see Figure 4.8, is characterized by (1 1 1) terraces separated by periodically broken steps (kinked steps) originating from (0 0 1) and (−1 1 1) monolayers. As in the example before, the (5 6 8) monolayer that forms a very open set of atoms describing the kink corners, shown in Figure 4.7 by red balls, does not evidence the surface morphology.

Another complication can arise for crystal surfaces defined by **large Miller indices**. For example, fcc single crystal surfaces corresponding to $(h\ k\ l) = (2m\ 1\ 1)$

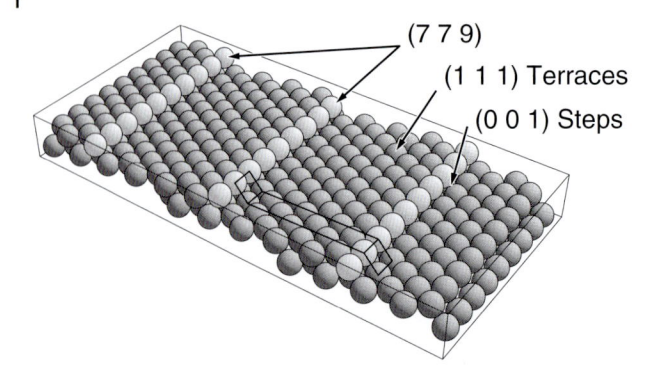

Figure 4.7 Structure of the stepped (7 7 9) surface of an fcc crystal. Atom balls along the step lines (defining the (7 7 9) monolayer) are emphasized in light gray. Steps and terraces are labeled accordingly and illustrated by line frames.

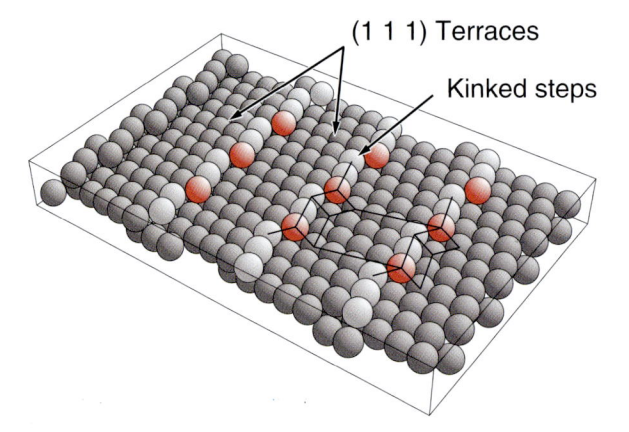

Figure 4.8 Structure of the kinked (5 6 8) surface of an fcc crystal. Atom balls along the kink lines are emphasized in light gray and red (the latter defining the (5 6 8) monolayer). Kinks and terraces are illustrated by line frames.

and $m > 1$ in sc notation can form alternating (1 0 0) terraces of widths given by m and $m+1$ atom rows, respectively. The terraces are separated by (1 1 1) single height steps, see right part of Figure 4.9 for $m = 3$ corresponding to an fcc(6 1 1) surface. Removing all atoms from the smaller of the two terraces results in a stepped surface of identical (1 0 0) terraces of widths given by 2m atom rows that are separated by (1 1 1)-oriented double height steps as shown in the left part of Figure 4.9. The appearance of steps (and kinks) of **multiple atom height** will be discussed in more detail in the next section. Other surface structures have also been discussed in [89–92].

From the example shown in Figure 4.7 it is clear that **step edges** at single crystal surfaces are formed by periodic atom rows along general lattice vectors $\underline{R}_N$ (**step edge vectors**), which connect neighboring atoms referring to rather small

(111) Double height   **(111) Single height**

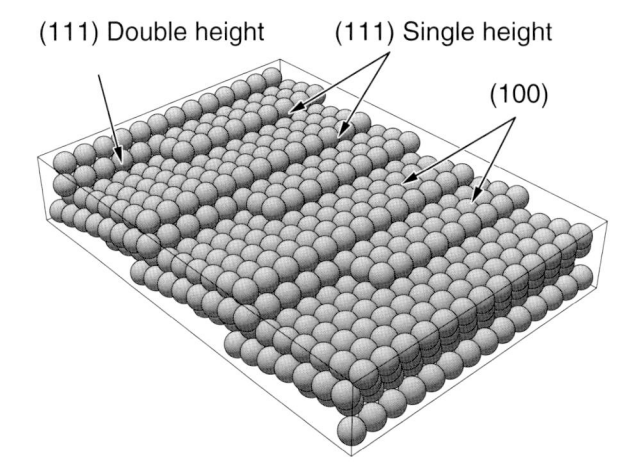

(100)

Figure 4.9 Structure of the stepped (6 1 1) surface of a crystal with an fcc lattice, left part with double height steps and identical terraces, right part with single height steps and alternating narrow and wide terraces.

interatomic distances. Thus, the periodicity of surfaces containing steps along $\underline{R}_N$ is described by netplanes parallel to $\underline{R}_N$. As a consequence, corresponding monolayer normal vectors, pointing along reciprocal lattice vectors $\underline{G}_{(h\,k\,l)}$ according to Eq. (3.6), must be perpendicular to $\underline{R}_N$. Thus, representations

$$\underline{R}_N = p_1\,\underline{R}_1 + p_2\,\underline{R}_2 + p_3\,\underline{R}_3, \quad \underline{G}_{(h\,k\,l)} = h\,\underline{G}_1 + k\,\underline{G}_2 + l\,\underline{G}_3 \qquad (4.1)$$

with integer p_i and h, k, l yield

$$(\underline{R}_N\,\underline{G}_{(h\,k\,l)}) = (p_1\,\underline{R}_1 + p_2\,\underline{R}_2 + p_3\,\underline{R}_3)\,(h\,\underline{G}_1 + k\,\underline{G}_2 + l\,\underline{G}_3)$$
$$= 2\pi\,(p_1\,h + p_2\,k + p_3\,l) = 0 \qquad (4.2)$$

where the orthogonality relation of real and reciprocal lattice vectors (2.96) has been applied. Relation (4.2) can be used to find $(h\,k\,l)$-indexed surfaces for a given edge vector $\underline{R}_N$ and also to verify edge vectors $\underline{R}_N$ at a stepped $(h\,k\,l)$ surface.

As an example, we consider the **fcc lattice**, defined by Eq. (2.102) and represented by simple cubic lattice vectors $\underline{R}_1, \underline{R}_2, \underline{R}_3$ in Cartesian coordinates. Then, vectors $\underline{R}_N$ connecting the nearest neighbors are given by

$$\underline{R}_N = \pm a/2\,(0, 1, \pm 1), \quad \pm a/2\,(1, 0, \pm 1), \quad \pm a/2\,(1, \pm 1, 0) \qquad (4.3)$$

Relation (4.2) together with Eq. (4.3) and using $(h\,k\,l)$ in sc notation results in six linear Diophantine equations

$$h \pm k = 0, \quad h \pm l = 0, \quad k \pm l = 0 \qquad (4.4)$$

with solutions

$$(h\,k\,l) = (m \pm m\ n), \quad (m\ n \pm m), \quad (m\ n \pm n), \quad m, n\ \text{integer} \qquad (4.5)$$

These Miller index triplets characterize orientations of all fcc surfaces with steps formed by rows of nearest neighbor atoms (which also include the atomically flat surfaces given by ($\pm$1 $\pm$1 $\pm$1), ($\pm$1 0 0), (0 $\pm$1 0), and (0 0 $\pm$1)).

If vectors $\underline{R}_N$ connecting the second nearest neighbors, given by

$$\underline{R}_N = \pm a\,(1,0,0),\ \pm a\,(0,1,0),\ \pm a\,(0,0,1) \tag{4.6}$$

are considered, an analogous procedure yields Miller index triplets

$$(h\ k\ l) = (0\ m\ n),\quad (m\ 0\ n),\quad (m\ n\ 0),\quad m, n\ \text{integer} \tag{4.7}$$

defining orientations of all fcc surfaces with steps formed by rows of second nearest neighbor atoms (which also include the atomically flat surfaces given by ($\pm$1 0 0), (0 $\pm$1 0), (0 0 $\pm$1)).

4.3 Miller Index Decomposition

According to Eq. (3.8), surfaces with large Miller index values ($h\ k\ l$) correspond to rather open monolayers of low atom density. They can be characterized in many cases morphologically by combinations of **terraces** with ($h_t\ k_t\ l_t$) orientation separated by **steps** with ($h_s\ k_s\ l_s$) orientation, as discussed in the previous section. Here, the Miller index triplets ($h\ k\ l$), ($h_t\ k_t\ l_t$), and ($h_s\ k_s\ l_s$) are connected by an **additivity theorem**, which is discussed in the following.

Starting from a **monoatomic** single crystal with its primitive lattice described by lattice vectors $\underline{R}_{o1}, \underline{R}_{o2}, \underline{R}_{o3}$, a **stepped surface** looks like that shown in Figure 4.10, which sketches the stepped (3 3 5) surface of an fcc crystal. Here, step-adapted lattice vectors $\underline{R}_1, \underline{R}_2, \underline{R}_3$ can be constructed where $\underline{R}_1$ and $\underline{R}_2$ describe the periodicity of the terrace monolayers, with $\underline{R}_1$ pointing along the step edges, and $\underline{R}_3$ along the connection between the lower and upper edge of each step. Let us assume further that terraces are n_t vector lengths $\underline{R}_2$ "wide" and the steps are n_s vector lengths $\underline{R}_3$ "high" ($n_t = 3$ and $n_s = 1$ in Figure 4.10). Then the atoms at two adjacent step edges, A, B, and at C in Figure 4.10, determine a plane with a normal vector defining the ($h\ k\ l$) direction of the stepped surface while $\underline{R}_1$ and $\underline{R}_2$ refer to ($h_t\ k_t\ l_t$) of the terrace and $\underline{R}_3$ and $\underline{R}_1$ to ($h_s\ k_s\ l_s$) of the step side (($h\ k\ l$) = (3 3 5), ($h_t\ k_t\ l_t$) = (1 1 1), and ($h_s\ k_s\ l_s$) = (0 0 2) in Figure 4.10). As a result, the **reciprocal lattice vector** $\underline{G}_{(h\ k\ l)}$ of the **stepped surface**, which is perpendicular to the plane through atoms A, B, and C, is determined by

$$\underline{G}_{(h\ k\ l)} = 2\pi/\beta\,(\underline{AB} \times \underline{AC}) = 2\pi/\beta\,\{\underline{R}_1 \times (n_t\underline{R}_2 - n_s\underline{R}_3)\}$$

$$= 2\pi/\beta\,\{n_t\,(\underline{R}_1 \times \underline{R}_2) + n_s(\underline{R}_3 \times \underline{R}_1)\} = n_t\,\underline{G}_{(h\ k\ l)t} + n_s\,\underline{G}_{(h\ k\ l)s}$$

$$\beta = (\underline{R}_1 \times \underline{R}_2)\,\underline{R}_3 \tag{4.8}$$

Thus, after the three reciprocal lattice vectors have been decomposed into their Miller index combinations, one obtains the **additivity theorem for stepped**

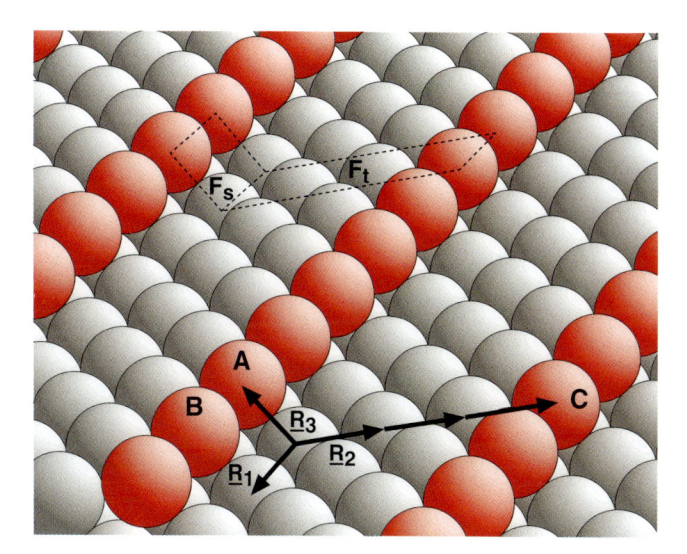

Figure 4.10 Scheme of a Miller index decomposition for the stepped (3 3 5) surface of an fcc crystal. The atom balls along the step lines, forming a (3 3 5) monolayer are emphasized in red. The step-adapted lattice vectors $\underline{R}_1$, $\underline{R}_2$, $\underline{R}_3$ are sketched accordingly. The elementary terrace and step sections of area F_t and F_s, respectively, are framed by dashed lines.

surfaces

$$(h\ k\ l) = n_t\ (h_t\ k_t\ l_t) + n_s\ (h_s\ k_s\ l_s) \tag{4.9}$$

The **scalar factors** n_t and n_s in this equation have a simple geometric meaning. The elementary terrace section defined as the periodic repeat cell along the terrace, sketched by dashed lines in Figure 4.10, has an area F_t, where

$$F_t = n_t\ |\underline{R}_1 \times \underline{R}_2| = n_t\ (V_{el}/2\pi)\ |G_{(h\ k\ l)t}| \tag{4.10}$$

while the area F_s of the repeat cell of step side, also sketched by dashed lines in Figure 4.10, is given by

$$F_s = n_s\ |\underline{R}_3 \times \underline{R}_1| = n_s\ (V_{el}/2\pi)\ |G_{(h\ k\ l)s}| \tag{4.11}$$

where $|\underline{R}_i \times \underline{R}_j|$ are the unit cell areas of the corresponding terrace and step planes. Thus, the factors n_t and n_s in Eqs. (4.8) and (4.9), respectively, define the relative sizes of the periodic repeat cells along the terraces and the steps with respect to their corresponding unit cell areas. The definition of terraces and steps requires that the terrace area F_t be larger than that of the separating step F_s. Thus, if the elementary cells $|\underline{R}_1 \times \underline{R}_2|$ and $|\underline{R}_3 \times \underline{R}_1|$ are of comparable size the scalar factor n_t in Eq. (4.9) will be larger than n_s and $(h_t\ k_t\ l_t)$ will be the dominant component of the $(h\ k\ l)$ triplet. In this spirit, the $(h\ k\ l)$-oriented surface will be called **vicinal surface** with the $(h_t\ k_t\ l_t)$ (terrace) surface representing the **vicinal partner**. Thus, the stepped fcc(3 3 5) surface shown in Figure 4.10, which decomposes to

$$(3\ 3\ 5) = 3\ (1\ 1\ 1) + 1\ (0\ 0\ 2)$$

is vicinal, with (1 1 1) being the vicinal partner.

The additivity theorem (4.9) is the basis of the so-called **step notation** [93, 94] of stepped vicinal surfaces according to which an $(h\,k\,l)$ surface is, in its general form, denoted as

$$(h\,k\,l) \equiv [p_1\,(h_t\,k_t\,l_t) \times p_2\,(h_s\,k_s\,l_s)], \quad p_1 = n_t + 1, \quad p_2 = n_s \quad (4.12)$$

Here, the terrace width of $n_t\,\underline{R}_2$ used above corresponds to $(n_t + 1)$ rows of terrace atoms used in the definition of the step notation. This definition was initially proposed for surfaces of **cubic crystals** (fcc and bcc) with Miller indices of simple cubic notation and single steps ($n_s = 1$) [93] whereas the additivity theorem is more general and independent of the lattice type. Further, each of the Miller index triplets $(h\,k\,l)$, $(h_t\,k_t\,l_t)$, and (h_s, k_s, l_s) in notation Eq. (4.12) is assumed to be scaled such that its indices do not have a common divisor. For example, (2 2 0) is written as (1 1 0). Examples of the additivity theorem (4.9) for crystals with simple cubic (sc), fcc, and bcc lattices together with the corresponding step notations are given in Table 4.2.

Table 4.2 Decomposition of Miller indices of vicinal stepped surfaces of crystals with sc, fcc, and bcc lattices.

$(h\,k\,l) = n_t\,(h_t\,k_t\,l_t) + n_s\,(h_s\,k_s\,l_s)$	Step notation
fcc (sc)	
(7 7 5) = 6 (1 1 1) + (1 1 −1)	[7 (1 1 1) × (1 1 −1)]
(3 3 5) = 3 (1 1 1) + (0 0 2)	[4 (1 1 1) × (0 0 1)]
(9 1 1) = 4 (2 0 0) + (1 1 1)	[5 (1 0 0) × (1 1 1)]
(p+2 p+2 p) = (p+1) (1 1 1) + (1 1 −1)	[(p+2) (1 1 1) × (1 1 −1)]
(p+2 p p) = p (1 1 1) + (2 0 0)	[(p+1) (1 1 1) × (1 0 0)]
(2p+1 1 1) = p (2 0 0) + (1 1 1)	[(p+1) (1 0 0) × (1 1 1)]
fcc (gen)	
(5 5 1) = 4 (1 1 0) + (1 1 1)	[5 (1 1 0) × (1 1 1)]
(4 3 2) = 3 (1 1 1) + (1 0 −1)	[4 (1 1 1) × (1 0 −1)]
bcc (sc)	
(5 5 2) = 5 (1 1 0) + (0 0 2)	[6 (1 1 0) × (0 0 1)]
(6 6 10) = 5 (1 1 2) + (1 1 0)	[6 (1 1 2) × (1 1 0)]
(8 1 1) = 4 (2 0 0) + (0 1 1)	[5 (2 0 0) × (0 1 1)]
(p p 2) = p (1 1 0) + (0 0 2)	[(p+1) (1 1 0) × (0 0 1)]
(p+1 p+1 2p) = p (1 1 2) + (1 1 0)	[(p+1) (1 1 2) × (1 1 0)]
(2p 1 1) = p (2 0 0) + (0 1 1)	[(p+1) (1 0 0) × (0 1 1)]
bcc (gen)	
(4 1 1) = 4 (1 0 0) + (0 1 1)	[5 (1 0 0) × (0 1 1)]
(1 1 2) = 2 (0 0 1) + (1 1 0)	[3 (0 0 1) × (1 1 0)]
sc (gen)	
(9 1 1) = 9 (1 0 0) + (0 1 1)	[10 (1 0 0) × (0 1 1)]

The table includes corresponding step notations, see text. Labels (sc) and (gen) refer to simple cubic and generic index notation. Constant p can assume any positive integer value.

Surfaces, for which the decomposition Eq. (4.9) suggests **multiple-atom-height steps**, $n_s > 1$, can give rise to more complex structural behavior depending on local binding. For strong nearest neighbor binding, like in metals with fcc and bcc lattices, these surfaces still form single-height steps with minimal variation of their terrace widths even if $n_s > 1$. Here, the multiple-atom-height step region, which has terraces of "width" $n_t \underline{R}_2$ and steps of "height" $n_s \underline{R}_3$, is partitioned in n_s additional **subterraces**,

$$((p+1)\,n_s - n_t) \quad \text{terraces A of width } p\,\underline{R}_2 \text{ and}$$
$$(n_t - p n_s) \quad \text{terraces B of width } (p+1)\,\underline{R}_2, \quad p = [n_t/n_s] \quad (4.13)$$

where [x] denotes the integer truncation function. (The mathematics behind this partitioning is spelled out in Appendix E.1.) As an example, the (15 15 23) surface of a crystal with an fcc lattice, see Figure 4.11, is decomposed in (1 1 1) terraces and (0 0 1) steps according to

$$(15\ 15\ 23) = 15\,(1\ 1\ 1) + 4\,(0\ 0\ 2)$$

with four subterraces, one type A of width $3\,\underline{R}_2$ and three type B of widths $4\,\underline{R}_2$ filling the initial multiple-step region with $4\,\underline{R}_3$ "high" steps and $15\,\underline{R}_2$ "wide" terraces.

The partitioning of multiple-atom-height step regions into subterraces A, B separated by single-height steps according to Eq. (4.13) yields subterrace sequences which in general are not regular and can be evaluated by number theoretical methods as described in Appendix E.1. However, these exact mathematical sequences may not be observed at real crystal surfaces where diffusion processes also determine the different widths of subterraces with single-height steps for a given surface orientation $(h\,k\,l)$.

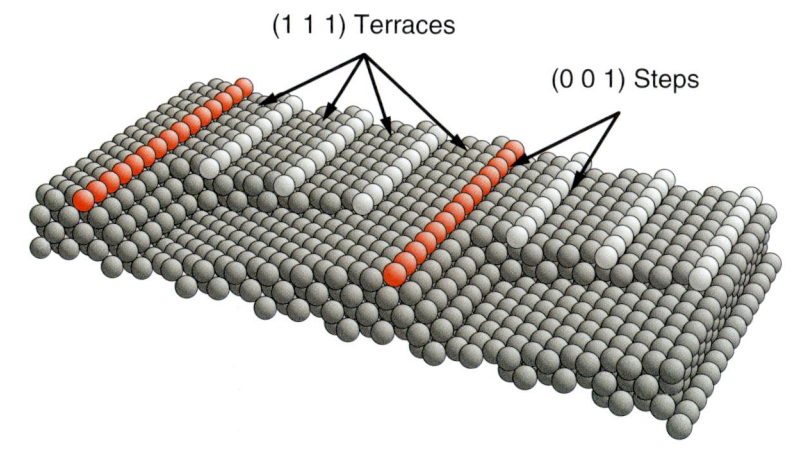

Figure 4.11 Structure of the stepped (15 15 23) surface of an fcc crystal, with multiple-atom-height steps (front) as well as with single-height steps and subterraces of variable width (back).

Surfaces with large Miller index values $(h\,k\,l)$ can also be characterized morphologically in some cases by combinations of terraces with $(h_t\,k_t\,l_t)$ orientation separated by **kinked steps** with $(h_{s1}\,k_{s1}\,l_{s1})$ and $(h_{s2}\,k_{s2}\,l_{s2})$ orientation. Analogous to stepped surfaces the Miller index triplets $(h\,k\,l)$, $(h_t\,k_t\,l_t)$, $(h_{s1}\,k_{s1}\,l_{s1})$, and $(h_{s2}\,k_{s2}\,l_{s2})$ are connected by an **additivity theorem**, which is proven in the following.

Starting from a **monoatomic** single crystal with its primitive lattice described by $\underline{R}_{o1}$, $\underline{R}_{o2}$, and $\underline{R}_{o3}$, a **kinked surface** looks like that shown in Figure 4.12 which sketches the kinked fcc(11 13 19) surface. Here, kink-adapted lattice vectors $\underline{R}_1$, $\underline{R}_2$, and $\underline{R}_3$ can be constructed where $\underline{R}_1$ and $\underline{R}_2$ describe the periodicity of the terrace monolayers, with the two vectors pointing along the two kink directions and $\underline{R}_3$ along the connection between the lower and upper corner of a kink.

Let us assume further that the two kink edges are m_1, m_2 vector lengths $\underline{R}_1$, $\underline{R}_2$ "long" ($m_1 = 3$ and $m_2 = 1$ in Figure 4.12) and the terrace width between kinked steps is described by a vector $\underline{R}_t = n_1\,\underline{R}_1 + n_2\,\underline{R}_2$ connecting the lower corner of one kink with the upper corner of an adjacent kink ($n_1 = 6$ and $n_2 = 3$ in Figure 4.12). In addition, the kinks are assumed to be n_s vector lengths $\underline{R}_3$ "high" ($n_s = 1$ in Figure 4.12). Then the atoms at corners of adjacent kink lines, A, B, and C

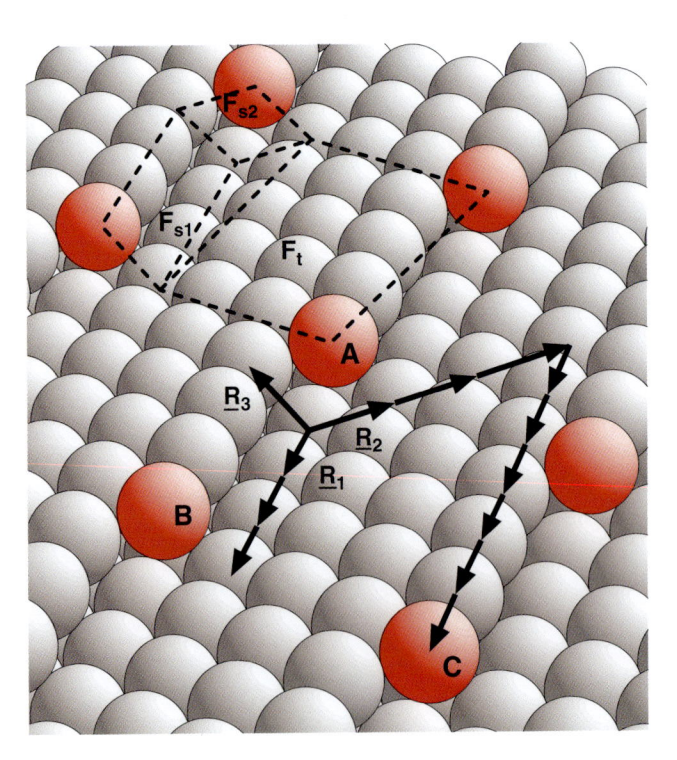

Figure 4.12 Scheme of a Miller index decomposition for the kinked (11 13 19) surface of an fcc crystal. The atom balls at the kink centers, forming a (11 13 19) monolayer are emphasized in red. The kink-adapted lattice vectors $\underline{R}_1$, $\underline{R}_2$, $\underline{R}_3$ are sketched accordingly. The elementary terrace and two kink sections of area F_t, F_{s1}, and F_{s2} are framed by dashed lines.

in Figure 4.12, determine a plane with a normal vector defining the $(h\,k\,l)$ direction of the kinked surface. Further, $\underline{R}_1$ and $\underline{R}_2$ refer to $(h_t\,k_t\,l_t)$ of the terrace, $\underline{R}_3$ and $\underline{R}_1$ to $(h_{s1}\,k_{s1}\,l_{s1})$ of one kink side, and $\underline{R}_2$ and $\underline{R}_3$ to $(h_{s2}\,k_{s2}\,l_{s2})$ of the other kink side $((h\,k\,l) = (11\,13\,19)$, $(h_t\,k_t\,l_t) = (1\,1\,1)$, $(h_{s1}\,k_{s1}\,l_{s1}) = (0\,0\,2)$, $(h_{s2}\,k_{s2}\,l_{s2}) = (-1\,1\,1)$ in Figure 4.12). As a result, the **reciprocal lattice vector** $\underline{G}_{(h\,k\,l)}$ of the **kinked surface**, which is perpendicular to the plane through atoms A, B, and C, is determined by

$$\underline{G}_{(h\,k\,l)} = 2\pi/\beta\,(\underline{AB} \times \underline{AC})$$

$$= 2\pi/\beta\,(m_1\underline{R}_1 - m_2\underline{R}_2) \times (n_1\underline{R}_1 + n_2\underline{R}_2 - m_2\underline{R}_2 - n_s\underline{R}_3)$$

$$= 2\pi/\beta\,\{(m_1n_2 + m_2n_1 - m_1m_2)(\underline{R}_1 \times \underline{R}_2)$$

$$+ m_1n_s\,(\underline{R}_3 \times \underline{R}_1) + m_2n_s\,(\underline{R}_2 \times \underline{R}_3)\}$$

$$= (m_1n_2 + m_2n_1 - m_1m_2)\,\underline{G}_{(h\,k\,l)t}$$

$$+ m_1n_s\,\underline{G}_{(h\,k\,l)s1} + m_2n_s\,\underline{G}_{(h\,k\,l)s2}$$

$$\beta = (\underline{R}_1 \times \underline{R}_2)\,\underline{R}_3 \tag{4.14}$$

Thus, after the four reciprocal lattice vectors have been decomposed into their Miller index combinations, one obtains the **additivity theorem for kinked surfaces**

$$(h\,k\,l) = p_t\,(h_t\,k_t\,l_t) + p_{s1}\,(h_{s1}\,k_{s1}\,l_{s1}) + p_{s2}\,(h_{s2}\,k_{s2}\,l_{s2})$$

$$p_t = (m_1n_2 + m_2n_1 - m_1m_2), \quad p_{s1} = m_1n_s, \quad p_{s2} = m_2n_s \tag{4.15}$$

The **scalar factors** in this equation have a simple geometric meaning. The elementary terrace section defined as the periodic repeat cell along the terrace, sketched by dashed lines in Figure 4.12, has an area F_t, where

$$F_t = p_t\,|\underline{R}_1 \times \underline{R}_2| = p_t\,(V_{el}/2\pi)\,|\underline{G}_{(h\,k\,l)t}| \tag{4.16}$$

while the areas F_{s1}, F_{s2} of the two kink step sides, also sketched by dashed lines, are given by

$$F_{s1} = p_{s1}\,|\underline{R}_3 \times \underline{R}_1| = p_{s1}\,(V_{el}/2\pi)\,|\underline{G}_{(h\,k\,l)s1}| \tag{4.17}$$

$$F_{s2} = p_{s2}\,|\underline{R}_2 \times \underline{R}_3| = p_{s2}\,(V_{el}/2\pi)\,|\underline{G}_{(h\,k\,l)s2}| \tag{4.18}$$

where $|\underline{R}_i \times \underline{R}_j|$ are the unit cell areas of the corresponding terrace and step planes. Thus, the factors p_t, p_{s1}, p_{s2} in Eqs. (4.14) and (4.15) define the relative sizes of the periodic repeat cells along the terraces and the two types of steps along the kink line with respect to their corresponding unit cell areas. The definition of terraces and steps requires that the terrace area F_t is larger than that of the separating steps F_{s1} and F_{s2}. Thus, if the elementary cells $|\underline{R}_1 \times \underline{R}_2|$, $|\underline{R}_3 \times \underline{R}_1|$, and $|\underline{R}_2 \times \underline{R}_3|$ are of comparable size the scalar factor p_t in Eq. (4.15) will be larger than p_{s1} and p_{s2} and $(h_t\,k_t\,l_t)$ will be the dominant component of the $(h\,k\,l)$ triplet. In this spirit, the $(h\,k\,l)$ oriented surface will be called **vicinal surface** with the $(h_t\,k_t\,l_t)$ (terrace) surface representing the **vicinal partner**. Thus, the kinked fcc(11 13 19) surface shown in Figure 4.12 which decomposes to

$$(11\ 13\ 19) = 12(1\ 1\ 1) + 3(0\ 0\ 2) + 1(-1\ 1\ 1) \tag{4.19}$$

is vicinal with (1 1 1) being the vicinal partner.

The additivity theorem (4.15) for kinked surfaces can be written alternatively as

$$(h\ k\ l) = p_t\,(h_t\ k_t\ l_t) + n_s\,(h'_s\ k'_s\ l'_s) \tag{4.20}$$

with p_t according to Eq. (4.15) and

$$(h'_s\ k'_s\ l'_s) = m_1\,(h_{s1}\ k_{s1}\ l_{s1}) + m_2\,(h_{s2}\ k_{s2}\ l_{s2}) \tag{4.21}$$

where relations (4.20) and (4.21) are identical to the additivity theorem (4.9) for stepped surfaces. Therefore, the kinked surface can also be understood as a stepped surface whose terraces are separated by "steps" which themselves are characterized in their orientation by stepped surfaces. As an example, the kinked fcc(11 13 19) surface shown in Figure 4.12 with a decomposition according to Eq. (4.19) can also be interpreted as a combination of the "stepped" surface

$$(11\ 13\ 19) = 12(1\ 1\ 1) + 1(-1\ 1\ 7)$$

with the stepped surface

$$(-1\ 1\ 7) = 3(0\ 0\ 2) + 1(-1\ 1\ 1)$$

The distinction between kinked and stepped surfaces becomes questionable in cases where the kink line sections are very short. As an illustration, Figure 4.13 shows the fcc(10 2 0) surface that allows an interpretation as a kinked surface with (2 0 0) terraces and (1 1 1)/(1 1 −1) kink steps (sections a–c) according to

$$(10\ 2\ 0) = 4(2\ 0\ 0) + 1(1\ 1\ 1) + 1(1\ 1\ -1)$$

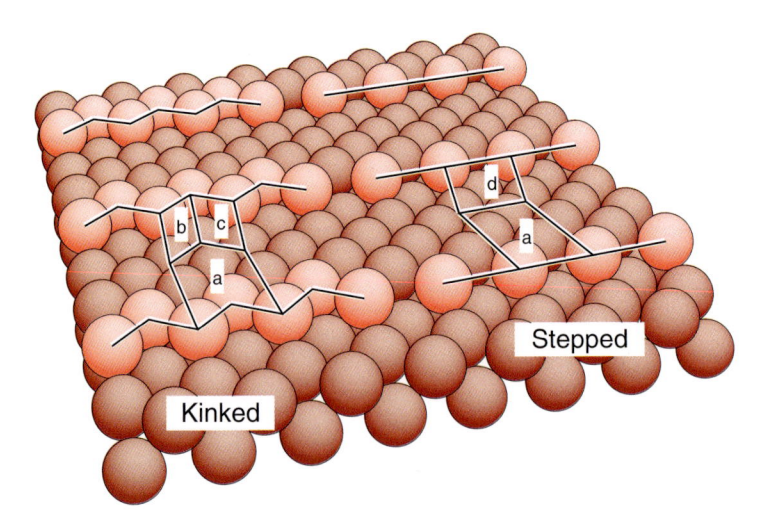

Figure 4.13 Structure of the (10 2 0) surface of an fcc crystal. Step and kink atoms are emphasized by lighter color. The kink decomposition is shown to the left (repeat cells labeled a, b, c and outlined by black lines). The step decomposition is shown to the right (repeat cells labeled a, d).

but also as a stepped surface with (2 0 0) terraces separated by (2 2 0) steps (sections a and d) according to

$$(10\ 2\ 0) = 4(2\ 0\ 0) + 1(2\ 2\ 0)$$

The additivity theorem (4.15) is the basis of the so-called **microfacet notation** [94, 95] of vicinal surfaces according to which an $(h\ k\ l)$ surface is, in its general form, denoted as

$$(h\ k\ l) = a_\lambda\ (h_t\ k_t\ l_t) + b_\mu\ (h_{s1}\ k_{s1}\ l_{s1}) + c_\nu\ (h_{s2}\ k_{s2}\ l_{s2}) \qquad (4.22)$$

This notation was initially proposed for crystals with **cubic lattices** (fcc and bcc) and Miller indices of simple cubic notation and single steps only, whereas the additivity theorem is more general and independent of the lattice type. Further, each of the Miller index triplets $(h\ k\ l)$, $(h_t k_t\ l_t)$, (h_{s1}, k_{s1}, l_{s1}), and (h_{s1}, k_{s1}, l_{s1}) in Eq. (4.22) is assumed to be scaled such that its indices do not have a common divisor. For example, (12 8 4) is written as (3 2 1). In addition, parameters a, b, c and λ, μ, ν are chosen as independent numbers (resulting in indexed number quantities like "3_4"), where λ, μ, ν denote the true decomposition given by Eq. (4.15) with

$$\lambda = p_t, \quad \mu = p_{s1}, \quad \nu = p_{s2} \qquad (4.23)$$

according to Eq. (4.15) while parameters a, b, c are scaled further to guarantee the additivity of the Miller indices in Eq. (4.22) that can be expressed formally by

$$a = (p_t/\chi)\ \gcd(h_t, k_t, l_t), \qquad b = (p_{s1}/\chi)\ \gcd(h_{s2}, k_{s2}, l_{s2})$$
$$c = (p_{s2}/\chi)\ \gcd(h_{s2}, k_{s2}, l_{s2}), \qquad \chi = \gcd(h, k, l) \qquad (4.24)$$

with $\gcd(n_1, n_2, n_3)$ denoting the greatest common divisor of the three integers n_1, n_2, n_3, see Appendix E.1.

As an example, for a crystal with an fcc lattice the additivity theorem for the (20 16 14) indexed surface (sc notation) reads

$$(20\ 16\ 14) = 15\ (1\ 1\ 1) + 2\ (2\ 0\ 0) + (1\ 1\ -1)$$

while the corresponding microfacet notation reads

$$(10\ 8\ 7) = (15/2)_{15}\ (1\ 1\ 1) + 2_2\ (1\ 0\ 0) + (1/2)_1\ (1\ 1\ -1)$$

Using generic Miller indices or simple cubic Miller indices with the correct numerical constraints, see Eqs. (3.20b) and (3.25b), yields

$$a = \lambda = p_t, \quad b = \mu = p_{s1}, \quad c = \nu = p_{s2} \qquad (4.25)$$

making the indexed numbers unnecessary and resulting in a notation according to the additivity theorem (4.15). Examples of the additivity theorem for crystals with fcc and bcc lattices together with their corresponding microfacet notations are given in Table 4.3.

Surfaces, for which the decomposition Eq. (4.15) suggests **multiple-atom-height kinks**, $n_s > 1$, can result in much more complex structural behavior depending on local binding. This is analogous to stepped surfaces described

Table 4.3 Decomposition of Miller indices of vicinal kinked surfaces of crystals with fcc and bcc lattices.

$(h\,k\,l) = a\,(h_t\,k_t\,l_t) + b\,(h_{s1}\,k_{s1}\,l_{s1}) + c\,(h_{s2}\,k_{s2}\,l_{s2})$	Microfacet notation
fcc (sc)	
(17 11 9) = 10 (1 1 1) + 3 (2 0 0) + (1 1 −1)	10_{10} (1 1 1) + 6_3 (1 0 0) + 1_1 (1 1 −1)
(11 3 1) = 4 (2 0 0) + 2 (1 1 1) + (1 1 −1)	8_4 (1 0 0) + 2_2 (1 1 1) + 1_1 (1 1 −1)
(17 15 1) = 7 (2 2 0) + (1 1 1) + (2 0 0)	14_7 (1 1 0) + 1_1 (1 1 1) + 2_1 (1 0 0)
(2p+7 2p+1 2p−1) = 2p (1 1 1) + 3 (2 0 0) + (1 1 −1)	$(2p)_{2p}$ (1 1 1) + 6_3 (1 0 0) + 1_1 (1 1 −1)
(2p+1 3 1) = (p−1) (2 0 0) + 2 (1 1 1) + (1 1 −1)	$(2p-2)_{(p-1)}$ (1 0 0) + 2_2 (1 1 1) + 1_1 (1 1 −1)
(2p +1 2p−1 1) = (p−1) (2 2 0) + (1 1 1) + (2 0 0)	$(2p-2)_{(p-1)}$ (1 1 0) + 2_1 (1 0 0) + 1_1 (1 1 1)
fcc (gen)	
(10 13 14) = 10 (1 1 1) + 3 (0 1 1) + (0 0 1)	10_{10} (1 1 1) + 3_3 (0 1 1) + 1_1 (0 0 1)
(2 6 7) = 4 (0 1 1) + 2 (1 1 1) + (0 0 1)	4_4 (0 1 1) + 2_2 (1 1 1) + 1_1 (0 0 1)
bcc (sc)	
(8 7 3) = 6 (1 1 0) + 2 (1 0 1) + (0 1 1)	6_6 (1 1 0) + 2_2 (1 0 1) + 1_1 (0 1 1)
(15 10 3) = 10 (1 1 0) + 3 (1 0 1) + (2 0 0)	10_{10} (1 1 0) + 3_3 (1 0 1) + 2_1 (1 0 0)
(18 16 4) = 15 (1 1 0) + 3 (1 0 1) + (0 1 1)	$(15/2)_{15}$ (1 1 0) + $(3/2)_3$ (1 0 1) + $(1/2)_1$ (0 1 1)
(2p+2 2p+1 3) = 2p (1 1 0) + 2 (1 0 1) + (0 1 1)	$(2p)_{2p}$ (1 1 0) + 2_2 (1 0 1) + 1_1 (0 1 1)
(2p+5 2p 3) = 2p (1 1 0) + 3 (1 0 1) + (2 0 0)	$(2p)_{2p}$ (1 1 0) + 3_3 (1 0 1) + 2_1 (1 0 0)
bcc (gen)	
(1 2 6) = 6 (0 0 1) + 2 (0 1 0) + (1 0 0)	6_6 (0 0 1) + 2_2 (0 1 0) + 1_1 (1 0 0)
(1 3 15) = 15 (0 0 1) + 3 (0 1 0) + (1 0 0)	15_{15} (0 0 1) + 3_3 (0 1 0) + 1_1 (1 0 0)

The table includes corresponding microfacet notations, see text. Labels (sc) and (gen) refer to the Miller index notations, simple cubic and generic, see above. Constant p can assume any positive integer value.

earlier. For strong nearest neighbor binding, as in metals with fcc and bcc lattices, these surfaces still form kinks with single atom steps even if $n_s > 1$. As an example, the (37 25 17) surface of a crystal with an fcc lattice, see Figure 4.14, decomposes into (1 1 1) terraces and (1 1 −1)/(2 0 0) kinks of double step height, which is clear from the decomposition

$$(37\ 25\ 17) = 21\,(1\ 1\ 1) + 4\,(1\ 1\ -1) + 6\,(2\ 0\ 0)$$

with $m_1 = 2$, $m_2 = 3$, $n_s = 2$, $n_1 = 7$, $n_2 = 3$ according to Eq. (4.15). Here, single step height kink lines contain two sections differing in length and separate terraces of different width. Further, the corresponding kink lines are structurally more complex as indicated by the black lines in Figure 4.14.

So far, the discussion of the shape of vicinal surfaces was restricted to monoatomic single crystals with only one atom in the primitive unit cell

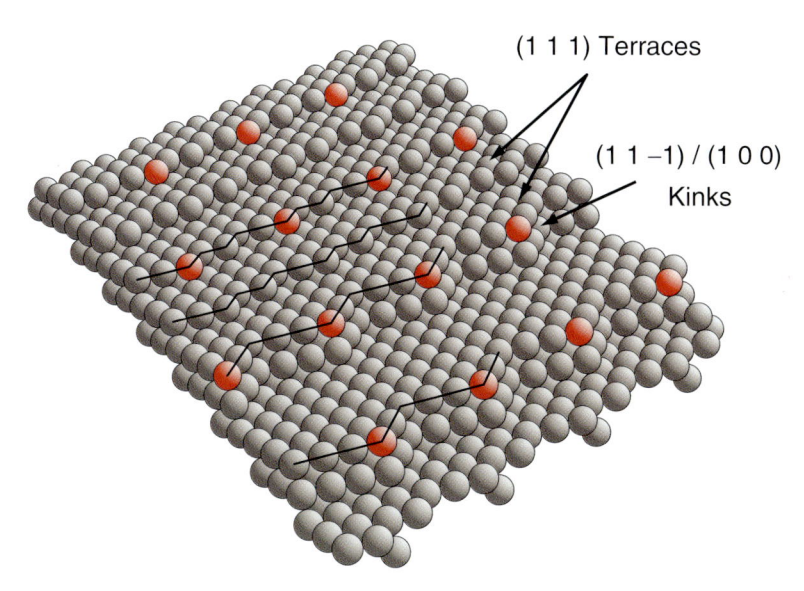

Figure 4.14 Structure of the kinked fcc(37 25 17) surface with alternating single-height kinks (back) and two lines of double-height kinks (front). The red atom balls define the (37 25 17) monolayer. Kink edges are emphasized by black lines.

(primitive crystals). A generalization to **polyatomic crystals** with more than one atom in the primitive unit cell and/or different elements is straight forward since, according to Section 2.2.1, any general crystal can be decomposed formally into a set of primitive crystals with lattices that are identical to that of the general crystal. Thus, vicinal surfaces of polyatomic crystals can be considered as superpositions of those of their primitive component crystals. However, the detailed local structure at the surfaces may be rather complicated depending on the crystal type. As an illustration Figure 4.15 shows an example of moderate complexity, the kinked (15 11 9) surface of cubic MgO. This surface can be decomposed in (1 1 1) terraces and (1 1 −1)/(2 0 0) kinks, where the terrace sections alternate between the two atom types leading to a highly polar "zebra-striped" surface.

A more complex example is given in Figure 4.16 showing the structure of the stepped (0 1 8) surface of cubic **strontium titanate**, $SrTiO_3$, (perovskite lattice). This surface exhibits alternating (0 0 1) oriented terraces of binary TiO_2 and SrO units. However, it must be emphasized that the structures of the MgO(15 11 9) and $SrTiO_3$(0 1 8) surfaces shown in Figures 4.15 and 4.16, respectively, are to some extent academic. The corresponding real crystal surfaces that can be measured are very likely to be modified by interatomic binding effects leading to local relaxation of atom positions and reconstruction at the surface. This will be discussed in greater detail in Chapter 5.

Altogether, the two **decomposition theorems** (4.9) and (4.15) can be used to characterize general $(h\ k\ l)$-indexed surfaces of single crystals by surfaces of high

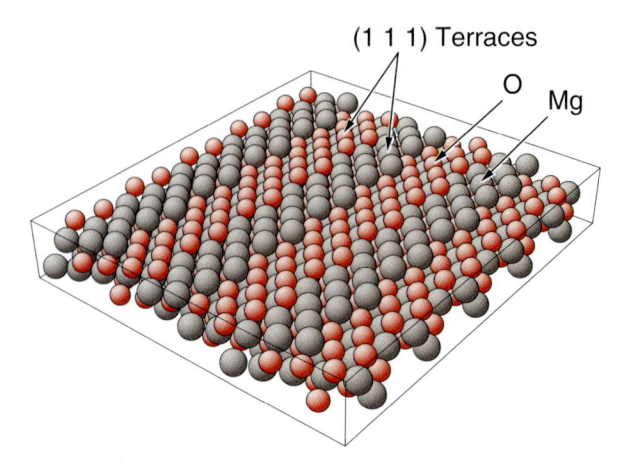

Figure 4.15 Structure of the kinked (15 11 9) surface of cubic MgO with alternating (1 1 1) terraces of Mg and O atoms. The atoms are shown in different color and labeled accordingly.

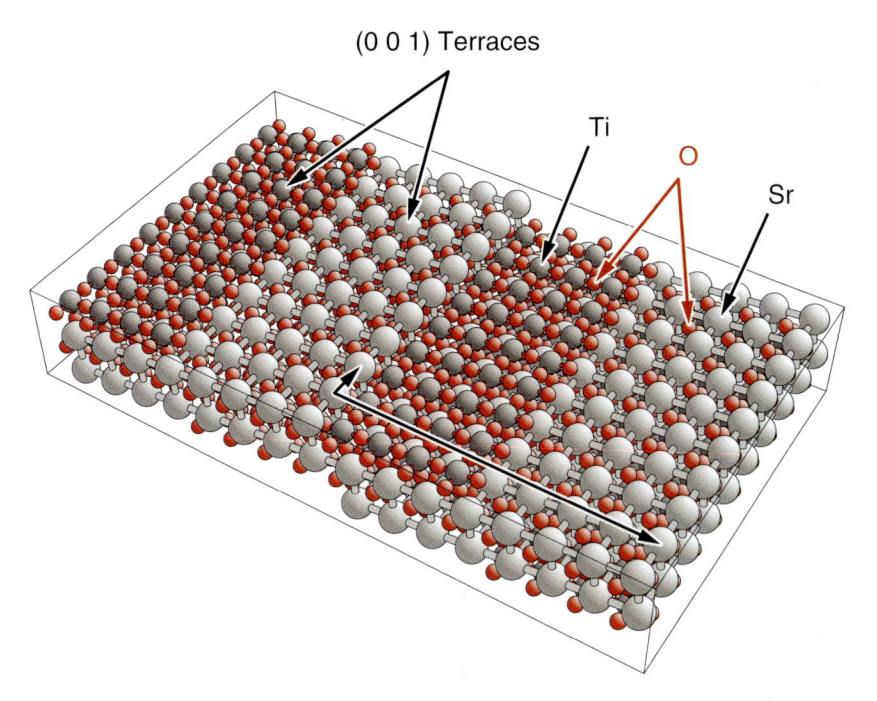

Figure 4.16 Structure of the stepped (0 1 8) surface of a SrTiO$_3$ crystal (perovskite lattice, cubic-P). The atoms are shown in different color and labeled accordingly. The surface-adapted lattice vectors (unlabeled black arrows) illustrate the surface periodicity.

atom density, usually corresponding to low Miller index values, which describe the surface morphology by combinations of dense terraces separated by steps and/or kinks. This can be achieved in the most general case by replacing the initial reciprocal lattice vectors $\underline{G}_{o1}, \underline{G}_{o1}, \underline{G}_{o1}$, which describe vector $\underline{G}_{(hkl)}$ of the $(h\,k\,l)$-indexed surface according to

$$\underline{G}_{(h\,k\,l)} = h\,\underline{G}_{o1} + k\,\underline{G}_{o2} + l\,\underline{G}_{o3} \tag{4.26}$$

by a set of transformed reciprocal lattice vectors $\underline{G}_1, \underline{G}_2, \underline{G}_3$. The latter also form a basis of the reciprocal lattice but at the same time refer to normal directions of high-density monolayers. If the transformed vectors are represented by

$$\underline{G}_i = h_i\,\underline{G}_{o1} + k_i\,\underline{G}_{o2} + l_i\,\underline{G}_{o3}, \quad i = 1, 2, 3 \tag{4.27}$$

with corresponding Miller indices $(h_i\,k_i\,l_i)$ then the transformation reads

$$\begin{pmatrix} \underline{G}_1 \\ \underline{G}_2 \\ \underline{G}_3 \end{pmatrix} = \begin{pmatrix} h_1 & k_1 & l_1 \\ h_2 & k_2 & l_2 \\ h_3 & k_3 & l_3 \end{pmatrix} \cdot \begin{pmatrix} \underline{G}_{o1} \\ \underline{G}_{o2} \\ \underline{G}_{o3} \end{pmatrix}, \quad \begin{pmatrix} \underline{G}_{o1} \\ \underline{G}_{o2} \\ \underline{G}_{o3} \end{pmatrix} = \begin{pmatrix} h_1 & k_1 & l_1 \\ h_2 & k_2 & l_2 \\ h_3 & k_3 & l_3 \end{pmatrix}^{-1} \cdot \begin{pmatrix} \underline{G}_1 \\ \underline{G}_2 \\ \underline{G}_3 \end{pmatrix} \tag{4.28}$$

Together with Eq. (4.26) this leads to

$$\underline{G}_{(h\,k\,l)} = (h\ k\ l) \cdot \begin{pmatrix} \underline{G}_{o1} \\ \underline{G}_{o2} \\ \underline{G}_{o3} \end{pmatrix} = (h\ k\ l) \cdot \begin{pmatrix} h_1 & k_1 & l_1 \\ h_2 & k_2 & l_2 \\ h_3 & k_3 & l_3 \end{pmatrix}^{-1} \cdot \begin{pmatrix} \underline{G}_1 \\ \underline{G}_2 \\ \underline{G}_3 \end{pmatrix} \stackrel{\text{def}}{=} (a_1\ a_2\ a_3) \cdot \begin{pmatrix} \underline{G}_1 \\ \underline{G}_2 \\ \underline{G}_3 \end{pmatrix} \tag{4.29}$$

where a_i, $i = 1, 2, 3$, are integer-valued coefficients that can be calculated by solving the system of linear Diophantine equations

$$\begin{pmatrix} a_1 \\ a_2 \\ a_3 \end{pmatrix} = \begin{pmatrix} h_1 & h_2 & h_3 \\ k_1 & k_2 & k_3 \\ l_1 & l_2 & l_3 \end{pmatrix}^{-1} \cdot \begin{pmatrix} h \\ k \\ l \end{pmatrix} \tag{4.30}$$

with solutions

$$a_1 = \frac{1}{\Delta}[h(k_2 l_3 - k_3 l_2) + k(l_2 h_3 - l_3 h_2) + l(h_2 k_3 - h_3 k_2)] \tag{4.31a}$$

$$a_2 = \frac{1}{\Delta}[h(k_3 l_1 - k_1 l_3) + k(l_3 h_1 - l_1 h_3) + l(h_3 k_1 - h_1 k_3)] \tag{4.31b}$$

$$a_3 = \frac{1}{\Delta}[h(k_1 l_2 - k_2 l_1) + k(l_1 h_2 - l_2 h_1) + l(h_1 k_2 - h_2 k_1)] \tag{4.31c}$$

$$\Delta = h_3(k_1 l_2 - k_2 l_1) + k_3(l_1 h_2 - l_2 h_1) + l_3(h_1 k_2 - h_2 k_1) \tag{4.31d}$$

where for generic Miller indices $\Delta = 1$ while for simple cubic Miller indices of fcc and bcc lattices $\Delta = 4$ and $\Delta = 2$, respectively, must be used, see Exercise 4.17. Thus, if the Miller indices $(h\,k\,l)$ of the surface are to be analyzed and those, $(h_i\,k_i\,l_i)$, of the corresponding high atom density surfaces are known, relations (4.31) provide the decomposition coefficients a_i. This suggests a simple **trial-and-error** method to characterize the structure of a general $(h\,k\,l)$ surface by terraces, steps, and kinks of surfaces with high atom density.

In a first step, a set of reciprocal lattice vectors $\underline{G}$ with Miller indices $(h\ k\ l)$, representing surfaces of high atom density, is evaluated. Of these, three reciprocal lattice vectors $\underline{G}_1$, $\underline{G}_2$, $\underline{G}_3$, and Miller indices $(h_1\ k_1\ l_1)$, $(h_2\ k_2\ l_2)$, and $(h_3\ k_3\ l_3)$ according to Eq. (4.27), are chosen such that they form a reciprocal lattice basis, that is,

$$\det\begin{pmatrix} h_1 & k_1 & l_1 \\ h_2 & k_2 & l_2 \\ h_3 & k_3 & l_3 \end{pmatrix} = \Delta \qquad (4.32)$$

where $\Delta = 1$ for generic Miller indices, $\Delta = 4$ and $\Delta = 2$ for sc Miller indices in fcc and bcc lattices, respectively. Then for each vector triplet $\underline{G}_1$, $\underline{G}_2$, $\underline{G}_3$ the decomposition coefficients a_1, a_2, a_3 are evaluated using Eq. (4.31) where only vector triplets with coefficients $a_i \geq 0$ need to be considered. For these solutions the Miller indices $(h_1\ k_1\ l_1)$, $(h_2\ k_2\ l_2)$, $(h_3\ k_3\ l_3)$ and coefficients a_i are rearranged such that $a_1 \geq a_2 \geq a_3 \geq 0$ after which three different scenarios of surface structures can be distinguished

1) For $\mathbf{a_2 = a_3 = 0}$ the $(h\ k\ l) = (h_1\ k_1\ l_1)$ oriented surface is characterized as a surface of **high atom density**.
2) For $\mathbf{a_1 \geq a_2 > 0}$ and $\mathbf{a_3 = 0}$ the $(h\ k\ l)$ oriented surface is **stepped** with $(h_1\ k_1\ l_1)$ representing the terrace surface (vicinal partner), with terraces of "width" a_1 and step "heights" a_2. Here, the value $a_2 = 1$ yields single atom steps while $a_2 > 1$ refers to multiple atom steps where the latter may lead to subterraces with single atom steps as discussed earlier.
3) For $\mathbf{a_1 \geq a_2 \geq a_3 > 0}$ the $(h\ k\ l)$ oriented surface is **kinked** with $(h_1\ k_1\ l_1)$ representing the terrace surface (vicinal partner). If $\mathbf{g = gcd(a_2, a_3)}$, then two cases can be distinguished,
    a. $\mathbf{g = 1}$ there are continuous kink lines of a_2 and a_3 atom vectors long sections and adjacent terraces are separated by single atom steps. Here, $a_3 = 1$ yields single atom kinks while $a_3 > 1$ refers to multiple atom kinks where the latter may result in more complex single atom kinks.
    b. $\mathbf{g > 1}$ there are continuous kink lines of a_2/g and a_3/g atom vectors long sections and adjacent terraces are separated by multiple atom steps that are g atoms high, which may result in more complex single atom steps with single atom kinks.

The decomposition is most evident for $(h\ k\ l)$ oriented surfaces where the decomposition coefficients a_i are rather different with large terraces, corresponding to a_1 distinctly larger than a_2, a_3, and, in the case of kinked surfaces, large kink lines with small kink edges, corresponding to $a_2 > 1$ and $a_3 = 1$. This must be considered when choosing meaningful vector triplets $\underline{G}_1$, $\underline{G}_2$, $\underline{G}_3$ and Miller indices $(h_1\ k_1\ l_1)$, $(h_2\ k_2\ l_2)$, $(h_3\ k_3\ l_3)$ for the decomposition. An example is the fcc(16 10 8) surface that decomposes to

$$(16\ 10\ 8) = 9(1\ 1\ 1) + 3(2\ 0\ 0) + (1\ 1\ -1)$$

On the other hand, decompositions with very similar values a_i may not yield a clear structural description. Here, an example is the fcc(21 13 3) surface shown in Figure 4.17 which decomposes to

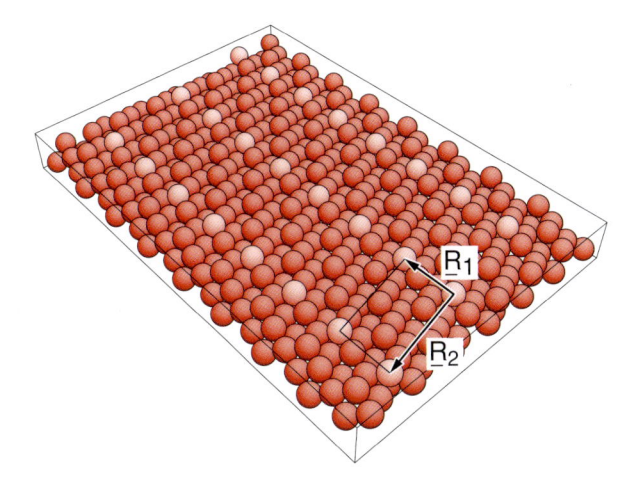

Figure 4.17 Structure of the (21 13 3) surface of an fcc crystal. The atoms of the topmost monolayer are emphasized by lighter color. The surface unit cell is outlined by black lines with surface lattice vectors $\underline{R}_1$, $\underline{R}_2$ labeled accordingly.

$$(21\ 13\ 3) = 5(2\ 2\ 0) + 4(2\ 0\ 0) + 3(1\ 1\ 1)$$

showing (1 1 1) terraces with a fairly complex kink structure. Surfaces of this type may not be called *vicinal*.

As examples of general decompositions, we consider crystals with an **fcc lattice**, where high-density surfaces are given by the Miller index families {1 1 1}, {2 0 0}, {2 2 0} with altogether 26 $(h\ k\ l)$ members. Here, useful examples of decompositions are

$$\begin{aligned}
(h\ k\ l) &= (k+l)/2\,(1\ 1\ 1) + (h-l)/2\,(1\ 1\ -1) + (h-k)/2\,(1\ -1\ 1) \\
&= (k+l)/2\,(1\ 1\ 1) + (k-l)/2\,(1\ 1\ -1) + (h-k)/2\,(2\ 0\ 0) \\
&= l\,(1\ 1\ 1) + (h-l)/2\,(2\ 0\ 0) + (k-l)/2\,(0\ 2\ 0) \\
&= h/2\,(2\ 0\ 0) + k/2\,(0\ 2\ 0) + l/2\,(0\ 0\ 2) \qquad \{\text{even } (h\ k\ l) \text{ only}\} \\
&= l\,(1\ 1\ 1) + (k-l)/2\,(2\ 2\ 0) + (h-k)/2\,(2\ 0\ 0) \qquad (4.33)
\end{aligned}$$

Crystals of **bcc lattices** offer high-density surfaces described by the Miller index families {1 1 0}, {2 0 0}, {2 1 1} with altogether 42 $(h\ k\ l)$ members. Here, useful examples of decompositions are

$$\begin{aligned}
(h\ k\ l) &= (-h+k+l)/2\,(0\ 1\ 1) + (h-k+l)/2\,(1\ 0\ 1) \\
&\quad + (h+k-l)/2\,(1\ 1\ 0) \\
&= k\,(1\ 1\ 0) + l\,(1\ 0\ 1) + (h-k-l)/2\,(2\ 0\ 0) \\
&= -k\,(1\ -1\ 0) + (h+k-l)/2\,(2\ 0\ 0) + l\,(1\ 0\ 1) \\
&= (h-k)\,(2\ 1\ 1) + (-h+2k)\,(1\ 1\ 0) + (-h+k+l)/2\,(0\ 0\ 2) \qquad (4.34)
\end{aligned}$$

The characterization of general $(h\ k\ l)$-indexed surfaces according to the above recipe allows one to distinguish between surfaces of high atom density, stepped, and kinked vicinal surfaces. While the underlying Miller index decomposition is not unique it is most general and applies to surfaces of crystals with **any Bravais lattice type**. An alternative distinction between stepped and kinked surfaces has been proposed for crystals with highly symmetric cubic and hexagonal close-packed (hcp) lattices [96]. Here, a stepped surface is defined by the atoms of its terrace edges forming linear arrays with nearest neighbor distances separating the atoms. This definition is rather intuitive but may not be applicable to crystals with general Bravais lattices.

4.4 Chiral and Achiral Surfaces

There is an additional structural property, **handedness** or **chirality** (kheir (χειρ) is Greek for "hand"), which can be used to discriminate between surfaces of single crystals [96] but is of much more general relevance [97]. For example, chiral molecules have been found to be optically active in the presence of circularly polarized light [97], and large organic biomolecules can react quite differently with their environment depending on their chiral components [98]. Further, chiral crystal surfaces have attracted much interest since their interaction with large (chiral) adsorbates has been found, in some cases, to differ dramatically depending on their chiral orientation (enantioselective adsorption) [99, 100]. This will be discussed in Section 6.7.

The formal **definition** of a general three-dimensional **chiral** object is that it cannot be superimposed onto its mirror image. This definition has also been used to describe the symmetry properties of molecules: applying a mirror operation creates an image molecule, which may or may not be brought to coincide with the initial species by simple rotations and translation. If there is no coincidence possible the molecule will be called **chiral** and in the coincidence case it will be called **achiral**. The two mirror partners of a chiral molecule are also known as **enantiomers**. As an example, Figure 4.18 shows the two enantiomers of bromochlorofluoromethane, BrClFCH, where mirroring creates two different species, a right-handed R-BrClFCH ("R-" for "rectus," latin for right) and a left-handed S-BrClFCH molecule ("S-" for "sinister," latin for left) using a nomenclature according to the stereochemical rules [101]. Here, the peripheral atoms see different arrangements of atom neighbors. For example, rotating R-BrClFCH clockwise about its C–H axis moves the bromine toward the chlorine atom whereas rotating S-BrClFCH about its C–H axis to move the bromine toward chlorine requires an anticlockwise rotation. The different orientation is also clear from the interatomic vectors $\underline{R}_1$, $\underline{R}_2$, $\underline{R}_3$, pointing from the central carbon to fluorine, chlorine, and bromine, and shown at the bottom of Figure 4.18. These vectors form a right-handed triplet while their mirror images $\underline{R}'_1$, $\underline{R}'_2$, $\underline{R}'_3$ form a left-handed triplet.

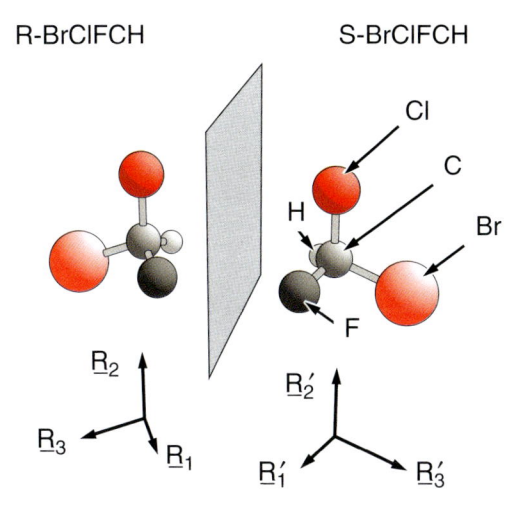

Figure 4.18 Balls-and-sticks models of the two enantiomers of bromochlorofluoromethane, BrClFCH, with a mirror plane in between. The right- and left-handed vector triplets $\underline{R}_1$, $\underline{R}_2$, $\underline{R}_3$, and $\underline{R}'_1$, $\underline{R}'_2$, $\underline{R}'_3$ referring to the R-BrClFCH and S-BrClFCH species, respectively, are shown at the bottom.

The example shows that the concept of chirality is connected mathematically with the handedness of vector triplets in three-dimensional space. A non-coplanar vector triplet $\underline{R}_1$, $\underline{R}_2$, $\underline{R}_3$ is called **right-handed** if the corresponding volume product, $V = (\underline{R}_1 \times \underline{R}_2)\,\underline{R}_3$, see Appendix F, assumes a **positive** value, whereas the triplet is considered to be **left-handed** if V is **negative**. The three point symmetry operations, mirroring, inversion, and rotation, affect the handedness of a vector triplet differently depending on the operation.

1) A **mirror operation** $\sigma(\underline{r}_o, \underline{m})$ with respect to a plane of normal vector $\underline{m}$ through the origin $\underline{r}_o = \underline{0}$ is, according to Eq. (3.89), defined by a transformation

$$\underline{r}' = \sigma(\underline{0}, \underline{m})\underline{r} = \underline{r} - 2(\underline{r}\,\underline{m})\,\underline{m} \tag{4.35}$$

As a result, the vector triplet $\underline{R}_1$, $\underline{R}_2$, $\underline{R}_3$ is transformed to $\underline{R}'_1$, $\underline{R}'_2$, $\underline{R}'_3$ where, as proven in Appendix F,

$$V' = (\underline{R}'_1 \times \underline{R}'_2)\,\underline{R}'_3 = -(\underline{R}_1 \times \underline{R}_2)\,\underline{R}_3 = -V \tag{4.36}$$

This shows that mirroring **changes** the **handedness** of vector triplets.

2) An **inversion** $i(\underline{r}_o)$ with respect to the origin $\underline{r}_o = \underline{0}$ is, according to Eq. (3.64), defined by a transformation

$$\underline{r}' = i(\underline{0})\underline{r} = -\underline{r} \tag{4.37}$$

Therefore, the vector triplet $\underline{R}_1$, $\underline{R}_2$, $\underline{R}_3$ is transformed to $\underline{R}'_1$, $\underline{R}'_2$, $\underline{R}'_3$ where the volume product with respect to the inversion center yields

$$V' = (\underline{R}'_1 \times \underline{R}'_2)\,\underline{R}'_3 = [(-\underline{R}_1) \times (-\underline{R}_2)]\,(-\underline{R}_3) = -V \tag{4.38}$$

This shows that inversion **changes** also the **handedness** of vector triplets.

3) A (clockwise) **rotation** $C_\varphi(\underline{r}_o, \underline{e})$ by an angle φ about an axis along $\underline{e}$ through the rotation center at the origin $\underline{r}_o = \underline{0}$ is, according to Eqs. (3.73) and (3.74) and using a representation

$$\underline{r} = x_1 \underline{e}_1 + x_2 \underline{e}_2 + x_3 \underline{e}_3 \quad x_i = \underline{r}\,\underline{e}_i, \quad i = 1, 2, 3 \tag{4.39}$$

where $\underline{e}_1, \underline{e}_2, \underline{e}_3 = \underline{e}$ are Cartesian unit vectors with respect to the rotation axis along $\underline{e}_3$, defined by a transformation

$$\underline{r}' = C_\varphi \underline{r} = x_1' \underline{e}_1 + x_2' \underline{e}_2 + x_3' \underline{e}_3 \tag{4.40}$$

with

$$\begin{pmatrix} x_1' \\ x_2' \\ x_3' \end{pmatrix} = \begin{pmatrix} \cos\varphi & -\sin\varphi & 0 \\ \sin\varphi & \cos\varphi & 0 \\ 0 & 0 & 1 \end{pmatrix} \cdot \begin{pmatrix} x_1 \\ x_2 \\ x_3 \end{pmatrix} \tag{4.41}$$

Thus, the vector triplet $\underline{R}_1, \underline{R}_2, \underline{R}_3$ is transformed to $\underline{R}_1', \underline{R}_2', \underline{R}_3'$ where the volume product with respect to the rotation center at the origin $\underline{r}_o = \underline{0}$ is yields

$$V' = (\underline{R}_1' \times \underline{R}_2')\,\underline{R}_3' = (\underline{R}_1 \times \underline{R}_2)\,\underline{R}_3 = V \tag{4.42}$$

as can be proven by simple calculus using Eqs. (4.39) to (4.41). This shows that rotation **does not change** the **handedness** of vector triplets.

Altogether, inversion and mirroring change the handedness of the corresponding vector triplets while rotation does not. Thus, applying a combination of different rotations with a mirror operation or with an inversion will always change the handedness of vector triplets describing the atom positions of a molecule and may lead to a molecule of different conformation. In this case the molecule is called **chiral**. On the other hand, a molecule with mirror or inversion symmetry will not change its shape when mirroring or inversion with rotations are applied and the molecule is called **achiral**. Therefore, **chirality** can be based on the behavior of a system with respect to mirroring or inversion and, hence, the existence of corresponding **mirror planes** or **inversion centers**. This equivalence is also clear from the fact that the inversion operation $i(\underline{r}_o)$ can always be represented by a combination of mirroring $\sigma(\underline{r}_o, \underline{m})$ and a 180° rotation $C_{180}(\underline{r}_o, \underline{m})$

$$i(\underline{r}_o) = \sigma(\underline{r}_o, \underline{m})\, C_{180}(\underline{r}_o, \underline{m}) = C_{180}(\underline{r}_o, \underline{m})\, \sigma(\underline{r}_o, \underline{m}) \tag{4.43}$$

which can be proven using the above transformations. In the following text, we will focus on the definition of chirality based on mirror symmetry, which is common practice in the literature.

It should be mentioned in passing that, in general, chirality is associated with atom centers in a molecule, so-called **chiral centers**, where mirror operations change the handedness of the atom arrangement near the center. In the BrClFCH molecule shown in Figure 4.18, the carbon center acts as the chiral center distinguishing between R-BrClFCH and S-BrClFCH. However, larger molecules may contain several chiral centers. As an example, Figure 4.19 shows the tartaric acid (TA) molecule [102], $C_4O_6H_6$, including two chiral carbon centers, denoted as C*

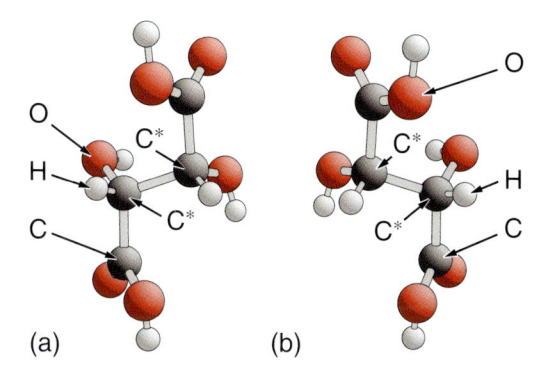

Figure 4.19 Balls-and-sticks models of the two enantiomers of tartaric acid (TA) from crystal data [102], (a) (R,R)-TA and (b) (S,S)-TA. The atom balls labeled C* refer to the two chiral carbon centers.

in the figure, which can be both left- and right-handed. Hence, there are four possible isomeric species. In (R,R)-TA and (S,S)-TA, see Figure 4.19, both carbon centers are left-handed or both right-handed according to the Cahn–Ingold–Prelog rules [101] and the two species are enantiomers connected by mirror imaging. In the other two, (L,R)-TA and (R,L)-TA, one of the carbon centers is left-handed and the other right-handed such that they are structurally identical and achiral (usually referred to as *meso tartaric acid* and diastereoisomeric with respect to the chiral species).

The concept of **chirality** can also be applied to extended systems such as **bulk single crystals**. For example, according to the basic definition, a primitive crystal that contains inversion centers by definition will always be achiral. Surfaces of ideal single crystals are terminated by $(h\ k\ l)$ oriented monolayers that are described by netplane-adapted lattice vectors $\underline{R}_1$ and $\underline{R}_2$, where a stacking vector $\underline{R}_3$ connects adjacent parallel monolayers, see Section 4.1. Here, a **surface** can be considered to be **chiral** if it does not exhibit mirror symmetry along any plane that is perpendicular to the surface. These surfaces always have chiral partners that can be obtained by applying a mirror operation to the initial surface with the mirror plane perpendicular to the surface. The mirroring transforms the netplane-adapted lattice vectors $\underline{R}_1$ and $\underline{R}_2$ such that the vector product $\underline{R}_1 \times \underline{R}_2$ changes its sign but not its absolute value. This vector $\underline{R}_1 \times \underline{R}_2$ defines (up to a constant factor) the reciprocal lattice vector $\underline{G}_{(h\ k\ l)}$ and, hence, the Miller indices of the surface, see Eq. (3.6). Therefore, the **chiral partner** of the $(h\ k\ l)$ surface is defined by Miller indices $(-h\ -k\ -l)$.

The two chiral partners of a $(h\ k\ l)$-oriented surface may also be assigned a **handedness**, which is most evident for kinked surfaces of primitive crystals. It was shown previously, see Eq. (4.15), that the reciprocal lattice vector $\underline{G}_{(h\ k\ l)}$ of a kinked surface can be written as a sum of three contributions $p_t\ \underline{G}_{(h\ k\ l)t}$, $p_{s1}\ \underline{G}_{(h\ k\ l)s1}$, and $p_{s2}\ \underline{G}_{(h\ k\ l)s2}$, characterizing its terrace orientation as well as those of the two steps defining each kink. Sorting the three contributions according to their lengths yields the three vectors $\underline{G}_i$, $i = 1-3$, pointing out of the surface where $\underline{G}_1$

4 Ideal Single Crystal Surfaces

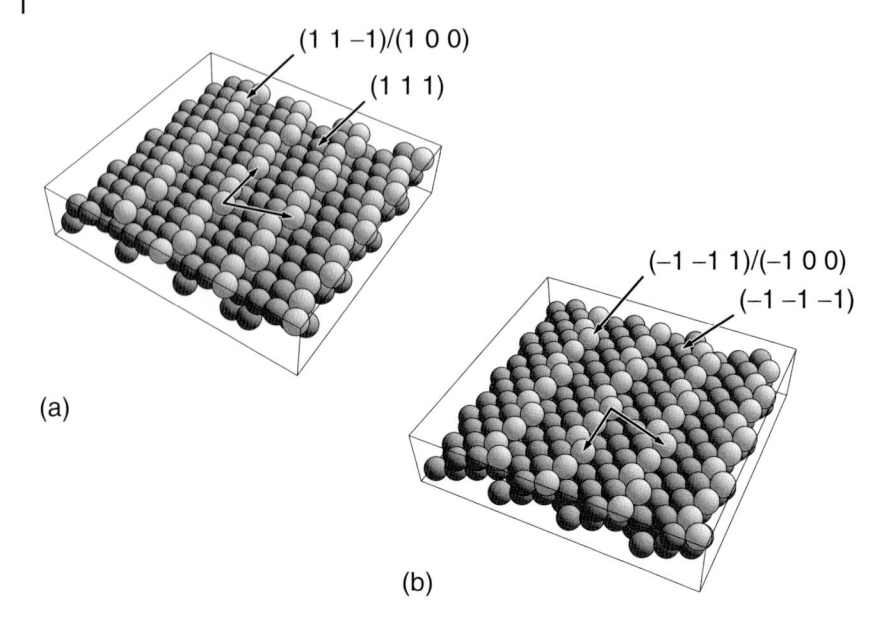

Figure 4.20 Structure of the (a) kinked fcc(11 9 5)^R surface with (1 1 1) terraces and (1 1 −1)/(1 0 0) kinks, (b) chiral partner surface fcc(−11 −9 −5)^S. The atoms along the kink lines are emphasized by light balls. The surface-adapted lattice vectors illustrate the surface periodicity.

is assumed to be of the smallest length and $\underline{G}_3$ of the largest. Then the $(h\,k\,l)$ surface is called **right-handed**, also denoted as $(h\,k\,l)^R$, if the vectors $\underline{G}_1, \underline{G}_2, \underline{G}_3$ form a right-handed system, quantified by the volume product

$$V_{ch} = (\underline{G}_1 \times \underline{G}_2)\,\underline{G}_3 \tag{4.44}$$

being **positive**, whereas the surface is called **left-handed**, denoted as $(h\,k\,l)^S$, if $\underline{G}_1, \underline{G}_2, \underline{G}_3$ form a left-handed system and V_{ch} is **negative**. This assignment is **unique** if the lengths of the three vectors $\underline{G}_i$ are all different and is compatible with the nomenclature proposed in the literature [103, 104].

As an example, Figure 4.20a shows the chiral (11 9 5) surface of an fcc crystal, described by (1 1 1) terraces with kinks of (1 1 −1) and (1 0 0) steps (confirming the additivity relation (11 9 5) = 7 (1 1 1) + 2 (1 1 −1) + (2 0 0)). The corresponding volume product V_{ch} equals to

$$V_{ch} = \alpha\,((2\,0\,0) \times (2\,2\,-2))\,(7\,7\,7) = 56\alpha, \quad \alpha > 0$$

where α is a global positive constant. Hence, the surface is right-handed and may be written as (11 9 5)^R. The chiral partner surface (−11 −9 −5) shown in Figure 4.20b (a mirrored copy of the surface section in Figure 4.20a) is left-handed and may be termed (−11 −9 −5)^S. The kink lines, emphasized by light balls in Figure 4.20, show the difference between the two surfaces quite clearly. In contrast, fcc bulk crystals are intrinsically achiral in three dimensions since they contain multiple mirror symmetry. However, none of the mirror planes is perpendicular to the (11 9 5) surface which is why the corresponding surface is chiral.

There are cases where, after sorting, two of the vectors $\underline{G}_i$ are equal in length. Then the above sorting process does not lead to a unique solution and the volume product V_{ch} can become positive or negative depending on the sequence chosen for the vectors $\underline{G}_i$. This can appear for general Bravais lattices where the concept of dense monolayers, forming terraces separated by steps and kinks, is less evident and the concept of chirality may not be applicable. Here, sorting the reciprocal lattice vectors $\underline{G}_{(h\,k\,l)t}$, $\underline{G}_{(h\,k\,l)s1}$, $\underline{G}_{(h\,k\,l)s2}$ themselves according to their length can provide a unique assignment. In cases where two vectors $\underline{G}_i$ are equal in length and also their scalar prefactors agree with each other, the assignment of a handedness of the surface becomes unclear. This happens, for example, for stepped surfaces with open step edges such that the edges may also be interpreted as kink lines. A simple example is the achiral fcc (4 1 0) surface shown in Figure 4.21 whose Miller indices can be decomposed as

$$(4\,1\,0) \equiv (8\,2\,0) = 3(2\,0\,0) + (2\,2\,0)$$

suggesting a stepped surface with (1 0 0) oriented terraces separated by (1 1 0) steps or as

$$(4\,1\,0) \equiv (8\,2\,0) = 3(2\,0\,0) + (1\,1\,1) + (1\,1\,-1)$$

suggesting a kinked surface with (1 0 0) oriented terraces separated by kink lines of symmetric (1 1 1) and (1 1 −1) step sections.

As stated earlier, an $(h\ k\ l)$ surface is considered to be **achiral** only if there is at least one mirror plane of the crystal which is perpendicular to the surface. Thus, all parallel $(h\ k\ l)$ **monolayers** that are stacked from the surface toward the bulk must **share** one or several **mirror planes** perpendicular to the surface. As an example, Figure 4.22 shows the achiral (3 3 1)

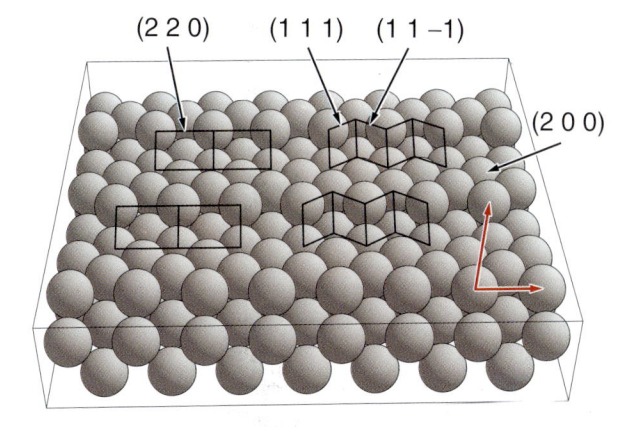

Figure 4.21 Structure of the fcc(4 1 0) surface described by (1 0 0) terraces with (1 1 0) steps (left part) or with (1 1 −1)/(1 1 −1) kinks (right part). The step and kink areas are outlined by black lines. The surface-adapted lattice vectors in red illustrate the overall surface periodicity.

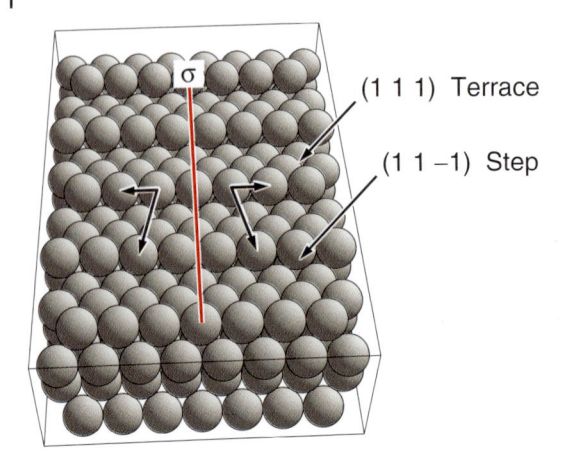

Figure 4.22 Structure of the achiral stepped fcc(3 3 1) surface with (1 1 1) terraces and (1 1 −1) steps. The mirror plane perpendicular to the surface is indicated by a red line labeled σ. The netplane-adapted lattice vectors (left- and right-handed) illustrate the surface periodicity.

surface of an fcc crystal, described by (1 1 1) terraces with (1 1 −1) steps ((3 3 1) = 2 (1 1 1) + (1 1 −1) according to the additivity theorem). The mirror plane perpendicular to the steps and, thus, perpendicular to the surface is indicated in Figure 4.22 by the red line labeled σ. In addition, the isolated topmost monolayer (formed by the atoms at the step edges) as well as all the underlying isolated monolayers contain a mirror line on the mirror plane labeled σ.

In Section 3.6.4 it was shown that netplanes containing mirror lines correspond to either primitive rectangular (including square) or centered rectangular (including hexagonal) lattices in two dimensions. This **two-dimensional** symmetry applies to each separate monolayer near the (*h k l*) surface. However, the overall **three-dimensional** symmetry and morphology near the surface, determining the surface chirality, is also influenced by the netplane-adapted vector $\underline{R}_3$ connecting adjacent monolayers. For example, the fictitious monoatomic triclinic crystal with lattice vectors

$$\underline{R}_1 = a\,(1,\,0\,,\,0), \quad \underline{R}_2 = a\,(0,\,1,\,0), \quad \underline{R}_3 = a\,(1/\sqrt{2},\,1/\sqrt{3},\,1/2) \quad (4.45)$$

yields a (0 0 1) oriented surface as shown in Figure 4.23 in a parallel projection perpendicular to the surface. Here, each of the (0 0 1) monolayers with lattice vectors $\underline{R}_1$, $\underline{R}_2$ given by Eq. (4.45) is of square symmetry including mirror lines. However, the (0 0 1) surface combining all monolayers has no mirror symmetry and is chiral.

In addition to chiral surfaces, there is often an infinite but discrete set of **achiral** (*h k l*) surfaces of a crystal depending on its symmetry. This set can be determined by simple geometric considerations. Achiral (*h k l*) surfaces have been defined by the existence of at least one mirror plane of the corresponding bulk crystal pointing perpendicular to the surface. This means, in particular, that the **mirror plane normal** vector $\underline{m}$ must point **parallel** to the surface. Hence, vector $\underline{m}$ must

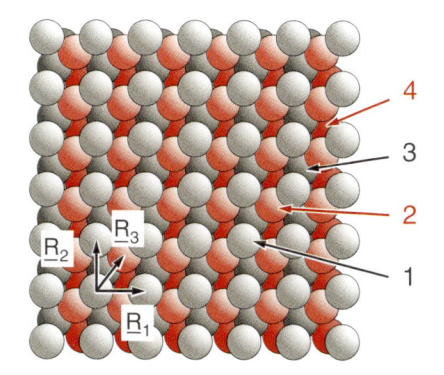

Figure 4.23 Structure of the chiral (0 0 1) surface of a fictitious primitive triclinic crystal, see text. The netplane-adapted lattice vectors $\underline{R}_1$, $\underline{R}_2$, $\underline{R}_3$ illustrate the monolayer and bulk periodicity. The atoms of the topmost four monolayers are painted differently and labeled accordingly.

be perpendicular to the surface normal vector pointing along the reciprocal lattice vector $\underline{G}_{(h\,k\,l)}$. This can be used to find the Miller indices of all possible achiral $(h\,k\,l)$ surfaces of a single crystal.

Consider an $(h\,k\,l)$ surface of an ideal single crystal with lattice vectors $\underline{R}_{o1}$, $\underline{R}_{o2}$, $\underline{R}_{o3}$. Then the normal vector $\underline{m}$ of any mirror plane defined by $\sigma(\underline{r}_o, \underline{m})$ in the crystal can be represented by

$$\underline{m} = x_1 \underline{R}_{o1} + x_2 \underline{R}_{o2} + x_3 \underline{R}_{o3} \tag{4.46}$$

The corresponding reciprocal lattice of the crystal is given by vectors $\underline{G}_{o1}$, $\underline{G}_{o2}$, $\underline{G}_{o3}$ according to Eq. (2.95) and an $(h\,k\,l)$ surface is defined by its normal vector along

$$\underline{G}_{(h\,k\,l)} = h\underline{G}_{o1} + k\underline{G}_{o2} + l\underline{G}_{o3} \tag{4.47}$$

Thus, the **condition** of an **achiral** surface with vector $\underline{m}$ perpendicular to $\underline{G}_{(h\,k\,l)}$ results in a linear equation

$$(\underline{G}_{(h\,k\,l)} \underline{m}) = x_1 h + x_2 k + x_3 l = 0 \tag{4.48}$$

where the orthogonality relation of real and reciprocal lattice vectors (2.96) has been used. Therefore, all achiral $(h\,k\,l)$ surfaces of a single crystal can be obtained by considering normal vectors $\underline{m}$ of all mirror planes of the crystal and then selecting $(h\,k\,l)$ according to Eq. (4.48) for each vector $\underline{m}$. This procedure is most general and applies to single crystals with any lattice.

As examples of finding all achiral surfaces we consider **cubic crystals** that may be simple, face-, or body-centered. The simple cubic lattice with lattice vectors

$$\underline{R}_1^{sc} = a\,(1,\,0,\,0), \quad \underline{R}_2^{sc} = a\,(0,\,1,\,0), \quad \underline{R}_3^{sc} = a\,(0,\,0,\,1) \tag{4.49}$$

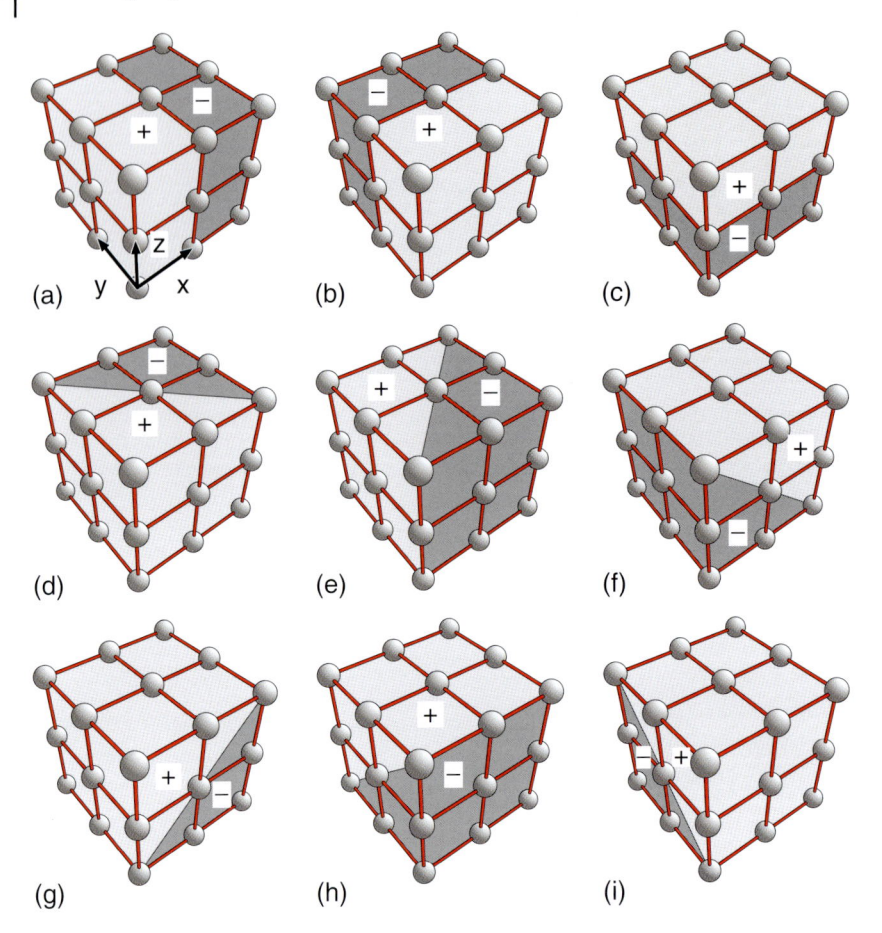

Figure 4.24 The nine mirror planes of a primitive simple cubic crystal, (a–i). Crystal atoms of nearest neighbors are connected by sticks. Mirror planes are indicated by boundaries between light and dark regions labeled + and −. All mirror planes go through the center of the cube. Cartesian coordinates are included in (a).

in Cartesian coordinates offers **nine** mirror planes, as shown in Figure 4.24, which are described by normal vectors $\underline{m}$ with

$$\underline{m} = (1, 0, 0), \quad (0, 1, 0), \quad (0, 0, 1), \quad \text{and}$$
$$\underline{m} = \alpha(1, \pm1, 0), \; \alpha(1, 0, \pm1), \; \alpha(0, 1, \pm1), \quad \alpha = 1/\sqrt{2} \quad (4.50)$$

According to Eq. (4.48), this results in **nine** different sets of Miller indices $(h\,k\,l)$ describing normal directions of achiral surfaces of simple cubic crystals, which are listed in Table 4.4.

Fcc and bcc lattices that describe many metal single crystals, share all mirror planes with those of the sc lattice. Therefore, achiral surfaces of the corresponding crystals are characterized by all sets of Miller indices (in simple cubic notation) given in Table 4.4. This also applies to the polyatomic crystals MgO and NaCl,

4.4 Chiral and Achiral Surfaces

Table 4.4 Possible sets of Miller indices $(h\ k\ l)$ describing orientations of achiral surfaces of cubic crystals.

Set	Constraint (4.48)	$(h\ k\ l)$
1	$h = 0$	(0 m n)
2	$k = 0$	(m 0 n)
3	$l = 0$	(m n 0)
4, 5	$h \pm k = 0$	(m ±m n)
6, 7	$h \pm l = 0$	(m n ±m)
8, 9	$k \pm l = 0$	(m n ±n)

Parameters m, n are integer-valued with at least one being non-zero.

described by fcc lattices, or to CsCl, described by an simple cubic lattice. Note that the Miller indices in simple cubic notation given in Table 4.4 can be applied to fcc and bcc crystals without the constraints discussed in Section 3.4 since they describe directions only.

The **achiral** surfaces of **fcc** crystals listed in Table 4.4 can all be connected with flat high-density surfaces, determined by low Miller index directions (1 1 1), (0 0 1), and (0 1 1) with their symmetry equivalents. They also appear for stepped surfaces composed of high-density terraces and steps as discussed in Sections 4.2 and 4.3. This is clear from the additivity theorem of Miller indices (Eq. (4.9)) for stepped surfaces, which allows decomposing the indices of all sets in Table 4.4. As examples, consider sets 1 and 4 with positive indices m, n. Here, the additivity theorem (note the fcc constraints for Miller indices in simple cubic notation required for quantitative evaluations) yields

$$\text{set } 1 : (0\ 2m\ 2n) = m\,(0\ 2\ 0) + n\,(0\ 0\ 2)$$
$$= (m - n)\,(0\ 2\ 0) + n\,(0\ 2\ 2) \qquad m \geq n$$
$$= (n - m)\,(0\ 0\ 2) + m\,(0\ 2\ 2) \qquad m \leq n \qquad (4.51a)$$

$$\text{set } 4 : (2m\ \pm 2m\ 2n) = (m + n)\,(1\ \pm 1\ 1) + (m - n)\,(1\ \pm 1 - 1) \quad m \geq n$$
$$= 2m\,(1\ \pm 1\ 1) + (n - m)\,(0\ 0\ 2) \qquad m \leq n$$
$$(2m + 1\ \pm (2m + 1)\ 2n + 1) =$$
$$= (m + n + 1)\,(1\ \pm 1\ 1) + (m - n)\,(1\ \pm 1 - 1), \quad m \geq n$$
$$= (2m + 1)\,(1\ \pm 1\ 1) + (n - m)\,(0\ 0\ 2), \qquad m \leq n$$
$$(4.51b)$$

with the other sets leading to analogous results. As an illustration, Figure 4.22 shows a model of the achiral stepped fcc(3 3 1). This may suggest that all stepped $(h\ k\ l)$ surfaces of fcc crystals are achiral, which can be proven mathematically for surfaces with steps formed by atom rows with smallest or second smallest interatomic distance, see Section 4.2. Larger interatomic distances result in kinked surfaces that are chiral.

The **achiral** surfaces of **bcc** crystals listed in Table 4.4 can also be connected with flat high-density surfaces, determined here by low Miller index directions (1 1 0), (1 0 0), (2 1 1) including their symmetry equivalents. However, in contrast to fcc crystals stepped ($h\,k\,l$) surfaces of bcc crystals may be either achiral or chiral. Examples of achiral stepped surfaces from Table 4.4 (with bcc constraints for Miller indices in simple cubic notation and m, n > 0) are

$$\text{set 1}: (0\ m\ m+2n) = m\,(0\ 1\ 1) + n\,(0\ 0\ 2) \tag{4.52a}$$

$$\text{set 4}: (m \pm m\ 2n) = m\,(1 \pm 1\ 0) + n\,(0\ 0\ 2) \tag{4.52b}$$

In contrast, Figure 4.25 shows a model of the stepped chiral (1 2 3) surface of a bcc iron crystal (according to the additivity theorem (1 2 3) = 2 (0 1 1) + (1 0 1)) together with its chiral (−1 −2 −3) partner surface. In this figure the black lines perpendicular to the steps at both surfaces illustrate the missing mirror symmetry that results in chirality.

The three-dimensional lattice of a **hexagonal** crystal can be described in Cartesian coordinates by lattice vectors (obtuse representation, see Section 2.2.2.1)

$$\underline{R}_1 = a\,(1,\ 0,\ 0),\quad \underline{R}_2 = a\,(-1/2,\ \sqrt{3}/2,\ 0),\quad \underline{R}_3 = c\,(0,\ 0,\ 1) \tag{4.53}$$

and includes **seven** mirror planes as shown in Figure 4.26. The corresponding normal vectors $\underline{m}$ can be described by

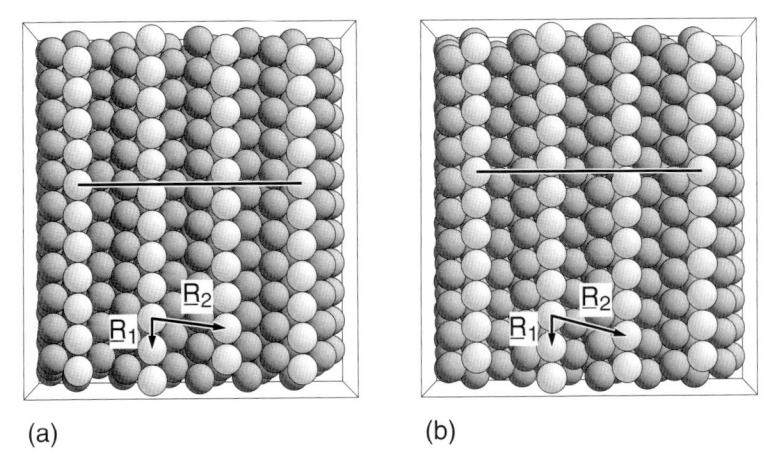

$$\underline{m} = (\cos\varphi,\ \sin\varphi,\ 0),\ \varphi = 0°,\ 30°,\ 60°,\ 90°,\ 120°,\ 150°,\ \text{and}$$
$$\underline{m} = (0,\ 0,\ 1) \tag{4.54}$$

Figure 4.25 Structure of the (a) perfect stepped (1 2 3) surface of bcc iron with (0 1 1) terraces and (1 0 1) steps, and (b) its chiral partner surface given by (−1 −2 −3). The atom balls along the step lines are emphasized by light color. Black lines perpendicular to the steps indicate the chirality of the surfaces, see text. The netplane-adapted lattice vectors $\underline{R}_1$ and $\underline{R}_2$ illustrate the surface periodicity.

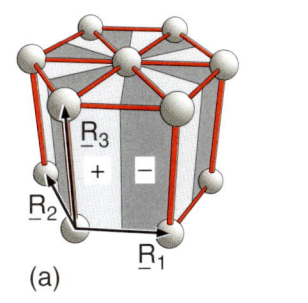

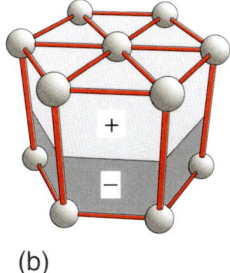

Figure 4.26 The seven mirror planes of a hexagonal crystal, (a) six vertical planes, (b) one horizontal plane. Crystal atoms of nearest neighbors are connected by sticks. Mirror planes are indicated by boundaries between light and dark regions labeled + and −. All mirror planes go through the center of the hexagonal prism. The lattice vectors are included in (a).

or by seven lattice directions along

$$\underline{R} = \underline{R}_1, \quad \underline{R} = 2\underline{R}_1 + \underline{R}_2, \quad \underline{R} = \underline{R}_1 + \underline{R}_2, \quad \underline{R} = \underline{R}_1 + 2\underline{R}_2$$
$$\underline{R} = \underline{R}_2, \quad \underline{R} = -\underline{R}_1 + \underline{R}_2, \quad \underline{R} = \underline{R}_3 \quad (4.55)$$

This results, according to Eq. (4.48), in **seven** different sets of Miller indices $(h\,k\,l)$, $(l\,m\,n\,q)$ describing normal directions of achiral single crystal surfaces, which are listed in Table 4.5.

As an illustration, Figure 4.27 shows a model of the perfect stepped $(1\,0\,-1\,5)$ surface of a **hexagonal cobalt** crystal that is achiral. According to the additivity theorem, the corresponding Miller–Bravais indices can be decomposed to yield

$$(1\,0\,-1\,5) = 5\,(0\,0\,0\,1) + (1\,0\,-1\,0)$$

which refers to set 5 of Table 4.5. Here, the crystal lattice is given by hexagonal lattice vectors $\underline{R}_1, \underline{R}_2, \underline{R}_3$ according to Eq. (4.53) with $c/a = 1.623$ for cobalt, quite

Table 4.5 Possible sets of Miller indices $(h\,k\,l)$, $(l\,m\,n\,q)$ describing orientations of achiral surfaces of hexagonal crystals.

Set	Constraint (4.48)	$(h\,k\,l)$ generic	$(l\,m\,n\,q)$ Miller–Bravais
1	$h = 0$	(0 m n)	(0 m −m n)
2	$2h + k = 0$	(m −2m n)	(m −2m m n)
3	$h + k = 0$	(m −m n)	(m −m 0 n)
4	$h + 2k = 0$	(2m −m n)	(2m −m −m n)
5	$k = 0$	(m 0 n)	(m 0 −m n)
6	$h − k = 0$	(m m n)	(m m −2m n)
7	$l = 0$	(m n 0)	(m n −m −n 0)

The Miller indices are given in generic and in Miller–Bravais (four-index) notation. Parameters m, n are integer-valued with at least one being non-zero.

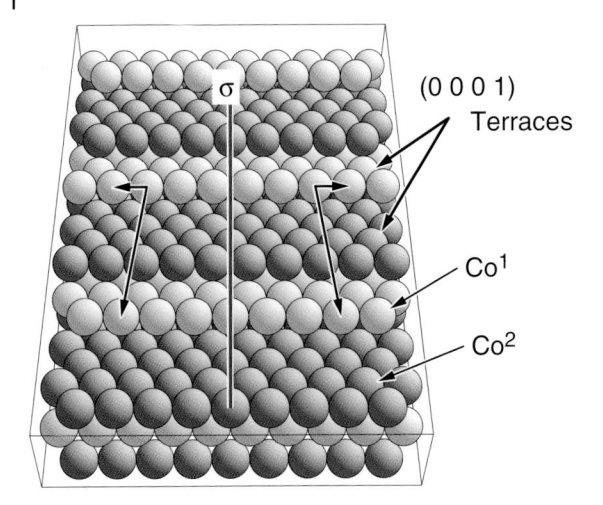

Figure 4.27 Structure of the stepped (1 0 −1 5) surface of hexagonal (hcp) cobalt with (0 0 0 1) terraces. The atoms Co^1 and Co^2, referring to the two atoms in the primitive hcp unit cell, are distinguished by light and dark gray. The mirror plane perpendicular to the surface is indicated by a red line labeled σ. The netplane-adapted lattice vectors (left- and right-handed) illustrate the surface periodicity.

close to the value $\sqrt{(8/3)} = 1.633$ for the ideal hcp crystal. Further, the primitive unit cell contains two atoms located at

$$\underline{r}_1 = 0, \quad \underline{r}_2 = 2/3\,\underline{R}_1 + 1/3\,\underline{R}_2 + 1/2\,\underline{R}_3 \tag{4.56}$$

These atoms, denoted by Co^1 and Co^2 in Figure 4.27, form alternating hexagonal (0 0 0 1) terraces of different widths at the (1 0 −1 5) surface and are separated by (1 0 −1 0) steps. The hexagonal shape of the terraces and their relative positioning leads to mirror planes perpendicular to the steps and, thus, to an achiral (1 0 −1 5) surface.

4.5
Exercises

4.1 Determine the densest (close-packed) surfaces of (a) sc, (b) fcc, (c) bcc, (d) hcp, (e) diamond, and (f) CsCl crystals.

4.2 How many differently terminated surfaces for given $(h\,k\,l)$ are there for perfect crystals of Ni, GaAs, NaCl, CsCl, graphite? Give the maximum number of terminations and determine Miller indices of corresponding surface orientations. Find orientations with less than the maximum number of terminations.

4.3 Determine Miller indices of polar and nonpolar surfaces of NaCl and CsCl crystals. Hint: Polar surfaces of these crystals are monoatomic.

4.4 Analyze the surfaces of an fcc crystal with sc Miller indices by their structure (assume m > 1)

(a) (0 1 m)
(b) (1 1 m)
(c) (m−1 m m+1)
(d) (m m m+2)
(e) (m m m+4)
(f) (7 8 11)
(g) (1 31 108)

Characterize terraces, steps, and kinks by their orientations and widths/heights.

4.5 Consider the rutile TiO_2 crystal defined in Exercise 2.23. Analyze the ideal surfaces of the bulk truncated crystal with orientations

(a) (0 0 1), (b) (1 0 0), (c) (0 1 1), (d) (1 1 1)

Determine for each orientation the number and structure of different terminations. Find point symmetry elements of the TiO_2 bulk crystal, which also appear at the surface.

4.6 Which Miller index values (h k l) of the simple cubic lattice are not strictly valid for numerical evaluations of fcc and bcc lattices when sc indexing is used? Characterize the netplanes and surface structures described by Miller indices (in sc notation)
(a) (h k l) = (2m 2m 2p+1) for crystals with an fcc lattice.
(b) (h k l) = (2m 2m 2p+1) for crystals with a bcc lattice.
Discuss example surfaces.

4.7 Determine the conditions required for surfaces of bcc crystals to possess steps consisting of

(a) atom rows with smallest interatomic distance. Show that the corresponding Miller indices can be represented by (m n ±(m+n)) and (m n ±(m−n)) in sc notation.
(b) atom rows with second smallest interatomic distance. Show that the corresponding Miller indices can be represented by (0 m n), (m 0 n), and (m n 0) in sc notation.

4.8 Determine Miller indices of a crystal with a bcc lattice and a surface consisting of six (nearest neighbor) atom distances wide terraces with single atom steps.

4.9 Give an example of a kinked surface of a silicon crystal with (1 1 1) terraces.

4.10 Visualize the facet edge of a stepped surface of an fcc crystal where (1 1 1) and (1 0 0) indexed surfaces join (Miller indices in sc notation).

4.11 Determine neighbor shells (first to fifth neighbors) for atoms of the first, second, ... surface layer of the (1 0 0), (1 1 0), and (1 1 1) surfaces of (a) fcc and (b) bcc crystals.
From which surface layer on are the third neighbor shells complete (reflecting those of the bulk)?

4.12 Decompose the Miller index triplets $(h\ k\ l)$ of netplanes into those of densely packed netplanes and give the formal decomposition relations for

(a) fcc (7 9 9)
(b) bcc (1 1 10)
(c) hcp (0 0 0 1)
(d) hcp (5 1 −6 0)
(e) sc (7 8 11)

Characterize the surfaces with Miller index orientations $(h\ k\ l)$ given in (a–e).

4.13 Give alternative netplane decompositions for the (4 3 1) surface of a crystal with an fcc lattice (sc notation). Visualize the decompositions.

4.14 Build a surface of a crystal with an fcc lattice that consists of alternating six and seven atom distances long kinks and determine the corresponding sc Miller indices. Which general irregularities of the kink sequences can arise for crystals with an fcc lattice?

4.15 Show that the stepped (1 2 3) surface of a crystal with a bcc lattice is chiral. Discuss the geometric structure of the two chiral partner surfaces.

4.16 Show that the hexagonal graphite crystal, defined in Exercise 2.7, allows all achiral surfaces given in Table 4.5.

4.17 Prove that in the Miller index decomposition for Miller indices of fcc and bcc lattices using the sc notation, a scaling factor $\Delta = 4$ and $\Delta = 2$, respectively, is required in relations (4.31).

4.18 Molecules with $n > 1$ chiral centers allow 2^n different arrangements of their left- and right-handed centers distinguishing between isomeric species. Of these, pairs of molecules are enantiomers if their chiral centers are complementary in handedness. (The members of other pairs are called *diastereomers*.)

(a) Show that there are up to $2^{(n-1)}$ different enantiomer pairs.
(b) For which chiral arrangements can the molecules with $n > 1$ chiral centers become achiral?

4.19 Determine the chirality (left- or right-handedness, denoted X = L, S in $(h\ k\ l)^X$) of the following kinked surfaces of cubic bulk substrate (Miller indices in sc notation).

(a) sc: $(6\ 2\ 1)^X$, $(7\ 5\ 1)^X$, $(10\ 3\ 1)^X$, $(10\ 7\ 1)^X$, $(13\ 3\ 1)^X$.

(b) fcc: $(5\ 3\ 1)^X$, $(6\ 4\ 3)^X$, $(8\ 5\ 4)^X$, $(8\ 7\ 4)^X$, $(10\ 8\ 5)^X$, $(17\ 1\ 3)^X$.

(c) bcc: $(11\ 10\ 3)^X$, $(12\ 7\ 1)^X$, $(16\ 3\ 1)^X$, $(16\ -3\ -1)^X$, $(16\ 1\ 3)^X$.

5
Real Crystal Surfaces

Atoms at **real crystal surfaces** appearing in nature experience a different local binding environment (connected with different atom coordination) as compared to atom sites inside the bulk crystal. This leads to **structures** of real surfaces, which **differ** from those of simple **bulk truncation** discussed for ideal single crystal surfaces. The differences may be rather small, examples are many elemental metal surfaces, but can also be quite substantial for semiconductor or oxide surfaces. Real surfaces can be restructured locally by bond changes including making and breaking of bonds, which may result in an overall **disordered** structure. In many other cases, surfaces will still exhibit a two-dimensionally periodic atom arrangement. However, the periodicity, specific atom positions, and the placement of atom layers may be different from those of bulk layers. These effects are usually described by surface **relaxation** and **reconstruction**, where details as well as nomenclature have been treated differently in the literature [94]. However, the basic concepts discussed in this section are universal. Real crystal surfaces are often covered by adsorbates, which introduce additional structural features as will be treated in Chapter 6.

5.1
Surface Relaxation

The effect of surface **relaxation** is the simplest modification observed for real surfaces. It assumes that the $(h\,k\,l)$ surface of a substrate, whose bulk lattice is given by a netplane-adapted lattice vectors $\underline{R}_1, \underline{R}_2, \underline{R}_3$, is terminated by overlayers forming $(h\,k\,l)$ monolayers, identical to those in the bulk. However, **relative positions** of the **overlayer atoms** near the surface, expressed by inter-layer distances and lateral shifts, **deviate** from corresponding positions in the **bulk**. This is described in the simplest case by complete overlayers shifting slightly with respect to their bulk positions. In most cases that are observed in experiments [22, 23] these shifts happen **perpendicular** to the surface, either toward (**inwards** relaxation) or away from the substrate (**outwards** relaxation).

Crystallography and Surface Structure: An Introduction for Surface Scientists and Nanoscientists,
Second Edition. Klaus Hermann.
© 2017 Wiley-VCH Verlag GmbH & Co. KGaA. Published 2017 by Wiley-VCH Verlag GmbH & Co. KGaA.

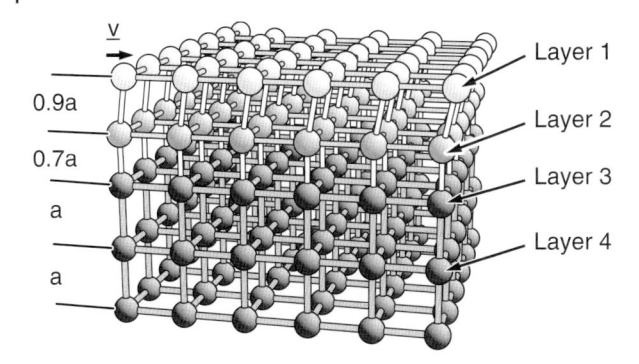

Figure 5.1 Hypothetical (0 0 1) surface section of a crystal with a simple cubic lattice (lattice constant a) with the two topmost overlayers relaxed, see text. Layer indices and inter-layer distances are indicated.

Formally, atom positions of **relaxed overlayers** near the surface are described by

$$\underline{R}^{(m)} = \underline{r}_i + n_1 \underline{R}_1 + n_2 \underline{R}_2 + \underline{s}^{(m)}, \quad n_1, n_2 \text{ integer}, \quad i = 1, \ldots p \quad (5.1)$$

for layer m near the surface, where $\underline{r}_i$ refers to positions of atoms inside the unit cell of the bulk lattice, n_1, n_2 are integer-valued coefficients accounting for the overlayer (netplane) periodicity, and $\underline{s}^{(m)}$ is a shift vector corresponding to the absolute positioning of layer m. As an illustration, Figure 5.1 shows the (0 0 1) surface of a fictitious crystal with a simple cubic lattice (lattice constant a), where the topmost layer no. 1 is relaxed inward by 10% and shifted sideways by vector $\underline{v}$ and layer no. 2 is relaxed inward by 30%.

Shift vectors $\underline{s}^{(m)}$ are equal to ($n_3 \underline{R}_3$) for **bulk truncated** surfaces of ideal single crystals. Further, $\underline{s}^{(m)}$ is expected to approach the bulk value ($n_3 \underline{R}_3$) for layers positioned well **below** the surface. Relaxation occurs for most metal surfaces, where, so far, mainly monolayer shifts perpendicular to the surface have been considered [105] (typical shifts amount to 1–5% of the bulk inter-layer spacing), with only few examples of lateral shifts in cases of stepped surfaces [22, 23].

5.2
Surface Reconstruction

Real surfaces that differ structurally from simple bulk truncation other than by relaxation are described as **reconstructed** surfaces. Reconstruction of a single crystal surface may result in surface **disorder** or may yield a **periodic** surface structure including minor or sizable displacements of the atoms and different periodicity compared with the bulk. Further, **additional** or **fewer atoms** may exist in the layer unit cells compared with those of an ideal bulk truncation. In the periodic case, the (h k l) surface of the substrate is terminated by monolayers that exhibit a two-dimensional periodicity given by vectors $\underline{R}'_1$ and $\underline{R}'_2$. These vectors can differ from those of the corresponding (h k l) bulk netplanes, $\underline{R}_1$ and $\underline{R}_2$, to form **superlattices**. In addition, the building units (two-dimensional unit cells) of the surface monolayers may contain a number of atoms different from that of the bulk layers. These monolayers will be called **overlayers** in the following.

Surface reconstruction is usually **combined** with relaxation such that atom positions of monolayers near the surface are described mathematically by

$$\underline{R}^{(m)} = \underline{r}'_i + n_1 \underline{R}'_1 + n_2 \underline{R}'_2 + \underline{s}^{(m)} \tag{5.2}$$

for layer m near the surface where $\underline{r}'_i$ refers to atom positions inside the reconstructed overlayers (which may or may not include positions of the initial bulk crystal), n_1, n_2 are integer-valued coefficients accounting for the overlayer periodicity, and $\underline{s}^{(m)}$ is a shift vector that describes possible layer relaxation. The periodicity vectors $\underline{R}'_1$ and $\underline{R}'_2$ can be connected with those of the (h k l) bulk netplanes, $\underline{R}_1$ and $\underline{R}_2$, by linear **2 × 2 transformations**, written in matrix form as

$$\begin{pmatrix} \underline{R}'_1 \\ \underline{R}'_2 \end{pmatrix} = \begin{pmatrix} m_{11} & m_{12} \\ m_{21} & m_{22} \end{pmatrix} \cdot \begin{pmatrix} \underline{R}_1 \\ \underline{R}_2 \end{pmatrix} = \underline{\underline{M}} \cdot \begin{pmatrix} \underline{R}_1 \\ \underline{R}_2 \end{pmatrix} \tag{5.3}$$

assuming surface-adapted lattice vectors $\underline{R}_1, \underline{R}_2, \underline{R}_3$ describing the bulk periodicity. As a consequence, the **unit cell area** F' of a reconstructed overlayer is given by

$$F' = |\underline{R}'_1 \times \underline{R}'_2| = |(m_{11} \underline{R}_1 + m_{12} \underline{R}_2) \times (m_{21} \underline{R}_1 + m_{22} \underline{R}_2)| =$$
$$= |(m_{11} m_{22} - m_{12} m_{21})(\underline{R}_1 \times \underline{R}_2)| = |\det(\underline{\underline{M}})| F \tag{5.4}$$

where F is the unit cell area of the (h k l) bulk netplane. Thus, $|\det(\underline{\underline{M}})|$ gives the ratio of the unit cell area F' of the overlayer and that, F, of the corresponding bulk layers. The transformation matrix $\underline{\underline{M}}$ in Eq. (5.3), called **reconstruction matrix** and sometimes written as $(\mathbf{m_{11} \, m_{12} \, | \, m_{21} \, m_{22}})$ for convenience, allows a classification of reconstructed periodic surfaces into three categories,

1) Reconstruction with **commensurate** superlattices is described by a reconstruction matrix $\underline{\underline{M}}$ according to Eq. (5.3) containing only **integer-valued** elements m_{ij}. In this case the periodicity vectors $\underline{R}'_1, \underline{R}'_2$ of the overlayer are also general vectors of the (h k l) bulk netplane and the unit cell area of the overlayer is an integer multiple of that of the bulk netplane. This includes systems, for which matrix $\underline{\underline{M}}$ equals the **unit matrix** and the reconstructed overlayer is of the same periodicity as the (h k l) bulk netplane. As a simple example, Figure 5.2 compares the ideal (1 1 0) surface of fcc platinum with the so-called (1 × 2)-missing-row reconstructed surface [106] taken from the **Surface Structure Database**, SSD 78.77, where every second row of atoms of the topmost (1 1 0) layer is missing. This results in a reconstruction matrix

$$\underline{\underline{M}} = \begin{pmatrix} 1 & 0 \\ 0 & 2 \end{pmatrix}, \quad F' = 2F \tag{5.5}$$

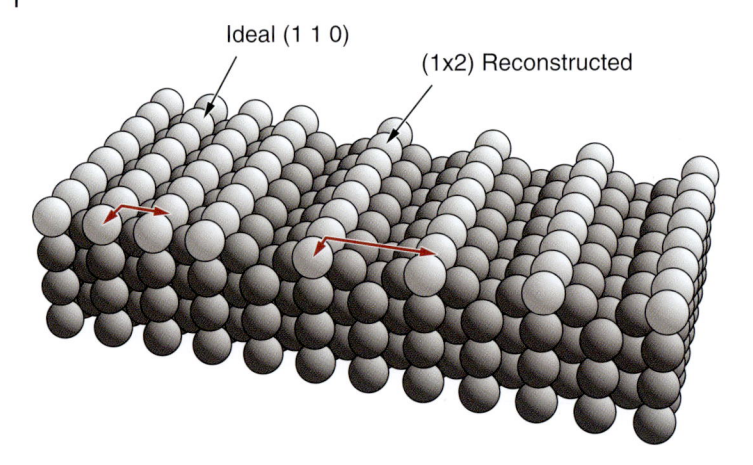

Figure 5.2 Ideal (left) and the (1×2) reconstructed Pt(1 1 0) surface (right). The layer periodicity vectors are indicated in red for both surface structures.

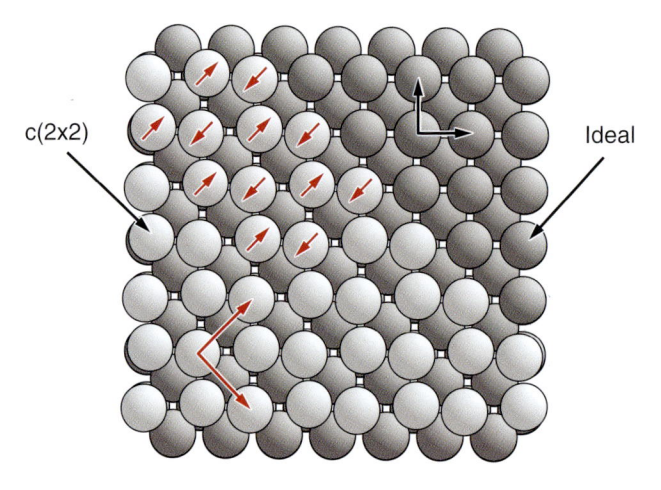

Figure 5.3 Ideal (upper right) and c(2×2)-reconstructed W(1 0 0) surface (lower left). The layer periodicity vectors are shown separately for the ideal substrate (black) and the reconstructed overlayer (red). Corresponding atom displacements are indicated by red arrows.

Note that in this example and the following ones we make use of the so-called **Wood notation** to denote the overlayer periodicity such as "(1 × 2)" or (centered) "c(2 × 2)." This notation will be discussed in detail in Section 6.3.

Another example is the centered (2 × 2) reconstruction of the (1 0 0) surface of bcc tungsten [22] taken from the Surface Structure Database, SSD 74.14, and illustrated by Figure 5.3. Here, the reconstruction matrix $\underline{\underline{M}}$ is given by

$$\underline{\underline{M}} = \begin{pmatrix} 1 & -1 \\ 1 & 1 \end{pmatrix}, \quad F' = 2\,F \tag{5.6}$$

In addition to the transformed periodicity, atom positions of the topmost overlayer are displaced by alternating lateral shifts as indicated by red arrows in Figure 5.3, which yields diagonal zigzag rows of tungsten atoms. Thus, this type of reconstruction may also be called **displacive**.

As more complicated examples, we mention surface structures, in which commensurate **reconstruction** is **combined** with major **repositioning** of individual atoms near the surface. An example is given by the symmetric dimer (2×1) [107] and the buckled dimer c(4×2) reconstructed (1 0 0) surfaces of silicon [108], see Figure 5.4, where alternating rows of surface atoms are shifted laterally as well as up and down forming surface **dimers** in order to optimize their Si–Si bonds.

The buckled dimer c(4×2) reconstruction [108] shown in Figure 5.4c is an example of a more general behavior of reconstructed overlayers, which, as a result of their coupling with the substrate, are not strictly planar with atoms shifted up and down resulting in **buckled surfaces**. In general, these perpendicular shifts can be described by **modulation functions** $\Delta z(\underline{r})$, where $\underline{r}$ denotes lateral positions along the surface. Due to the lateral periodicity of the overlayers $\Delta z(\underline{r})$ is also a periodic function. Thus, it may be represented by a **Fourier expansion** with respect to the overlayer periodicity. The modulation of atom positions is not restricted to reconstructed overlayers only

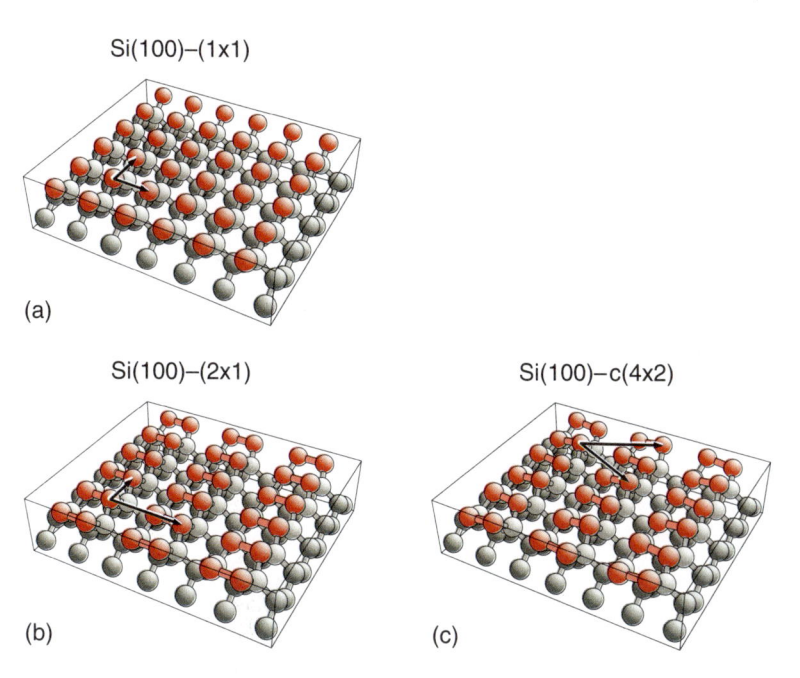

Figure 5.4 The Si(1 0 0) surface, (a) ideal unreconstructed (1×1), (b) reconstructed (2×1) with symmetric dimers in top layer, and (c) reconstructed c(4×2) with buckled dimers in top layer. The corresponding overlayers are shown in red with their lattice vectors sketched in black.

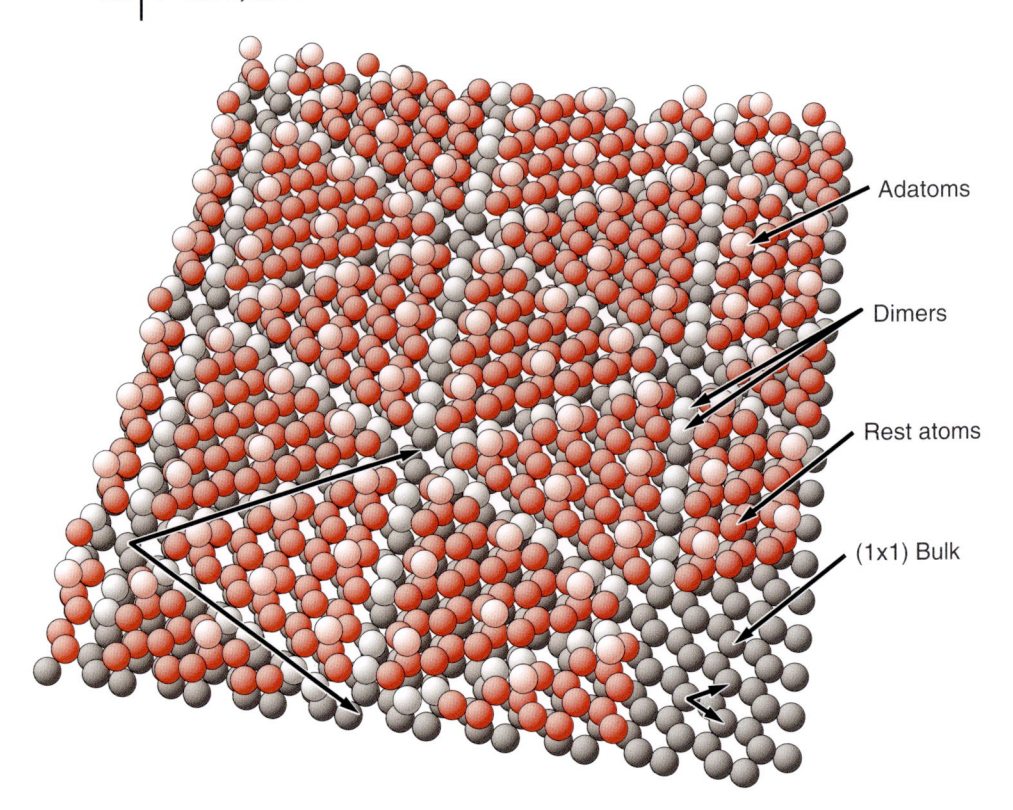

Figure 5.5 Structure of the reconstructed Si(1 1 1) – (7 × 7) surface according to the dimer–adatom–stacking-fault (DAS) model. The overlayer is removed at the bottom right to reveal the ideal bulk termination of Si(1 1 1). The different Si atom types are labeled accordingly. The periodicity vectors of the overlayer and of the ideal bulk termination are sketched in black.

but may also reach deeper into the substrate making the definition of corresponding modulation functions for substrate layers necessary. Further, the concept of modulated atom positions, described by appropriate modulation functions, can also be applied to all other types of reconstruction discussed below.

A very complex example is given by the **dimer–adatom–stacking-fault (DAS) model** of the (7 × 7) reconstructed (1 1 1) surface of silicon [109, 110], see Figure 5.5, where the topmost three monolayers of the surface reconstruct. Here, Si adatoms stick out of the surface, Si_2 dimers stabilize in long trenches that cross to form open holes, and the rest atoms together with the adatoms yield a mirror symmetry inside the surface unit cell (not existing in the bulk) described as a stacking fault.

2) Reconstruction with **coincidence superlattices**, sometimes also called **high-order** commensurate (HOC) or **scaled commensurate** lattices, is described by a reconstruction matrix $\underline{M}$ according to Eq. (5.3), which

contains **rational**- and **integer**-valued elements m_{ij} with at least one being rational. Thus, matrix $\underline{M}$ can be written as

$$\underline{M} = \begin{pmatrix} r_{11} & r_{12} \\ r_{21} & r_{22} \end{pmatrix}, \quad r_{ij} = \frac{p_{ij}}{q_{ij}} \quad \text{with} \quad p_{ij}, q_{ij} \text{ integer} \tag{5.7}$$

where $q_{ij} = 1$ corresponds to an integer r_{ij}. Together with $c_i = \mathrm{lcm}(q_{i1}, q_{i2})$ denoting the least common multiple (see Appendix E.1) of the two denominators q_{i1}, q_{i2}, $i = 1, 2$, this matrix can be written as a **reconstruction matrix**

$$\underline{M} = \begin{pmatrix} c_1 & 0 \\ 0 & c_2 \end{pmatrix}^{-1} \cdot \begin{pmatrix} r'_{11} & r'_{12} \\ r'_{21} & r'_{22} \end{pmatrix}, \quad r'_{ij} = c_i r_{ij} = \frac{c_i}{q_{ij}} p_{ij} \tag{5.8}$$

to yield

$$\begin{pmatrix} \underline{R}''_1 \\ \underline{R}''_2 \end{pmatrix} = \begin{pmatrix} c_1 & 0 \\ 0 & c_2 \end{pmatrix} \cdot \begin{pmatrix} \underline{R}'_1 \\ \underline{R}'_2 \end{pmatrix} = \begin{pmatrix} r'_{11} & r'_{12} \\ r'_{21} & r'_{22} \end{pmatrix} \cdot \begin{pmatrix} \underline{R}_1 \\ \underline{R}_2 \end{pmatrix} \tag{5.9}$$

where elements r'_{ij} and c_i are **integer**-valued. Thus, if the initial overlayer lattice $\underline{R}'_1, \underline{R}'_2$ in Eq. (5.3) is represented by a lattice with a larger unit cell given by scaled lattice vectors

$$\underline{R}''_1 = c_1 \underline{R}'_1, \quad \underline{R}''_2 = c_2 \underline{R}'_2 \tag{5.10}$$

then the resulting matrix $\underline{M}$ in Eq. (5.8) is replaced by an integer-valued matrix and the superlattice will be commensurate. This property of matrix $\underline{M}$ explains the nomenclature "scaled commensurate" for this type of reconstruction. The **distinction** between simple commensurate and coincidence superlattices may be considered somewhat artificial. However, we will keep this distinction to indicate that **coincidence** superlattices connect **non-primitive** unit cells of the overlayer, larger than corresponding primitive cells, with those of the substrate. In contrast, **(simple) commensurate** superlattices connect **primitive** unit cells of the overlayer with those of the substrate.

As an example of a coincidence superlattice, Figure 5.6 shows a postulated structure of the (1 1 1) oriented gold surface [111], referred to as Au(1 1 1) $-(\sqrt{3} \times 22)$rect, where the topmost gold layer forms a hexagonal lattice which is compressed unilaterally along $\underline{R}_2$ by 4.35% ($= 1/23$) such that 23 atom distances of the overlayer along $\underline{R}_2$ coincide with 22 atom distances of the substrate. As a consequence, the primitive unit cell of the surface is rectangular and given by scaled lattice vectors

$$\underline{R}''_1 = 2 \underline{R}_1 - \underline{R}_2, \quad \underline{R}''_2 = 22 \underline{R}_2 \tag{5.11}$$

which are orthogonal and where $|\underline{R}''_1| = \sqrt{3}\,|\underline{R}_1|$ (explaining the nomenclature "$(\sqrt{3} \times 22)$rect"). The overlayer atoms of the rows along $\underline{R}_2$ compensate their lateral **compressive stress** by gradually shifting their positions normal to the surface, as illustrated in Figure 5.6b. This results in a periodically **buckled surface**, where the buckling can be described by a modulation function $\Delta z(\underline{r})$ as discussed earlier.

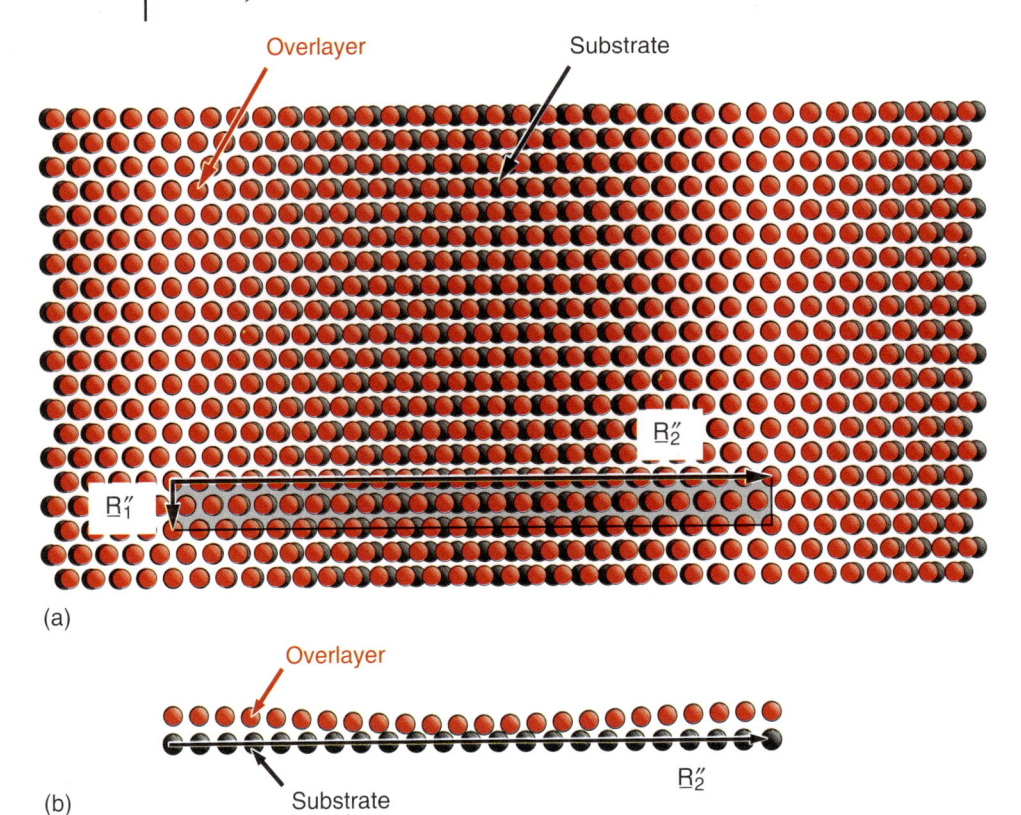

Figure 5.6 (a) Coincidence superlattice of the Au(1 1 1) – ($\sqrt{3} \times 22$)rect surface shown by its overlayer (red) and the topmost substrate layer (black) for a normal view. The common unit cell is emphasized in gray with scaled lattice vectors $\underline{R}''_1$, $\underline{R}''_2$ indicated. (b) Parallel view of the unit cell along $\underline{R}''_1$ illustrating the perpendicular displacement (buckling) of the overlayer atoms.

As another example, Figure 5.7 shows a fictitious surface with two graphene layers, that is, graphite monolayers with honeycomb structure, corresponding to a reconstruction matrix

$$\underline{\underline{M}} = \frac{1}{12} \begin{pmatrix} 11 & 2 \\ -2 & 13 \end{pmatrix} = \begin{pmatrix} 0.9167 & 0.1667 \\ -0.1667 & 1.0833 \end{pmatrix} \tag{5.12}$$

discussed below, see also Eq. (5.27). This surface forms a **coincidence lattice** with scaled lattice vectors $\underline{R}''_i = 12\, \underline{R}'_i$ according to Eq. (5.10) and shows a **hexagonal moiré pattern** whose periodicity vectors $\underline{R}^p_1$, $\underline{R}^p_2$ (not included in Figure 5.7) can be described by

$$\begin{pmatrix} \underline{R}^p_1 \\ \underline{R}^p_2 \end{pmatrix} = \begin{pmatrix} 3 & 5 \\ -5 & 8 \end{pmatrix} \cdot \begin{pmatrix} \underline{R}_1 \\ \underline{R}_2 \end{pmatrix} \tag{5.13}$$

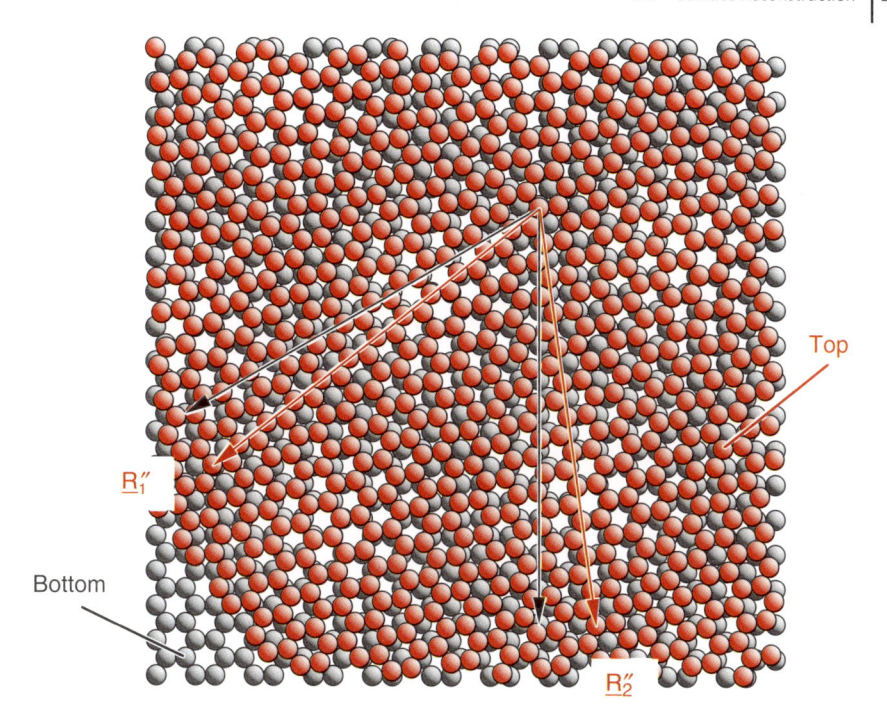

Figure 5.7 Coincidence superlattice of two graphene sheets. The scaled lattice vectors $\underline{R}_1'', \underline{R}_2''$ of the top layer are shown in red, those of the bottom layer in black.

as discussed in detail in Section 6.5. This leads to

$$\begin{pmatrix} \underline{R}_1'' \\ \underline{R}_2'' \end{pmatrix} = 12\,\underline{\underline{M}} \cdot \begin{pmatrix} 3 & 5 \\ -5 & 8 \end{pmatrix}^{-1} \cdot \begin{pmatrix} \underline{R}_1^p \\ \underline{R}_2^p \end{pmatrix} = \begin{pmatrix} 2 & -1 \\ 1 & 1 \end{pmatrix} \cdot \begin{pmatrix} \underline{R}_1^p \\ \underline{R}_2^p \end{pmatrix} \quad (5.14)$$

which shows that the coincidence lattice is also **commensurate** with the **moiré lattice** and the unit cell area given by the scaled lattice vectors $\underline{R}_1'', \underline{R}_2''$ is three times, that is, an integer multiple of, the periodicity cell area suggested by the moiré pattern. This is a more general result of coincidence lattices forming moiré patterns as shown in Section 6.5.

3) Reconstruction with **incommensurate** superlattices is described by a reconstruction matrix $\underline{\underline{M}}$ containing elements m_{ij} of which at least one is **irrational**, that is, cannot be represented by an integer or a rational number. In this case at least one of the periodicity vectors $\underline{R}_1', \underline{R}_2'$ in Eq. (5.3) cannot be described by lattice vectors of the corresponding $(h\,k\,l)$ substrate netplane using integer- or rational-valued linear combinations. Further, the **combined surface system** (overlayer with substrate layers) is **not strictly periodic** in two dimensions. As an example, Figure 5.8 shows a (1 0 0) surface of fcc gold [112] taken from the Surface Structure Database, SSD 79.80. Here, the topmost overlayer is reconstructed with a slightly distorted hexagonal structure, while the (1 0 0) monolayers of the substrate are of square geometry and also

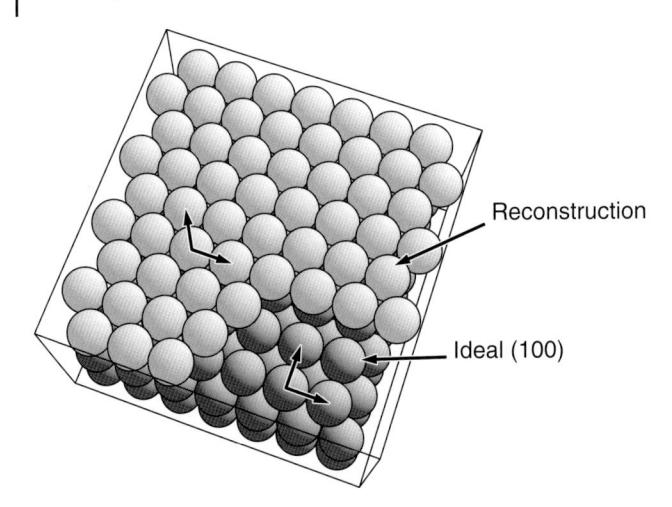

Figure 5.8 Hexagonal reconstructed Au(1 0 0) surface. Atoms of the top reconstructed and the underlying substrate layers are painted differently with periodicity vectors sketched accordingly.

slightly distorted near the surface. Assuming a perfectly hexagonal overlayer on a substrate with exact square lattice this results in a reconstruction matrix

$$\underline{\underline{M}} = \eta \begin{pmatrix} 1 & 0 \\ -1/2 & \sqrt{3}/2 \end{pmatrix} \tag{5.15}$$

where η (≈ 0.95) is the ratio of the lattice constants of the reconstructed layer and the initial (1 0 0) monolayer.

An esthetically pleasing class of incommensurate superlattices is given by surfaces with **rotational superlattices**. Here, the topmost overlayer retains its internal lattice (except for minor lateral distortions and buckling) but is rotated by an angle α with respect to the underlying substrate layer. Simple algebraic calculus shows that the reconstruction matrix $\underline{\underline{M}}$ of a rotational superlattice is given by

$$\underline{\underline{M}} = \frac{1}{\sin(\omega)} \begin{pmatrix} \sin(\omega - \alpha) & q^{-1} \sin(\alpha) \\ -q \sin(\alpha) & \sin(\omega + \alpha) \end{pmatrix}, \quad q = \frac{R_2}{R_1} \tag{5.16}$$

assuming an anticlockwise rotation of the overlayer by an angle α, where the angle between the lattice vectors $\underline{R}_1$ and $\underline{R}_2$ of the ($h\,k\,l$) bulk netplane is ω. These surface systems exhibit **spatial interference** patterns, so-called **moiré** patterns as shown above, which have been observed for many surfaces as will be discussed in great detail in Section 6.5.

While rotational superlattices are **incommensurate** in general a detailed mathematical analysis shows that, depending on the angles ω, α, and the ratio q of the vector lengths R_2 and R_1, they can also yield **coincidence superlattices**, which bridge incommensurate and HOC surface systems. As a conceptual example, we consider a simple cubic substrate with a

(0 0 1) oriented surface, where the monolayers of the substrate are described by square lattices with lattice vectors given by

$$\underline{R}_1 = a\,(1,\,0), \quad \underline{R}_2 = a\,(0,\,1) \tag{5.17}$$

(ignoring the third dimension normal to the surface). Further, we assume that the topmost monolayer (overlayer) is rotated with respect to the substrate layers by an angle α with $\tan(\alpha) = n_2/n_1$ where n_1, n_2 are positive integers. Then the reconstruction matrix (5.16) with $q = 1$, $\omega = 90°$ reads

$$\underline{\underline{M}} = \begin{pmatrix} \cos(\alpha) & \sin(\alpha) \\ -\sin(\alpha) & \cos(\alpha) \end{pmatrix} = \frac{1}{\sqrt{n_1^2 + n_2^2}} \begin{pmatrix} n_1 & n_2 \\ -n_2 & n_1 \end{pmatrix} \tag{5.18}$$

and the lateral lattice vectors of the overlayer, $\underline{R}'_1$, $\underline{R}'_2$ are given by

$$\begin{pmatrix} \underline{R}'_1 \\ \underline{R}'_2 \end{pmatrix} = \frac{1}{\sqrt{n_1^2 + n_2^2}} \begin{pmatrix} n_1 & n_2 \\ -n_2 & n_1 \end{pmatrix} \cdot \begin{pmatrix} \underline{R}_1 \\ \underline{R}_2 \end{pmatrix} \tag{5.19}$$

Thus, the lattice vectors (5.19) describe, in general, an **incommensurate** overlayer. However, for selected integer values n_1, n_2 with

$$n_1^2 + n_2^2 = N^2, \; N \text{ integer} \tag{5.20}$$

which correspond to a discrete set of angles α where

$$\cos(\alpha) = n_1/N, \; \sin(\alpha) = n_2/N \tag{5.21}$$

the reconstruction matrix (5.18) becomes

$$\underline{\underline{M}} = \begin{pmatrix} \cos(\alpha) & \sin(\alpha) \\ -\sin(\alpha) & \cos(\alpha) \end{pmatrix} = \frac{1}{N} \begin{pmatrix} n_1 & n_2 \\ -n_2 & n_1 \end{pmatrix} \tag{5.22}$$

and describes a **coincidence lattice** reconstruction, as discussed above. Thus, if the rotated overlayer lattice $\underline{R}'_1$, $\underline{R}'_2$ of Eq. (5.3) is represented by non-primitive lattice vectors $\underline{R}''_1$, $\underline{R}''_2$ with

$$\underline{R}''_1 = N\,\underline{R}'_1, \quad \underline{R}''_2 = N\,\underline{R}'_2 \tag{5.23}$$

the resulting matrix transformation between $(\underline{R}_1, \underline{R}_2)$ and $(\underline{R}''_1, \underline{R}''_2)$ based on Eqs. (5.19), (5.22), and (5.23) is integer-valued and the overlayer will be high-order commensurate. This is illustrated in Figure 5.9 for $n_1 = 4$, $n_2 = 3$ (hence $N = 5$ corresponding to $\alpha = 36.87°$). The supercell, given by $\underline{R}''_1$, $\underline{R}''_2$, is common to both the substrate and overlayer lattice, and contains 25 atoms per unit cell in each layer. However, it does not represent the primitive cell of the compound system, which is given by vectors $\underline{R}''_{o1}$, $\underline{R}''_{o2}$ with

$$\begin{pmatrix} \underline{R}''_{o1} \\ \underline{R}''_{o2} \end{pmatrix} = \begin{pmatrix} 2 & -1 \\ 1 & 2 \end{pmatrix} \cdot \begin{pmatrix} \underline{R}_1 \\ \underline{R}_2 \end{pmatrix} \tag{5.24}$$

as sketched in Figure 5.9 where only five atoms are included in the unit cell of each layer.

The above relations (5.17)–(5.23) are valid for any rotational overlayer on a substrate with a **square lattice** and corresponding values n_1, n_2, N result in

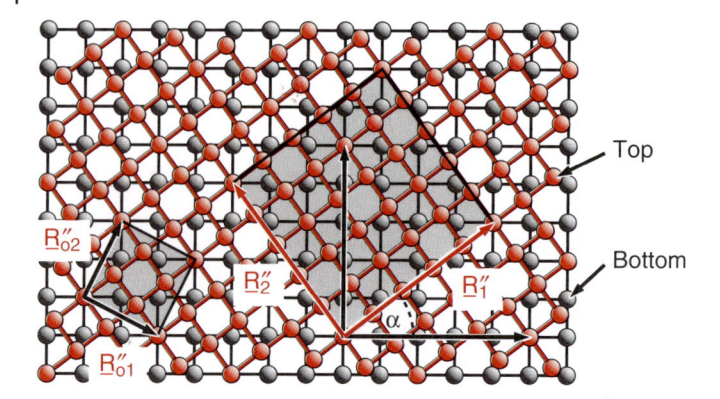

Figure 5.9 Commensurate rotational overlayer on a substrate with square lattice corresponding to $\alpha = 36.87°$, see text. The common unit cells, primitive cell with overlayer lattice vectors $\underline{R}''_{o1}$, $\underline{R}''_{o1}$ to the left, scaled cell with vectors $\underline{R}''_1$, $\underline{R}''_2$ to the right, are emphasized in gray.

coincidence lattices. This includes cases where n_1 and N become quite large while n_2 remains much smaller, leading to rotational superlattices with quite small rotation angles α. (Actually, solutions of the Pythagorean equation (5.20) can be generated explicitly by an algorithm discussed in Appendix E.4.) As an example, Figure 5.10 shows the superlattice corresponding to $n_1 = 84$, $n_2 = 13$, $N = 85$ (reflecting an angle $\alpha = 8.797°$) exhibiting a clear

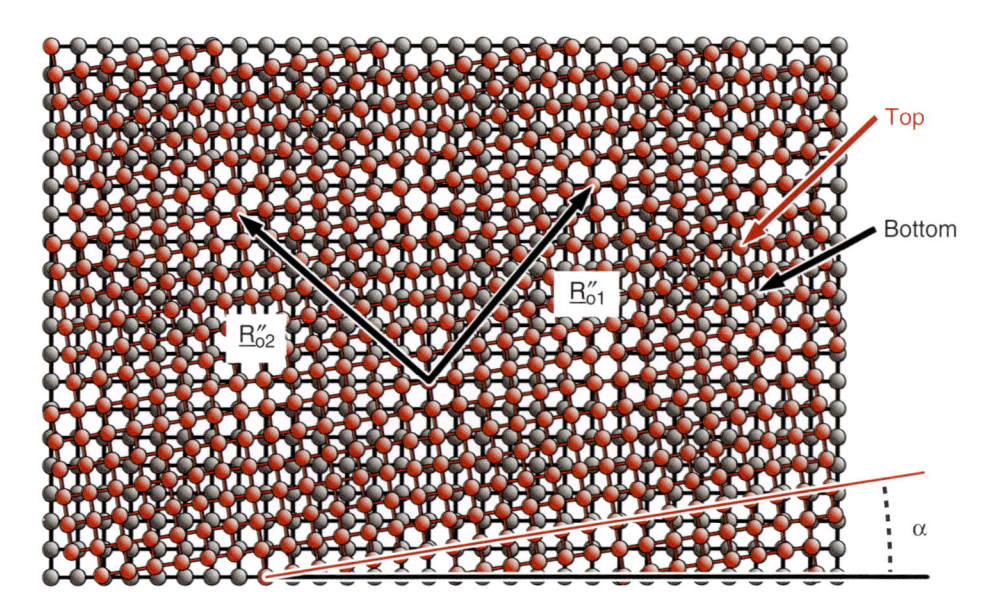

Figure 5.10 Commensurate rotational overlayer on substrate with square lattice corresponding to $\alpha = 8.797°$ sketched at the bottom, see text. The lattice vectors $\underline{R}''_{o1}$, $\underline{R}''_{o2}$ of the primitive common unit cell are indicated accordingly.

moiré pattern. Here, a supercell, defined by $\underline{R}_1''$, $\underline{R}_2''$ according to Eq. (5.23), with $85^2 = 7225$ overlayer atoms would be common to the substrate and the overlayer lattice. As before, a more detailed analysis evidences periodicity with much smaller lattice vectors $\underline{R}_{o1}''$, $\underline{R}_{o2}''$ where

$$\begin{pmatrix} \underline{R}_{o1}'' \\ \underline{R}_{o2}'' \end{pmatrix} = \begin{pmatrix} 6 & 7 \\ -7 & 6 \end{pmatrix} \cdot \begin{pmatrix} \underline{R}_1 \\ \underline{R}_2 \end{pmatrix} \qquad (5.25)$$

as sketched in Figure 5.10. Here, only 85 overlayer atoms are included in the primitive cell.

In cases of **truly incommensurate** rotational superlattices the reconstruction matrix $\underline{\underline{M}}$ according to Eq. (5.16) must contain irrational elements. However, these elements can always be approximated by rational numbers which, altogether, results in an approximate reconstruction matrix $\underline{\underline{M}}$, which describes a coincidence lattice as discussed above for substrates with square lattice. As an example, for hexagonal lattices, a reconstructed surface with two graphene layers, rotated by $\alpha = 8°$ with respect to each other, corresponds to a reconstruction matrix $\underline{\underline{M}}$ according to Eq. (5.16) (setting $q = 1, \omega = 60°$)

$$\underline{\underline{M}} = \frac{2}{\sqrt{3}} \begin{pmatrix} \sin(52°) & \sin(8°) \\ -\sin(8°) & \sin(68°) \end{pmatrix} = \begin{pmatrix} 0.909916 & 0.160703 \\ -0.160703 & 1.070620 \end{pmatrix} \qquad (5.26)$$

This matrix can be approximated by

$$\underline{\underline{M}} = \frac{1}{12} \begin{pmatrix} 11 & 2 \\ -2 & 13 \end{pmatrix} = \begin{pmatrix} 0.916667 & 0.166667 \\ -0.166667 & 1.083333 \end{pmatrix} \qquad (5.27)$$

describing reconstruction with a coincidence superlattice, as shown in Figure 5.7, which is visually indistinguishable from that of the incommensurate overlayer structure describing 8° rotation.

The transformation matrix $\underline{\underline{M}}$ defining transformations between lattice vectors of the ideal bulk-truncated and the reconstructed surface forms the basis of the **2 × 2 matrix notation** to characterize **reconstructed surfaces**. For a general single crystal surface of a substrate Sub with Miller indices $(h\,k\,l)$ and its topmost layer reconstructed according to reconstruction matrix $\underline{\underline{M}}$, the 2 × 2 matrix notation can be written as

$$\text{Sub}(h\,k\,l) - \underline{\underline{M}} = \text{Sub}(h\,k\,l) - \begin{pmatrix} m_{11} & m_{12} \\ m_{21} & m_{22} \end{pmatrix} \qquad (5.28)$$

with an alternative notation

$$\text{Sub}(h\,k\,l) - (m_{11}\,m_{12}\,|\,m_{21}\,m_{22}) \qquad (5.29)$$

where the latter in its one-line format is easier to write than Eq. (5.28). Examples are

- $\text{Pt}(1\,1\,0) - \begin{pmatrix} 1 & 0 \\ 0 & 2 \end{pmatrix}$ or $\text{Pt}(1\,1\,0) - (1\,0\,|\,0\,2)$

  describing the (1×2) reconstructed Pt(1 1 0) surface, see Figure 5.2,

- $W(1\,0\,0) - \begin{pmatrix} 1 & 1 \\ -1 & 1 \end{pmatrix}$ or $W(1\,0\,0) - (1\,1\,|\,-1\,1)$

  describing the c(2×2) reconstructed W(1 0 0) surface, see Figure 5.3,

- $Si(1\,0\,0) - \begin{pmatrix} 2 & -1 \\ 2 & 1 \end{pmatrix}$ or $Si(1\,0\,0) - (2\,-1\,|\,2\,1)$

  describing the c(4×2) reconstructed Si(1 0 0) surface, see Figure 5.4c,

- $C(0\,0\,0\,1) - \begin{pmatrix} 3 & 5 \\ -5 & 8 \end{pmatrix}$ or $C(0\,0\,0\,1) - (3\,5\,|\,-5\,8)$

  yielding an approximate description of a graphite C(0 0 0 1) surface with its topmost layer rotated by 8° at its top, see Figure 5.7.

These notations have been introduced some time ago [94] and have also been recommended by the International Union of Pure and Applied Chemistry (IUPAC) [86]. However, they are used by surface scientists much less frequently compared with the Wood notation, see Section 6.3.

5.3
Growth Processes

The geometric structure of crystal surfaces is determined, apart from static inter-atomic coupling, also by dynamic details of diffusion and nucleation that eventually result in thin film formation and crystal growth. These processes are completely analogous for thin film and adsorbate layer formation, and the corresponding structural details can be discussed on the same footing, see also Section 6.1. Overlayers (**adsorbate layers**) are composed of foreign atoms or molecules at a substrate surface which, depending on layer thickness, may also be called **heteroepitaxial (thin) films**. These are distinguished from overlayers whose atoms are of the same chemical type as those of the substrate surface, usually termed **homoepitaxial (thin) films** and which become important for crystal growth.

The growth of an overlayer has to be initiated by **nucleation centers** at the substrate surface where adsorbing atoms or molecules, from gas phase or diffusing at the substrate surface, can stabilize to form larger surface aggregates that may eventually lead to closed overlayers. These nucleation centers can be adsorbate particles that approach the surface and bind at preferred surface sites (e.g., near steps or kinks) or surface perturbations, such as lattice imperfections (e.g., dislocations, defects, or vacancies) and impurities. Their position and arrangement is governed by complex individual binding properties, surface diffusion, and thermodynamic behavior. Thus, there are only few general qualitative criteria as to structural details of nucleation centers. This aspect goes beyond the scope of this book and further details of the underlying physics can be found in [35].

The growth of adsorbate overlayers and thin films has been discussed for a long time where three basic **growth modes** have been considered [113] and verified by experiments [35]. These modes can be distinguished roughly by energetic quantities that consider binding between atoms within the overlayer as well as between overlayer and substrate atoms. In the following, we restrict our discussion of crystal growth to elemental overlayers and substrates, for example,

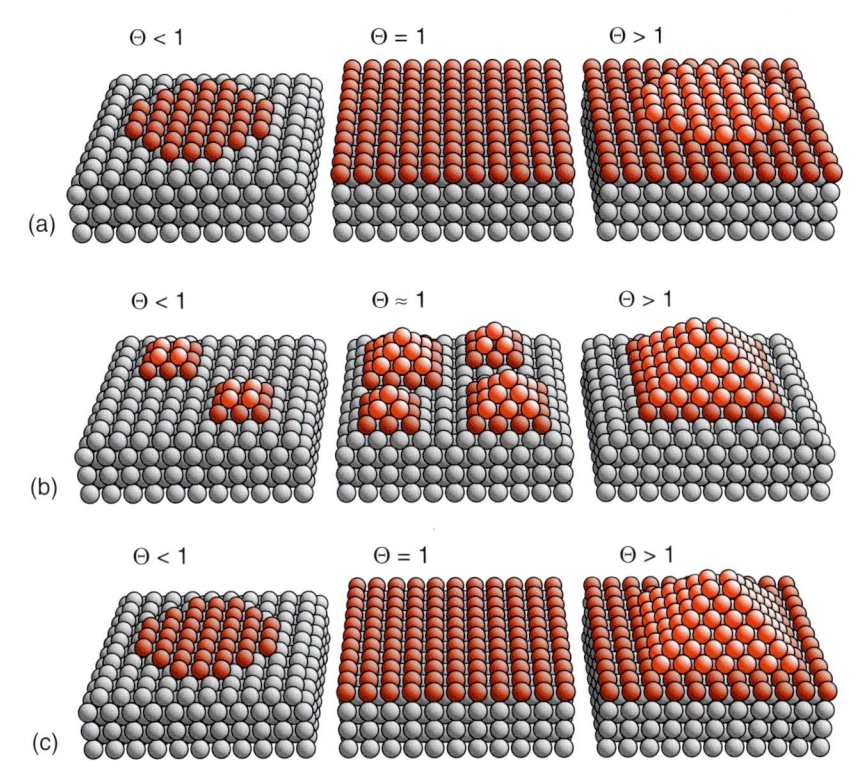

Figure 5.11 Schematic sketch of the three different growth modes, (a) Frank-Van-der-Merwe (layer-by-layer), (b) Volmer–Weber (three-dimensional clusters), and (c) Stranski–Krastanov (clusters above monolayer). Overlayer (substrate) atoms are shown in red (gray) where those above the first adsorbate layer are painted in lighter red. The amount of deposited species is denoted by a coverage Θ (monolayer coverage where $\Theta = 1$ refers to one overlayer atom per substrate atom).

elemental metal systems with growth modes shown schematically in Figure 5.11. However, it must be emphasized that growth mechanisms of molecular adsorbate overlayers can be described in a completely analogous way.

1) **Frank-Van-der-Merwe** (FM) growth mode (**layer-by-layer** growth, see Figure 5.11a). This mode assumes that binding between overlayer and substrate atoms dominates or is comparable in strength with binding between overlayer atoms. Here, adsorbing atoms stabilize at neighboring surface sites, forming monolayer islands at small coverage. With increasing coverage these islands grow until a complete monolayer film is obtained. At larger coverage (and assuming inter-layer binding to dominate), a second monolayer film starts to build above the first continuing the layer-by-layer growth process where each single layer is completed before the next starts to build. For heteroepitaxial growth, where the atoms of the overlayers differ from those of the substrate, the building process is often accompanied by

local stress or strain acting on each overlayer which, after equilibration, gives rise to layer-dependent structural variations.

As an example, ultra-thin films of iron, 1–20 monolayers (ML) thick, are found to grow in layer-by-layer mode on the (1 0 0) surface of the fcc copper single crystal substrate [114, 115] with different structural phases. For very thin Fe layers, 1–4 ML, the lateral lattice constants of the overlayers and of the Cu substrate agree while the distances between adjacent monolayers in the film are larger by 4% compared with those of the substrate, resulting in a centered tetragonal film lattice. Above 10 ML thickness the Fe films assume a bcc lattice structure, the generic bulk structure of iron, with a (1 1 0) surface termination [114].

2) **Volmer–Weber** (VW) growth mode (**three-dimensional cluster** growth, see Figure 5.11b). This mode assumes that binding between overlayer atoms dominates or is comparable in strength with binding between overlayer and substrate atoms. Here, adsorbing atoms stabilize at the substrate surface forming three-dimensional clusters, often in the shape of several layers thick islands, where the size of the clusters and/or their density at the surface increases with increasing adsorbate coverage. Only at very large coverage do the clusters combine to form a closed and often quite rough overlayer surface. Many noble metals grow on insulator or semiconductor substrate in the VW mode; for an overview see, for example, [116]. Examples are silver clusters growing on a mica substrate or gold growing on MgF_2 [116].

3) **Stranski–Krastanov** (SK) growth mode (**mixed layer-by-layer and cluster** growth, see Figure 5.11c). This mode combines the two previous modes where it is assumed that binding between overlayer atoms and between overlayer and substrate atoms are of comparable strength. Here, adsorbing atoms stabilize initially at the substrate surface forming monolayer islands that grow until one or a few complete monolayers (often called **wetting layers**) are reached. After this, additional adatoms bind on top of the wetting layers in three-dimensional clusters of increasing size and/or density.

Metals are found to often grow on metal and insulator substrates in the SK mode; for an overview see, for example, [117]. Examples are gold and silver growing on a tungsten substrate [118] or silver growing on a silicon substrate [119].

At vicinal metal surfaces with flat terraces separated by steps or kinks, sites near step or kink atoms can be assumed to act as nucleation centers for homoepitaxial growth. As a consequence, atoms from the gas phase, or adsorbed at terrace sites and diffusing along the terrace, will stabilize at regular surface sites next to step or kink atoms. This leads to a continuation of the terraces, possibly changing terrace widths, and causing additional step irregularities. Here, one can distinguish between two different growth scenarios. First, with increasing coverage, atoms may stabilize by forming single atom rows along step or kink lines where each row is completed before atoms adsorb at the sites of the next row. This is the one-dimensional equivalent of Frank-Van-der-Merwe growth and does not introduce new structural features except for kinks due to incomplete atom rows.

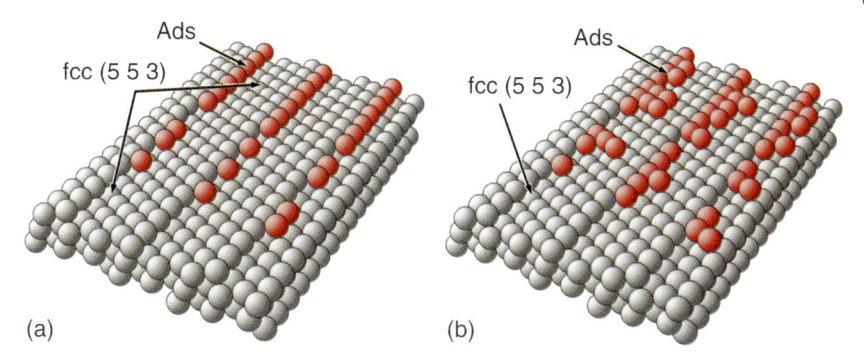

Figure 5.12 Schematic sketch of (a) Frank-Van-der-Merwe type and (b) Volmer–Weber type growth at a stepped (5 5 3) surface of an fcc metal substrate. Adsorbate (substrate) atoms (of the same element type) are shown in red (gray).

Alternatively, atoms may adsorb near step or kink lines also allowing for incomplete rows, which results in irregular step and kink structures and can be considered equivalent to Volmer-Weber growth. As an illustration, Figure 5.12 shows a stepped (5 5 3) surface of a fictitious fcc metal substrate where atoms of the same element type adsorb according to a FM type, Figure 5.12a, and a VW type growth scenario, see Figure 5.12b.

So far, heteroepitaxial growth processes at the substrate surface have been considered only for systems where the adsorbing species does not intermix with substrate atoms. Intermixing can happen if the adsorbates react chemically with substrate atoms, which results in a surface composition with an interface region where there is no clear phase separation between an adsorbate and a substrate. Examples are surface oxides or sulfides originating from oxygen and sulfur adsorption at metal substrates. Further, adsorbing metal atoms can mix with surface atoms of a metal substrate to form surface alloy layers which, in their composition, may not exist as bulk alloys [120]. Structural details of these systems

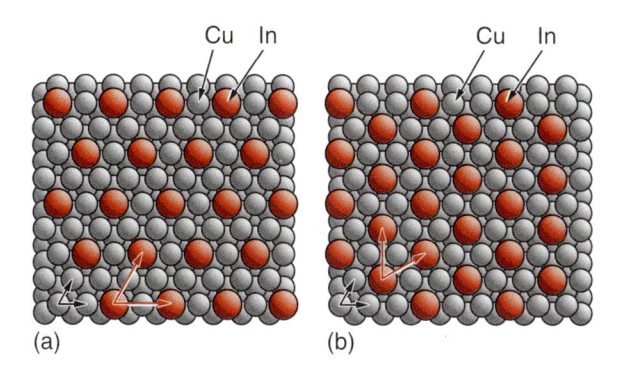

Figure 5.13 Cu(1 1 1) + In surface section with alloy formation at the topmost layer, (a) (2×2) overlayer with Cu_3In composition and (b) $(\sqrt{3} \times \sqrt{3})R30°$ overlayer with Cu_2In composition. Indium (copper) atoms are shown in red (gray). The lattice vectors of the overlayer and of the Cu substrate are sketched in red and black, respectively.

can be very complex and have to be treated on an individual basis, which goes beyond the scope of this book. As an illustration, Figure 5.13 shows experimental structures of indium adsorbed on the Cu(1 1 1) surface [121] where alloying of the topmost surface layer occurs according to a (2×2) overlayer, representing a Cu_3In surface alloy, see Figure 5.13a, as well as according to a $(\sqrt{3} \times \sqrt{3})R30°$ overlayer with Cu_2In composition, see Figure 5.13b. So far, Cu_2In and Cu_3In have not been observed as bulk alloys.

5.4
Faceting

Real surfaces of single crystals may be **rough** beyond simple buckling of their topmost layers and can combine **small** flat surface sections of different $(h\,k\,l)$-indexed orientation. This structural feature is called **faceting**. Facets are also found in crystallites and nanoparticles where they confine the particle surface and determine the global shape as discussed in Section 2.7. At extended single crystal surfaces, facet formation is often observed as a consequence of thermal equilibration after sputtering by atom or ion beams or as a result of etching and polishing. It originates from physical and chemical processes, where an $(h\,k\,l)$ surface of a flat single crystal can be stabilized energetically by introducing finite sections of differently oriented $(h'\,k'\,l')$ surface sections.

Other examples are **oxide crystals**, where surfaces with highly polar termination can lower their electrostatic energy by forming local facets with nonpolar termination. This has been proposed for the highly polar (1 1 1) surface of MgO (NaCl lattice, see Figure 1.1), where thermal treatment (annealing at high temperatures)

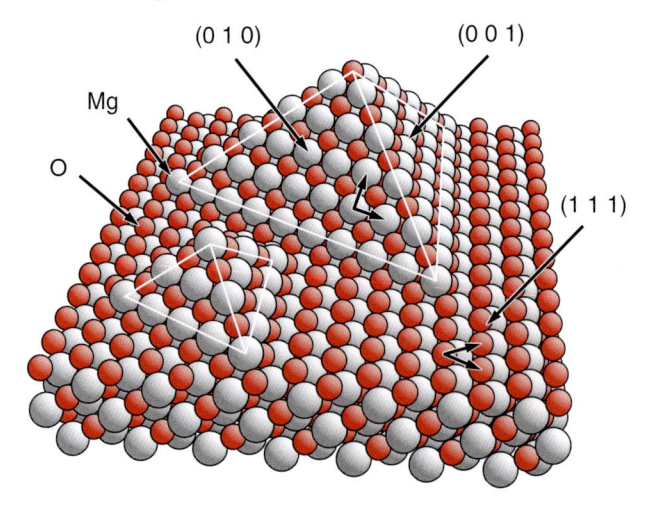

Figure 5.14 Ionic MgO(1 1 1) surface section with two pyramids terminated by facets of non-polar (0 0 1), (0 1 0), and (1 0 0) monolayers. The facet edges are emphasized by white lines. Corresponding netplane orientations are labeled with adapted lattice vectors indicated accordingly.

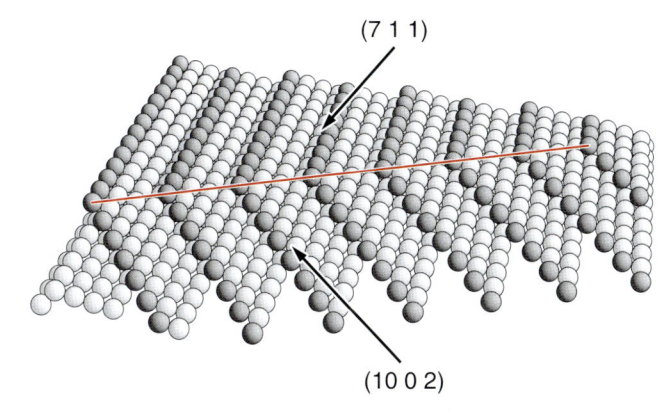

Figure 5.15 Facet edge separating stepped (7 1 1) and (10 0 2) surfaces of an fcc crystal. Step edges are indicated by darker balls and facet edge atoms are connected by a red line.

produces facets, whose sides resemble nonpolar (0 0 1), (0 1 0), and (1 0 0) terminated surfaces [122]. This is illustrated in Figure 5.14 showing a section of a MgO(1 1 1) surface with two pyramids terminated by nonpolar facets.

Another example is given in Figure 5.15, where two stepped surfaces of an fcc (7 1 1) crystal, describing (1 0 0) terraces with (1 1 1) steps, and fcc(10 0 2), describing (1 0 0) terraces with (0 0 1) steps, join to form a facet edge.

Facet edges will be **denoted** in the following by $(h\,k\,l)/(h'\,k'\,l')$, where $(h\,k\,l)$ and $(h'\,k'\,l')$ are the Miller indices of the two surface sections that join to form the edge. Further, facet edges are called **positive** if the two joining surface sections form a **roof-shaped** arrangement with respect to the underlying crystal bulk. In contrast, **negative** facet edges result from surface sections forming a **trough-shaped** arrangement with respect to the bulk. This is illustrated in Figure 5.16, where positive and negative edges of (0 0 1)/(1 1 1) facets of an fcc crystal surface are shown. Evidently, a faceted surface, which still gives the appearance to be flat on a larger scale, must contain both positive and negative facet edges.

The direction of a facet edge is defined by the **facet edge vector**, $\underline{R}_{facet}$, which points parallel to the cutting line of the corresponding two surface sections. If their orientations are defined by Miller indices $(h\,k\,l)$ and $(h'\,k'\,l')$, respectively, then vector $\underline{R}_{facet}$, common to both netplanes, is perpendicular to both reciprocal lattice vectors $\underline{G}_{(h\,k\,l)}$ and $\underline{G}_{(h'k'l')}$ as given by Eq. (3.6). Therefore, $\underline{R}_{facet}$ can be represented by the (scaled) vector product

$$\begin{aligned}
\underline{R}_{facet} &= \chi\,(\underline{G}_{(h\,k\,l)} \times \underline{G}_{(h'k'l')}) \\
&= \chi\,(h\,\underline{G}_{o1} + k\,\underline{G}_{o2} + l\,\underline{G}_{o3}) \times (h'\,\underline{G}_{o1} + k'\,\underline{G}_{o2} + l'\,\underline{G}_{o3}) \\
&= \chi\,\{(k\,l' - l\,k')\,(\underline{G}_{o2} \times \underline{G}_{o3}) + (l\,h' - h\,l')\,(\underline{G}_{o3} \times \underline{G}_{o1}) \\
&\quad + (h\,k' - k\,h')\,(\underline{G}_{o1} \times \underline{G}_{o2})\} \\
&= \chi\,(2\pi)^2/\beta\,\{(k\,l' - l\,k')\,\underline{R}_{o1} + (l\,h' - h\,l')\,\underline{R}_{o2} + (h\,k' - k\,h')\,\underline{R}_{o3}\} \\
\beta &= (\underline{R}_{o1} \times \underline{R}_{o2})\,\underline{R}_{o3} \quad\quad\quad\quad\quad\quad\quad\quad\quad\quad\quad\quad\quad\quad (5.30)
\end{aligned}$$

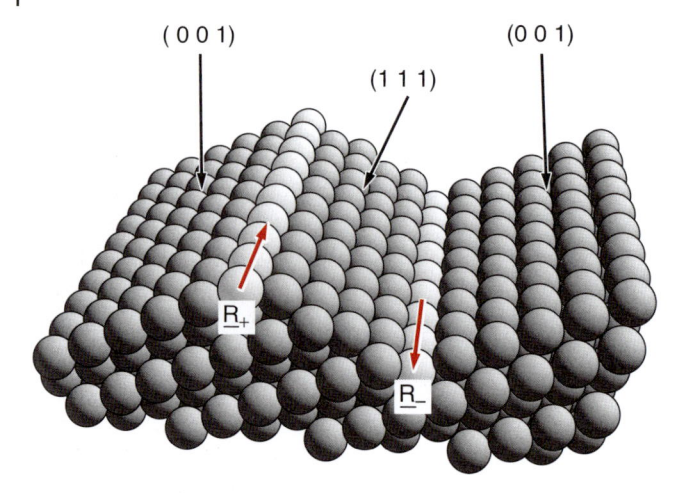

Figure 5.16 Structure of surface sections near positive and negative edges of (0 0 1)/(1 1 1) facets of a crystal with an fcc lattice. The positive (negative) edge is indicated by its facet edge vector $\underline{R}_+$ ($\underline{R}_-$) in red.

where the reciprocity between real space and reciprocal lattice vectors, discussed in Section 2.5, has been applied. Thus, fixing the scaling factor χ in Eq. (5.30) at

$$\chi = \frac{\beta}{(2\pi)^2} \qquad (5.31)$$

yields a facet edge vector $\underline{R}_{facet}$, which equals a **general lattice vector**. Vector $\underline{R}_{facet}$ is of smallest length along its direction if the Miller indices $(h\ k\ l)$ and $(h'\ k'\ l')$ have no common divisor greater than 1, that is, if $\gcd(h\ k\ l) = \gcd(h'\ k'\ l') = 1$. Relation 5.30 with 5.31 can also be expressed mathematically in a simpler **determinantal form** as

$$\underline{R}_{facet} = \det \begin{pmatrix} h & k & l \\ h' & k' & l' \\ \underline{R}_{o1} & \underline{R}_{o2} & \underline{R}_{o3} \end{pmatrix} \qquad (5.32)$$

Swapping the two top rows in matrix (5.32) changes only the sign of its determinant and, hence, inverts the direction of $\underline{R}_{facet}$. Thus, edge vectors $\underline{R}_{facet}$ of an $(h\ k\ l)/(h'\ k'\ l')$ facet and of its corresponding $(h'\ k'\ l')/(h\ k\ l)$ facet (one belongs to a positive and the other to a negative facet edge, see Figure 5.16) are always equal in length but opposite in direction.

The two surface sections joining at the facet edge form a **facet angle** φ_{facet} as illustrated in Figure 5.17. This angle can be evaluated by considering the scalar product of the corresponding normal vectors along $\underline{G}_{(h\ k\ l)}$ and $\underline{G}_{(h'k'l')}$ as

$$\cos \varphi_{facet} = (\underline{G}_{(h\ k\ l)}\, \underline{G}_{(h'k'l')})/(|\underline{G}_{(h\ k\ l)}|\, |\underline{G}_{(h'k'l')}|) \qquad (5.33)$$

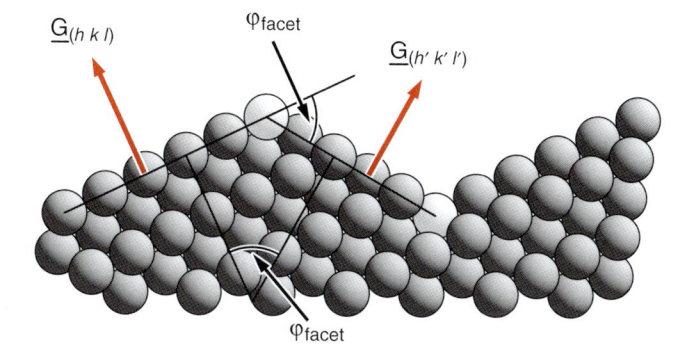

Figure 5.17 Structure of a facet edge separating $(h\,k\,l)$- and $(h'\,k'\,l')$-oriented surface sections. The facet angle φ_{facet} and the corresponding reciprocal lattice vectors $\underline{G}_{(h\,k\,l)}$ and $\underline{G}_{(h'k'l')}$ are labeled accordingly. Edge atoms are emphasized by light balls.

Table 5.1 Angles φ_{facet} and edge vectors $\underline{R}_{facet}$ of facets formed by selected $(h\,k\,l)$ surfaces of crystals with (a) fcc and (b) bcc lattices.

	Facet $(h\,k\,l)/(h'\,k'\,l')$	$\cos\varphi_{facet}$	φ_{facet} (°)	$\underline{R}_{facet}/a$
(a) Face-centered cubic lattice				
1	(1 1 1)/(−1 1 1)	1/3	70.53	(0, −1/2, 1/2)
2	(1 1 1)/(0 0 2)	$1/\sqrt{3}$	54.74	(1/2, −1/2, 0)
3	(1 1 1)/(0 2 2)	$\sqrt{(2/3)}$	35.26	(0, −1/2, 1/2)
4	(0 0 2)/(0 2 0)	0	90.00	(−1, 0, 0)
5	(0 0 2)/(0 2 2)	$1/\sqrt{2}$	45.00	(−1, 0, 0)
6	(1 1 1)/(1 1 3)	$5/\sqrt{33}$	29.50	(1/2, −1/2, 0)
7	(0 0 2)/(1 1 3)	$3/\sqrt{11}$	25.24	(−1/2, 1/2, 0)
8	(7 1 1)/(10 0 2)	$34/\sqrt{1326}$	20.98	(1/2, −1, 5/2)
(b) Body-centered cubic lattice				
1	(0 1 1)/(1 0 1)	1/2	60.00	(1/2, 1/2, −1/2)
2	(0 1 1)/(1 1 0)	1/2	60.00	(−1/2, 1/2, −1/2)
3	(1 0 1)/(1 1 0)	1/2	60.00	(−1/2, 1/2, 1/2)
4	(0 1 1)/(0 0 2)	$1/\sqrt{2}$	45.00	(1, 0, 0)
5	(0 1 1)/(1 1 2)	$\sqrt{(3/4)}$	30.00	(1/2, 1/2, −1/2)
6	(0 0 2)/(1 2 1)	$1/\sqrt{6}$	65.91	(−2, 1, 0)
7	(0 0 2)/(1 1 2)	$2/\sqrt{6}$	35.26	(−1, 1, 0)
8	(1 1 2)/(1 2 1)	5/6	33.56	(−3/2, 1/2, 1/2)

The facets $(h\,k\,l)/(h'\,k'\,l')$ are listed according to monolayer density with $\rho(h\,k\,l) \geq \rho(h'\,k'\,l')$. All Miller indices are given in sc notation. The edge vectors are defined by Cartesian coordinates (x, y, z), normalized to lattice constant a, and refer to the vectors of smallest length.

As examples, Table 5.1 lists angles φ_{facet} and edge vectors $\underline{R}_{facet}$ of facets formed by high-density $(h\,k\,l)$ surfaces of crystals with fcc and bcc lattices calculated using Eqs. (5.32) and (5.33).

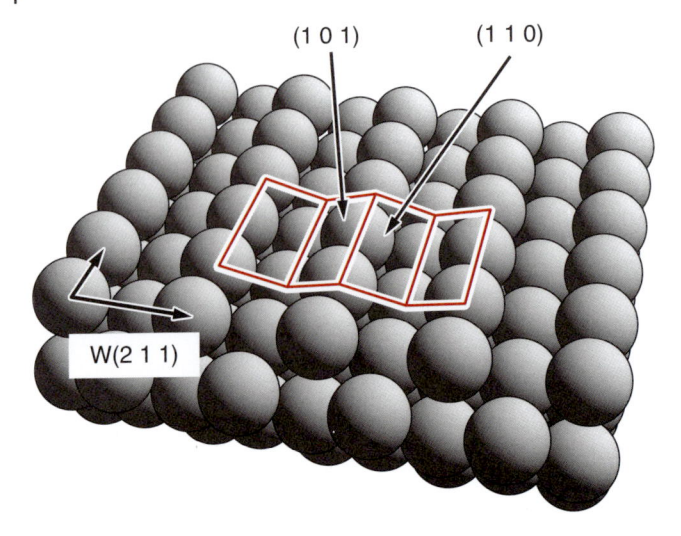

Figure 5.18 Perspective view of a W(2 1 1) surface with (1 0 1) and (1 1 0) facet stripes. The (2 1 1)-adapted lattice vectors are shown at the lower left. Joining facet areas are sketched by red lines and labeled accordingly.

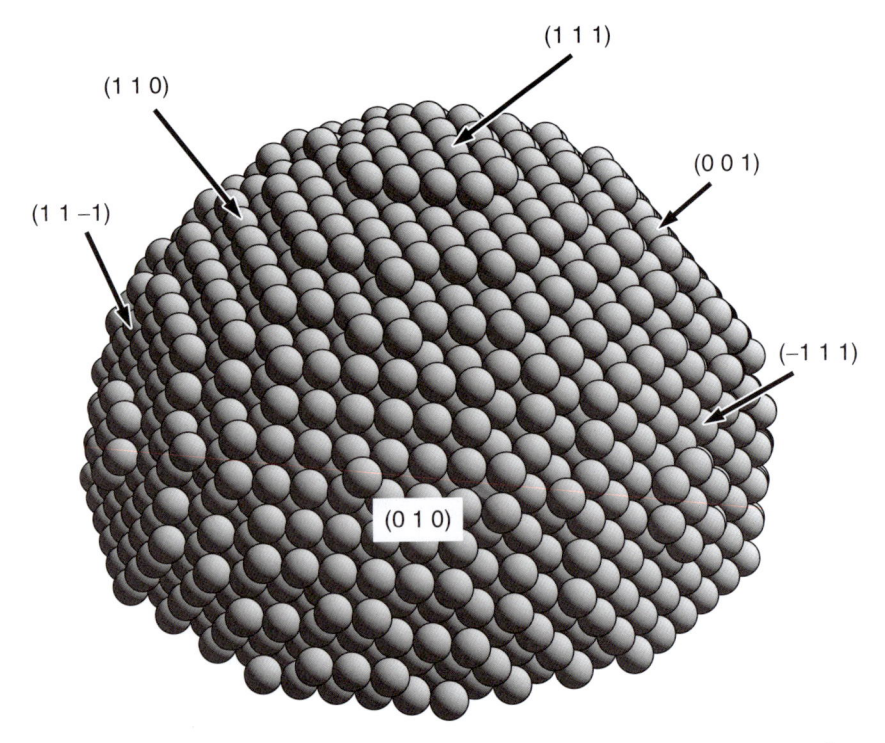

Figure 5.19 Spherical section of an fcc crystal exposing different (*h k l*) oriented facets of high atom density. The facets are labeled by their Miller indices.

Many **open surfaces** of single crystals expose small local planar sections of high atom density corresponding to low Miller index netplanes. Therefore, they are often considered to be **(micro) faceted**. As an illustration, the (2 1 1) oriented surface of bcc tungsten can conceptually be thought of as being stepped with (1 0 1) oriented terraces and (1 1 0) oriented steps. (The additivity theorem for Miller indices (4.9) of stepped surfaces yields (2 1 1) = (1 0 1) + (1 1 0).) But this surface may also be described as consisting of faceted stripes with (1 0 1) and (1 1 0) orientation reflecting the densest monolayers of the bcc lattice, see Figure 5.18. This is one simple example where **stepped** (or **kinked**) surfaces of single crystals may also be called **(micro) faceted**.

More complex examples of faceting, where many facet edges of different type can occur, are given, for example, by curved surfaces. These appear at crystalline spheres, cylinders, and tips, and are of great physical interest in connection with metal tips used in field emission or for scanning tunneling microscopy [34]. As an illustration, Figure 5.19 shows a spherical section of an fcc crystal, which may model the tip of a scanning tunneling microscope. This hemisphere exposes facets of different (h k l)-oriented surfaces of high density (labeled in the figure) with stepped and kinked transitions between them.

5.5
Exercises

5.1 Consider a (0 0 1)-oriented surface of a fictitious monoatomic crystal with an sc lattice described by Minkowski-reduced lattice vectors $\underline{R}_{o1}, \underline{R}_{o2}, \underline{R}_{o3}$. As a result of surface relaxation the inter-layer separation $d_{i,i+1}$ along $\underline{R}_{o3}$ is affected according to

$$d_{i,i+1} = R_{o3}(1 + q_i), \quad i = 1, 2 \ldots$$

where index i counts monolayers from the surface top. Discuss variations in the neighbor shells (up to fifth shell) of atoms of the three topmost surface layers assuming q_i values $q_1 = -0.2$, $q_2 = -0.1$, $q_3 = 0.05$, $q_i = 0.0$ for $i > 4$.

5.2 Consider the (0 0 1)-oriented surface of a Pd crystal (fcc lattice, lattice constant $a = 3.89$ Å). The topmost four monolayers are relaxed perpendicular to the surface with inter-layer distances $d_{i,i+1}$ varying according to $d_{12} = 1.0487\, d_o$, $d_{23} = 1.0025\, d_o$, $d_{34} = 0.9922\, d_o$, $d_{45} = 1.0025\, d_o$ (d_o denotes the bulk inter-layer spacing). Determine the geometric structure of neighbor shells (up to third nearest neighbors) of atoms of the three topmost surface layers.

5.3 Consider a Ni(0 0 1) surface (fcc lattice) with c(2 × 2) reconstruction of the topmost surface layer. Give alternative representations of the reconstructed layer (matrix definition).

5.4 Consider a Cu(1 1 1) and (0 0 1) surface (fcc lattice) with the surface layer rotated by 10°. Determine the approximate lateral lattice constant of the resulting superlattice.

5.5 Consider the superposition of two adjacent fcc monolayers that are rotated by small angles α with respect to each other. Determine the resulting geometry as a function of the rotation angle α and discuss corresponding superlattices for monolayers oriented (1 1 1) and (0 0 1).

5.6 Discuss the Cu(1 1 0) surface (fcc lattice) with missing row reconstructions, (2×1) and (1×2), of the first layer.
  (a) Determine corresponding reconstruction matrices.
  (b) Evaluate monolayer orientations of the corresponding microfacets.

5.7 Discuss the structure of a Si(0 0 1) surface (diamond lattice) with a missing row reconstruction of the first layer. Determine the reconstruction matrix. Calculate distances of neighbor shells (up to third neighbors) of atoms of the three topmost surface layers

5.8 Discuss the structure of a Si(1 1 0) surface (diamond lattice). Show that this surface allows only one unique termination.

5.9 Consider a Si(0 0 1) surface (diamond lattice),
  (a) without relaxation or reconstruction. Show that the two possible terminations differ only by a 90° rotation about the surface normal.
  (b) with a buckling $c(4\times 2)$ dimer reconstruction of the first layer. Determine neighbor shell radii of the atoms of the reconstruction layer.

5.10 Discuss the model of the (7×7) reconstructed Si(1 1 1) surface according to a LEED (low-energy electron diffraction) analysis by Tong et al. [110], see Figure 5.5. How many atoms does each of the elementary cells of the first three surface layers contain?

5.11 Consider rotational reconstruction with isotropic scaling of the top layer (overlayer) of a primitive simple cubic lattice at the (0 0 1), (0 1 1), and (1 1 1) surface. Determine the rotation angles that yield coincidence lattice overlayers. Which values do the scaling constants of the lattice vectors assume for a given rotation? Hint: Use results of Appendix E.4.

5.12 Consider rotational reconstruction with isotropic scaling of the top layer (overlayer) of a primitive hexagonal lattice. Determine rotation angles that yield coincidence lattice overlayers. Which values do the scaling constants of the lattice vectors assume for a given rotation? Hint: Use results of Appendix E.4.

5.13 Consider rotational reconstruction of the topmost overlayer of a primitive tetragonal crystal. Determine the constraints for the lattice constants a, c and for Miller indices $(h\,k\,l)$ to yield coincidence lattice overlayers.

5.14 Consider rotational reconstruction of the topmost overlayer of the (0 0 1), (1 0 0), and (1 0 1) oriented surface of a primitive tetragonal crystal. Determine the lattice constants a, c to yield coincidence lattice overlayers.

5.15 Consider a Pd(0 0 1) surface (fcc lattice) with a crystallite of palladium forming a pyramid of square base on top of it. (The internal structure of the crystallite is assumed to be identical to that of the bulk crystal.)
(a) Determine Miller indices of the four facet planes of the pyramid.
(b) Calculate the angle between two crossing facet planes of the pyramid.

5.16 Consider a Ni(1 1 1) surface (fcc lattice) with a crystallite of palladium forming a pyramid of triangular base on top of it. (The internal structure of the crystallite is assumed to be identical to that of the bulk crystal.)
(a) Determine Miller indices of the three facet planes of the pyramid.
(b) Calculate the angle between two crossing facet planes of the pyramid.

5.17 Consider fcc metal single crystals terminated by surfaces of orientations given by Miller indices (1 1 0), (1 1 3), (2 1 1). These surfaces may be interpreted as microfaceted. Determine Miller indices of corresponding facets.

5.18 Consider bcc metal single crystals terminated by surfaces of orientations given by Miller indices (0 0 1), (1 1 2), (0 1 3), (1 1 1). These surfaces may be interpreted as microfaceted. Determine Miller indices of corresponding facets.

5.19 Consider an fcc crystal sphere with an atom in its center and including all atoms up to a distance $r = 5a$ from the center (a = lattice constant).
(a) Characterize surface sections of the ball corresponding to densest monolayers.
(b) Discuss transitions between sections of low $(h\,k\,l)$ index monolayers as a result of the ball curvature.
(c) How many atoms does the ball contain?

5.20 Consider a lattice described by two equivalent lattice vector sets $\underline{R}_1, \underline{R}_2, \underline{R}_3$ and $\underline{R}'_1, \underline{R}'_2, \underline{R}'_3$ with

$$\begin{pmatrix} \underline{R}'_1 \\ \underline{R}'_2 \\ \underline{R}'_3 \end{pmatrix} = \underline{\underline{T}} \cdot \begin{pmatrix} \underline{R}_1 \\ \underline{R}_2 \\ \underline{R}_3 \end{pmatrix}$$

Then facet edge vectors $\underline{R}_{\text{facet}}$ can be represented by either of the lattice vector sets. Show that

$$\underline{R}_{\text{facet}} = \det \begin{pmatrix} H & K & L \\ H' & K' & L' \\ \underline{R}'_1 & \underline{R}'_2 & \underline{R}'_3 \end{pmatrix} = \det(\underline{\underline{T}}) \cdot \det \begin{pmatrix} h & k & l \\ h' & k' & l' \\ \underline{R}_1 & \underline{R}_2 & \underline{R}_3 \end{pmatrix}$$

where $(h\,k\,l)$, $(h'\,k'\,l')$ and $(H\,K\,L)$, $(H'\,K'\,L')$ are Miller indices referring to the two lattice vector sets. Hint: Use results of Sections 5.4 and 3.4.

6
Adsorbate Layers

6.1
Definition and Classification

Adsorption at single crystal surfaces of $(h\ k\ l)$ orientation can be described by foreign atoms and/or molecules – they will be called **adsorbates** or *adparticles* in this section – binding to a substrate and forming overlayers. Hence, adsorption processes are closely related to growth mechanisms at single crystal surfaces where different growth modes have been discussed in Section 5.3, and structural aspects are completely analogous in both types of systems. If the adparticles are identical in type to atoms of the substrate, adsorption increases only the substrate at its corresponding surface, possibly introducing additional surface reconstruction as discussed in Section 5.2. This will not lead to new structure details and can be ignored in the present section.

Structural properties of adsorbate overlayers depend strongly on the interaction between the adsorbates and the substrate surface, as well as between different adsorbates within the overlayer depending on the overlayer density. The latter is usually defined by an adsorbate **coverage** Θ which is given by the ratio of the number density of adsorbates in the overlayer and the atom density of the topmost substrate layer, where $\Theta = 1$ is defined as **monolayer coverage**. In cases of very weak adsorbate–substrate and adsorbate–adsorbate interactions at low coverage, adsorbate particles can diffuse easily on the substrate surface and corresponding overlayers cannot be expected to show any structural order. They will form completely **disordered two-dimensional** gas or liquid films. Examples are light rare gas atoms, like helium or neon, physisorbed at low coverage $\Theta \ll 1$ at metal surfaces, where interatomic coupling is governed by van der Waals type interactions [123]. These cases are not relevant for general crystallographic considerations.

If the adsorbate–substrate interaction becomes stronger while the adsorbate–adsorbate interaction is still weak, the adsorbates may bind only at specific geometric sites of the substrate surface. At low adsorbate coverage, $\Theta < 1$, not all equivalent surface sites will be populated by adsorbates and there is a disordered population of sites. Thus, corresponding adsorbate overlayers can be described structurally by fixed overlayer lattices, which are commensurate with the lattice of the substrate surface. However, not all overlayer lattice sites are occupied by adsorbates. These disordered systems are usually called **two-dimensional**

Crystallography and Surface Structure: An Introduction for Surface Scientists and Nanoscientists,
Second Edition. Klaus Hermann.
© 2017 Wiley-VCH Verlag GmbH & Co. KGaA. Published 2017 by Wiley-VCH Verlag GmbH & Co. KGaA.

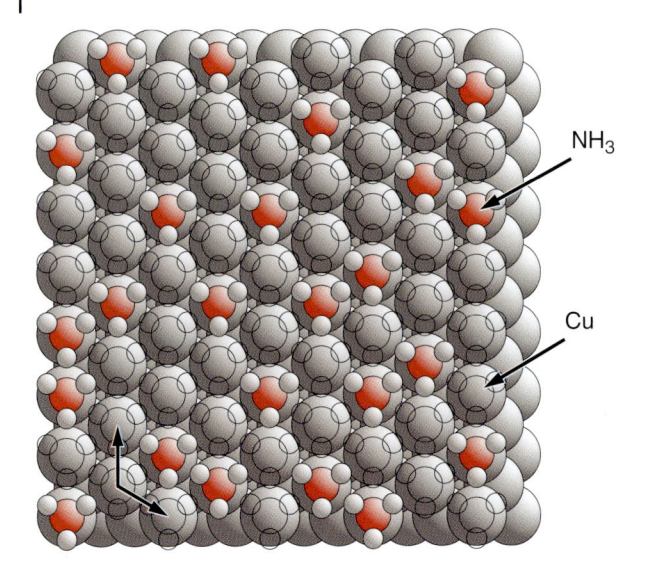

Figure 6.1 Structure of the Cu(1 1 1) + (disordered) - NH_3 adsorbate system. The lattice vectors of the NH_3 adsorbate lattice for a complete (1 × 1) overlayer are shown in black. The random population of the lattice sites by NH_3 molecules corresponds to a coverage $\Theta = 1/3$. Unoccupied lattice sites are indicated by open circles.

lattice gas systems. An example is given by the adsorption of ammonia on the Cu(1 1 1) surface at low coverage [124]. Figure 6.1 illustrates a possible structure of the Cu(1 1 1) + (disordered) − NH_3 system for an NH_3 coverage $\Theta = 1/3$. Here, NH_3 molecules stabilize always on top of copper atoms of the substrate surface (which forms a hexagonal lattice) where only 1/3 of the top sites are occupied.

If, on the other hand, the adsorbate–substrate interaction is weak while the adsorbate–adsorbate interaction becomes strong, then adsorbates may, at lower coverage, combine to two-dimensional **islands** or form three-dimensional **clusters** at the surface, which are randomly distributed. Their structural properties are influenced only to a small extent by those of the substrate surface and have to be treated individually. At higher adsorbate coverage, islands and clusters can increase in size and form defect-free periodic overlayers that are oriented randomly on the surface and whose lattices are, in general, not expected to be commensurate with that of the substrate. There are also cases of **partially disordered** adsorbate systems. Here, we mention only overlayers that are periodic in one dimension and disordered in the other forming periodic adsorbate rows on the surface that are positioned in a disordered fashion.

In addition, there are many adsorbate systems [22] where the adsorbates interact strongly with the substrate and also couple with each other at large enough coverage forming two-dimensionally **periodic overlayers**. Thus, the discussion of structural properties of these adsorbate systems is, from a crystallographic point of view, completely **analogous** to that of single crystal surfaces with topmost layers that are relaxed or reconstructed, see Sections 5.1 and 5.2. It is also strongly connected with structural aspects of epitaxial crystal growth discussed in Section 5.3. Differences arise only in that the atom types (elements) in adsorbate overlayers will differ from those of the substrate. On the other hand, heteroepitaxial thin films are, in their structural description, analogous to adsorbate overlayers and can be treated on the same footing. Altogether, we can distinguish between three cases.

1) Adsorbate overlayers can form **commensurate** overlayer lattices. Here, the lattice vectors of the overlayers, $\underline{R}'_1$ and $\underline{R}'_2$, are connected with those of the substrate, $\underline{R}_1$ and $\underline{R}_2$, by an **integer-valued** transformation matrix $\underline{\underline{M}}$ according to Eq. (5.3). Further, the adsorbates stabilize in one or several specific sites at the substrate surface as will be discussed below. An example from the Surface Structure Database, SSD 28.6.8.45, is the adsorption of CO on the Ni(1 1 0) surface, formally described as Ni(1 1 0) + p2mg(2 × 1) − 2CO [125], shown in Figure 6.2, where

$$\underline{\underline{M}} = \begin{pmatrix} 2 & 0 \\ 0 & 1 \end{pmatrix}$$

Here, the CO molecules stabilize in bridge sites between Ni atoms of the topmost substrate layer, where their molecular axes are tilted alternatingly to the left and right of the Ni ridges, with two CO molecules in each overlayer unit cell.

More complex cases of commensurate superlattices include large molecular adsorbates at reconstructed single crystal surfaces of metals where the adsorbates bind at preferred surface sites. As an example, C_{60} adsorbate molecules, so-called "buckyballs", form a 4 × 4 superlattice on the Cu(1 1 1) surface [126],

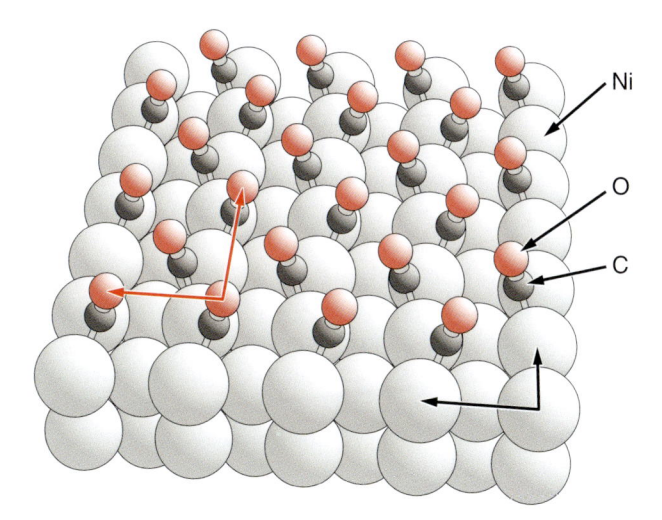

Figure 6.2 Structure of the Ni(1 1 0) + p2mg(2 × 1) − 2CO adsorbate system. The lattice vectors of the CO adsorbate layer and of the Ni substrate are sketched in red and black, respectively.

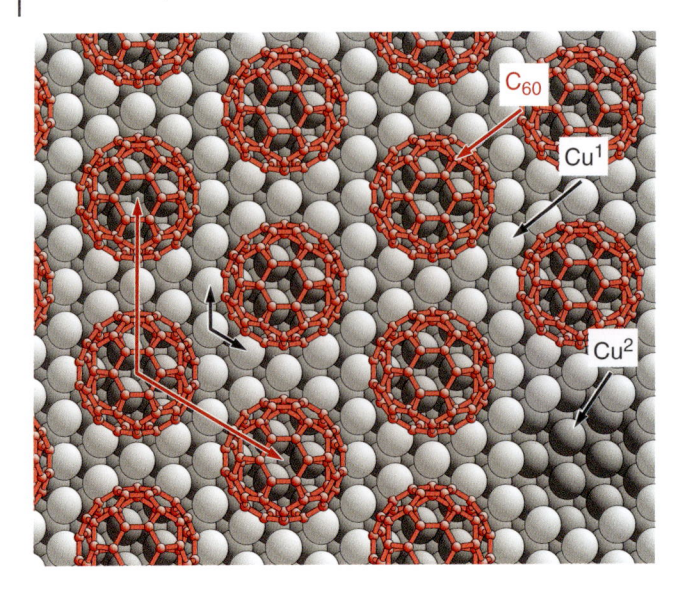

Figure 6.3 Structure of the Cu(1 1 1) + (4×4) − C_{60} adsorbate system. The lattice vectors of the C_{60} adsorbate layer and of the Cu substrate are sketched in red and black, respectively. Cu atoms of the first and second substrate surface layer, labeled Cu^1 and Cu^2, are shown in light and dark gray, the C_{60} adsorbates in red with bond sticks. One adsorbate is removed to illustrate the missing first layer Cu atoms.

which is described as Cu(1 1 1) + (4×4) − C_{60}, see Figure 6.3. Here, the topmost layer of the Cu substrate (first layer) is reconstructed forming a 4×4 overlayer with hexagonal holes described by seven missing Cu atoms. These holes act as binding sites for the C_{60} adsorbates that sit above threefold hollow sites of the second surface layer where they bind with three copper atoms of the layer.

2) Adsorbate overlayers can form **coincidence** lattices, sometimes also called **high-order** commensurate (**HOC**) or **scaled commensurate** lattices. Here, the transformation matrix $\underline{M}$, connecting lattice vectors of the overlayer with those of the substrate surface according to Eq. (5.3), contains integer and **rational** matrix elements m_{ij} with at least one rational. Coincidence lattices can appear for metal overlayers on a substrate of a different metal where the two metal lattice constants do not match each other. As an illustration, Figure 6.4 shows a fictitious adsorbate system with a hexagonal overlayer of face-centered cubic (fcc) metal B adsorbed at a (1 1 1) surface of an fcc metal A (only the topmost hexagonal substrate layer is shown). Here, the lattice constant of metal B is larger than that of metal A by 6.25% (= 1/16). This results in a coincidence lattice overlayer structure described by a transformation matrix

$$\underline{M} = \frac{1}{16} \begin{pmatrix} 17 & 0 \\ 0 & 17 \end{pmatrix} \tag{6.1}$$

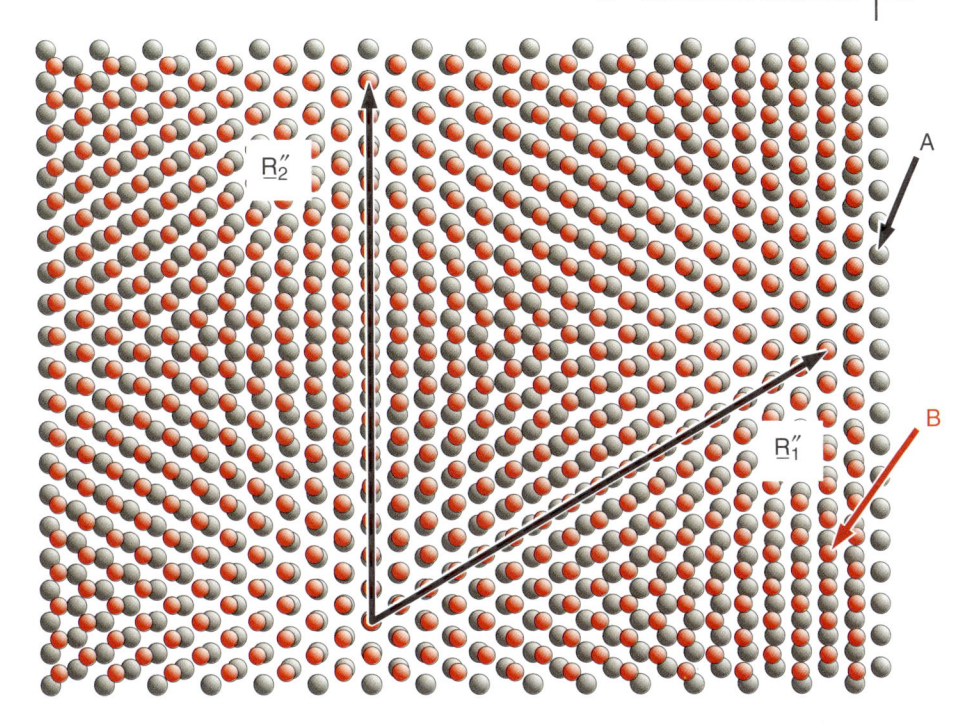

Figure 6.4 Structure of a fictitious adsorbate system with a hexagonal overlayer of fcc metal B (red balls) on a (1 1 1) surface of an fcc metal A (gray balls), see text. The substrate is represented by its topmost layer. The lattice constant of metal B is larger than that of metal A by 6.25% = 1/16. The lattice vectors of the common superlattice, $\underline{R}''_1, \underline{R}''_2$, are sketched accordingly.

In this system the overlayer of metal B forms a hexagonal coincidence lattice with the substrate. The common hexagonal supercell is given by multiples of the substrate lattice vectors, $\underline{R}''_1 = 17\,\underline{R}_1$ and $\underline{R}''_2 = 17\,\underline{R}_2$, also evidenced by the hexagonal interference pattern in Figure 6.4.

In real adsorbate overlayer systems, **coincidence lattices** are usually combined with a **modulation** of overlayer (and substrate) **atom positions** perpendicular and parallel to the surface as a result of local binding effects. Therefore, overlayer atom positions inside the unit cell of the coincidence lattice are described only **approximately** by lattice vectors defined by rational-valued transformation matrices $\underline{M}$ such as Eq. (6.1). An example is the adsorption of graphene, a graphite monolayer with honeycomb structure, on the (0 0 0 1) surface of ruthenium, which has been observed by LEED (low-energy electron diffraction) and STM (scanning tunneling microscopy) [127]. Here, the graphene overlayer forms, together with the topmost Ru layers, a coincidence superlattice, shown in Figure 6.5, where very small lateral distortions are combined with perpendicular warping, see Figure 6.5b. HOC overlayers will be discussed in more detail in Sections 6.4 and 6.5.

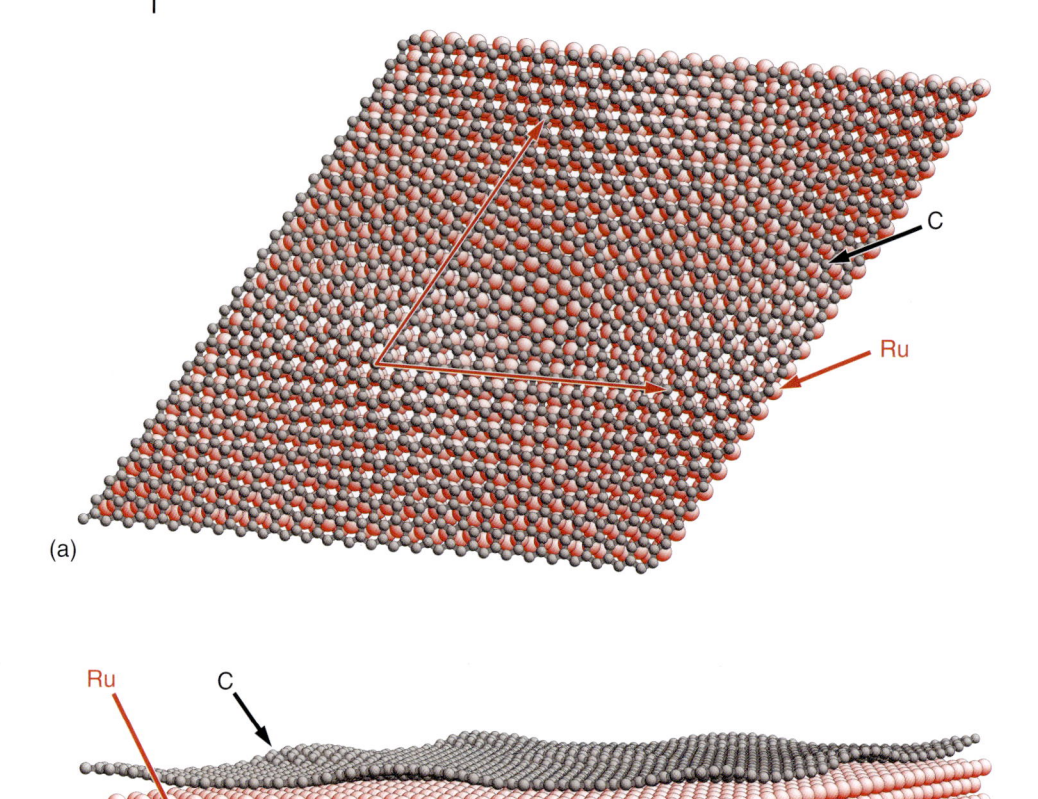

Figure 6.5 Graphene overlayer adsorbed on the Ru(0 0 0 1) surface. The Ru surface is shown by its topmost three layers. (a) View perpendicular to the surface. The superlattice periodicity, 12×12 for the Ru substrate and 13×13 for graphene from DFT simulations, is indicated by lattice vectors. (b) View almost parallel to the surface demonstrating the overlayer warping.

3) Adsorbate overlayers can form **incommensurate** overlayer lattices. Here, the transformation matrix $\underline{M}$, connecting the lattice vectors of the overlayer with those of the substrate surface according to Eq. (5.3), contains matrix elements m_{ij}, of which at least one is **irrational**. In this case the combined adsorbate–substrate system is not strictly periodic in two dimensions. An example from the Surface Structure Database, SSD 47.54.1, is the adsorption of Xe atoms adsorbed on the Ag (1 1 1) surface [128] shown in Figure 6.6. Here, both the Xe adsorbate overlayer and the topmost substrate layer form hexagonal lattices. However, their lattice constants are different and, in addition, the relative orientation of the two layers with respect to each other may vary. The orientation shown in Figure 6.6 is only one of many possibilities

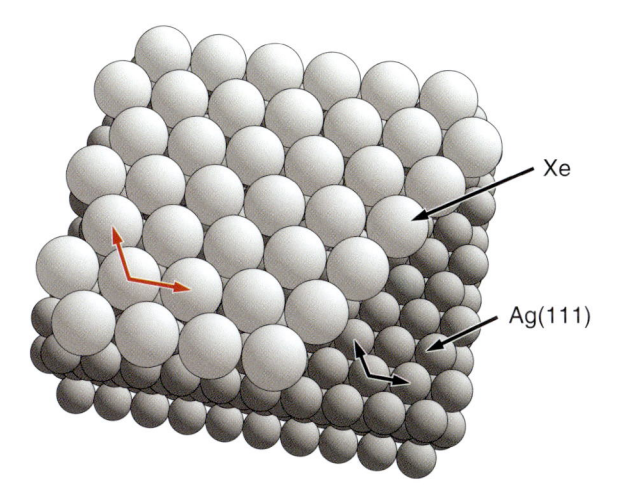

Figure 6.6 Structure of the incommensurate Ag(1 1 1) + Xe system. The periodicity vectors of the Xe adsorbate layer and of the Ag substrate are shown for one possible orientation only with overlayer vectors chosen to be parallel to those of the substrate.

where the overlayer can be rotated and shifted laterally. Analogous to the earlier examples, the Xe overlayer may not be completely flat with modulations due to local binding effects which are, however, expected to be small in the present case. Also, the slightest lateral adsorbate–substrate interaction may contract or expand the overlayer into a commensurate relationship, perhaps with a large coincidence supercell reflecting an HOC lattice. The latter corresponds formally to approximating all elements of the real-valued transformation matrix $\underline{\underline{M}}$ by rational numbers for which many mathematical algorithms have been proposed [129].

6.2
Adsorbate Sites

The structural characterization of an adsorbate covered surface also includes quantitative details about all surface sites where the adsorbates stabilize as well as about orientation and changed internal structure in cases of molecular adsorbates. These details are basically determined by the local binding of each adsorbate with its nearby atoms of the substrate surface, that is, those of the topmost substrate layers. The planar substrate surface of a single crystal with ideal bulk termination, that is, of two-dimensional lateral periodicity and defined by Miller indices $(h\ k\ l)$, is also characterized by its **lateral point symmetry** reflecting a two-dimensional point symmetry group, see Section 3.8. As a result, there are preferred lateral sites inside the elementary cell of the substrate surface, which are compatible with its symmetry, so-called **high-symmetry sites**. In many surface systems adsorbates are found to stabilize laterally at or near these high-symmetry sites with a directed perpendicular distance z_{ads}. The latter is usually measured from the plane through the topmost $(h\ k\ l)$ monolayer of the substrate where $z_{ads} > 0$ will be called **above** and $z_{ads} < 0$ **below** the surface. In the following, examples of high-symmetry sites at surfaces of different symmetry are discussed and illustrated by results from measured adsorbate systems. Additional sites can be inspected in Appendix A, which collects the most important high-symmetry sites of common surfaces. All following example structures are denoted by "SSD n.m" where n.m refers the corresponding structure to the entry number of the Surface Structure Database (NIST SSD, Version 5 or oSSD), see Section 7.2 and Appendix H.

At ideal substrate surfaces with **square** netplanes and appropriate point symmetry, see Section 3.8.4, the primitive periodicity cell offers three distinct lateral high-symmetry sites, shown in Figure 6.7, (a) the top site, (b) the fourfold hollow site, and (c) the twofold bridge site.

At the **top** site the adsorbate atom or molecule stabilizes directly above a substrate atom at the surface, see Figure 6.7, site (a). This allows strong directional binding of the adsorbate with the substrate, which can be a result of strong covalent bond formation at the surface. An example of periodically ordered **molecular** adsorbates is Ni(1 0 0) + c(2×2) − CO [130], SSD 28.6.8.8, shown in Figure 6.8, where the CO molecules adsorb in top sites with carbon pointing to the surface and $z_{ads}(C) = 1.70$ Å.

At the **fourfold hollow** site the adsorbate atom or molecule stabilizes laterally in the center between four substrate atoms at the surface that form the periodicity cell, see Figure 6.7, site (b). This allows for largest coordination at the surface, which can be due to major electrostatic interactions appearing for ionic adsorbates. An example of periodically ordered **atomic** adsorbates is

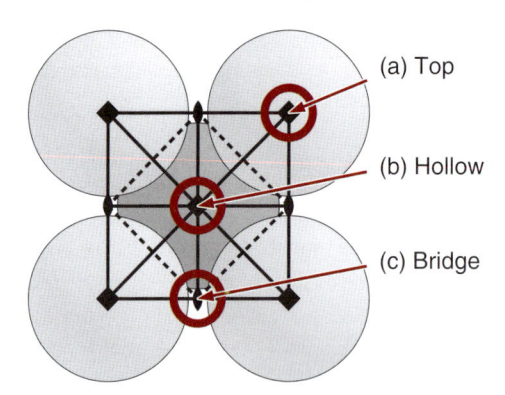

Figure 6.7 Primitive periodicity cell of a substrate with square lattice. The lateral high-symmetry sites, (a) top, (b) hollow, and (c) bridge, are shown by red circles. The gray balls indicate substrate atoms of different layers and all possible point symmetry elements of the cell are included by corresponding symbols, see Section 3.8.4.

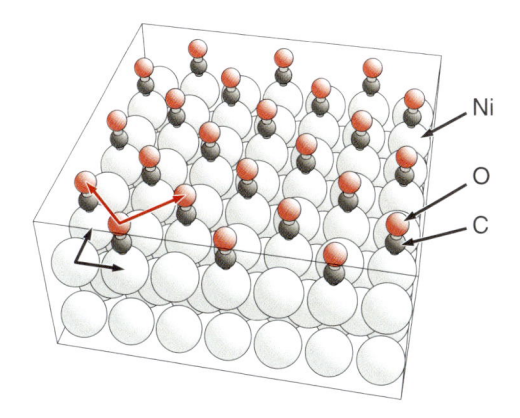

Figure 6.8 Structure of the Ni(1 0 0) + c(2 × 2) – CO adsorbate system. The lattice vectors of the CO adsorbate layer and of the Ni substrate are shown in red and black, respectively.

Cu(1 0 0) + c(2 × 2) – Cl [131], SSD 29.17.7, shown in Figure 6.19, where the chlorine atoms adsorb in the hollow sites with $z_{ads} = 1.59$ Å.

At the **twofold bridge** site the adsorbate atom or molecule stabilizes laterally between two adjacent substrate atoms at the surface, see Figure 6.7, site (c). An example of periodically ordered **atomic** adsorbates is Ir(1 0 0) + (1 × 2) – O [132], SSD 77.8.4, where the oxygen atoms adsorb in bridge sites with $z_{ads} = 1.30$ Å.

At ideal substrate surfaces with **hexagonal** netplanes and appropriate point symmetry, see Section 3.8.5, the primitive periodicity cell offers four distinct lateral high-symmetry sites, shown in Figure 6.9, (a) the top site, (b) the threefold hcp hollow site, (c) the threefold fcc hollow site, and (d) the twofold bridge site.

At the **top** site the adsorbate atom or molecule stabilizes directly above a substrate atom at the surface, see Figure 6.9, site (a), which allows strong directional binding of the adsorbate with the substrate. An example of periodically ordered **atomic** adsorbates is Cu(1 1 1) + (2 × 2) – Cs [133], SSD 29.55.1, where the Cs atoms adsorb in top sites with $z_{ads} = 3.01$ Å. Further, Rh(1 1 1) + ($\sqrt{3} \times \sqrt{3}$)R30° – CO [134], SSD 45.6.8.7a, is an example of periodically ordered **molecular** adsorbates where the CO molecules adsorb in top sites with carbon pointing to the surface and $z_{ads}(C) = 1.81$ Å.

Surfaces with hexagonal netplanes can offer two different **threefold hollow** sites depending on the structure of the substrate layers below the topmost layer. For cubic and hexagonal close-packed crystals these are the **hcp hollow** site, see Figure 6.9, site (b), where there is a substrate atom of the second surface layer directly underneath the adsorbate site, and the **fcc hollow** site, see Figure 6.9, site (c), where an atom of the third surface layer is underneath. An example of periodically ordered **atomic** adsorbates is Ni(1 1 1) + (2 × 2) – O [135], SSD 28.8.75a, where oxygen atoms adsorb in fcc hollow sites with $z_{ads} = 1.09$ Å. In addition, the CO molecules of the **molecular** adsorbate system Pd(1 1 1) + ($\sqrt{3} \times \sqrt{3}$)R30° – CO [136], SSD 46.6.8.13, shown

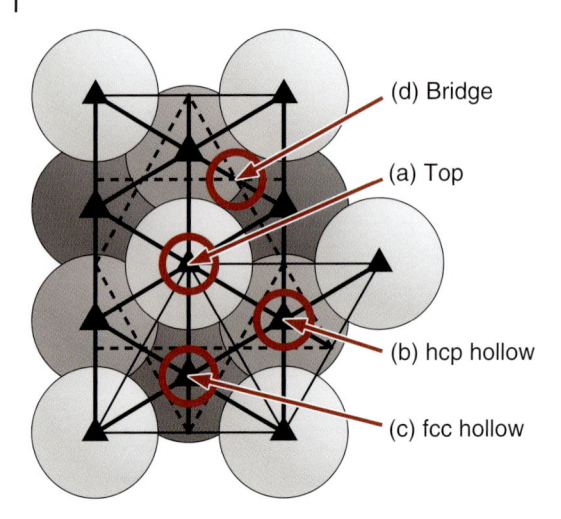

Figure 6.9 Primitive and rectangular periodicity cells of a substrate with hexagonal lattice based on cubic and hexagonal close-packed bulk structures. The lateral high-symmetry sites, (a) top, (b) hcp hollow, (c) fcc hollow, and (d) bridge, are shown by red circles. The gray balls indicate substrate atoms of different layers and the point symmetry elements of the cell are included by corresponding symbols, see Section 3.8.5.

in Figure 6.20, adsorb in fcc hollow sites with carbon pointing to the surface and $z_{ads}(C) = 1.25$ Å. Further, the NO molecules of Ni(1 1 1) + c(4×2) – 2NO [137], SSD 28.7.8.8, shown in Figure 6.10, adsorb in both fcc and hcp hollow sites with nitrogen pointing to the surface and $z_{ads}(N)$ varying between 1.18 and 1.32 Å.

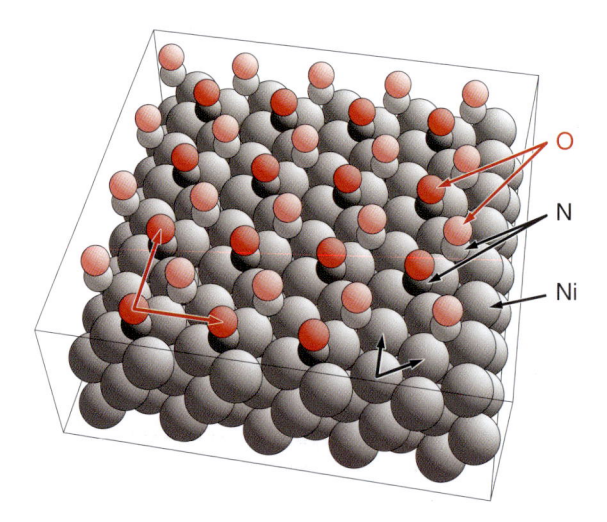

Figure 6.10 Structure of the Ni(1 1 1) + c(4×2) – 2NO adsorbate system. NO molecules in fcc hollow sites are painted darker that those in hcp hollow sites. The lattice vectors of the NO adsorbate layer and of the Ni substrate are shown in red and black, respectively.

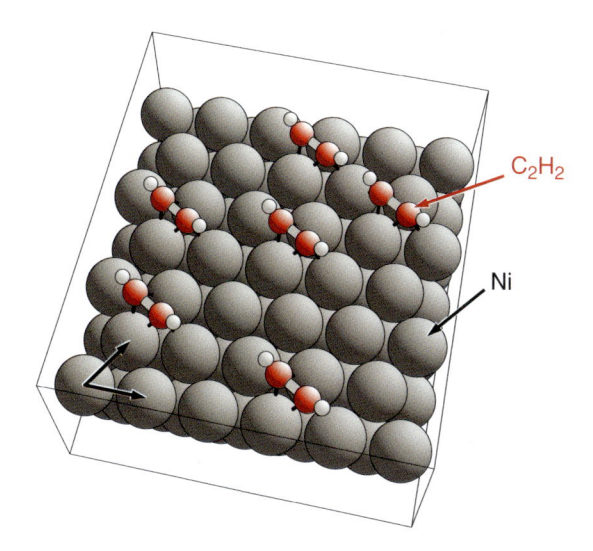

Figure 6.11 Structure of the Cu(1 1 1) + (disordered) – C_2H_2 adsorbate system. The lattice vectors of the Cu substrate are shown in black.

Actually, the topmost nickel layer is not exactly planar while the nitrogen centers are coplanar, resulting in different $z_{ads}(N)$ values.

At the **twofold bridge** site the adsorbate atom or molecule stabilizes laterally between two adjacent substrate atoms at the surface, see Figure 6.9, site (d). An example of **molecular** adsorption is Cu(1 1 1) + (disordered) – C_2H_2 [138], SSD 29.6.1.6, shown in Figure 6.11, where the two carbon atoms of the acetylene adsorbate bend slightly asymmetrically over bridge sites such that they approach adjacent fcc and hcp hollow sites where $z_{ads}(C)$ amounts to 1.38 Å (fcc) and 1.44 Å (hcp), respectively. Due to the threefold symmetry of the clean Cu(1 1 1) surface there are three equivalent bridge site structures with the orientation of the C_2H_2 adsorbate rotated by ±120°. Of these only one orientation is shown in Figure 6.11.

At ideal substrate surfaces with **primitive rectangular** netplanes and appropriate point symmetry, see Section 3.8.2, the primitive periodicity cell offers four distinct lateral high-symmetry sites, shown in Figure 6.12, (a) the top site, (b) the twofold long bridge site, (c) the twofold short bridge site, and (d) the fourfold hollow site, sometimes also called **center site**.

At the **top** site the adsorbate atom or molecule stabilizes directly above a substrate atom at the surface, see Figure 6.12, site (a). An example of periodically ordered **molecular** adsorbates is Cu(1 1 0) + (2 × 1) – CO [139], SSD 29.6.8.7, where the CO molecules adsorb in the top sites with carbon pointing to the surface and $z_{ads}(C) = 1.87$ Å.

At the **twofold long bridge** site the adsorbate atom or molecule stabilizes laterally between two adjacent substrate atoms at the surface, see Figure 6.12, site (b). An example of **atomic** adsorption was suggested for Ag(1 1 0) + (2 × 1) – O [140], SSD 47.8.4, shown in Figure 6.13, where the oxygen atoms stabilize between silver

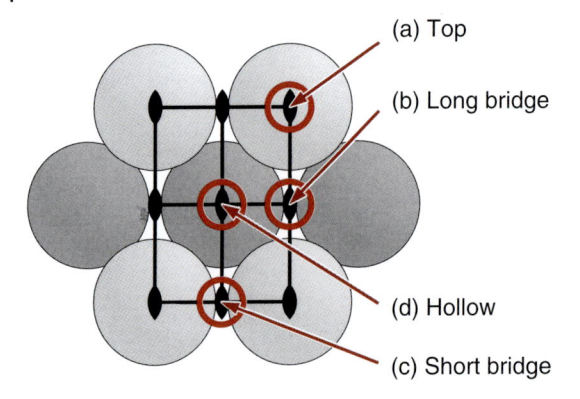

Figure 6.12 Primitive cell of a substrate with primitive rectangular lattice based on cubic bulk structures. The lateral high-symmetry sites, (a) top, (b) long bridge, (c) short bridge, and (d) hollow, are shown by red circles. The gray balls indicate substrate atoms of different layers and the point symmetry elements of the cell are included by corresponding symbols, see Section 3.8.2.

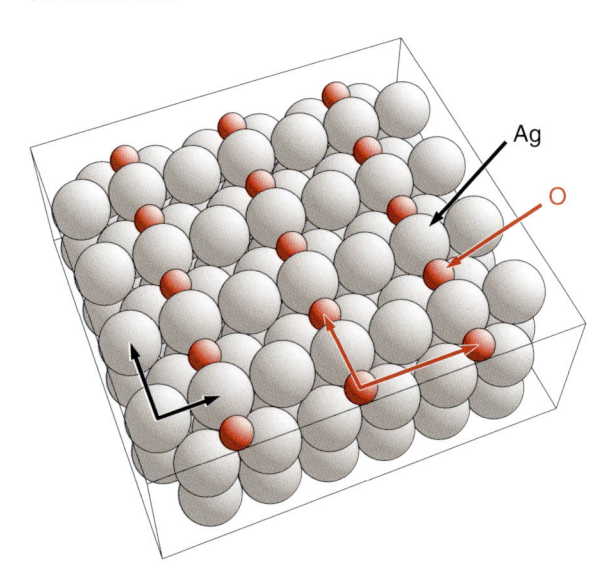

Figure 6.13 Structure of the Ag(1 1 0) + (2 × 1) – O adsorbate system. The lattice vectors of the O adsorbate layer and of the Ag substrate are shown in red and black, respectively.

atoms of the topmost substrate row with $z_{ads} = 0.2$ Å, that is, only slightly above the surface plane through the rows.

At the **twofold short bridge** site the adsorbate atom or molecule also stabilizes laterally between two adjacent substrate atoms at the surface but the substrate atoms are closer together than at the long bridge site, see Figure 6.12, site (c). An example of periodically ordered **atomic** adsorbates is Pt(1 1 0) + c(2 × 2) – Br [106], SSD 78.35.1, shown in Figure 6.14, where the bromine atoms adsorb in

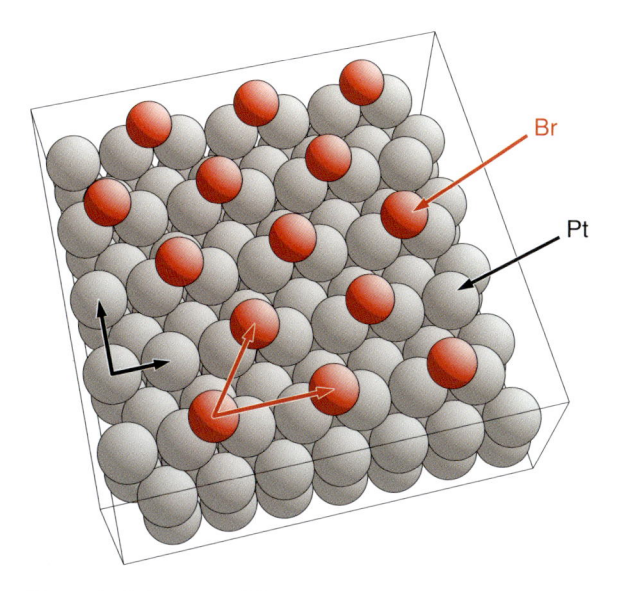

Figure 6.14 Structure of the Pt(1 1 0) + c(2 × 2) – Br adsorbate system. The lattice vectors of the Br adsorbate layer and of the Pt substrate are shown in red and black, respectively.

short bridge sites above the topmost substrate rows with $z_{ads} = 2.04$ Å. Further, the CO molecules of the **molecular** adsorbate system Ni(1 1 0) + p2mg(2 × 1) – 2CO [125], SSD 28.6.8.45, shown in Figure 6.2, adsorb quite near the short bridge sites. Here, their molecular axes are alternately tilted with respect to the surface normal as discussed below.

At the **fourfold hollow** site the adsorbate atom or molecule stabilizes laterally in the center between four substrate atoms at the surface, which form the periodicity cell, see Figure 6.12, site (d). An example of periodically ordered **atomic** adsorbates is Ni(1 1 0) + c(2 × 2) – S [141], SSD 28.16.57b, where the sulfur atoms adsorb in hollow sites with $z_{ads} = 0.77$ Å.

At ideal substrate surfaces with **centered rectangular** netplanes and appropriate point symmetry, see Section 3.8.3, the primitive periodicity cell offers two distinct lateral high-symmetry sites, shown in Figure 6.15, (a) the top site and (b) the fourfold hollow site. Further, there is an additional site, (c) the so-called threefold hollow site, which is not exactly of high symmetry but is treated on the same footing since it has been observed in a number of surface systems.

At the **top** site the adsorbate atom or molecule stabilizes directly above a substrate atom at the surface, see Figure 6.15, site (a). Examples for this type of site do not seem to exist in the literature.

At the **fourfold hollow** site the adsorbate atom or molecule stabilizes laterally between four substrate atoms at the surface, two adjacent and two at larger distance, see Figure 6.15, site (b). An example of periodically ordered **atomic** adsorbates is Fe(1 1 0) + (2 × 2) – S [142], SSD 26.16.4, where the sulfur atoms adsorb with $z_{ads} = 0.77$ Å.

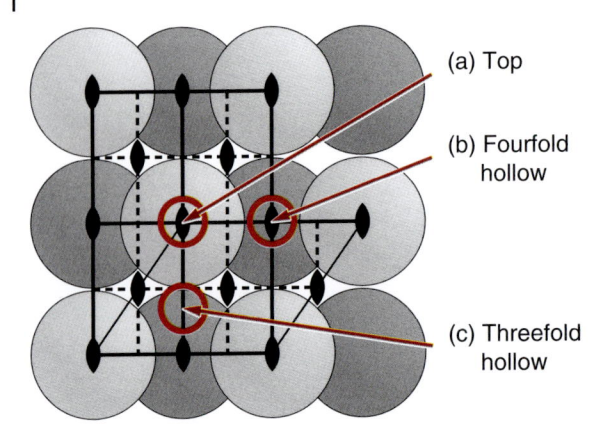

Figure 6.15 Primitive cell of a substrate with centered rectangular lattice based on cubic bulk structures. The lateral high-symmetry sites, (a) top, (b) fourfold hollow, and (c) threefold hollow, are shown by red circles. The gray balls indicate substrate atoms of different layers and the point symmetry elements of the cell are included by corresponding symbols, see Section 3.8.3.

At the **threefold hollow** site the adsorbate atom or molecule stabilizes laterally between three substrate atoms at the surface forming a triangle with two equal sides and one larger by only 15%, see Figure 6.15, site (c). An example of periodically ordered **atomic** adsorbates is W(1 1 0) + (2 × 1) − O [143], SSD 74.8.1, shown in Figure 6.16, where the oxygen atoms adsorb with $z_{ads} = 1.25$ Å.

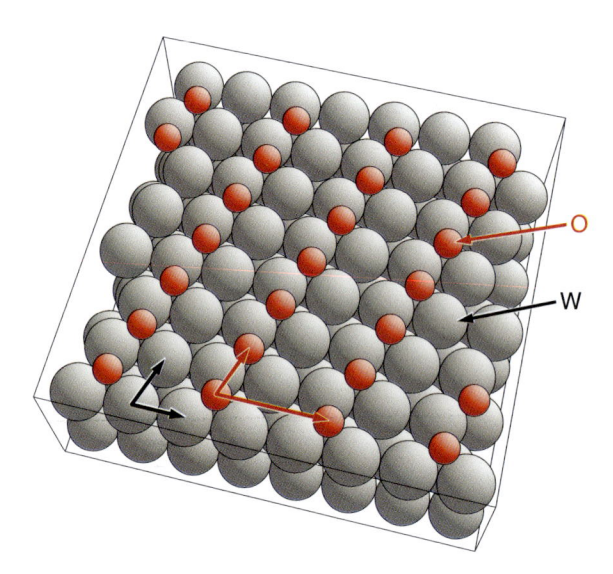

Figure 6.16 Structure of the W(1 1 0) + (2 × 1) − O adsorbate system. The lattice vectors of the O adsorbate layer and of the W substrate are shown in red and black, respectively.

6.2 Adsorbate Sites

In the example system Ni(1 1 1) + c(4×2) − 2NO [137] discussed above, see Figure 6.10, adsorption at both fcc and hcp hollow sites of the hexagonal substrate surface appears in the same structure. This illustrates that ordered adsorbate overlayers are not always connected with only one single adsorption site and that **mixed site adsorption** can occur. Another example is Pt(1 1 1) + c(4×2) − 2CO [144], SSD 78.6.8.4, shown in Figure 6.40, where CO molecules adsorb in both top and bridge sites of the hexagonal substrate surface with carbon pointing to the surface and $z_{ads}(C) = 1.85$ Å for the top site, 1.55 Å for the bridge site.

Apart from adsorption at sites near substrate atoms of the otherwise unreconstructed surface, adsorbates can also replace substrate atoms at the surface (**substitutional adsorption**). A simple example is Cu(1 0 0) + c(2×2) − Pd [145], SSD 29.46.2, where palladium and copper atoms of the topmost layer form a checkerboard structure resulting in surface alloying. Another example is Cu(1 1 0) + c(2×2) − Mn [146], SSD 29.25.8, shown in Figure 6.17, where manganese and copper atoms in alternating sequence form the topmost surface rows.

So far, adsorbate structure has been characterized by lateral surface sites and perpendicular distances z_{ads} of the adsorbate from the topmost substrate plane. For **molecular adsorbates** this needs to be supplemented by parameters that describe the **orientation** of the adsorbate relative to the surface and the **internal** molecular structure where the latter may be distorted if compared with the structure of the free molecule. For example, the information stating which part of an adsorbate molecule points to the substrate and the inclination of a molecular axis with respect to the surface normal of the substrate are essential for a complete structural characterization. Assuming no internal distortion of the adsorbed molecule, three angles are needed to specify the orientation of a nonsymmetrical molecule relative to a surface. As a simple example, the CO

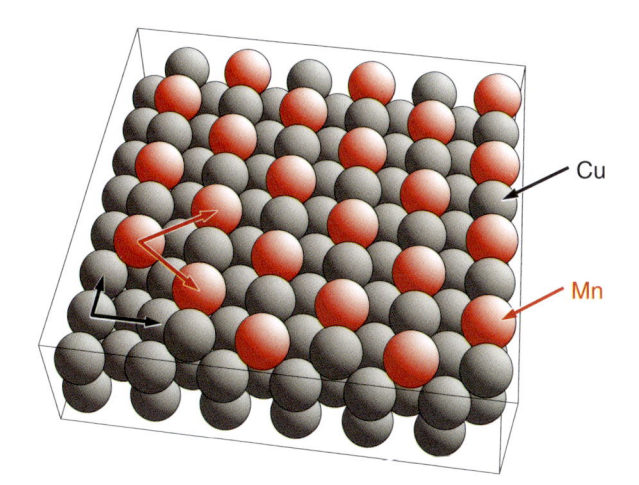

Figure 6.17 Structure of the Cu(1 1 0) + c(2×2) − Mn adsorbate system. The lattice vectors of the Mn/Cu atoms of the topmost layer and of the Cu substrate are shown in red and black, respectively.

molecule is found to adsorb in many cases with its molecular axis along the surface normal and with its carbon end pointing toward the substrate, see Ni(1 0 0) + c(2×2) − CO in Figure 6.8. However, the molecular axis of CO can also be tilted away from the surface normal as shown for Ni(1 1 0) + p2mg(2×1) − 2CO in Figure 6.2. The latter illustrates the more general result that adsorbates may adsorb differently depending on their coverage on the substrate surface due to adsorbate–adsorbate interaction. In the Ni(1 1 0) + p2mg(2×1) − 2CO system, adjacent CO adsorbate molecules, getting quite near to each other, try to minimize their mutual repulsion by tilting in different direction whereas in the Ni(1 0 0) + c(2×2) − CO system the adsorbates are further away and stabilize in equal orientation.

Further, adsorbate atoms or molecules are expected to influence the structure of the underlying **substrate** by inducing relaxation and reconstruction. This may create new adsorption sites that do not exist at the clean surface. As an example, the clean Cu(1 1 0) − (1×1) surface [147], SSD 29.65, exists as an unreconstructed bulk terminated structure with primitive rectangular lattice. In contrast, in the Cu(1 1 0) + (2×3) − 4N system [148], SSD 29.7.10, see Figure 6.18, the nitrogen adsorbate reconstructs the top copper layer to form a buckled nearly square lattice. Another example of large structural effects of the adsorbate on the substrate is the Si(1 1 1) surface. Without adsorbates this surface shows a complex (7×7) reconstruction [109, 110] described by the DAS (dimer-adatom-stacking-fault) model, see Figure 5.5. After hydrogen adsorption the reconstruction disappears completely, yielding a simple Si(1 1 1) + (1×1) − H adsorbate system [149], SSD 14.1.30, with an unreconstructed substrate and top-site-bonded hydrogen atoms.

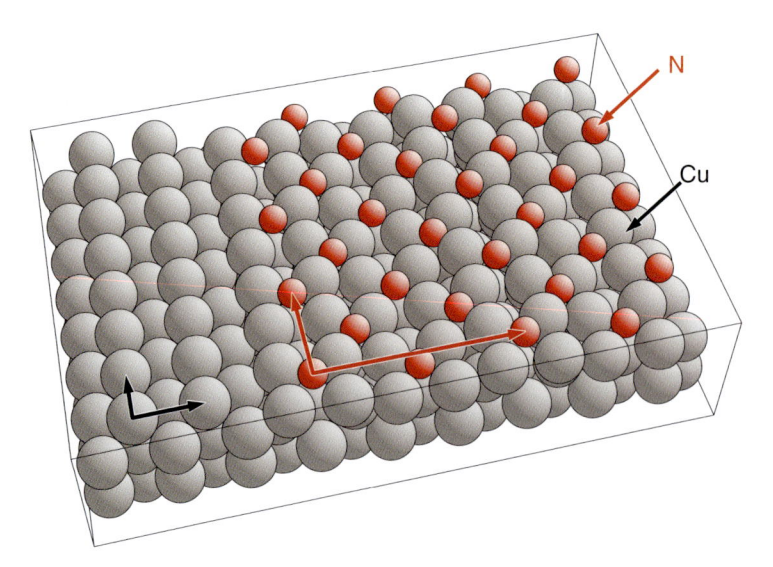

Figure 6.18 Structure of the Cu(1 1 0) + (2×3) − 4N adsorbate system. The lattice vectors of the unreconstructed substrate to the left are shown in black while those of the reconstructed surface to the right are shown in red.

6.3 Wood Notation of Surface Structure

As a simpler alternative to the matrix notation, the structure of ordered **reconstructed** single crystal surfaces as well as of ordered **adsorbate layers** is often characterized by the **Wood notation** [150]. Here, we adopt mainly the nomenclature used in [94]. It should be mentioned in passing that a number of attempts have been made to suggest alternative notation schemes that could give a unique description of surface structure of any complexity [44], analogous to notations used in bulk crystallography [33]. These include generalizations of the 2×2 matrix notation [100] or a scheme proposed by the authors of the NIST SSD [23], see Chapter 7. However, in contrast to the Wood notation, these schemes have never been widely accepted within the surface science community.

As a first example of the Wood notation the so-called (1×2)-missing-row reconstructed platinum surface was discussed in Section 5.2. This surface is denoted as Pt(1 1 0) − (1×2) where the periodicity of the topmost (1 1 0) surface layer is described as "(1×2)" and every second row of atoms of this layer is missing. The notation refers to a rectangular lattice of the topmost layer whose second lattice vector is enlarged by a factor 2 compared with that of the underlying substrate lattice while the first remains unchanged, see Figure 5.2. As a result, the corresponding reconstruction matrix $\underline{\underline{M}}$ is given by Eq. (5.5).

A slightly more complex example is the Cu(1 0 0) surface with a periodic chlorine overlayer of half the density of the topmost Cu layer [131] as shown in Figure 6.19. The example, taken from the Surface Structure Database, SSD 29.17.7, is commonly denoted as Cu(1 0 0) + c(2×2) − Cl following the Wood notation, where the periodicity of the Cl adsorbate layer is characterized by "c(2×2)". This corresponds to a centered rectangular overlayer with rectangular lattice vectors twice those of the underlying substrate layer. The overlayer periodicity can also be described by a primitive square lattice with lattice vectors increased by a factor $\sqrt{2}$ and rotated by 45° with respect to those of the substrate. Thus, the corresponding reconstruction matrix $\underline{\underline{M}}$ is given by

$$\underline{\underline{M}} = \begin{pmatrix} 1 & -1 \\ 1 & 1 \end{pmatrix}$$

and, within the Wood notation scheme, the overlayer structure can be written as Cu(1 0 0) + ($\sqrt{2} \times \sqrt{2}$)R45° − Cl. The appearance of irrational numbers, $\sqrt{2}$, in the notation still leads to an integer-valued reconstruction matrix characterizing the commensurate overlayer. This applies also to the Pd(1 1 1) + ($\sqrt{3} \times \sqrt{3}$)R30° − CO adsorbate structure [136], SSD 46.6.8.13, shown in Figure 6.20. Here, the lattices of both the substrate and the CO overlayer are hexagonal with the lattice vectors of the overlayer increased by a factor $\sqrt{3}$ and rotated by 30° with respect to those of the substrate. For this structure, using obtuse hexagonal lattice vectors,

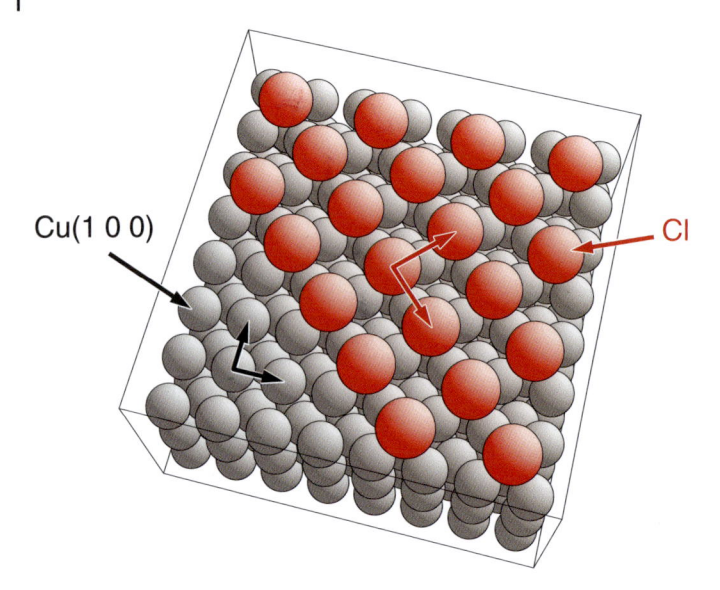

Figure 6.19 Structure of the Cu(1 0 0) + c(2 × 2) – Cl adsorbate system. The periodicity vectors of the Cl adsorbate layer and of the Cu substrate are shown in red and black, respectively.

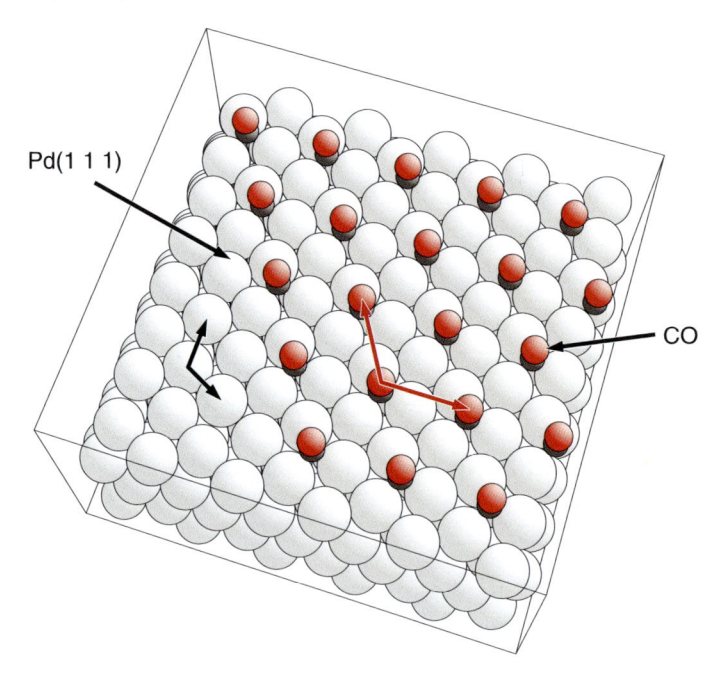

Figure 6.20 Structure of the Pd(1 1 1) + ($\sqrt{3} \times \sqrt{3}$)R30° – CO adsorbate system. The periodicity vectors of the CO adsorbate layer and of the Pd substrate are shown in red and black, respectively.

the transformation matrix $\underline{M}$ is given by

$$\underline{M} = \begin{pmatrix} 2 & 1 \\ -1 & 1 \end{pmatrix}$$

The **general** case of the **Wood notation** of a **reconstructed** surface is formally given by

$$\text{Sub}(h\ k\ l) - \kappa(\gamma_1 \times \gamma_2)R\alpha - \eta\text{Sub} \tag{6.2a}$$

and a general surface with an **adsorbate overlayer** is written as

$$\text{Sub}(h\ k\ l) + \kappa(\gamma_1 \times \gamma_2)R\alpha - \eta\text{Ovl} \tag{6.2b}$$

where it is assumed that the substrate "**Sub**" is described by stacking two-dimensionally periodic layers with periodicity vectors $\underline{R}_1$ and $\underline{R}_2$ representing $(h\ k\ l)$ Miller index planes. In addition, either the topmost substrate layer at the surface is reconstructed yielding a periodic **reconstruction layer** (which may actually include more than one substrate layer), formula (6.2a), or the surface is covered by a **periodic overlayer** "**Ovl**", formula (6.2b). In both cases, the periodicity vectors $\underline{R}'_1$ and $\underline{R}'_2$ of the topmost (over)layer are given by **linear combinations** of the substrate surface vectors $\underline{R}_1$ and $\underline{R}_2$ where for

a) $\kappa =$ "p" (**primitive**) vector $\underline{R}'_1$ equals the substrate surface vector $\underline{R}_1$ rotated anti-clockwise by an angle α along the surface plane and scaled by factor γ_1 to yield $|\underline{R}'_1| = \gamma_1 |\underline{R}_1|$. The same procedure is applied to $\underline{R}_2$ using angle α and scaling factor γ_2 to yield $\underline{R}'_2$.

b) $\kappa =$ "c" (**centered**) vectors $\underline{R}'_1, \underline{R}'_2$ describe a centered two-dimensional lattice starting from a primitive set $\underline{R}'_{1p}, \underline{R}'_{2p}$ constructed according to (a) followed by a linear transformation

$$\underline{R}'_1 = 1/2\,(\underline{R}'_{1p} + \underline{R}'_{2p}), \quad \underline{R}'_2 = 1/2\,(-\underline{R}'_{1p} + \underline{R}'_{2p}). \tag{6.3}$$

Further, the unit cell of the reconstructed substrate surface is assumed to contain $\eta \geq 1$ non-equivalent species "Sub" and in the adsorbate system the unit cell of the overlayer may contain $\eta \geq 1$ non-equivalent species "Ovl" of the same type. The general Wood notation (6.2) is often **simplified** by omitting κ if $\kappa =$ "p", omitting η if $\eta = 1$, and omitting "Rα" if $\alpha = 0°$. The qualifier κ has been used in a few cases to also give **additional information** about the overlayer lattice and its symmetry. An example is given by the Ni(1 1 0) + p2mg(2 × 1) − 2CO adsorbate system [125], see Figure 6.2, where the unit cell of the CO overlayer is described in its two-dimensional symmetry by symmetry group p2mg.

The most complex case, adsorption of a reconstructed overlayer at a reconstructed substrate surface, would be denoted as

$$\text{Sub}(h\ k\ l) - \kappa_o(\gamma_{o1} \times \gamma_{o2})R\alpha_o - \eta_o\text{Sub} + \kappa(\gamma_1 \times \gamma_2)R\alpha - \eta\text{Ovl} \tag{6.4}$$

where in addition to Eq. (6.2b) the substrate reconstruction would be described in analogy with that of the overlayer as "$\kappa_o(\gamma_{o1} \times \gamma_{o2})R\alpha_o - \eta_o\text{Sub}$". This added complication that has been observed only rarely will be ignored in the following.

The Wood notation can describe both **commensurate** and **incommensurate** overlayers. However, it is **not general** due to its restrictions in the overlayer periodicity which equates the angle between the lattice vectors $\underline{R}'_1$ and $\underline{R}'_2$ with that between $\underline{R}_1$ and $\underline{R}_2$ as discussed in Appendix C. The periodicity information of the Wood notation can be expressed alternatively by a **2 × 2 matrix transformation** according to Eq. (5.3), that is, given by

$$\begin{pmatrix} \underline{R}'_1 \\ \underline{R}'_2 \end{pmatrix} = \begin{pmatrix} m_{11} & m_{12} \\ m_{21} & m_{22} \end{pmatrix} \cdot \begin{pmatrix} \underline{R}_1 \\ \underline{R}_2 \end{pmatrix} = \underline{\underline{M}}_{p,c} \cdot \begin{pmatrix} \underline{R}_1 \\ \underline{R}_2 \end{pmatrix} \tag{6.5}$$

referring to the reconstruction matrix $\underline{\underline{M}}_{p,c}$, which may also be written as $(m_{11}\, m_{12}\,|\, m_{21}\, m_{22})$.

If the periodicity vectors $\underline{R}_1$ and $\underline{R}_2$ of the substrate form an angle ω, that is, $(\underline{R}_1\, \underline{R}_2) = R_1 R_2 \cos(\omega)$, then simple algebra, see Appendix C, yields

- for "**primitive**" overlayers denoted by "$\cdots - p(\gamma_1 \times \gamma_2)R\alpha - \cdots$"

$$\underline{\underline{M}}_p = \frac{1}{\sin(\omega)} \begin{pmatrix} \gamma_1 \sin(\omega - \alpha) & \gamma_1 q^{-1} \sin(\alpha) \\ -\gamma_2 q \sin(\alpha) & \gamma_2 \sin(\omega + \alpha) \end{pmatrix}, \quad q = \frac{R_2}{R_1} \tag{6.6a}$$

- for "**centered**" overlayers denoted by "$\cdots - c(\gamma_1 \times \gamma_2)R\alpha - \cdots$"

$$\underline{\underline{M}}_c = \frac{1}{2\sin(\omega)} \begin{pmatrix} \gamma_1 \sin(\omega - \alpha) - \gamma_2 q \sin(\alpha) & \gamma_1 q^{-1} \sin(\alpha) + \gamma_2 \sin(\omega + \alpha) \\ -\gamma_1 \sin(\omega - \alpha) - \gamma_2 q \sin(\alpha) & -\gamma_1 q^{-1} \sin(\alpha) + \gamma_2 \sin(\omega + \alpha) \end{pmatrix} \tag{6.6b}$$

In the following, **examples** of Wood notations are listed together with corresponding transformation matrices describing the overlayer periodicity for different types of Bravais lattices characterized by the parameters q and ω.

- Overlayers on a substrate with **square** netplane (e.g., fcc(1 0 0), bcc(1 0 0), diamond(0 0 1), zincblende(0 0 1)), where $R_1 = R_2$, $\omega = 90°$, see Figure 6.21. Example notations are

(a) $\quad p(a \times b) = (a \times b):\quad \underline{\underline{M}}_p = \begin{pmatrix} a & 0 \\ 0 & b \end{pmatrix}\quad a,\, b \text{ integer}$ (6.7a)

(b) $\quad c(4 \times 2): \quad \underline{\underline{M}}_c = \begin{pmatrix} 2 & 1 \\ -2 & 1 \end{pmatrix}$ (6.7b)

(c) $\quad c(2 \times 2) = (\sqrt{2} \times \sqrt{2})R45°: \quad \underline{\underline{M}}_c = \underline{\underline{M}}_p = \begin{pmatrix} 1 & 1 \\ -1 & 1 \end{pmatrix}$ (6.7c)

(d) $\quad p(a\sqrt{2} \times b\sqrt{2})R45°: \quad \underline{\underline{M}}_p = \begin{pmatrix} a & a \\ -b & b \end{pmatrix}$ (6.7d)

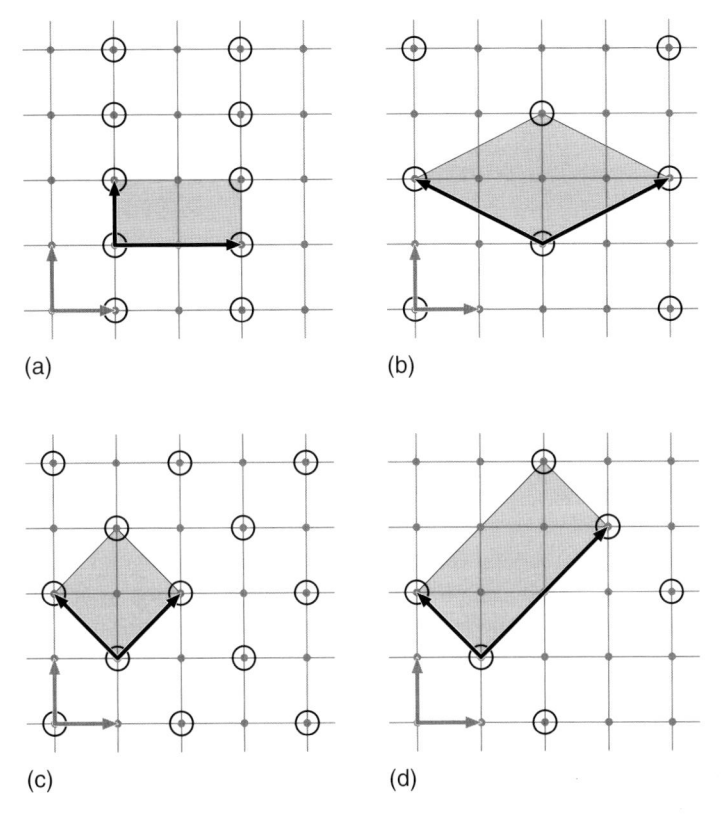

Figure 6.21 Overlayer unit cells on a substrate with square lattice, (a) (2×1), (b) c(4×2), (c) c(2×2), and (d) p(2√2× √2)R45°, see text. Overlayer and substrate lattice vectors are shown in black and gray, respectively. The overlayer unit cell is emphasized in gray.

- Overlayers on a substrate with **rectangular** netplane (e.g., fcc(1 1 0), diamond(1 1 0), zincblende(1 1 0)), where $R_1 \neq R_2$, $\omega = 90°$, see Figure 6.22. Example notations are

(a) $\quad p(a \times b) = (a \times b) : \quad \underline{\underline{M}}_p = \begin{pmatrix} a & 0 \\ 0 & b \end{pmatrix}, \quad$ a, b integer $\quad$ (6.8a)

(b) $\quad p(2 \times 2) = (2 \times 2) : \quad \underline{\underline{M}}_p = \begin{pmatrix} 2 & 0 \\ 0 & 2 \end{pmatrix} \quad$ (6.8b)

(c) $\quad c(2 \times 2) : \quad \underline{\underline{M}}_c = \begin{pmatrix} 1 & 1 \\ -1 & 1 \end{pmatrix} \quad$ (6.8c)

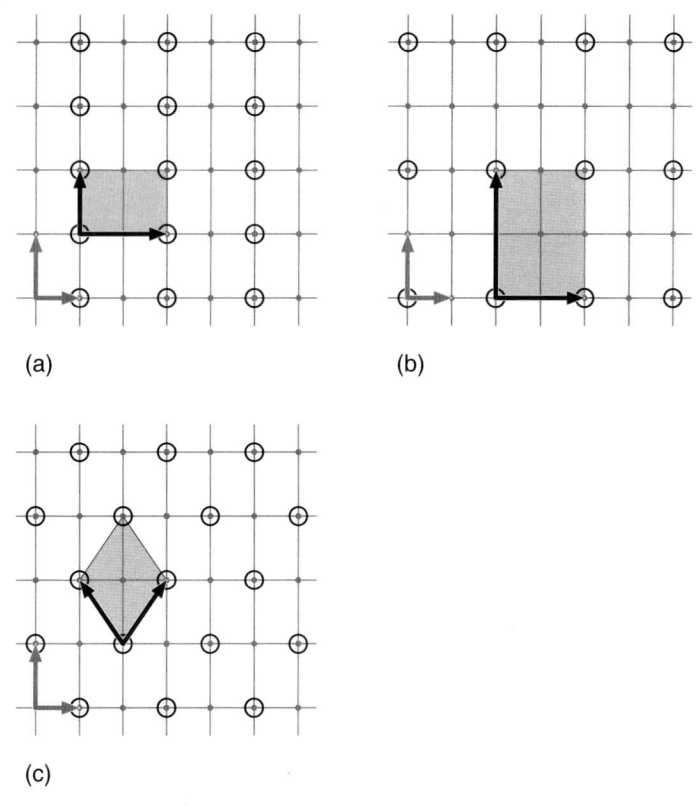

Figure 6.22 Overlayer unit cells on a substrate with rectangular lattice, (a) p(2×1), (b) p(2×2), and (c) c(2×2), see text. Overlayer and substrate lattice vectors are shown in black and gray, respectively. The overlayer unit cell is emphasized in gray.

- Overlayers on a substrate with **hexagonal** netplane (e.g., fcc(1 1 1), diamond(1 1 1), zincblende(1 1 1), graphite(0 0 0 1)), where $R_1 = R_2$, $\omega = 120°$ (obtuse representation), see Figure 6.23. Example notations are

(a) $\quad p(a \times b) = (a \times b):\quad \underline{\underline{M}}_p = \begin{pmatrix} a & 0 \\ 0 & b \end{pmatrix},\quad$ a, b integer $\quad$ (6.9a)

(b) $\quad c(2 \times 2)$ (also called (2×1)): $\quad \underline{\underline{M}}_p = \begin{pmatrix} 1 & -1 \\ 1 & 1 \end{pmatrix}\quad$ (6.9b)

(c) $\quad c(4 \times 2):\quad \underline{\underline{M}}_p = \underline{\underline{M}}_c = \begin{pmatrix} 0 & -2 \\ 2 & 1 \end{pmatrix}\quad$ (6.9c)

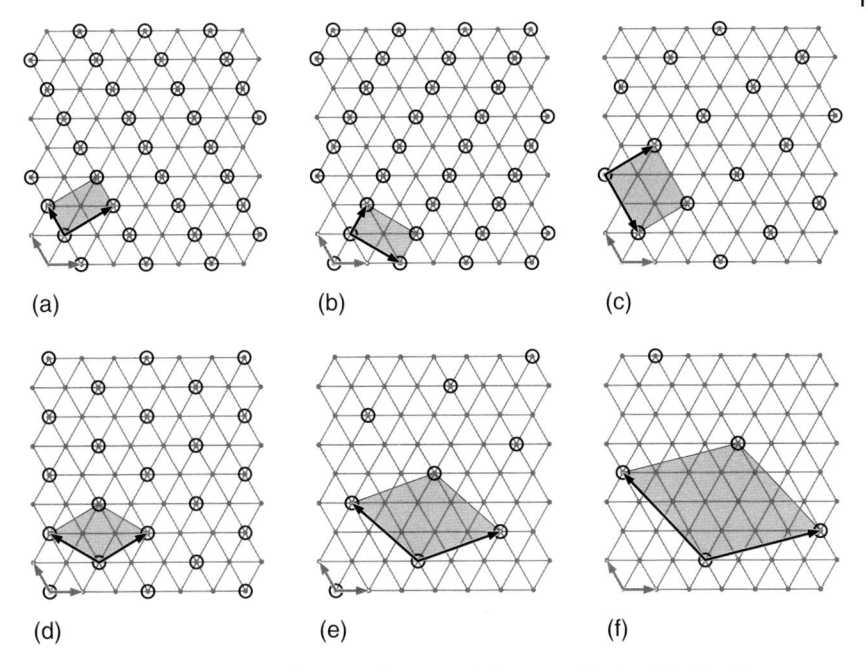

Figure 6.23 Overlayer unit cells on a substrate with hexagonal lattice, (a) p(2 × 1), (b) c(2 × 2), (c) c(4 × 2), (d) ($\sqrt{3} \times \sqrt{3}$)R30°, (e) ($\sqrt{7} \times \sqrt{7}$)R19.1°, and (f) ($\sqrt{13} \times \sqrt{13}$)R13.9°, see text. Overlayer and substrate lattice vectors are shown in black and gray, respectively. The overlayer unit cell is emphasized in gray.

(d) $\quad p(\sqrt{(a^2 - ab + b^2)} \times \sqrt{(a^2 - ab + b^2)})R\alpha \quad$ with

$$\cos\alpha = (a - b/2) / \sqrt{(a^2 - ab + b^2)}: \quad \underline{\underline{M}}_p = \begin{pmatrix} a & b \\ -b & a-b \end{pmatrix} \quad (6.9d)$$

Specific cases are

(d1) $\quad (a, b) = (2, 1)$

$$(\sqrt{3} \times \sqrt{3})R30°: \quad \underline{\underline{M}}_p = \begin{pmatrix} 2 & 1 \\ -1 & 1 \end{pmatrix} \quad (6.9e)$$

(d2) $\quad (a, b) = (3, 1)$

$$(\sqrt{7} \times \sqrt{7})R19.1°: \quad \underline{\underline{M}}_p = \begin{pmatrix} 3 & 1 \\ -1 & 2 \end{pmatrix} \quad (6.9f)$$

(d3) $\quad (a, b) = (4, 1)$

$$(\sqrt{13} \times \sqrt{13})R13.9°: \quad \underline{\underline{M}}_p = \begin{pmatrix} 4 & 1 \\ -1 & 3 \end{pmatrix} \quad (6.9g)$$

6.4
High-Order Commensurate (HOC) Overlayers

As mentioned in Section 6.1 adsorbate overlayers can form a high-order commensurate (**HOC**) lattice structure, which is formally defined by a transformation matrix $\underline{M}$, connecting lattice vectors of the overlayer $\underline{R}'_{o1}$, $\underline{R}'_{o2}$ with those of the substrate surface, $\underline{R}_{o1}$ and $\underline{R}_{o2}$, according to

$$\begin{pmatrix} \underline{R}'_{o1} \\ \underline{R}'_{o2} \end{pmatrix} = \begin{pmatrix} r_{11} & r_{12} \\ r_{21} & r_{22} \end{pmatrix} \cdot \begin{pmatrix} \underline{R}_{o1} \\ \underline{R}_{o2} \end{pmatrix} = \underline{M} \cdot \begin{pmatrix} \underline{R}_{o1} \\ \underline{R}_{o2} \end{pmatrix}, \quad r_{ij} = p_{ij}/q_{ij}, \ p_{ij}, \ q_{ij} \text{ integer}$$
(6.10)

where all elements r_{ij} are **rational** or integer-valued with at least one being rational. Thus, the transformation matrix can be written as

$$\underline{M} = \begin{pmatrix} n_1 & 0 \\ 0 & n_2 \end{pmatrix}^{-1} \cdot \begin{pmatrix} a_{11} & a_{12} \\ a_{21} & a_{22} \end{pmatrix} \stackrel{\text{def.}}{=} \underline{B}^{-1} \cdot \underline{A}, \quad n_i, a_{ij} \text{ integer}$$
(6.11)

with

$$n_i = \frac{q_{i1} \, q_{i2}}{\gcd(q_{i1}, q_{i2})}, \quad a_{ij} = n_i \frac{p_{ij}}{q_{ij}}, \quad b_{ij} = n_i \delta_{ij}, \quad i, j = 1, 2$$
(6.12)

(Function $\gcd(x, y)$ denotes the greatest common divisor, see Appendix E.1.) This means that matrix $\underline{M}$ can always be represented by a product of an inverted matrix $\underline{B}$ and a matrix $\underline{A}$ where both matrices are integer-valued. As a consequence, relations (6.10) with (6.11) can be rewritten as

$$\begin{pmatrix} \underline{R}'_1 \\ \underline{R}'_2 \end{pmatrix} \stackrel{\text{def.}}{=} \begin{pmatrix} n_1 \underline{R}'_{o1} \\ n_2 \underline{R}'_{o2} \end{pmatrix} = \underline{B} \cdot \begin{pmatrix} \underline{R}'_{o1} \\ \underline{R}'_{o2} \end{pmatrix} = \underline{A} \cdot \begin{pmatrix} \underline{R}_{o1} \\ \underline{R}_{o2} \end{pmatrix}$$
(6.13)

Therefore, in an HOC lattice structure one can always find a superlattice with superlattice vectors $\underline{R}'_1$ and $\underline{R}'_2$ of the overlayer, which are commensurate with those of the substrate $\underline{R}_{o1}$ and $\underline{R}_{o2}$. The corresponding supercell describing the periodicity shared by the two vector sets, $\underline{R}'_1, \underline{R}'_2$ and $\underline{R}_{o1}, \underline{R}_{o2}$, has an area

$$F_s = |\underline{R}'_1 \times \underline{R}'_2| = |\det(\underline{B})| \, |\underline{R}'_{o1} \times \underline{R}'_{o2}| = |\det(\underline{A})| \, |\underline{R}_{o1} \times \underline{R}_{o2}|$$
(6.14)

with

$$|\det(\underline{B})| = |n_1 n_2|$$
(6.15)

The supercell spanned by $\underline{R}'_1, \underline{R}'_2$ may not be the smallest possible cell, the **primitive** supercell of the HOC lattice structure. This is clear by a simple numerical example. Let us assume a transformation matrix $\underline{M}$ and its representation by Eqs. (6.11) and (6.12) where

$$\underline{M} = \frac{1}{13} \begin{pmatrix} 14 & -5 \\ 5 & 14 \end{pmatrix} = \underline{B}^{-1} \cdot \underline{A} = \begin{pmatrix} 13 & 0 \\ 0 & 13 \end{pmatrix}^{-1} \cdot \begin{pmatrix} 14 & -5 \\ 5 & 14 \end{pmatrix}$$
(6.16)

6.4 High-Order Commensurate (HOC) Overlayers

which according to Eq. (6.14) leads to a supercell size $F_s = 169 \, |\underline{R}'_{o1} \times \underline{R}'_{o2}|$. However, an alternative representation of matrix $\underline{M}$ given by

$$\underline{M} = \frac{1}{13} \begin{pmatrix} 14 & -5 \\ 5 & 14 \end{pmatrix} = \underline{B}'^{-1} \cdot \underline{A}' = \begin{pmatrix} 3 & 2 \\ -2 & 3 \end{pmatrix}^{-1} \cdot \begin{pmatrix} 4 & 1 \\ -1 & 4 \end{pmatrix} \quad (6.17)$$

leads to a supercell size $F_s = 13 \, |\underline{R}'_{o1} \times \underline{R}'_{o2}|$, which is smaller by a factor 13 and represents the primitive supercell. This is illustrated for a rectangular substrate lattice $\underline{M}$ according to Eq. (6.16) in Figure 6.24 where the supercell corresponding to the initial matrix $\underline{A}$ in Eq. (6.16) and the primitive supercell corresponding to matrix $\underline{A}'$ in Eq. (6.17) are sketched.

The reduction of superlattice vector sets R'_1, R'_2, determined initially by Eqs. (6.11)–(6.13), to yield a primitive set R''_1, R''_2 (providing the smallest supercell) can be achieved by number theoretical methods as described in detail in Appendix E.5. The basic idea is that if R'_1, R'_2 can be reduced then there must be an alternative set of reduced (primitive) superlattice vectors R''_1, R''_2, which represent the initial superlattice vectors with

$$\begin{pmatrix} \underline{R}'_1 \\ \underline{R}'_2 \end{pmatrix} = \underline{T} \cdot \begin{pmatrix} \underline{R}''_1 \\ \underline{R}''_2 \end{pmatrix}, \quad \underline{T} \text{ integer} \quad (6.18)$$

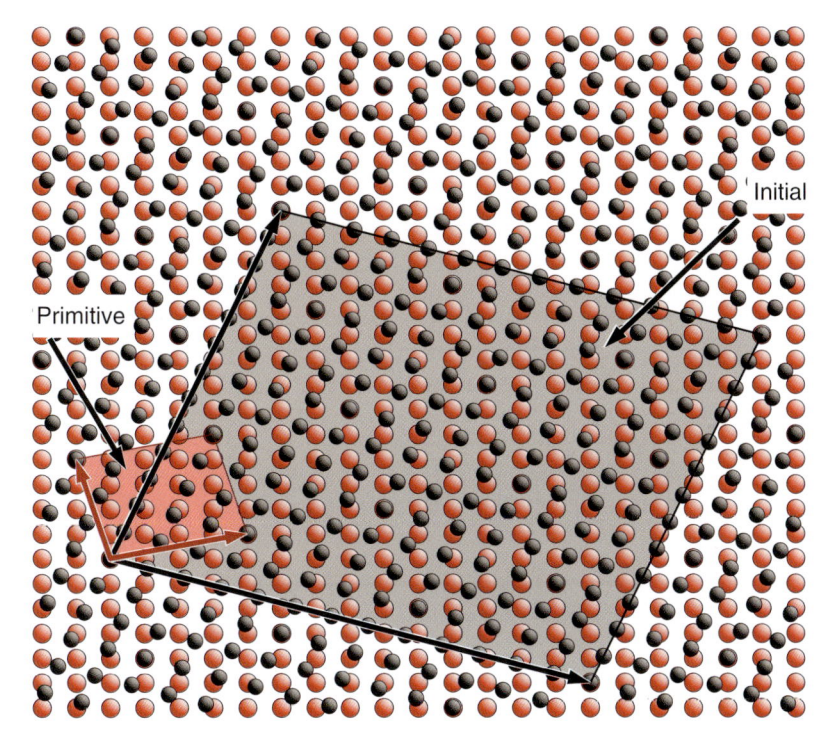

Figure 6.24 Fictitious rectangular substrate with HOC overlayer structure according to Eq. (6.16). The initial and primitive supercells with superlattice vectors are outlined in gray and red, respectively.

As a result, the reduced vectors $\underline{R}_1''$, $\underline{R}_2''$ yield according to Eq. (6.13)

$$\begin{pmatrix} \underline{R}_1'' \\ \underline{R}_2'' \end{pmatrix} = \underline{\underline{T}}^{-1} \cdot \begin{pmatrix} \underline{R}_1' \\ \underline{R}_2' \end{pmatrix} = \underline{\underline{T}}^{-1} \cdot \underline{\underline{B}} \cdot \begin{pmatrix} \underline{R}_{o1}' \\ \underline{R}_{o2}' \end{pmatrix} = \underline{\underline{T}}^{-1} \cdot \underline{\underline{A}} \cdot \begin{pmatrix} \underline{R}_{o1} \\ \underline{R}_{o2} \end{pmatrix} \qquad (6.19)$$

In addition, the vectors $\underline{R}_1''$, $\underline{R}_2''$ also represent general lattice vectors of both the overlayer and the substrate surface. This means that the transformation matrices $\underline{\underline{A}}'$, $\underline{\underline{B}}'$ with

$$\underline{\underline{A}}' = \underline{\underline{T}}^{-1} \cdot \underline{\underline{A}}, \quad \underline{\underline{B}}' = \underline{\underline{T}}^{-1} \cdot \underline{\underline{B}} \qquad (6.20)$$

must be integer-valued and leave the transformation matrix $\underline{\underline{M}}$ unchanged, that is,

$$\underline{\underline{M}} = \underline{\underline{B}}^{-1} \cdot \underline{\underline{A}} = \underline{\underline{B}}'^{-1} \cdot \underline{\underline{A}}' \qquad (6.21)$$

As a result, the size of the reduced supercell of the overlayer is determined by

$$F_s^{\text{red}} = |\underline{R}_1'' \times \underline{R}_2''| = |\det(\underline{\underline{A}}')| |\underline{R}_{o1} \times \underline{R}_{o2}| = \frac{|\det(\underline{\underline{A}})|}{|\det(\underline{\underline{T}})|} |\underline{R}_{o1} \times \underline{R}_{o2}|$$

$$= |\det(\underline{\underline{B}}')| |\underline{R}_{o1}' \times \underline{R}_{o2}'| = \frac{|\det(\underline{\underline{B}})|}{|\det(\underline{\underline{T}})|} |\underline{R}_{o1}' \times \underline{R}_{o2}'| = \frac{F_s}{|\det(\underline{\underline{T}})|} \qquad (6.22)$$

such that $|\det(\underline{\underline{T}})|$ can be understood as a reduction factor of the cell sizes corresponding to the two representations. Since the determinants of $\underline{\underline{A}}$, $\underline{\underline{A}}'$, $\underline{\underline{B}}$, $\underline{\underline{B}}'$, and $\underline{\underline{T}}$ are all integer-valued, relation (6.22) shows further that the integer $\det(\underline{\underline{T}})$ must be a common divisor of both integers $\det(\underline{\underline{A}})$ and $\det(\underline{\underline{B}})$ with an upper limit given by

$$1 \leq |\det(\underline{\underline{T}})| \leq g \stackrel{\text{def.}}{=} \gcd(|\det(\underline{\underline{A}})|, |\det(\underline{\underline{B}})|) \qquad (6.23)$$

Thus, relations (6.22) and (6.23) show that for $g = 1$, and hence $|\det(\underline{\underline{T}})| = 1$, the initial supercell vectors $\underline{R}_1'$, $\underline{R}_2'$ cannot be reduced further and $\underline{R}_1'$, $\underline{R}_2'$ yields already the primitive cell size. On the other hand, a value of $g > 1$ indicates that reduction is possible with a largest reduction factor given by $|\det(\underline{\underline{T}})| = g$. This can be used to quantify the cell size reduction in going from the initial HOC lattice representation (6.13) to the primitive representation (6.19). Further, the reduced transformation matrices $\underline{\underline{A}}'$, $\underline{\underline{B}}'$ for $|\det(\underline{\underline{T}})| = g$ can be evaluated by a number theoretical algorithm as described in Appendix E.5. In the numerical example (6.16), the corresponding values are $g = \gcd(221, 169) = 13$, yielding a supercell size that is 13 times larger than that given by the matrices of Eq. (6.17), which confirms the above result.

Recent experimental studies on the adsorption of C_{60} fullerene molecules ("buckyballs") at the hexagonal Pb(1 1 1) surface have shown an ordered C_{60} overlayer with HOC lattice structure shown in Figure 6.25 and described as Pb(1 1 1) + ($\sqrt{403/7} \times \sqrt{403/7}$)R22.85° – C_{60} [151]. This corresponds to a transformation matrix $\underline{\underline{M}}$ with

$$\underline{\underline{M}} = \frac{1}{7} \begin{pmatrix} 14 & 9 \\ -9 & 23 \end{pmatrix} = \underline{\underline{B}}^{-1} \cdot \underline{\underline{A}} = \begin{pmatrix} 7 & 0 \\ 0 & 7 \end{pmatrix}^{-1} \cdot \begin{pmatrix} 14 & 9 \\ -9 & 23 \end{pmatrix} \qquad (6.24)$$

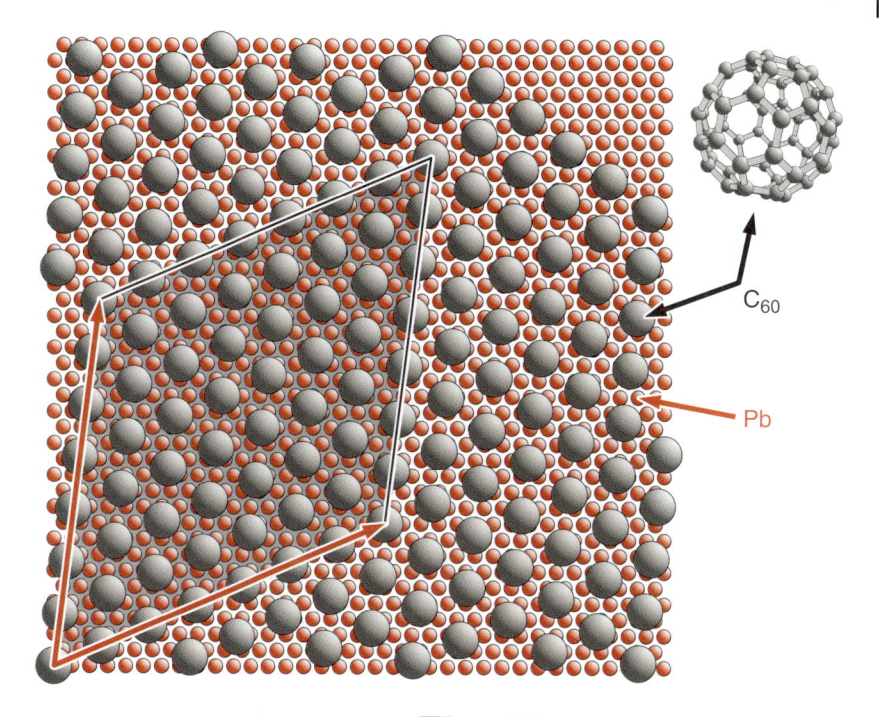

Figure 6.25 Structure of the Pb(1 1 1) + ($\sqrt{403}/7 \times \sqrt{403}/7$)R22.85° – C_{60} adsorbate system. The lattice vectors of the primitive superlattice are sketched in red and the supercell is outlined in black.

Here, $\gcd(|\det(\underline{A})|, |\det(\underline{B})|) = \gcd(403, 49) = 1$. Thus, representation (6.24) cannot be reduced further and the superlattice vectors sketched in Figure 6.25 reflect the primitive cell of the HOC lattice structure.

The Pb(1 1 1) + C_{60} adsorbate system is an example of a more general group of HOC lattice structures, **hexagonal** overlayers on hexagonal substrate, which has also been discussed in the literature [152]. Using an acute representation for the substrate lattice vectors $\underline{R}_{o1}$ and $\underline{R}_{o2}$ with $R_{o1} = R_{o2} = R$, any commensurate hexagonal overlayer with lattice vectors $\underline{R}'_{o1}, \underline{R}'_{o2}$ can be described by a transformation

$$\begin{pmatrix} \underline{R}'_{o1} \\ \underline{R}'_{o2} \end{pmatrix} = \begin{pmatrix} m & n \\ -n & m+n \end{pmatrix} \cdot \begin{pmatrix} \underline{R}_{o1} \\ \underline{R}_{o2} \end{pmatrix}, \quad m, n \text{ integer} \qquad (6.25)$$

since the resulting vectors $\underline{R}'_{o1}, \underline{R}'_{o2}$ are of equal length

$$R'_{o1} = R'_{o2} = (\sqrt{m^2 + n^2 + mn})\, R \qquad (6.26)$$

and form an angle of 60° as can be shown by simple vector calculus. As a consequence, HOC lattice structures involving hexagonal overlayer and substrate lattices can be defined by transformations

$$\begin{pmatrix} m' & n' \\ -n' & m'+n' \end{pmatrix} \cdot \begin{pmatrix} \underline{R}'_{o1} \\ \underline{R}'_{o2} \end{pmatrix} = \begin{pmatrix} m & n \\ -n & m+n \end{pmatrix} \cdot \begin{pmatrix} \underline{R}_{o1} \\ \underline{R}_{o2} \end{pmatrix}, \quad m, n, m', n' \text{ integer}$$

(6.27)

or

$$\begin{pmatrix} \underline{R}'_{o1} \\ \underline{R}'_{o2} \end{pmatrix} = \begin{pmatrix} m' & n' \\ -n' & m'+n' \end{pmatrix}^{-1} \cdot \begin{pmatrix} m & n \\ -n & m+n \end{pmatrix} \cdot \begin{pmatrix} \underline{R}_{o1} \\ \underline{R}_{o2} \end{pmatrix} = \begin{pmatrix} a & b \\ -b & a+b \end{pmatrix} \cdot \begin{pmatrix} \underline{R}_{o1} \\ \underline{R}_{o2} \end{pmatrix}$$

(6.28)

with

$$a = [m(m'+n') + nn']/\Delta, \quad b = (nm' - mn')/\Delta, \quad \Delta = m'^2 + n'^2 + m'n'$$

(6.29)

Here, the corresponding vector lengths are

$$R'_{o1} = R'_{o2} = \left(\sqrt{a^2 + b^2 + ab}\right) R = \sqrt{\frac{m^2 + n^2 + mn}{m'^2 + n'^2 + m'n'}} R$$

(6.30)

Relations (6.28) and (6.29) allow determining a joint superlattice with vectors $\underline{R}'_1, \underline{R}'_2$ where

$$\begin{pmatrix} \underline{R}'_1 \\ \underline{R}'_2 \end{pmatrix} = \begin{pmatrix} m' & n' \\ -n' & m'+n' \end{pmatrix} \cdot \begin{pmatrix} \underline{R}'_{o1} \\ \underline{R}'_{o2} \end{pmatrix} = \begin{pmatrix} m & n \\ -n & m+n \end{pmatrix} \cdot \begin{pmatrix} \underline{R}_{o1} \\ \underline{R}_{o2} \end{pmatrix}$$

(6.31)

and

$$R'_1 = R'_2 = \left(\sqrt{m^2 + n^2 + mn}\right) R$$

(6.32)

According to the preceding discussion, the superlattice cell spanned by vectors $\underline{R}'_1, \underline{R}'_2$ cannot be reduced further in size if the integer-valued determinants in Eq. (6.31) are coprime, that is, if

$$\gcd(m^2 + n^2 + mn, m'^2 + n'^2 + m'n') = 1$$

(6.33)

For the Pb(1 1 1) + C_{60} adsorbate system the present general formalism yields

$$m = 14, \ n = 9, \ m' = 7, \ n' = 0 \quad \text{and hence } \gcd(403, 49) = 1$$

(6.34)

which verifies the result obtained above.

HOC lattice structures formed by overlayers on substrate where both lattices are **square** can be treated analogous to the hexagonal case. Here, commensurate overlayers with square lattice vectors $\underline{R}'_{o1}, \underline{R}'_{o2}$ of the overlayer and $\underline{R}_{o1}, \underline{R}_{o2}$ of the substrate ($R_{o1} = R_{o2} = R$) can be described by a transformation

$$\begin{pmatrix} \underline{R}'_{o1} \\ \underline{R}'_{o2} \end{pmatrix} = \begin{pmatrix} m & n \\ -n & m \end{pmatrix} \cdot \begin{pmatrix} \underline{R}_{o1} \\ \underline{R}_{o2} \end{pmatrix}, \quad m, n \text{ integer}$$

(6.35)

since the resulting vectors $\underline{R}'_{o1}, \underline{R}'_{o2}$ are of equal length

$$R'_{o1} = R'_{o2} = (\sqrt{m^2 + n^2}) R$$

(6.36)

and form an angle of 90° as can be shown by simple vector calculus. Thus, HOC lattice structures involving square overlayer and substrate lattices can be defined by transformations

$$\begin{pmatrix} m' & n' \\ -n' & m' \end{pmatrix} \cdot \begin{pmatrix} \underline{R}'_{o1} \\ \underline{R}'_{o2} \end{pmatrix} = \begin{pmatrix} m & n \\ -n & m \end{pmatrix} \cdot \begin{pmatrix} \underline{R}_{o1} \\ \underline{R}_{o2} \end{pmatrix}, \quad m, n, m', n' \text{ integer} \quad (6.37)$$

or

$$\begin{pmatrix} \underline{R}'_{o1} \\ \underline{R}'_{o2} \end{pmatrix} = \begin{pmatrix} m' & n' \\ -n' & m' \end{pmatrix}^{-1} \cdot \begin{pmatrix} m & n \\ -n & m \end{pmatrix} \cdot \begin{pmatrix} \underline{R}_{o1} \\ \underline{R}_{o2} \end{pmatrix} = \begin{pmatrix} a & b \\ -b & a \end{pmatrix} \cdot \begin{pmatrix} \underline{R}_{o1} \\ \underline{R}_{o2} \end{pmatrix} \quad (6.38)$$

with

$$a = (mm' + nn')/\Delta, \quad b = (nm' - mn')/\Delta, \quad \Delta = m'^2 + n'^2 \quad (6.39)$$

where the corresponding vector lengths are

$$R'_{o1} = R'_{o2} = \left(\sqrt{a^2 + b^2}\right) R = \sqrt{\frac{m^2 + n^2}{m'^2 + n'^2}} R \quad (6.40)$$

As before, relations (6.38) and (6.39) allow to determine a joint superlattice with vectors $\underline{R}'_1, \underline{R}'_2$ where

$$\begin{pmatrix} \underline{R}'_1 \\ \underline{R}'_2 \end{pmatrix} = \begin{pmatrix} m' & n' \\ -n' & m' \end{pmatrix} \cdot \begin{pmatrix} \underline{R}'_{o1} \\ \underline{R}'_{o2} \end{pmatrix} = \begin{pmatrix} m & n \\ -n & m \end{pmatrix} \cdot \begin{pmatrix} \underline{R}_{o1} \\ \underline{R}_{o2} \end{pmatrix} \quad (6.41)$$

and

$$R'_1 = R'_2 = \left(\sqrt{m^2 + n^2}\right) R \quad (6.42)$$

The superlattice cell spanned by vectors $\underline{R}'_1, \underline{R}'_2$ cannot be reduced further in size if the integer-valued determinants in Eq. (6.41) are coprime, that is, if

$$\gcd(m^2 + n^2, m'^2 + n'^2) = 1 \quad (6.43)$$

6.5
Interference Lattices

Interference lattices and quasi-periodic long-range order at adsorbate covered surfaces expressed by one- and two-dimensional **moiré patterns** have been known for some time but have attracted considerable attention more recently in connection with graphene, a graphite monolayer with a honeycomb structure, adsorbing at metal surfaces. (In the following text, graphene will be named "**Gra**" in all structure formulas using the Wood notation.) As an example, experimental LEED and STM studies [127] suggest that graphene adsorbing on the hexagonal (0 0 0 1) surface of ruthenium forms a commensurate phase with a coincidence lattice described by a supercell of 25×25 carbon honeycombs on a 23×23 supercell of the hexagonal Ru(0 0 0 1) substrate. This was simulated by DFT (density-functional theory) calculations with different (smaller than observed) coincidence lattices [127] where Figure 6.5 shows results for the

Ru(0 0 0 1) + (12/13 × 12/13) − Gra structure. Here, the lateral long-range order is evident in the view perpendicular to the substrate surface, see Figure 6.5a. In addition, the lateral structure is combined with perpendicular warping of the graphene layer as shown in Figure 6.5b, which is characterized by local distortion away from the substrate when carbon gets near the ruthenium surface.

Boron nitride (h-BN) monolayers adsorbed on the (1 1 1) surface of rhodium have also been found to form coincidence lattices. These are described by 13 × 13 supercells of hexagonal B_3N_3 rings on top of 12 × 12 supercells of the Rh(1 1 1) substrate surface [153]. Analogous to the previous example, the coincidence lattice structure includes warping of the boron nitride overlayer. Interestingly, the warping of this overlayer leads to local indentation toward the substrate when adsorbate atoms get near the rhodium surface, which is contrary to the previous example. However, in this section structural details of overlayer warping will be ignored and focus will be put on the lateral structure.

Another example is the observed rotational superlattice formed by a silver monolayer on the Ni(1 1 1) substrate surface which can be described as Ni(1 1 1) + (1.167 × 1.167)R2.4° − Ag [154] in Wood notation. This structure can be represented approximately by an HOC lattice structure with a transformation matrix

$$\underline{\underline{M}} = \frac{1}{39} \begin{pmatrix} 46 & -2 \\ 2 & 44 \end{pmatrix} = \begin{pmatrix} 5 & 2 \\ -2 & 7 \end{pmatrix}^{-1} \cdot \begin{pmatrix} 6 & 2 \\ -2 & 8 \end{pmatrix} \quad (6.44)$$

reflecting Ni(1 1 1) + (1.155 × 1.155)R2.2° − Ag which is shown in Figure 6.26 (only the overlayer and the topmost substrate layer are included). The figure illustrates nicely that the combination of the Ag adsorbate and the Ni substrate lattices yields a structure with a clear lateral interference pattern (moiré pattern) of local areas (**moirons**) that form a hexagonal lattice with lattice vectors $\underline{R}_{M1}$ and $\underline{R}_{M2}$. The general formalism to describe these two-dimensional interference lattices quantitatively will be discussed in the following.

6.5.1
Basic Formalism

The basis of the formalism [155, 156] is a lateral surface structure combining a two-dimensional periodic substrate layer described by lattice vectors $\underline{R}_{o1}$ and $\underline{R}_{o2}$, with an overlayer whose periodicity is given by lattice vectors $\underline{R}'_{o1}$, $\underline{R}'_{o2}$ with

$$\begin{pmatrix} \underline{R}'_{o1} \\ \underline{R}'_{o2} \end{pmatrix} = \underline{\underline{M}} \cdot \begin{pmatrix} \underline{R}_{o1} \\ \underline{R}_{o2} \end{pmatrix} \quad (6.45)$$

where the 2×2 matrix $\underline{\underline{M}}$ quantifies the linear transformation as discussed in Section 6.1. For real-valued $\underline{\underline{M}}$ we define an integer approximant matrix $\underline{\underline{M}}_I$ whose elements a_{ij} are the integers nearest to m_{ij}, see Appendix E.5. Therefore, $\underline{\underline{M}}_I$ characterizes the commensurate structure closest to that given by matrix $\underline{\underline{M}}$.

As discussed in Appendix G, any functions $f^S(\underline{r})$, $f^O(\underline{r})$, describing lateral spatial properties of the **substrate** surface and of the **overlayer** by themselves can be

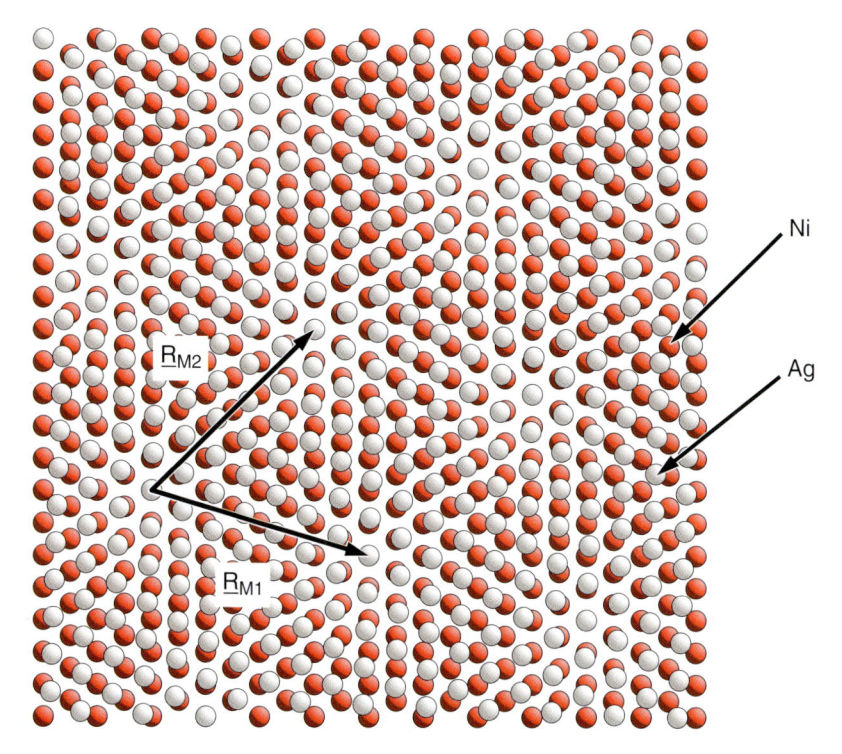

Figure 6.26 Structure of the Ni(1 1 1) + (1.155 × 1.155)R2.2° − Ag adsorbate system. The lattice vectors $\underline{R}_{M1}$, $\underline{R}_{M2}$ of the moiré pattern are labeled accordingly.

represented by the corresponding infinite Fourier series as

$$f^S(\underline{r}) = \sum_{j,k} c_{j,k}^S \exp[i\,(j\underline{G}_{o1} + k\underline{G}_{o2})\underline{r}] = \sum_{j,k} c_{j,k}^S \exp\left[i(j\ k)\begin{pmatrix}\underline{G}_{o1}\\ \underline{G}_{o2}\end{pmatrix}\underline{r}\right] \quad (6.46a)$$

$$f^O(\underline{r}) = \sum_{j',k'} c_{j',k'}^O \exp\left[i(j'\ k')\begin{pmatrix}\underline{G}'_{o1}\\ \underline{G}'_{o2}\end{pmatrix}\underline{r}\right], \quad j,\,k,\,j',\,k'\ \text{integer} \quad (6.46b)$$

with, in general, complex valued coefficients $c_{j,k}^S$ and $c_{j',k'}^O$, respectively, (which, at the real three-dimensional surface, may depend on a third coordinate perpendicular to the surface) and reciprocal lattice vectors $\underline{G}_{o1}$, $\underline{G}_{o2}$ and $\underline{G}'_{o1}$, $\underline{G}'_{o2}$ derived from orthogonality relations (2.96), that is,

$$(\underline{G}_{oi}\underline{R}_{oj}) = (\underline{G}'_{oi}\underline{R}'_{oj}) = 2\pi\delta_{ij}, \quad j,\,k = 1,\,2 \quad (6.47)$$

as discussed in Section 2.5. Note that in Eq. (6.46) we use a matrix notation for scalar products with vectors

$$p\underline{v}_1 + q\underline{v}_2 \stackrel{\text{def.}}{=} (p\ q)\begin{pmatrix}\underline{v}_1\\ \underline{v}_2\end{pmatrix} \quad (6.48)$$

which proves to be useful for the following.

The real space transformation (6.45) together with orthogonality relations (6.47) leads to a transformation of the reciprocal space lattice vectors of the substrate and the overlayer given by

$$\begin{pmatrix} \underline{G}'_{o1} \\ \underline{G}'_{o2} \end{pmatrix} = \underline{m} \cdot \begin{pmatrix} \underline{G}_{o1} \\ \underline{G}_{o2} \end{pmatrix} \quad \text{with} \quad \underline{m} = (\underline{M}^{-1})^+ = (\underline{M}^+)^{-1} = \underline{M}^{-1+} \quad (6.49)$$

where $\underline{M}^+$ denotes the transposed and $\underline{M}^{-1}$ the inverted matrix $\underline{M}$ with $\underline{M}^{-1+}$ being the combination.

According to Eqs. (6.46) the **superposition** of the property functions $f^S(\underline{r})$ of the substrate and $f^O(\underline{r})$ of the overlayer describing spatial properties of the combined system can be written as

$$f(\underline{r}) = f^S(\underline{r}) + f^O(\underline{r})$$

$$= \sum_{j,k} c_{j,k}^S \exp\left[i(j\ k) \begin{pmatrix} \underline{G}_{o1} \\ \underline{G}_{o2} \end{pmatrix} \underline{r}\right] + \sum_{j',k'} c_{j',k'}^O \exp\left[i(j'\ k') \begin{pmatrix} \underline{G}'_{o1} \\ \underline{G}'_{o2} \end{pmatrix} \underline{r}\right]$$

$$j, k, j', k' \text{ integer} \quad (6.50)$$

This superposition is **nonperiodic** for real-valued transformation matrices $\underline{M}$. In contrast, for **commensurate** lateral surface structures, the matrix $\underline{M}$ is integer-valued, thus agreeing with its **integer approximant** $\underline{M}_I$. Then in reciprocal space

$$\begin{pmatrix} \underline{G}'_{o1} \\ \underline{G}'_{o2} \end{pmatrix} = \underline{m} \cdot \begin{pmatrix} \underline{G}_{o1} \\ \underline{G}_{o2} \end{pmatrix} = \underline{m}_I \cdot \begin{pmatrix} \underline{G}_{o1} \\ \underline{G}_{o2} \end{pmatrix} \quad \text{with} \quad \underline{m}_I = \underline{M}_I^{-1+} \quad (6.51)$$

and we can write

$$(j\ k) \cdot \begin{pmatrix} \underline{G}_{o1} \\ \underline{G}_{o2} \end{pmatrix} = (j\ k) \cdot \underline{m}_I^{-1} \cdot \underline{m}_I \cdot \begin{pmatrix} \underline{G}_{o1} \\ \underline{G}_{o2} \end{pmatrix} \stackrel{\text{def.}}{=} (j'\ k') \cdot \begin{pmatrix} \underline{G}'_{o1} \\ \underline{G}'_{o2} \end{pmatrix} \quad (6.52)$$

where

$$\begin{pmatrix} j' \\ k' \end{pmatrix} = (\underline{m}_I^{-1})^+ \cdot \begin{pmatrix} j \\ k \end{pmatrix} = \underline{M}_I \cdot \begin{pmatrix} j \\ k \end{pmatrix} \quad (6.53)$$

Thus, each substrate component in Eq. (6.46a) can be written as

$$f_{j,k}^S(\underline{r}) = c_{j,k}^S \exp\left[i(j\ k) \begin{pmatrix} \underline{G}_{o1} \\ \underline{G}_{o2} \end{pmatrix} \underline{r}\right] = \frac{c_{j,k}^S}{c_{j',k'}^O} f_{j',k'}^O(\underline{r}) \quad (6.54)$$

where $f_{j',k'}^O(\underline{r})$ is an overlayer component in Eq. (6.46b) referring to indices j', k' that can be different from j, k. Thus, for commensurate structures, each substrate component (6.54) can be combined with a corresponding overlayer component and Fourier expansion (6.50) becomes altogether periodic with the overlayer lattice vectors $\underline{R}'_{o1}$, $\underline{R}'_{o2}$ describing its periodicity.

The basic mathematical idea of an interference lattice in a combined substrate/overlayer system is that all or an infinite subset of components $f_{j,k}^S(\underline{r})$ of the substrate expansion (6.50) interfere with the corresponding components $f_{j',k'}^O(\underline{r})$ of the overlayer that exhibit identical or very similar spatial variation.

If each component $f^S_{j,k}(\underline{r})$ interferes with a corresponding component $f^O_{j',k'}(\underline{r})$ (**basic** or **first order interference**) the superposition component can, together with Eqs. (6.49) and (6.53), be written as

$$f_{j,k}(\underline{r}) = f^S_{j,k}(\underline{r}) + f^O_{j',k'}(\underline{r})$$

$$= c^S_{j,k} \exp\left[i(j\ k)\begin{pmatrix}\underline{G}_{o1}\\ \underline{G}_{o2}\end{pmatrix}\underline{r}\right] + c^O_{j',k'} \exp\left[i(j\ k)\,\underline{\underline{m}}_I^{-1}\cdot\underline{\underline{m}}\cdot\begin{pmatrix}\underline{G}_{o1}\\ \underline{G}_{o2}\end{pmatrix}\underline{r}\right]$$

$$= c^S_{j,k} \exp\left[i(j\ k)\begin{pmatrix}\underline{G}_{o1}\\ \underline{G}_{o2}\end{pmatrix}\underline{r}\right] \kappa_{j,k}(\underline{r}) \qquad (6.55)$$

with a modulation factor $\kappa_{j,k}(\underline{r})$ given by

$$\kappa_{j,k}(\underline{r}) = 1 + \frac{c^O_{j',k'}}{c^S_{j,k}} \exp\left[i(j\ k)\begin{pmatrix}\underline{G}_{M1}\\ \underline{G}_{M2}\end{pmatrix}\underline{r}\right] \qquad (6.56)$$

and

$$\begin{pmatrix}\underline{G}_{M1}\\ \underline{G}_{M2}\end{pmatrix} = (\underline{\underline{m}}_I^{-1}\cdot\underline{\underline{m}} - \underline{\underline{1}})\cdot\begin{pmatrix}\underline{G}_{o1}\\ \underline{G}_{o2}\end{pmatrix} \qquad (6.57)$$

Thus, the superposition component $f_{j,k}(\underline{r})$ of Eq. (6.55) is a periodic function modulated by a factor $\kappa_{j,k}(\underline{r})$ which itself is a periodic function with lattice vectors $\underline{G}_{M1}$ and $\underline{G}_{M2}$ in reciprocal space. This corresponds to a lattice in real space with lattice vectors $\underline{R}_{M1}$ and $\underline{R}_{M2}$ (**moiré lattice vectors**) according to

$$\begin{pmatrix}\underline{R}_{M1}\\ \underline{R}_{M2}\end{pmatrix} = (\underline{\underline{m}}_I^{-1}\cdot\underline{\underline{m}} - \underline{\underline{1}})^{-1+}\cdot\begin{pmatrix}\underline{R}_{o1}\\ \underline{R}_{o2}\end{pmatrix} = (\underline{\underline{M}}_I - \underline{\underline{M}})^{-1}\cdot\underline{\underline{M}}\cdot\begin{pmatrix}\underline{R}_{o1}\\ \underline{R}_{o2}\end{pmatrix} = \underline{\underline{P}}\cdot\begin{pmatrix}\underline{R}_{o1}\\ \underline{R}_{o2}\end{pmatrix}$$
(6.58)

with a transformation matrix $\underline{\underline{P}}$ (**moiré matrix**) given by

$$\underline{\underline{P}} = (\underline{\underline{M}}_I - \underline{\underline{M}})^{-1}\cdot\underline{\underline{M}} = [\underline{\underline{1}} - \underline{\underline{M}}_M]^{-1}\cdot\underline{\underline{M}}_M, \quad \text{where} \quad \underline{\underline{M}}_M = \underline{\underline{M}}_I^{-1}\cdot\underline{\underline{M}}$$
(6.59)

If the structure of the combined substrate/overlayer system is approaching commensurability, which is expressed mathematically by the transformation matrix $\underline{\underline{M}}$ in Eqs. (6.45) and (6.58) getting near its integer approximant $\underline{\underline{M}}_I$, then matrix $\underline{\underline{P}}$ becomes rather large and the moiré lattice vectors $\underline{R}_{M1}$ and $\underline{R}_{M2}$ in Eq. (6.58) will be considerably larger than the lattice vectors $\underline{R}_{o1}$ and $\underline{R}_{o2}$ of the substrate. This geometry describes a long-range modulation whose periodicity is perceived as a periodic interference pattern associated with moiré patterns as will be illustrated in the following.

Moiré patterns have been found for graphene overlayers on a number of hexagonal metal substrate surfaces where an isotropically scaled ($\gamma\times\gamma$) overlayer structure has been discussed in the literature, for references see, for example, [155]. Here, the transformation matrix $\underline{\underline{M}}$ connecting lattice vectors $\underline{R}'_{o1}$ and $\underline{R}'_{o2}$ of the

Table 6.1 Scaling factors γ and moiré factors λ for scaled ($\gamma \times \gamma$) overlayers of graphene on different metal substrates.

Substrate	γ	λ
Pt(1 1 1) [157]	0.89	8.09
Ir(1 1 1) [158, 159]	0.91	10.11
Ru(0 0 0 1) [127]	0.92	11.50
Rh(1 1 1) [157, 160]	0.93	13.28
Cu(1 1 1) [161]	0.97	32.33
Ni(1 1 1) [162]	1.00	∞

The references denote experimental studies where the moiré patterns have been identified.

graphene layer with $\underline{R}_{o1}$ and $\underline{R}_{o2}$ of the metal substrate according to Eq. (6.45) is given by

$$\underline{\underline{M}} = \begin{pmatrix} \gamma & 0 \\ 0 & \gamma \end{pmatrix} = \gamma \underline{\underline{1}}, \quad \gamma = \frac{a_{Gra}}{a_{Me}} \qquad (6.60)$$

where γ is a scaling factor determined by the ratio of the lattice constants a_{gra} of the graphene overlayer and a_{Me} of the metal substrate Table 6.1 lists experimental γ values for different metal substrates where moiré patterns have been observed. This shows that for graphene γ values always lie near $\gamma = 1$ such that the integer approximant of $\underline{\underline{M}}$ equals the unit matrix $\underline{\underline{M}}_I = \underline{\underline{1}}$. This leads to a moiré matrix $\underline{\underline{P}}$ according to Eq. (6.59)

$$\underline{\underline{P}} = \begin{pmatrix} \lambda & 0 \\ 0 & \lambda \end{pmatrix} = \lambda \underline{\underline{1}}, \quad \lambda = \frac{\gamma}{|1 - \gamma|} \qquad (6.61)$$

with λ denoting the **moiré factor**. As an example, Figure 6.27 shows a simulation of a graphene overlayer on the Pt(1 1 1) surface where $\gamma = 0.89$. Here, a hexagonal moiré lattice corresponding to a lattice constant $a_{moiré} = 8.09\, a_{Pt}$ is obtained, which confirms the moiré pattern observed by experiment [157].

The moiré matrix $\underline{\underline{P}}$ of Eq. (6.61) shows that, in general, isotropic overlayer scaling always results in moiré lattice vectors $\underline{R}_{M1}$ and $\underline{R}_{M2}$ that point along the directions of the substrate lattice vectors $\underline{R}_{o1}$ and $\underline{R}_{o2}$, respectively. Thus, the moiré lattice is of the same Bravais lattice type as that of the substrate. Further, Eq. (6.58) together with Eq. (6.61) shows that moiré lattice vectors diverge in their lengths when γ approaches $\gamma = 1$, which corresponds to lattice constants of the overlayer and substrate lattices getting very close. In this limit the moiron arrangement will become very open with extremely large moirons (giant scaling) that may be too large to be observable. This is consistent with experimental findings for graphene overlayers on Ni(1 1 1) substrate [162] where the surface lattice constants differ by less that 1%.

High-order interference is obtained by an infinite subset of substrate components $f_{j,k}^{S,d}(\underline{r})$, given by

$$f_{j,k}^{S,d}(\underline{r}) = c_{j'',k''}^{S} \exp\left[i(j''\ k'') \begin{pmatrix} \underline{G}_{o1} \\ \underline{G}_{o2} \end{pmatrix} \underline{r}\right] \text{ with } \begin{pmatrix} j'' \\ k'' \end{pmatrix} = \underline{\underline{d}} \cdot \begin{pmatrix} j \\ k \end{pmatrix} \qquad (6.62)$$

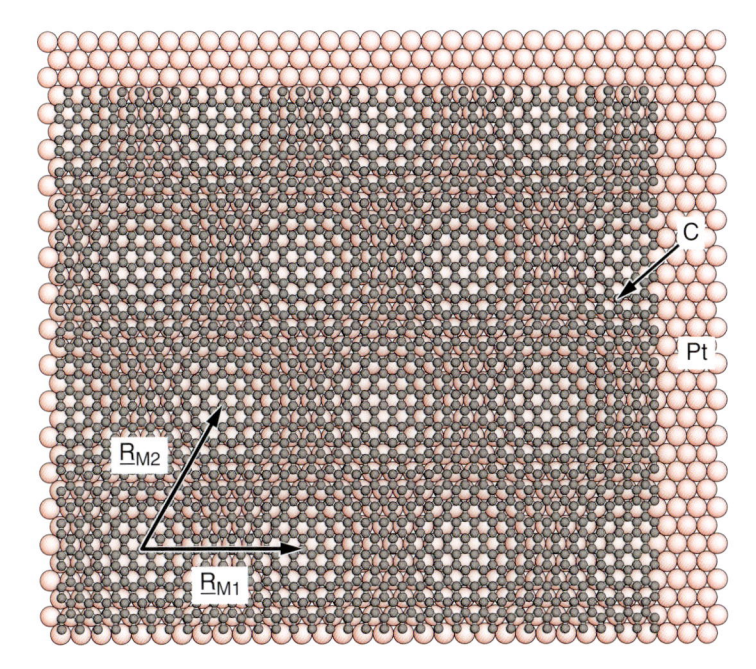

Figure 6.27 Moiré pattern of a graphene overlayer on the Pt(1 1 1) surface for a view perpendicular to the surface with moiré lattice vectors $\underline{R}_{M1}$, $\underline{R}_{M2}$ included.

in the Fourier expansion Eq. (6.46a), interfering with the corresponding overlayer components $f^O_{j',k'}(\underline{r})$. Here, the integer-valued matrix $\underline{\underline{d}}$ selects the infinite subset and can thus be considered an **interference order matrix** where $\underline{\underline{d}} = \underline{\underline{1}}$ (or more generally $\det(\underline{\underline{d}}) = 1$) refers to first order interference discussed earlier. Then analogous to Eq. (6.55) the superposition components can be written as

$$f^d_{j,k}(\underline{r}) = f^{S,d}_{j,k}(\underline{r}) + f^O_{j',k'}(\underline{r})$$

$$= c^S_{j'',k''} \exp\left[i(j\ k) \cdot \underline{\underline{d}}^+ \cdot \begin{pmatrix} G_{o1} \\ G_{o2} \end{pmatrix} \underline{r}\right] \kappa^d_{j,k}(\underline{r}) \qquad (6.63)$$

with a modulation factor $\kappa^d_{j,k}(\underline{r})$ where

$$\kappa^d_{j,k}(\underline{r}) = 1 + \frac{c^O_{j',k'}}{c^S_{j'',k''}} \exp\left[i(j\ k) \begin{pmatrix} \underline{G}^d_{M1} \\ \underline{G}^d_{M2} \end{pmatrix} \underline{r}\right] \qquad (6.64)$$

and

$$\begin{pmatrix} \underline{G}^d_{M1} \\ \underline{G}^d_{M2} \end{pmatrix} = (\underline{\underline{m}}_I^{-1} \cdot \underline{\underline{m}} - \underline{\underline{d}}^+) \cdot \begin{pmatrix} G_{o1} \\ G_{o2} \end{pmatrix} \qquad (6.65)$$

As before, the superposition component $f^d_{j,k}(\underline{r})$ of Eq. (6.63) is a periodic function modulated by a factor $\kappa^d_{j,k}(\underline{r})$, which itself is a periodic function with a lattice $\underline{G}^d_{M1}, \underline{G}^d_{M2}$ in reciprocal space that corresponds to a lattice with lattice vectors $\underline{R}^d_{M1}, \underline{R}^d_{M2}$ (**high-order moiré lattice vectors**) in real space according to

$$\begin{pmatrix} \underline{R}^d_{M1} \\ \underline{R}^d_{M2} \end{pmatrix} = (\underline{m}_I^{-1} \cdot \underline{m} - \underline{d}^+)^{-1+} \cdot \begin{pmatrix} \underline{R}_{o1} \\ \underline{R}_{o2} \end{pmatrix}$$

$$= (\underline{M}_I - \underline{M} \cdot \underline{d})^{-1} \cdot \underline{M} \cdot \begin{pmatrix} \underline{R}_{o1} \\ \underline{R}_{o2} \end{pmatrix} = \underline{P}^d \cdot \begin{pmatrix} \underline{R}_{o1} \\ \underline{R}_{o2} \end{pmatrix} \tag{6.66}$$

with a transformation matrix $\underline{P}^d$ (**high-order moiré matrix**) given by

$$\underline{P}^d = (\underline{M}_I - \underline{M} \cdot \underline{d})^{-1} \cdot \underline{M} = [\underline{1} - \underline{M}_M \cdot \underline{d}]^{-1} \cdot \underline{M}_M \quad \text{where} \quad \underline{M}_M = \underline{M}_I^{-1} \cdot \underline{M} \tag{6.67}$$

If the structure of the combined substrate/overlayer system is approaching high-order commensurability, expressed formally by the transformation matrix ($\underline{M} \cdot \underline{d}$) in Eq. (6.66) getting near its integer approximant $\underline{M}_I$, then matrix $\underline{P}^d$ is diverging and the moiré lattice vectors $\underline{R}^d_{M1}$ and $\underline{R}^d_{M2}$ in Eq. (6.66) will be considerably larger than the lattice vectors $\underline{R}_{o1}$ and $\underline{R}_{o2}$ of the substrate. The resulting long-range modulation can explain periodic high-order moiré patterns. Since these patterns refer to interference of only a subset of Fourier components representing the substrate periodicity their visual perception may not be as pronounced as that of first order moiré patterns. Further, high-order interference in two dimensions will, in general, depend on a 2×2 order matrix $\underline{d}$ rather than on a scalar quantity unless the matrix is restricted to an integer multiple d of the unit matrix as assumed elsewhere [163].

As an example, high-order moiré patterns have been found in experiment for graphene overlayers on the Ir(1 1 1) surface [158, 159] where the overlayer is rotated by almost 30° with respect to the hexagonal substrate surface. This system, characterized in Wood notation as Ir(1 1 1) + (0.9 × 0.9)R29.5° − Gra, can be approximated by an HOC lattice structure with a transformation matrix (using the acute representation of the hexagonal lattice)

$$\underline{M} = \frac{1}{151} \begin{pmatrix} 80 & 78 \\ -78 & 158 \end{pmatrix} = \begin{pmatrix} 5 & 9 \\ -9 & 14 \end{pmatrix}^{-1} \cdot \begin{pmatrix} -2 & 12 \\ -12 & 10 \end{pmatrix}$$

describing Ir(1 1 1) + ($\sqrt{(124/151)} \times \sqrt{(124/151)}$)R29.5° − Gra which is shown in Figure 6.28 (only the overlayer and the topmost substrate layer are included). Here, the hexagonal interference pattern is clearly visible with moiré lattice vectors $\underline{R}_{M1}$ and $\underline{R}_{M2}$ included in Figure 6.28. This pattern can be described by high-order interference [163] assuming a simplified "second" order matrix $\underline{d}$ and an appropriate integer approximant $\underline{M}_I$ with

$$\underline{M} = \begin{pmatrix} 0.530 & 0.517 \\ -0.530 & 1.046 \end{pmatrix} \approx \frac{1}{2}\begin{pmatrix} 1 & 1 \\ -1 & 2 \end{pmatrix} = \frac{1}{2}\underline{M}_I = \underline{M}_I \cdot \underline{d}^{-1}, \quad \underline{d} = \begin{pmatrix} 2 & 0 \\ 0 & 2 \end{pmatrix} \tag{6.68}$$

which yields for the moiré matrix according to Eq. (6.67) after some calculus

$$\underline{P}^d = \underline{d}^{-1} \cdot (\underline{M}_I \cdot \underline{d}^{-1} - \underline{M})^{-1} \cdot \underline{M} = -\begin{pmatrix} 10 & 2 \\ -2 & 12 \end{pmatrix}$$

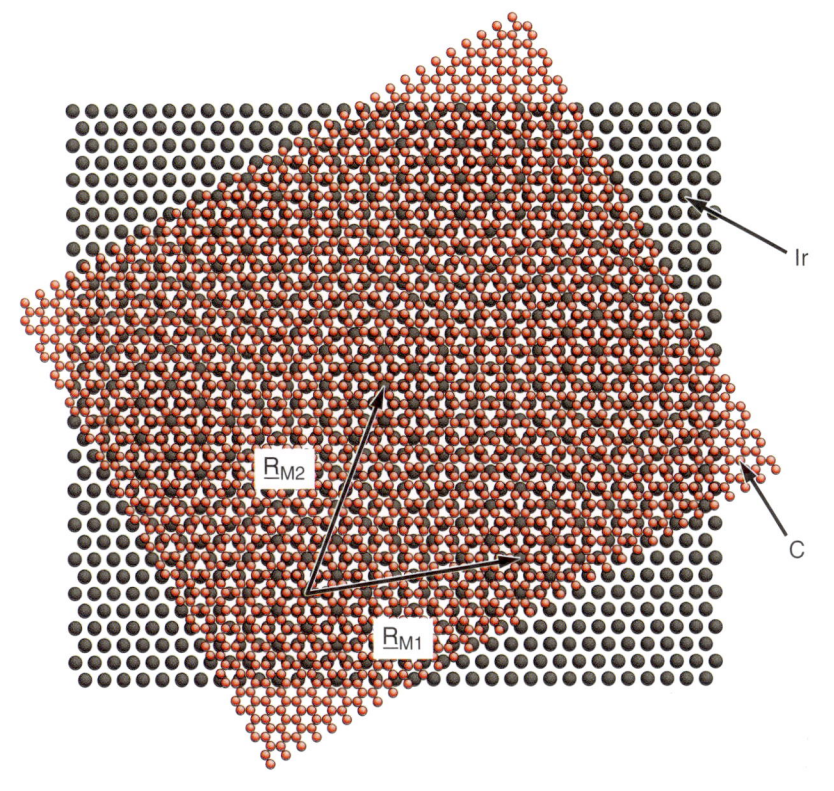

Figure 6.28 High-order moiré pattern of a graphene overlayer on the Ir(1 1 1) surface for a perpendicular view with high-order moiré lattice vectors $\underline{R}_{M1}$, $\underline{R}_{M2}$ included.

and results, in agreement with experiment [159], in a hexagonal moiré lattice with vectors $\underline{R}_{M1}$ and $\underline{R}_{M2}$ of lengths

$$|\underline{R}_{M1}| = |\underline{R}_{M1}| = \sqrt{124}\,|\underline{R}_{o1}| = \sqrt{124}\,a_{Ir} \tag{6.69}$$

where a_{Ir} is the lattice constant of the hexagonal Ir(1 1 1) substrate. Further, the moiré lattice is rotated by 8.948° with respect to the substrate lattice, which confirms the previous analysis [163].

If matrix $\underline{\underline{M}}$ in Eq. (6.45) describes a coincidence lattice of an HOC lattice structure, see Section 6.4, that is, if $\underline{\underline{M}}$ can be represented by two integer matrices $\underline{\underline{A}}$, $\underline{\underline{B}}$ with

$$\underline{\underline{M}} = \underline{\underline{B}}^{-1} \cdot \underline{\underline{A}} \tag{6.70}$$

then the moiré matrix (6.67) describing interference yields

$$\underline{\underline{P}}^d = (\underline{\underline{M}}_I - \underline{\underline{B}}^{-1} \cdot \underline{\underline{A}} \cdot \underline{\underline{d}})^{-1} \cdot (\underline{\underline{B}}^{-1} \cdot \underline{\underline{A}}) = (\underline{\underline{B}} \cdot \underline{\underline{M}}_I - \underline{\underline{A}} \cdot \underline{\underline{d}})^{-1} \cdot \underline{\underline{A}} \tag{6.71}$$

with $\underline{\underline{d}} = \underline{\underline{1}}$ referring to basic interference. As a consequence, the transformation between the moiré lattice vectors, $\underline{R}^d_{M1}$ and $\underline{R}^d_{M2}$, and those of the substrate lattice,

$\underline{R}_{o1}$ and $\underline{R}_{o2}$, given by Eq. (6.66), can be written as

$$\begin{pmatrix} \underline{R}_1 \\ \underline{R}_2 \end{pmatrix} = \underline{\underline{A}} \cdot \begin{pmatrix} \underline{R}_{o1} \\ \underline{R}_{o2} \end{pmatrix} = (\underline{\underline{B}} \cdot \underline{\underline{M}}_I - \underline{\underline{A}} \cdot \underline{\underline{d}}) \cdot \begin{pmatrix} \underline{R}_{M1}^d \\ \underline{R}_{M2}^d \end{pmatrix} \tag{6.72}$$

where $\underline{R}_1$ and $\underline{R}_2$ are the superlattice vectors describing the HOC supercell according to Eq. (6.13). Since the matrices $\underline{\underline{A}}$, $\underline{\underline{B}}$, $\underline{\underline{M}}_I$, and $\underline{\underline{d}}$ appearing on the right-hand side of Eq. (6.72) are all integer-valued the superlattice vectors $\underline{R}_1$ and $\underline{R}_2$ are given by integer-valued linear combinations of vectors $\underline{R}_{M1}$ and $\underline{R}_{M2}$. Therefore, the moiré lattice with vectors $\underline{R}_{M1}$ and $\underline{R}_{M2}$ can serve as a basis to define the coincidence lattice of the HOC structure.

As a simple example, graphene adsorbed on the hexagonal Ru(0 0 0 1) surface, which has been discussed earlier, was found to form an HOC coincidence lattice structure where 25 graphene cells fit on 23 Ru unit cells. This yields a lattice transformation (6.10) with a transformation matrix

$$\underline{\underline{M}} = \frac{23}{25} \begin{pmatrix} 1 & 0 \\ 0 & 1 \end{pmatrix} \quad \text{and} \quad \underline{\underline{M}}_I = \begin{pmatrix} 1 & 0 \\ 0 & 1 \end{pmatrix} \tag{6.73}$$

Hence, according to Eqs. (6.58) and (6.59) the moiré lattice vectors $\underline{R}_{M1}$ and $\underline{R}_{M2}$ are given by

$$\begin{pmatrix} \underline{R}_{M1} \\ \underline{R}_{M2} \end{pmatrix} = \frac{23}{2} \begin{pmatrix} 1 & 0 \\ 0 & 1 \end{pmatrix} \begin{pmatrix} \underline{R}_{o1} \\ \underline{R}_{o2} \end{pmatrix} \tag{6.74}$$

while the superlattice vectors $\underline{R}_1$ and $\underline{R}_2$ of the overlayer defined by Eq. (6.13) are

$$\begin{pmatrix} \underline{R}_1 \\ \underline{R}_2 \end{pmatrix} = 23 \begin{pmatrix} 1 & 0 \\ 0 & 1 \end{pmatrix} \begin{pmatrix} \underline{R}_{o1} \\ \underline{R}_{o2} \end{pmatrix} = 2 \begin{pmatrix} \underline{R}_{M1} \\ \underline{R}_{M2} \end{pmatrix} \tag{6.75}$$

Thus, the superlattice vectors $\underline{R}_1$ and $\underline{R}_2$ are twice as large as their moiré counterparts, $\underline{R}_{M1}$ and $\underline{R}_{M2}$, as shown in Figure 6.29, which confirms relation (6.72).

6.5.2
Interference and Wood Notation

Further quantification of moiré lattices is obtained for surfaces where the overlayer structure can be characterized using the Wood notation (6.2), see Section 6.3. In the present section only the primitive notation "$\cdots - p(\gamma_1 \times \gamma_2) R\alpha - \cdots$" (or "$\cdots - (\gamma_1 \times \gamma_2) R\alpha - \cdots$" in short) will be considered. According to Eq. (6.6a), the 2×2 transformation matrix of an overlayer structure characterized as "$\cdots - (\gamma_1 \times \gamma_2) R\alpha - \cdots$" is given by

$$\underline{\underline{M}} = \frac{1}{\sin(\omega)} \begin{pmatrix} \gamma_1 \sin(\omega - \alpha) & \gamma_1 q^{-1} \sin(\alpha) \\ -\gamma_2 q \sin(\alpha) & \gamma_2 \sin(\omega + \alpha) \end{pmatrix}, \quad q = \frac{R_2}{R_1} \tag{6.76}$$

where q, ω refer to the type of Bravais lattice. If the nearby commensurate overlayer structure described by the integer approximant $\underline{\underline{M}}_I$ corresponds to a Wood

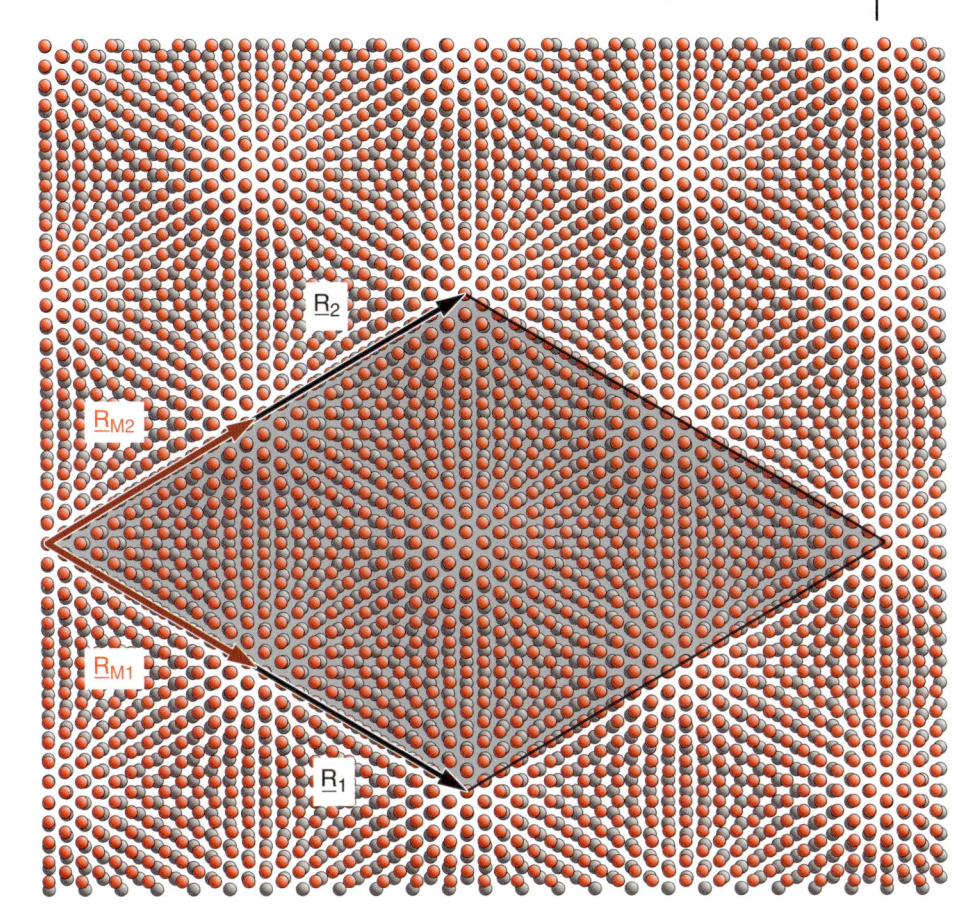

Figure 6.29 Graphene overlayer adsorbed on the Ru(0 0 0 1) surface. The Ru substrate is shown by its topmost layer with gray atom balls while the honeycomb elements of the graphene are simplified by red balls. The supercell is emphasized in gray with superlattice vectors $\underline{R}_1$, $\underline{R}_2$ and moiré lattice vectors $\underline{R}_{M1}$, $\underline{R}_{M2}$ labeled accordingly.

notation "$\cdots - (\gamma_{I1} \times \gamma_{I2}) R\alpha_I - \cdots$" based on the same Bravais lattice, that is, if

$$\underline{\underline{M}}_I = \frac{1}{\sin(\omega)} \begin{pmatrix} \gamma_{I1} \sin(\omega - \alpha_I) & \gamma_{I1} q^{-1} \sin(\alpha_I) \\ -\gamma_{I2} q \sin(\alpha_I) & \gamma_{I2} \sin(\omega + \alpha_I) \end{pmatrix} \qquad (6.77)$$

then the transformation matrix $\underline{\underline{M}}_M$ in Eq. (6.67) can be represented in Wood notation as

$$\text{``} \cdots - (\gamma_{M1} \times \gamma_{M2}) R\alpha_M - \cdots \text{''}, \gamma_{Mi} = \frac{\gamma_i}{\gamma_{Ii}}, \quad i = 1, 2, \quad \alpha_M = \alpha - \alpha_I \quad (6.78)$$

with a transformation matrix $\underline{\underline{M}}_M$ given by

$$\underline{\underline{M}}_M = \frac{1}{\sin(\omega)} \begin{pmatrix} \gamma_{M1} \sin(\omega - \alpha_M) & \gamma_{M1} q^{-1} \sin(\alpha_M) \\ -\gamma_{M2} q \sin(\alpha_M) & \gamma_{M2} \sin(\omega + \alpha_M) \end{pmatrix} \qquad (6.79)$$

If we further assume that the order matrix for high-order interference can be restricted to a multiple of the unit matrix, that is, to

$$\underline{\underline{d}} \cong d\underline{\underline{1}} = d \begin{pmatrix} 1 & 0 \\ 0 & 1 \end{pmatrix} \qquad (6.80)$$

then the (high-order) moiré matrix $\underline{\underline{P}}^d$ according to Eq. (6.67) can be written as

$$\underline{\underline{P}}^d = [\underline{\underline{1}} - \underline{\underline{M}}_M \cdot \underline{\underline{d}}]^{-1} \cdot \underline{\underline{M}}_M = \frac{1}{\Delta \sin(\omega)}$$

$$\cdot \begin{pmatrix} \gamma_{M1}[\sin(\omega - \alpha_M) - d\gamma_{M2}\sin(\omega)] & q^{-1}\gamma_{M1}\sin(\alpha_M) \\ -q\gamma_{M2}\sin(\alpha_M) & \gamma_{M2}[\sin(\omega + \alpha_M) - d\gamma_{M1}\sin(\omega)] \end{pmatrix} \qquad (6.81)$$

where

$$\Delta = 1 + d^2\gamma_{M1}\gamma_{M2} - d(\gamma_{M1} + \gamma_{M2})\cos(\alpha_M) - d(\gamma_{M2} - \gamma_{M1})\cot(\omega)\sin(\alpha_M) \qquad (6.82)$$

Using Eqs. (6.81) and (6.82) together with Eq. (6.66) one can derive parameters that are quite intuitive and can be actually measured in experiments performed for the overlayer structures. A comparison of the lengths of the moiré lattice vectors $\underline{R}^d_{M1}$, $\underline{R}^d_{M2}$ with those of the substrate surface, $\underline{R}_{o1}$ and $\underline{R}_{o2}$ yields the **moiré factors** λ_i with

$$\lambda_i = \frac{R^d_{Mi}}{R_{oi}} = \frac{\gamma_{Mi}}{\Delta}\sqrt{1 + d^2\gamma^2_{Mj} - 2d\gamma_{Mj}\cos(\alpha_M)}, \quad (i,j) = (1,2), (2,1) \qquad (6.83)$$

and the **moiré angles** φ_i defined by the angles between the moiré lattice vectors and their substrate counterparts, are given by

$$\cos(\varphi_i) = \frac{(\underline{R}^d_{M1}\underline{R}_{oi})}{R^d_{Mi}R_{oi}} = \frac{\cos(\alpha_M) - d\gamma_{Mj}}{\sqrt{1 + d^2\gamma^2_{Mj} - 2d\gamma_{Mj}\cos(\alpha_M)}}, \quad (i,j) = (1,2), (2,1) \qquad (6.84)$$

Special cases are systems where the integer approximant matrix $\underline{\underline{M}}_I$ equals the unit matrix which refers to $\gamma_{Ii} = 1$, $\alpha_I = 0°$ resulting in

$$\lambda_i = \frac{R^d_{Mi}}{R_{oi}} = \frac{\gamma_i}{\Delta}\sqrt{1 + d^2\gamma^2_j - 2d\gamma_j\cos(\alpha)}, \quad (i,j) = (1,2), (2,1)$$

$$\Delta = 1 + d^2\gamma_1\gamma_2 - d(\gamma_1 + \gamma_2)\cos(\alpha) - d(\gamma_2 - \gamma_1)\cot(\omega)\sin(\alpha)$$

$$\cos(\varphi_i) = \frac{(\underline{R}^d_{M1}\underline{R}_{oi})}{R^d_{Mi}R_{oi}} = \frac{\cos(\alpha) - d\gamma_j}{\sqrt{1 + d^2\gamma^2_j - 2d\gamma_j\cos(\alpha)}}, \quad (i,j) = (1,2), (2,1) \qquad (6.85)$$

for the moiré parameters as discussed in detail in [155]. This simplifies further for isotropic scaling combined with rotation, where $\gamma_1 = \gamma_2 = \gamma$, to yield

$$\lambda_1 = \lambda_2 = \lambda = \frac{R^d_{Mi}}{R_{oi}} = \frac{\gamma}{\sqrt{\Delta}}, \quad \Delta = 1 + d^2 \gamma^2 - 2d\gamma \cos(\alpha)$$

$$\cos(\varphi_1) = \cos(\varphi_2) = \cos(\varphi) = \frac{(R^d_{Mi} \, R_{oi})}{R^d_{Mi} \, R_{oi}} = \frac{\cos(\alpha) - d\gamma}{\sqrt{\Delta}} \quad (6.86)$$

As a first example, we consider the graphene overlayer on the Pt(1 1 1) surface discussed earlier. This system which has been observed in experiments [157] can be described in Wood notation as Pt(1 1 1) + (0.89 × 0.89) – Gra, reflecting an isotropically scaled overlayer with $\gamma = 0.89$ and no rotation, $\alpha = 0°$. Here, the integer approximant structure is (1 × 1), which corresponds to $\alpha_1 = 0°$ such that Eq. (6.86) is applicable and yields a hexagonal first order moiré lattice with

$$\lambda = \frac{\gamma}{|1 - \gamma|} = 8.09, \quad \varphi_i = 0° \quad (6.87)$$

which is shown in Figure 6.27 confirming the result in the preceding discussion.

Another example where Eq. (6.86) can be applied is a rotated graphene overlayer adsorbed on top of graphene. For this system, a rotation angle $\alpha = 3.5°$ has been claimed by experiment [164]. Thus, in Wood notation the overlayer structure is described as Gra + (1 × 1)R3.5° – Gra and exhibits a hexagonal first order moiré pattern where according to Eq. (6.86)

$$\lambda = \left(2\left|\sin\left(\frac{\alpha}{2}\right)\right|\right)^{-1} = 16.37, \quad \cos(\varphi) = \frac{\cos(\alpha) - 1}{\sqrt{2(1 - \cos(\alpha))}}, \quad \varphi = 90° + \frac{\alpha}{2} = 91.75° \quad (6.88)$$

which is shown in Figure 6.30.

The combination of rotated and scaled overlayers has been found in experiments for graphene adsorbed on hexagonal boron nitride (h-BN) [165]. In this system, the lattice constant of the graphene overlayer is smaller by 3% compared with that of the h-BN lattice and the overlayer is rotated by a rather small angle of 3° yielding a h-BN + (0.97 × 0.97)R3° – Gra structure in Wood notation. This can be approximated by an HOC lattice structure with a transformation matrix

$$\underline{\underline{M}} = \frac{1}{17}\begin{pmatrix} 16 & 1 \\ -1 & 17 \end{pmatrix} = \begin{pmatrix} 17 & 0 \\ 0 & 17 \end{pmatrix}^{-1} \cdot \begin{pmatrix} 16 & 1 \\ -1 & 17 \end{pmatrix}$$

representing h-BN + (0.971 × 0.971)R3° – Gra. The resulting hexagonal first order moiré pattern according to Eq. (6.86) leads to $\lambda = 16.52$ and $\varphi = 63.00°$. This is

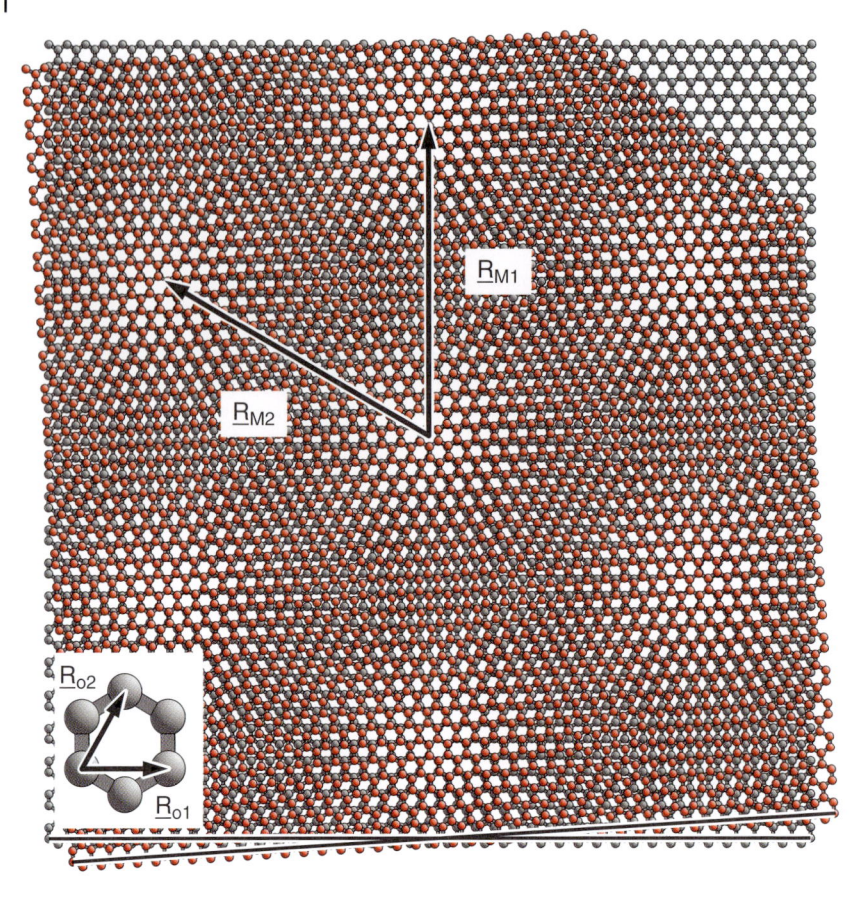

Figure 6.30 Graphene overlayer adsorbed on a graphene layer for a rotation angle $\alpha = 3.5°$ (indicated by lines at the bottom) with moiré vectors $\underline{R}_{M1}$, $\underline{R}_{M2}$ included. The lattice vectors $\underline{R}_{o1}$, $\underline{R}_{o2}$ of the underlying graphene monolayer are sketched at the lower left where a magnification by a factor 5 is applied for better visibility.

illustrated in Figure 6.31a where in the simulation the graphene C_6 and B_3N_3 honeycombs are replaced by gray and red balls for better visibility. For comparison, Figure 6.31b shows a corresponding simulation of the overlayer structure h-BN + (0.971 × 0.971)R0° − Gra, that is, without rotation, defined by a transformation matrix

$$\underline{\underline{M}} = \frac{1}{35}\begin{pmatrix} 34 & 0 \\ 0 & 34 \end{pmatrix} = \begin{pmatrix} 35 & 0 \\ 0 & 35 \end{pmatrix}^{-1} \cdot \begin{pmatrix} 34 & 0 \\ 0 & 34 \end{pmatrix}$$

The resulting moiré pattern is dramatically changed, representing a much larger moiré lattice, $\lambda = 34.00$ and $\varphi = 0.00°$, with well separated moirons compared to those of the overlayer structure with a small rotation. Further, the rotation of the overlayer by only 3° leads to a moiron lattice rotated by a sizeable 63°, which may be described as a giant rotation. This is a more general result found for overlayer

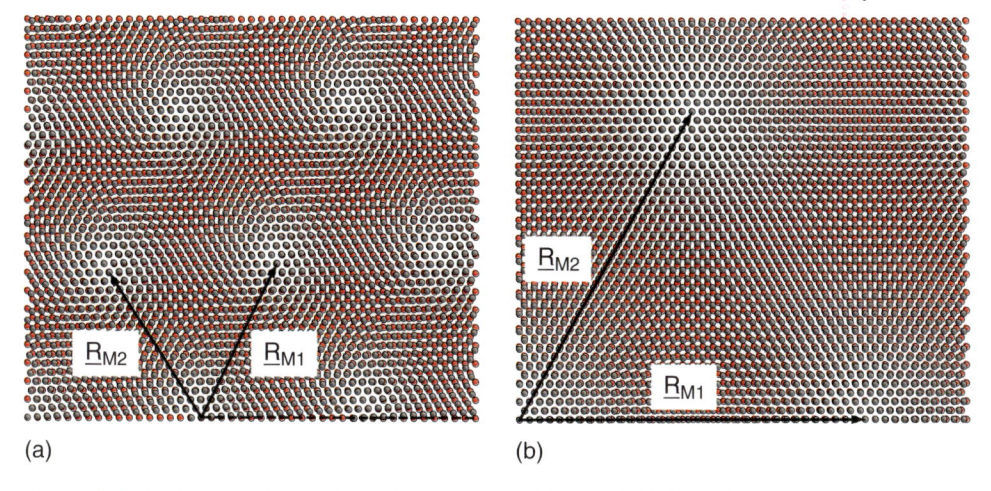

Figure 6.31 Graphene overlayer adsorbed on a hexagonal boron nitride layer for a rotation angle of (a) $\alpha = 3.0°$ and (b) $\alpha = 0.0°$. The graphene C_6 and B_3N_3 honeycombs are simulated by gray and red balls for better visibility. Moiré vectors $\underline{R}_{M1}$, $\underline{R}_{M2}$ are indicated accordingly.

structures with small rotation angles α and where the lattice mismatch is small, reflected by scaling factors $\gamma_i \approx 1$. In fact, it can be shown by a simple Taylor expansion that for even smaller values of α the moiré angle φ in Eq. (6.86) can be approximated by

$$\varphi \approx \frac{\alpha}{|1-\gamma|} \tag{6.89}$$

which substantiates the general result.

Another interesting example of a rotated and scaled overlayer is the Ir(1 1 1) + ($\sqrt{(124/151)} \times \sqrt{(124/151)}$)R29.5° – Gra adsorbate system [159] discussed earlier, see Figure 6.28, which exhibits a second order moiré pattern. Here, the integer approximant $\underline{\underline{M}}_I$, given by Eq. (6.68) can be written as

$$\underline{\underline{M}}_I = \begin{pmatrix} 1 & 1 \\ -1 & 2 \end{pmatrix} = \frac{1}{\sin(60°)} \begin{pmatrix} \sqrt{3}\sin(30°) & \sqrt{3}\sin(30°) \\ -\sqrt{3}\sin(30°) & \sqrt{3}\sin(90°) \end{pmatrix}$$

reflecting a ($\sqrt{3} \times \sqrt{3}$) R30° structure in Wood notation according to Eq. (6.77). Using Eq. (6.78) this results in scaling factors $\gamma_M = \sqrt{(124/453)} = 0.5232$ and rotation angles $\alpha_M = -0.4187°$. Thus, applying Eqs. (6.82)–(6.84) for second order interference, $d = 2$, yields moiré factors $\lambda_i = \sqrt{124}$ and moiré angles $\varphi_i = 8.948°$, which confirms the previous analysis.

An important result of the present formalism is that for isotropically scaled overlayers with scaling factors $\gamma_{M1} = \gamma_{M2} = \gamma_M$ the moiré factors λ_1 and λ_2 given by Eq. (6.83) are equal and do not depend explicitly on the type of Bravais lattice. Likewise, the moiré angles φ_1 and φ_2 defined by Eq. (6.84) agree with each other and are independent of the Bravais lattice type. Thus, for isotropically scaled and rotated overlayers the type of Bravais lattice of a possible moiré lattice is always that of the initial substrate. As an illustration, Figure 6.32 shows moiré patterns

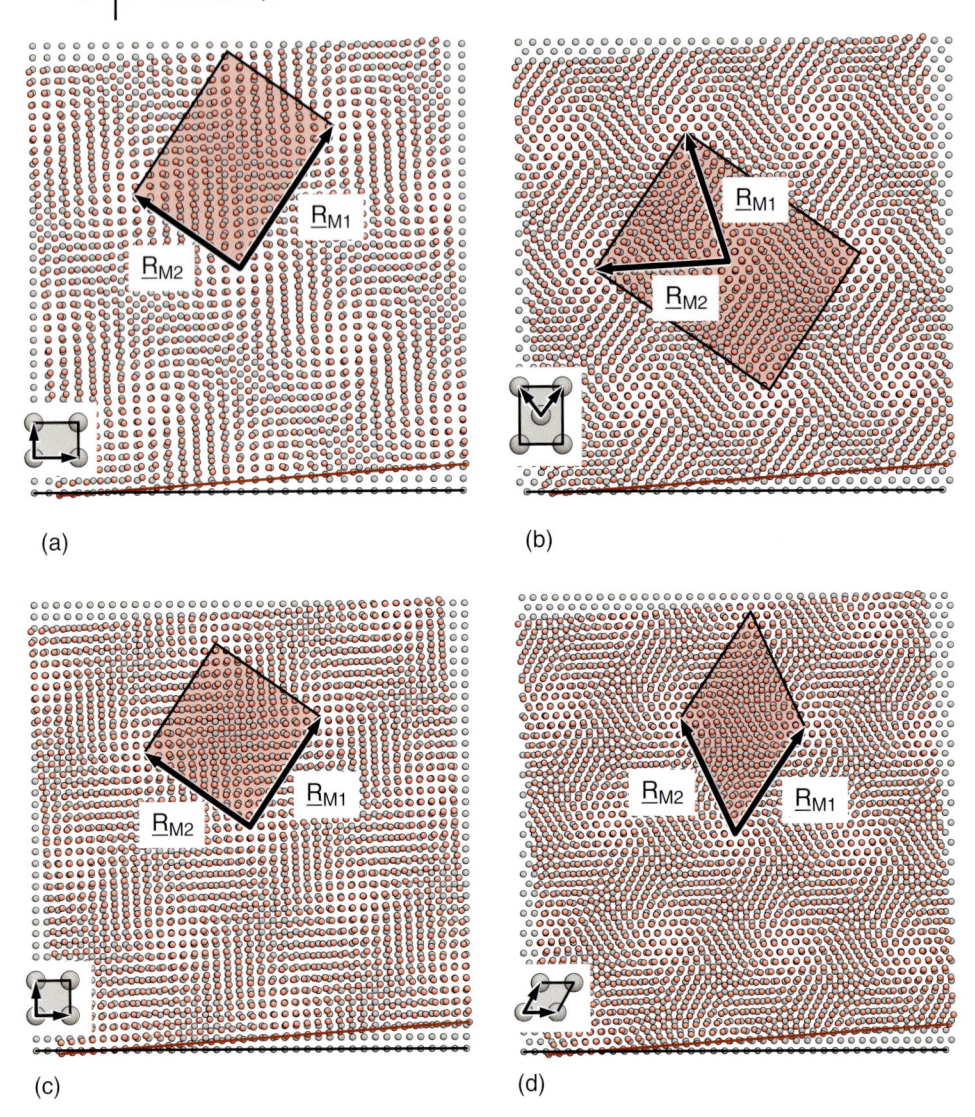

Figure 6.32 Moiré patterns of overlayers with different Bravais lattices, (a) primitive rectangular (p-rect., $R_{o1}/R_{o2} = 1.3$), (b) centered rectangular (c-rect., $\angle(\underline{R}_{o1}, \underline{R}_{o2}) = 75°$), (c) square, and (d) hexagonal (hex.) reflecting a $(0.95 \times 0.95)R4°$ overlayer structure. Moiré unit cells are emphasized in red with moiré vectors $\underline{R}_{M1}$, $\underline{R}_{M2}$ labeled accordingly. Substrate unit cells and lattice vectors $\underline{R}_{o1}$, $\underline{R}_{o2}$ are included at the lower left insets with a magnification factor 3 applied for better visibility.

for fictitious surface structures with different Bravais lattices where the patterns result from rotation by $\alpha = 4°$ and isotropic scaling by $p = 0.95$. This corresponds in all examples to a moiré angle $\varphi = 55.71°$ and a moiré factor $\lambda = 11.252$. For all four Bravais lattice types with symmetry, that is, for primitive and centered rectangular, square, and hexagonal lattices, the moiré lattice type agrees with that of the substrate and overlayer lattice. While this theoretical result is quite clear its experimental verification seems to exist only for hexagonal surface systems so far.

6.5.3
Anisotropic Scaling, Stretching, and Shifting

While the Bravais lattice type of moiré lattices resulting from rotated and isotropically scaled overlayers always agrees with that of the overlayer structure, anisotropic scaling with scaling factors $\gamma_1 \neq \gamma_2$ can lead to moiré patterns that do not reflect the initial Bravais lattice. As an illustration, Figure 6.33 shows two moiré patterns of a fictitious (distorted) hexagonal overlayer on a hexagonal substrate which refer to $(0.94 \times 0.98)R5°$ (Figure 6.33a) and to $(0.98 \times 0.94)R(-5°)$ (Figure 6.33b) in Wood notation. In both cases, the moiré pattern is characterized by distorted ellipsoidal moirons forming an almost rectangular lattice. In addition, the moirons are frizzy at their boundaries exhibiting a spiral structure where the spirals rotate clockwise in the left pattern and anti-clockwise in the right pattern of Figure 6.33. Thus, the two patterns may be considered chiral pairs.

Anisotropically scaled overlayers without rotation, reflecting $(\gamma_1 \times \gamma_2)$ overlayers in Wood notation with (1×1) as the integer approximant, that is, $\underline{\underline{M}}_I = \underline{\underline{1}}$, are

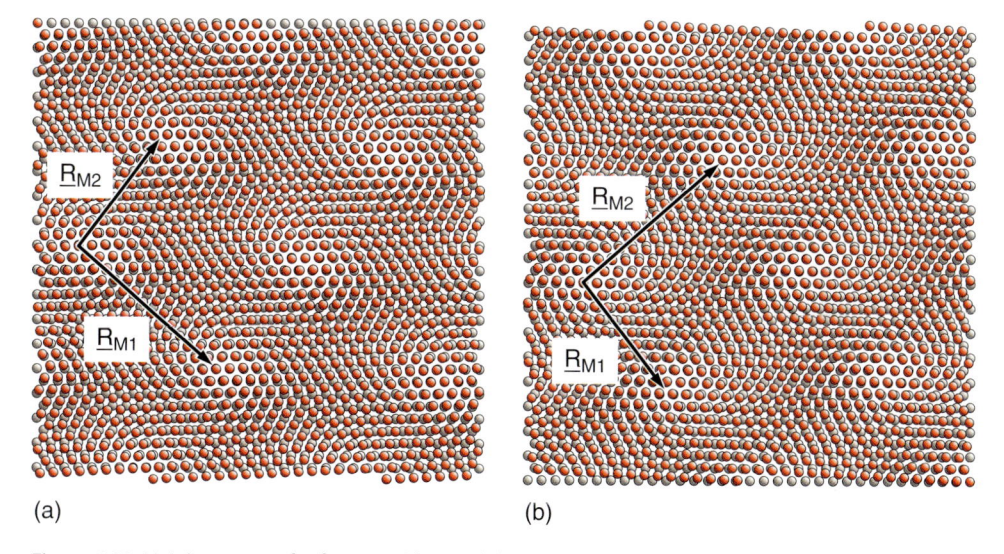

Figure 6.33 Moiré patterns of a fictitious (distorted) hexagonal overlayer on hexagonal substrate described as (a) $(0.94 \times 0.98)R5°$ and (b) $(0.98 \times 0.94)R(-5°)$. Moiré vectors $\underline{R}_{M1}$, $\underline{R}_{M2}$ are labeled accordingly.

characterized by transformation matrices $\underline{\underline{M}}$ according to Eq. (6.45) with

$$\underline{\underline{M}} = \begin{pmatrix} \gamma_1 & 0 \\ 0 & \gamma_2 \end{pmatrix} \tag{6.90}$$

Thus, they can lead to periodic first order moiré patterns with lattice vectors $\underline{R}_{M1}$ and $\underline{R}_{M2}$ defined by Eq. (6.58) with a moiré matrix $\underline{\underline{P}}$ according to Eq. (6.59) given as

$$\underline{\underline{P}} = \begin{pmatrix} \frac{\gamma_1}{1-\gamma_1} & 0 \\ 0 & \frac{\gamma_2}{1-\gamma_2} \end{pmatrix} \tag{6.91}$$

yielding moiré factors λ_i and moiré angles φ_i according to Eqs. (6.83) and (6.84)

$$\lambda_i = \frac{R_{Mi}}{R_{oi}} = \frac{\gamma_i}{|1-\gamma_i|}, \quad \varphi_i = 0°, \quad i = 1, 2 \tag{6.92}$$

Here, the limit $\gamma_1 = 1$, $\gamma_2 \neq 1$, corresponding to a $(1 \times \gamma_2)$ overlayer, leads to a diverging moiré factor λ_1. This means that the moiré lattice vector $\underline{R}_{M1}$ tends to infinity such that the interference along $\underline{R}_{M1}$ disappears while that along $\underline{R}_{M2}$ remains. As a result, the moiré pattern is periodic only in one dimension, exhibiting moirons as parallel infinite stripes (**one-dimensional moirons**) rather than confined regions where the stripes are perpendicular to $\underline{R}_{o2}$. This has been observed in experiments for the gold surface whose reconstruction has been characterized as Au(1 1 1) – ($\sqrt{3} \times 22$)rect [111]. As discussed in Section 5.2, the topmost surface layer experiences a lateral compression such that 23 atoms of the overlayer fit in a cell with 22 atoms of the substrate along $\underline{R}_{o2}$. This leads to a transformation matrix $\underline{\underline{M}}$ and to a moiré matrix $\underline{\underline{P}}$ with

$$\underline{\underline{M}} = \begin{pmatrix} 1 & 0 \\ 0 & \frac{22}{23} \end{pmatrix}, \quad \underline{\underline{P}} = \begin{pmatrix} \infty & 0 \\ 0 & 22 \end{pmatrix}$$

which confirms the observed arrangement of striped moirons, shown in Figure 5.6.

One-dimensional moiré patterns can also be found in overlayer structures where the overlayer lattice is **stretched** with respect to the substrate lattice. The lattice vectors $\underline{R}'_{o1}$, $\underline{R}'_{o2}$ of stretched overlayers are described by a transformation

$$\underline{R}'_{oi} = \underline{R}_{oi} + \gamma \, (\underline{R}_{oi} \, \underline{e}) \, \underline{e}, \quad i = 1, 2 \tag{6.93}$$

where γ is a stretch factor and $\underline{e}$, $\underline{e}'$ are unit vectors with $\underline{e}$ defining the stretch direction and $\underline{e}'$ being perpendicular to $\underline{e}$. Overlayer structures resulting from stretching cannot be described by Wood notation since the angle between $\underline{R}'_{o1}$ and $\underline{R}'_{o2}$ differs in general from that between $\underline{R}_{o1}$ and $\underline{R}_{o2}$. However, transformation (6.93) can be written as

$$\begin{pmatrix} \underline{R}'_{o1} \\ \underline{R}'_{o2} \end{pmatrix} = \underline{\underline{M}} \cdot \begin{pmatrix} \underline{R}_{o1} \\ \underline{R}_{o2} \end{pmatrix} \tag{6.94}$$

with

$$\underline{\underline{M}} = \frac{1}{(x_1 x'_2 - x_2 x'_1)} \begin{pmatrix} (1+\gamma) x_1 x'_2 - x_2 x'_1 & -\gamma x_1 x'_1 \\ \gamma x_2 x'_2 & x_1 x'_2 - (1+\gamma) x_2 x'_1 \end{pmatrix} \tag{6.95}$$

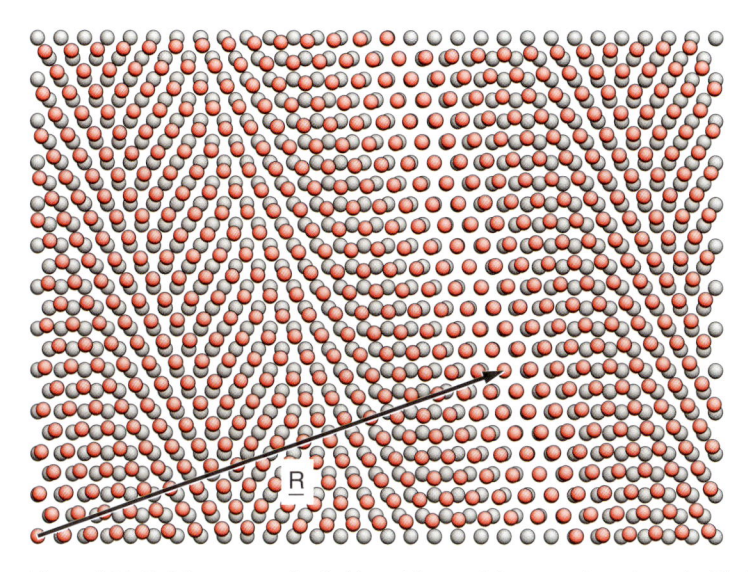

Figure 6.34 Moiré patterns of a fictitious (distorted) hexagonal overlayer (red balls) on hexagonal substrate (gray balls) where the overlayer is stretched by a factor $\gamma = 8/7$ along vector $\underline{R}$ included in the figure.

where

$$x_i = (\underline{R}_{oi}\, \underline{e}), \quad x'_i = (\underline{R}_{oi}\, \underline{e}'), \quad i = 1, 2 \qquad (6.96)$$

After some calculation this leads to

$$\underline{\underline{1}} - \underline{\underline{M}} = \frac{\gamma}{(x_1 x'_2 - x_2 x'_1)} \begin{pmatrix} x_1 & 0 \\ 0 & x_2 \end{pmatrix} \cdot \begin{pmatrix} -x'_2 & x'_1 \\ -x'_2 & x'_1 \end{pmatrix}, \quad \det(\underline{\underline{1}} - \underline{\underline{M}}) = 0 \qquad (6.97)$$

As a result, the moiré matrix $\underline{\underline{P}}$ according to Eq. (6.59) with $\underline{\underline{M}}_I = \underline{\underline{1}}$ becomes singular, thus, excluding a two-dimensional moiré pattern. However, stretching along direction $\underline{e}$ can lead to spatial variations that are perceived as one-dimensional moiré stripes but the detailed analysis is rather complex. As an illustration, Figure 6.34 shows a fictitious (distorted) hexagonal overlayer on hexagonal substrate where the overlayer (red balls) is stretched by a factor $\gamma = 8/7$ along vector $\underline{e} = (2\,\underline{R}_{o1} + \underline{R}_{o2})/\sqrt{7}$ with respect to the substrate layer (gray balls). Thus, the moiron stripes repeat periodically along $\underline{e}$ with a periodicity length given by vector $\underline{R} = 8\,(2\,\underline{R}_{o1} + \underline{R}_{o2})$ while the stripes are not exactly perpendicular to $\underline{e}$.

Moiré lattices are also affected by lateral shifts of the overlayer. A global shift vector $\underline{s}$ translating the otherwise rigid overlayer can always be represented by lattice vectors $\underline{R}_{o1}$ and $\underline{R}_{o2}$ of the substrate, that is, by

$$\underline{s} = x_1\,\underline{R}_{o1} + x_2\,\underline{R}_{o2} = (x_1\ x_2) \begin{pmatrix} \underline{R}_{o1} \\ \underline{R}_{o2} \end{pmatrix} \qquad (6.98)$$

where the shift does not influence the periodicity of the overlayer structure but changes the origin of the corresponding lattice. As a result, the origin of the moiré lattice vectors $\underline{R}^d_{M1}, \underline{R}^d_{M2}$ given by Eq. (6.66) is shifted by

$$\underline{S}_M = x_1 \underline{R}^d_{M1} + x_2 \underline{R}^d_{M2} = (x_1 \ x_2) \begin{pmatrix} \underline{R}^d_{M1} \\ \underline{R}^d_{M2} \end{pmatrix} = (x_1 \ x_2) \ \underline{P}^d \cdot \begin{pmatrix} \underline{R}_{o1} \\ \underline{R}_{o2} \end{pmatrix} \quad (6.99)$$

Since moiré lattice vectors $\underline{R}_{M1}$ and $\underline{R}_{M2}$, in general, do not point along the directions of the substrate lattice vectors $\underline{R}_{o1}$ and $\underline{R}_{o2}$, vectors $\underline{s}$ and $\underline{S}_M$ may also point in different directions. Further, the lengths of the corresponding shift vectors may be quite different. As an illustration, we assume an overlayer structure to represent a combination of isotropic scaling and rotation described as $(\gamma \times \gamma)R\alpha$ in Wood notation. Thus, the lattice vectors of the moiré lattice, $\underline{R}^d_{M1}$ and $\underline{R}^d_{M2}$, are connected with those of the substrate lattice, $\underline{R}_{o1}$ and $\underline{R}_{o2}$, where moiré parameters λ, φ are given by Eq. (6.86). Then the ratio of the shift lengths is given by

$$\left(\frac{S_M}{s}\right)^2 = \frac{x_1^2 \, (R^d_{M1})^2 + x_2^2 \, (R^d_{M2})^2 + x_1 x_2 \, (\underline{R}^d_{M1} \underline{R}^d_{M2})}{x_1^2 \, (R_{01})^2 + x_2^2 \, (R_{02})^2 + x_1 x_2 \, (\underline{R}_{01} \underline{R}_{02})}$$

$$= \frac{x_1^2 \, (\lambda R_{01})^2 + x_2^2 \, (\lambda R_{02})^2 + x_1 x_2 \, (\lambda \underline{R}_{01} \lambda \underline{R}_{02})}{x_1^2 \, (R_{01})^2 + x_2^2 \, (R_{02})^2 + x_1 x_2 \, (\underline{R}_{01} \underline{R}_{02})} = \lambda^2 \quad (6.100)$$

and the angle between the shift vectors $\underline{s}$ and $\underline{S}_M$ yields with Eq. (6.86) after some calculus

$$\cos(\angle (\underline{S}_M, \underline{s})) = \frac{(\underline{S}_M \, \underline{s})}{S_M \, s} = \cos(\varphi) \quad (6.101)$$

Thus, for $(\gamma \times \gamma)R\alpha$ overlayer structures the shift vectors of the moiré lattice and of the substrate lattice, $\underline{S}_M$ and $\underline{s}$, respectively, are connected by the same relationships (6.86) as the lattice vectors themselves. This is evident from Figure 6.35, which shows shift directions of the overlayer and of

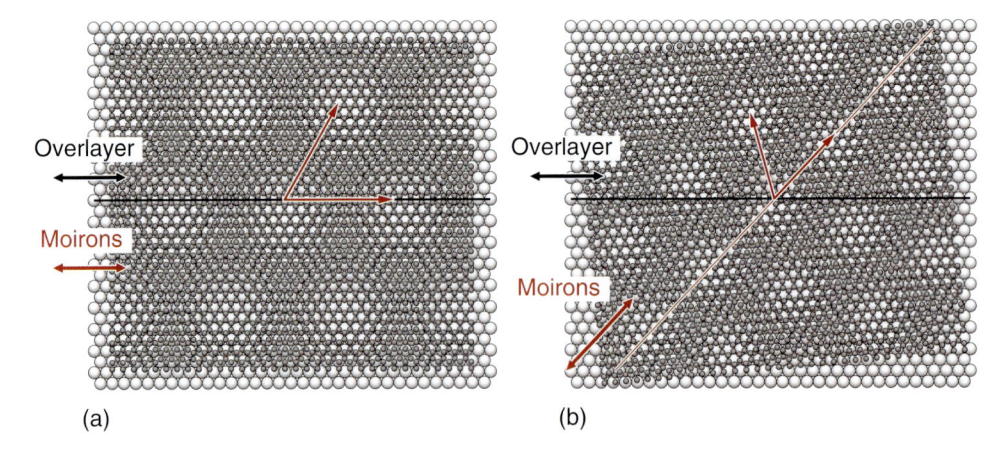

Figure 6.35 Graphene overlayer adsorbed on the Ir(1 1 1) surface for structures (a) Ir(1 1 1) + (0.906 × 0.906) − Gra and (b) Ir(1 1 1) + (0.906 × 0.906)R5° − Gra (fictitious). The moiré vectors $\underline{R}_{M1}$, $\underline{R}_{M2}$ are shown in red. Shift directions of the overlayer and of the moiré lattice are denoted by black and red double headed arrows

the moiré lattice for two overlayer structures. In Figure 6.35a, referring to Ir(1 1 1) + (0.906 × 0.906)R0° − Gra [159], the graphene overlayer is scaled but not rotated such that the moiré lattice vectors are parallel to those of the iridium substrate. Thus, the moiré lattice and the overlayer both shift in parallel, that is, horizontally in Figure 6.35a. In Figure 6.35b, referring to (fictitious) Ir(1 1 1) + (0.906 × 0.906)R5° − Gra, the graphene overlayer is scaled and rotated leading to moiré lattice vectors that point in directions different from those of the substrate lattice vectors. Therefore, the moiré lattice shifts in a diagonal direction while the overlayer shifts horizontally. However, in both cases the moiron shifts are much larger than the corresponding overlayer shifts, where according to Eq. (6.100) amplification factors $S_M/s = 9.61$ (without rotation) and 7.22 (with rotation) are found.

Moiré lattices of shifted overlayers can be observed at stepped metal surfaces. As an illustration, Figure 6.36 shows the simulation of a moiré pattern for graphene spreading over a monoatomic step on the Ir(1 1 1) surface [155, 166]. Here, the Ir atoms at adjacent terraces experience a lateral shift of $-\underline{s} = (\underline{R}_{o1} - 2\,\underline{R}_{o2})/3$ (shown by a red arrow denoted "$-\underline{s}$" in the figure inset at the bottom right), corresponding to a relative shift $\underline{s}$ of the graphene layer. This results, according to Eqs. (6.98) and (6.99), in a parallel shift $\underline{S}_M = (\underline{R}_{M1} - 2\,\underline{R}_{M2})/3$ (shown by a red arrow denoted "$\underline{S}_M$" in the figure) of the moiron arrangement between the terraces.

6.6
Symmetry and Domain Formation

As pointed out in Section 6.1, structure concepts discussed for the topmost layers of relaxed or reconstructed surfaces of real crystals can be used analogously to characterize adsorbate layers at surfaces. This applies in particular to **symmetry** properties. In the case of reconstructed or adsorbate layers at surfaces – both will be called *overlayers* in the following – those with **commensurate superlattices** are expected to exhibit **highest symmetry**, where **two aspects** are important. **First**, the overlayer and the substrate surface share a common periodicity (i.e., translational symmetry), that of the superlattice, with lattice vectors $\underline{R}'_1$, $\underline{R}'_2$. These lattice vectors are given by integer-valued combinations of the two-dimensional lattice vectors of the substrate surface, $\underline{R}_1$ and $\underline{R}_2$, according to Eq. (5.3). Further, the area of the superlattice unit cell is an integer multiple of that of the substrate surface. **Second**, point symmetry elements, common to the overlayer and the substrate surface, will combine to form the two-dimensional space group of the superlattice representing the joint overlayer/substrate surface. This space group may be identical to that of the overlayer (and can never include additional symmetry elements). However, it can also represent lower symmetry compared with that of each subsystem given by the two corresponding space groups.

Formally, the combination of two-dimensional space groups of overlayer and substrate lattices can be based on the notion that all **two-dimensional** symmetry

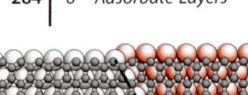

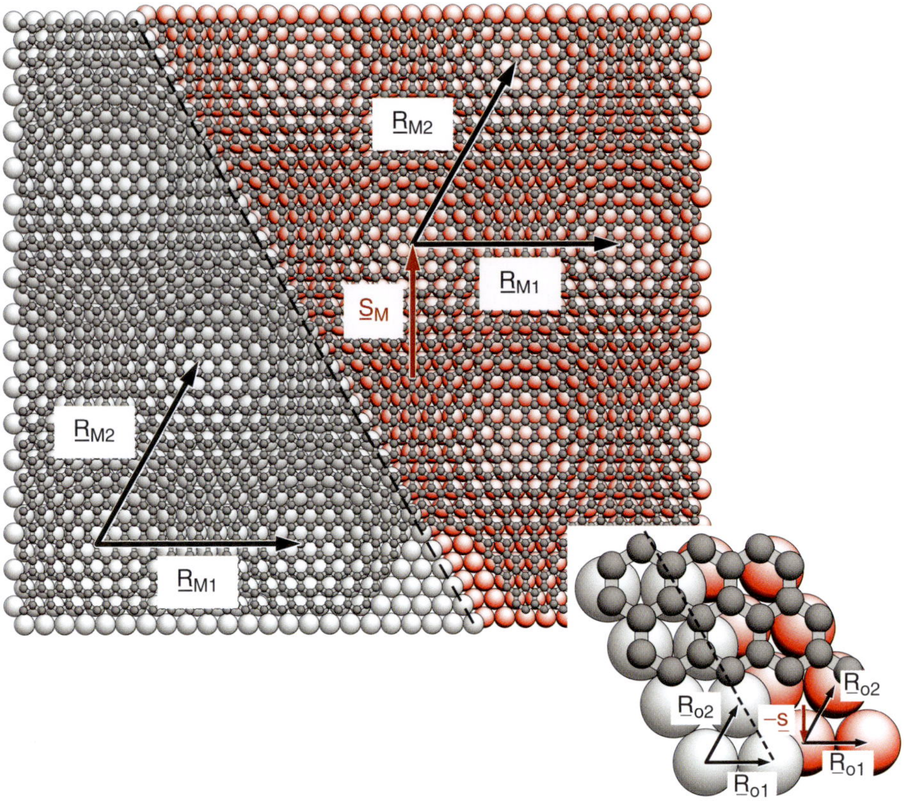

Figure 6.36 Graphene overlayer adsorbed at a stepped Ir(1 1 1) surface. Adjacent terraces are shown in red (lower terrace) and light gray (upper terrace) with moiré vectors $\underline{R}_{M1}$, $\underline{R}_{M2}$ included. The inset at the bottom right shows an enlarged local area near the step edge with lattice vectors $\underline{R}_{o1}$, $\underline{R}_{o2}$ at the two terraces. The vector labeled "$-\underline{s}$" denotes the lateral atom shift between upper and lower terrace.

elements appearing in an atom layer can be translated into **three-dimensional** ones at the surface. Thus, the rotation axes in two-dimensional atom layers reflect rotation axes pointing perpendicular to the layers in the three-dimensional space. Likewise, mirror lines in two-dimensional layers correspond to mirror planes perpendicular to the layers in three dimensions. Finally, glide lines in two-dimensional layers translate to glide planes perpendicular to the layers in three dimensions. Then the **joint** two-dimensional **space group** of the overlayer and substrate layers **collects** the two-dimensional equivalents of all those three-dimensional symmetry elements that are shared by both layers.

As an example, we consider the adsorption of oxygen on the (1 1 0) surface of fcc rhodium, for which experimental studies [167], SSD 45.8.7, have shown an Rh(1 1 0) + p2mg(2×1) − 2O overlayer structure shown in Figure 6.37. Here, oxygen atoms are placed at tilted bridge sites along the topmost rows of Rh atoms

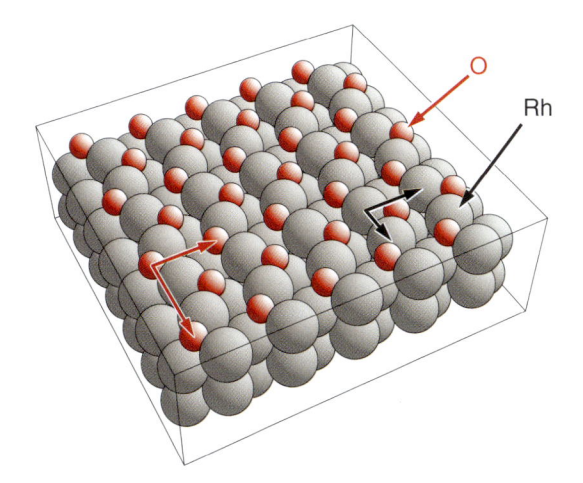

Figure 6.37 Structure of the Rh(1 1 0) + p2mg(2 × 1) - 2O adsorbate system. O adsorbate and Rh substrate atoms are shown in red and gray. Lattice vectors of the adsorbate layer and substrate are shown in red and black.

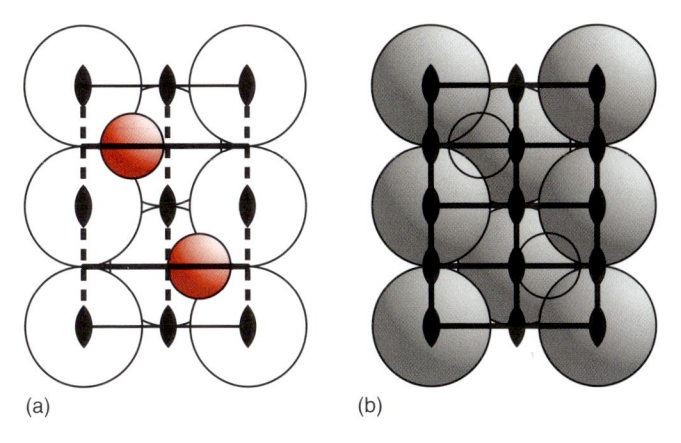

Figure 6.38 Surface symmetry elements of the Rh(1 1 0) + p2mg(2 × 1) - 2O adsorbate system. (a) O adsorbate symmetry elements, space group p2mg. (b) Rh(1 1 0) substrate symmetry elements, space group p2mm. The sketch includes two unit cells of the substrate to reflect the adsorbate periodicity. All symmetry elements are denoted according to ITC (International Tables of Crystallography).

in an alternating fashion (left and right tilt). As a result, the unit cell of the oxygen overlayer, containing two oxygen atoms, is rectangular with lattice vectors along and perpendicular to the substrate atom rows, where $\underline{R}'_1 = 2\,\underline{R}_1$ and $\underline{R}'_2 = \underline{R}_2$ define the superlattice. Further, the point symmetry elements of the overlayer are those of space group p2mg sketched in Figure 6.38a. In contrast, point symmetry elements of the substrate lattice, shown in Figure 6.38b, form the space group p2mm (for group notations, see Section 3.8.6), where the unit cell area is half that of the superlattice cell. (The sketch of Figure 6.38a includes two unit cells of the substrate

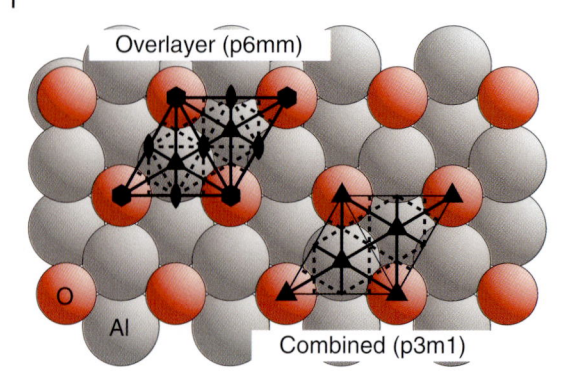

Figure 6.39 Structure of the Al(1 1 1) + (1 × 1) - O adsorbate system. The figure includes symmetry patterns of space group p6mm of the separate overlayer and substrate (top left) as well as of space group p3m1 of the combined overlayer/substrate surface (bottom right). All symmetry elements are denoted according to ITC.

to reflect the superlattice periodicity.) All twofold rotation centers of the oxygen overlayer as well as the two mirror lines also appear in the symmetry sketch of the substrate. Further, the glide lines of the overlayer with periodicity vectors $\underline{R}'_1 = 2\,\underline{R}_1$ are consistent with the parallel mirror lines of the substrate along $\underline{R}_1$. Thus, all symmetry elements of space group p2mg exist in both subsystems and this space group describes all point symmetry elements of the combined overlayer/substrate surface. Therefore, in this example the space group of the **overlayer** happens to be **identical** to that of the **combined system**.

An example where the space group of the **combined** system represents **lower symmetry** compared with that of the **overlayer** superlattice is the Al(1 1 1) + (1 × 1) − O adsorbate system [168], SSD 13.8.19, shown in Figure 6.39. Here, the unit cell of the separate oxygen overlayer includes point symmetry elements described by the hexagonal space group p6mm, see upper left pattern of Figure 6.39, which is identical to the symmetry of the unit cells of each underlying aluminum substrate layer. However, the substrate layer cells are **shifted** laterally with respect to those of the adsorbate overlayer. As a result, there are **fewer** point symmetry elements in the combined overlayer/substrate system than those in the adsorbate and substrate layers themselves and the symmetry of the combined system is described by space group p3m1, see lower right pattern of Figure 6.39.

Table 6.2 gives an overview of **allowed space groups** for all **combinations** of commensurate overlayer and substrate lattices sharing point symmetry elements (including glide lines). Here, the **rows** denote the two-dimensional **substrate lattice** at the surface and **columns** refer to space groups of the **combined** overlayer/substrate **superlattice.** The table shows that in numerous cases the symmetry of the substrate lattice allows **different** superlattice symmetries depending on the lattice vectors $\underline{R}'_1, \underline{R}'_2$ of the superlattice. A full analysis of compatible overlayer symmetries can be obtained by an interactive computational tool, **LEEDpat** [169], which allows finding all possible overlayer space groups for a given commensurate superlattice and a substrate space group.

6.6 Symmetry and Domain Formation | 287

Table 6.2 Allowed space groups of combined overlayer/substrate superlattices.

(a) Space groups no. 1–2 (oblique), 3, 4, 6–8 (p-rectangular), 5, 9 (c-rectangular) of the combined overlayer/substrate superlattice

Substrate lattice	Combined overlayer/substrate superlattice								
	(1) p1	(2) p2	(3) p1m1	(4) p1g1	(5) c1m1	(6) p2mm	(7) p2mg	(8) p2gg	(9) c2mm
(1) p1	any	—	—	—	—	—	—	—	—
(2) p2	any	any	—	—	—	—	—	—	—
(3) p1m1, m = 0	any	—	0	0	0	—	—	—	—
(4) p1g1, g = 0	any	—	—	0	0	—	—	—	—
(5) c1m1, m = 0	any	—	0	0	0	—	—	—	—
(6) p2mm, m = 0, 90	any	any	0, 90	0, 90	0, 90	0, 90	0, 90	0, 90	0, 90
(7) p2mg, m = 0, g = 90	any	any	0	0, 90	0	—	m = 0, g = 90	0, 90	—
(8) p2gg, g = 0, 90	any	any	—	0, 90	—	—	—	0, 90	—
(9) c2mm, m = 0, 90	any	any	0, 90	0, 90	0, 90	0, 90	0, 90	0, 90	0, 90
(10) p4	any	any	—	—	—	—	—	—	—
(11) p4mm, m = 0, 45	any	any	0, 45	0, 45	0, 45	0, 45	0, 45	0, 45	0, 45
(12) p4gm, g = 0	any	any	45	0, 45	45	45	45	0, 45	45
(13) p3	any	—	—	—	—	—	—	—	—
(14) p3m1, m = 90	any	—	90	90	90	—	—	—	—
(15) p31m, m = 0	any	—	0	0	0	—	—	—	—
(16) p6	any	any	—	—	—	—	—	—	—
(17) p6mm, m = 0, 90	any	any	0, 90	0, 90	0, 90	0, 90	0, 90	0, 90	0, 90

(continued overleaf)

Table 6.2 (Continued).

(b) Space groups no. 10–12 (square), 13–17 (hexagonal) of the combined overlayer/substrate superlattice

Substrate lattice	Combined overlayer/substrate superlattice							
	(10) p4	(11) p4mm	(12) p4gm	(13) p3	(14) p3m1	(15) p31m	(16) p6	(17) p6mm
(1) p1	—	—	—	—	—	—	—	—
(2) p2	—	—	—	—	—	—	—	—
(3) p1m1, m = 0	—	—	—	—	—	—	—	—
(4) p1g1, g = 0	—	—	—	—	—	—	—	—
(5) c1m1, m = 0	—	—	—	—	—	—	—	—
(6) p2mm, m = 0, 90	—	—	—	—	—	—	—	—
(7) p2mg, m = 0, g = 90	—	—	—	—	—	—	—	—
(8) p2gg, g = 0, 90	—	—	—	—	—	—	—	—
(9) c2mm, m = 0, 90	—	—	—	—	—	—	—	—
(10) p4	any	—	—	—	—	—	—	—
(11) p4mm, m = 0, 45	any	0, 45	0, 45	—	—	—	—	—
(12) p4gm, g = 0	any	—	0	—	—	—	—	—
(13) p3	—	—	—	any	—	—	—	—
(14) p3m1, m = 90	—	—	—	any	0	90	—	—
(15) p31m, m = 0	—	—	—	any	90	0	—	—
(16) p6	—	—	—	any	—	—	any	—
(17) p6mm, m = 0, 90	—	—	—	any	0, 90	0, 90	any	0, 90

In the tables, lattice vectors $\underline{R}_1$, $\underline{R}_2$ of the substrate are oriented such that $\underline{R}_1$ points along the horizontal x axis, except for centered rectangular lattices where $(\underline{R}_1 + \underline{R}_2)$ is assumed to point along x, see Figure 3.49. The table entries "any", 0, 45, and 90 denote allowed azimuthal angles (in degrees) of mirror (m) and glide (g) lines or planes relative to the substrate lattice while "—" refers to incompatible symmetries. Symmetry degenerate orientations, such as 120° rotations for hexagonal lattices, are not included in the list. The numbering sequence follows the scheme used in the ITC [33].

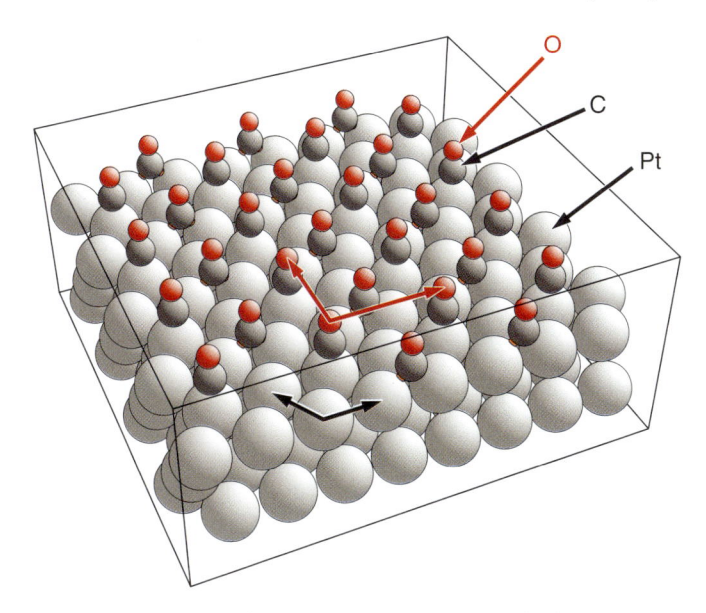

Figure 6.40 Structure of the Pt(1 1 1) + c(4×2) - 2CO adsorbate system in a perspective view. The periodicity vectors of the CO adsorbate layer and of the Pt substrate are shown in red and black, respectively.

In addition to the compatibility rules given in Table 6.2, **overlayers** of the same structure can be **oriented differently** with respect to a substrate surface with symmetry, where the resulting overlayer/substrate systems are energetically degenerate. At real crystal surfaces this degeneracy gives rise to **domain formation** where finite patches of the differently oriented overlayers coexist and can be observed by diffraction or imaging experiments [34, 94]. As an example, we consider the adsorption of carbon monoxide on the (1 1 1) surface of fcc platinum, where experimental studies [144], SSD 78.6.8.4, have yielded a Pt(1 1 1) + c(4×2) − 2CO overlayer structure, as shown by a perspective view in Figure 6.40. Here, CO molecules are placed at top and bridge sites above the hexagonal Pt surface layer, where the unit cell of the CO overlayer, containing two molecules, is rectangular with lattice vectors $\underline{R}'_1 = 2\underline{R}_1$ and $\underline{R}'_2 = \underline{R}_1 + 2\underline{R}_2$ defining the superlattice. The point symmetry elements of the overlayer are those of space group p2mm while the substrate symmetry is described by p3m1. This yields a combined two-dimensional symmetry pattern represented by space group p1m1. The symmetry is more evident in the view normal to the surface shown in Figure 6.41. This figure shows the three energetically degenerate orientations of CO overlayers which are rotated by 120° with respect to each other as a consequence of the threefold rotational symmetry of the substrate lattice. Thus, the substrate symmetry gives rise to three different **rotational domains** on the real surface.

Rotational domains of overlayers can appear whenever the substrate surface symmetry includes a rotation axis that is not shared by the overlayer. Domain formation can also be found for other symmetry elements of the substrate surface. A mirror plane of the substrate can induce two **mirror domains** of overlayers

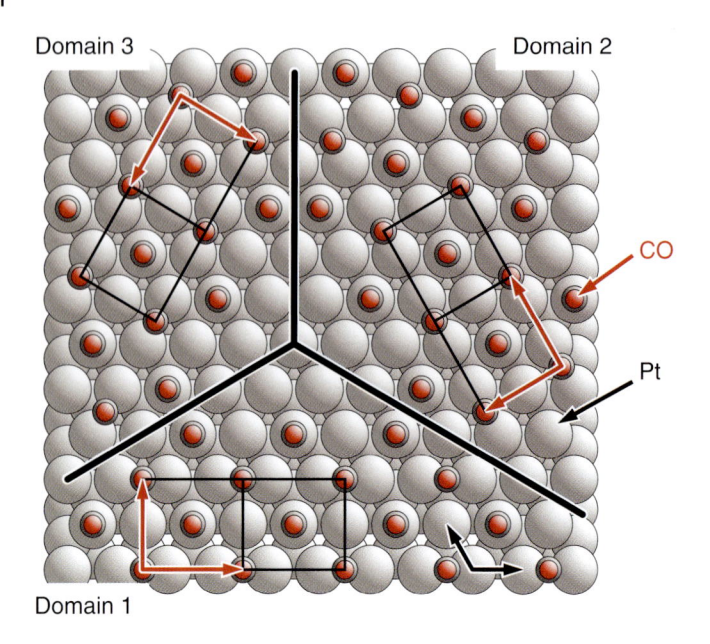

Figure 6.41 Structure of rotational domains of the Pt(1 1 1) + c(4 × 2) - 2CO adsorbate system with the three equivalent rotational domains. The domains are separated by black lines.

as suggested by experiments for the Ni(1 1 0) + c(2 × 2) – CN adsorbate system [170], SSD 28.6.7.2, shown in Figure 6.42. Here, CN molecules are found to adsorb as tilted species at bridging sites along the topmost rows of Ni atoms with the tilt pointing always in the same direction laterally. This offers a second, energetically degenerate, geometric configuration where the tilt occurs to the other side, which can give rise to two mirror domains. Actually, in the present system the two domains can be connected by a twofold rotation perpendicular to the surface and can, therefore, also be considered as twofold rotational domains.

Further, a glide line of the substrate surface can induce two **glide line domains** of overlayers that may also appear for the Ni(1 1 0) + c(2 × 2) – CN adsorbate system [170] as illustrated in Figure 6.43. Here, the glide line operation creates a second, energetically degenerate structure of the CN adsorbate with its tilt pointing to the other side with respect to the surface normal. This is combined with a shift by half an overlayer lattice vector along the glide line indicated by a red arrow in Figure 6.43.

In addition to point symmetry elements of the substrate surface its translational symmetry can also induce different domains, so-called **translational domains**. As an example, we consider the adsorption of atomic hydrogen on the (1 1 0) surface of rhodium where experiments found a Rh(1 1 0) + (1 × 3) – H adsorbate structure [171], SSD 45.1.5, as illustrated by Figure 6.44. Here, hydrogen atoms adsorb in threefold sites at the slopes of the (1 1 0) troughs of the rhodium surface. The hydrogen rows can be moved laterally in their perpendicular direction by a lattice

6.6 Symmetry and Domain Formation | 291

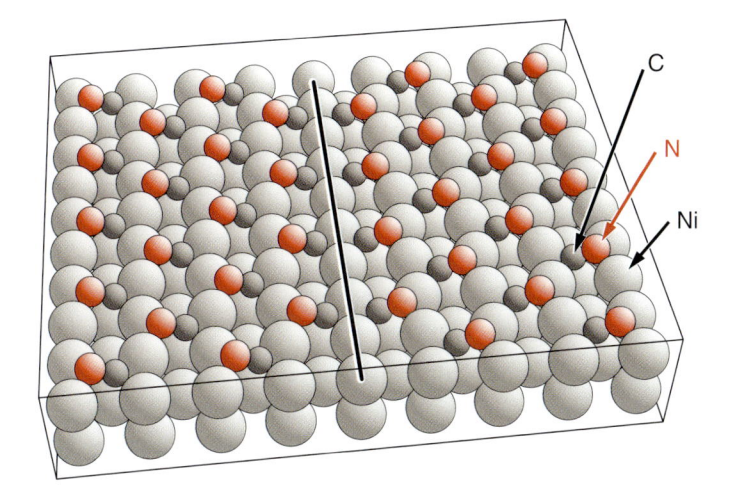

Figure 6.42 Structure of mirror domains of the Ni(1 1 0) + c(2 × 2) - CN adsorbate system. The two domains are separated by a black line.

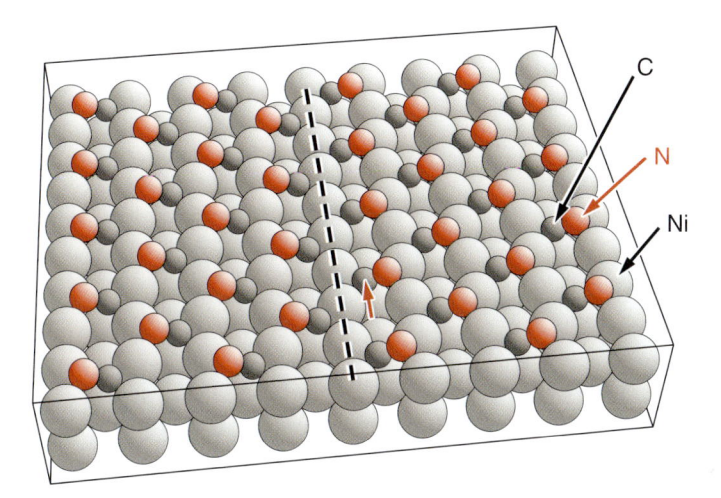

Figure 6.43 Structure of glide line domains of the Ni(1 1 0) + c(2 × 2) - CN adsorbate system. The domains are separated by a black dashed line. The shift vector along the glide line is indicated by a red arrow.

vector of the substrate lattice, indicated by a red arrow in Figure 6.44, to yield altogether three energetically degenerate configurations. This results in three different translational domains.

Another example of translational domains is given by the adsorption of atomic oxygen on the (1 1 0) surface of silver, where experiments show a Ag(1 1 0) + (2 × 1) − O adsorbate structure [172], as illustrated by Figure 6.45. Here, oxygen atoms adsorb between silver atoms of the topmost rows of the (2 × 1) reconstructed substrate forming rows of alternating Ag and O atoms. These rows

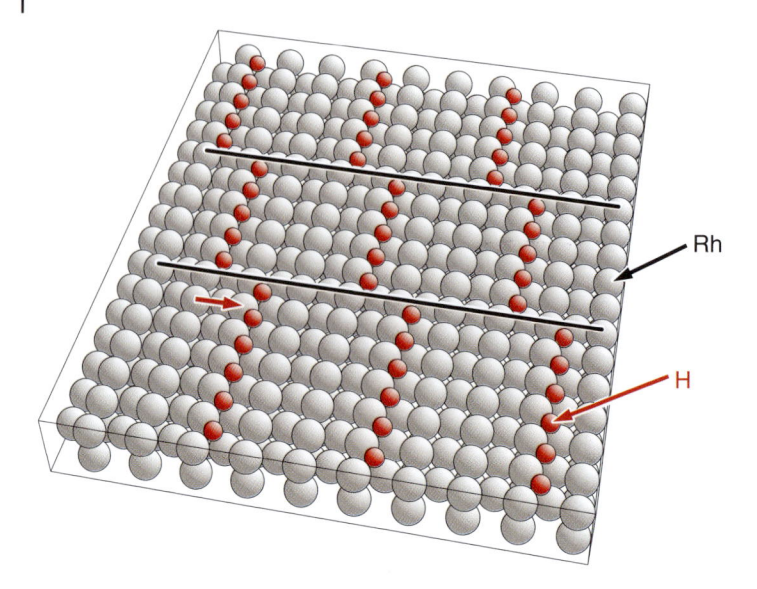

Figure 6.44 Structure of translational domains of the Rh(1 1 0) + (1 × 3) - H adsorbate system. The domains are separated by a black line with the shift vector indicated in red.

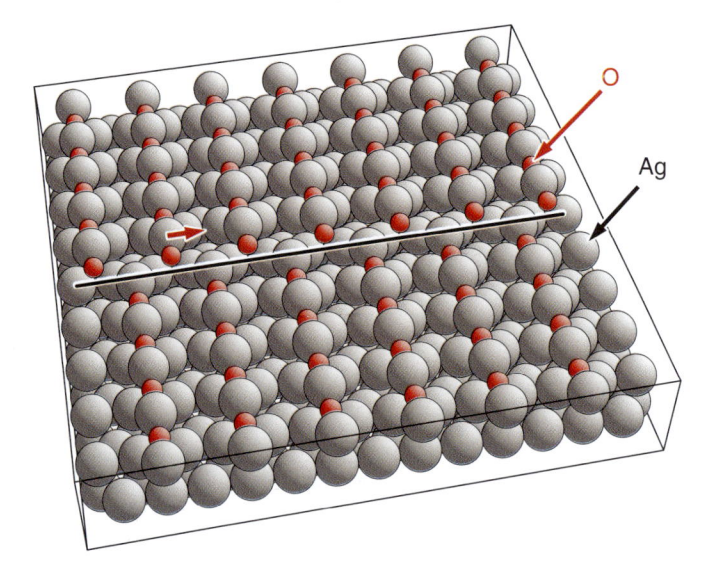

Figure 6.45 Structure of translational (antiphase) domains of the Ag(1 1 0) + (2 × 1) - O adsorbate system. The domains are separated by a black line with the shift vector indicated in red.

may be shifted laterally in their perpendicular direction by a lattice vector of the substrate lattice, illustrated by a red arrow in Figure 6.45, which yields an energetically degenerate configuration. The resulting two structures form different translational domains. While the two domains are completely equivalent the lateral shift between them, which reflects a structural phase inversion, can be observed in electron scattering experiments [34, 94] since it gives rise to scattering phase changes between adjacent domains. Therefore, these translational domains are sometimes also called **anti-phase domains**.

6.7
Adsorption at Surfaces and Chirality

Structural aspects of adsorption involving chiral adsorbates as well as chiral substrate surfaces have attracted wide scientific interest because these systems have been found to show exciting physical and chemical properties which are also important for practical applications. Examples are enantioselective catalytic reactions that are used by the chemical industry to produce different drugs. In many cases, chiral adsorption systems refer to organic molecular adsorbates and/or kinked metal surfaces [173–175] where a large variety of structural elements have been observed. These can be rather complex and have to be studied on a case-by-case basis, which goes far beyond the scope of this book. Therefore, this section focuses only on a few general guidelines and basic results that can be illustrated by simple examples.

The chirality of adsorbate systems can be classified into four different groups depending on the chirality of the overlayer and the substrate taken separately. Achiral or chiral overlayers may combine with achiral or chiral substrate surfaces. Here, achiral overlayers on achiral substrate represent the simplest group while chiral overlayers on chiral substrate are the most complex.

The adsorption of **single achiral** molecules on an **achiral** surface can create either an achiral or a chiral adsorbate system depending on the adsorption site and on the orientation of the adsorbate at the surface. If the substrate and the adsorbate share a mirror plane perpendicular to the surface the combined adsorbate system is achiral. This is illustrated in Figure 6.46 for ammonia adsorption on the Cu(1 1 1) surface where experiments for Cu(1 1 1) + (disordered) $-$ NH_3 [124], SSD 29.7.1.3, yield at rather low coverage a disordered overlayer structure with NH_3 stabilizing with its nitrogen end at top sites of the metal substrate while positions of the hydrogen centers could not be determined. In Figure 6.46a it is assumed that the NH_3 is structurally analogous to the free molecule with its threefold rotation axis perpendicular to the surface. The three hydrogen centers are rotated such that one of the three molecular mirror planes, indicated by a red line in Figure 6.46a, coincides with a substrate mirror plane, parallel to the horizontal black line. As a consequence, a mirror operation with respect to the common mirror plane reproduces the adsorbate system and the system is **achiral**. It should be mentioned in passing that the adsorption of single atoms at mirror-symmetry sites of an achiral surface will always result in an achiral adsorbate system whereas other adsorption sites of the atom result in chirality.

In Figure 6.46b the upper half shows the NH_3 adsorbate at the top site where, the three hydrogen centers are rotated about the molecular rotation axis anticlockwise by 15°, indicated by the label S-NH_3. As a consequence, the adsorbate does not share a mirror plane with the substrate surface. This is clear from comparing the

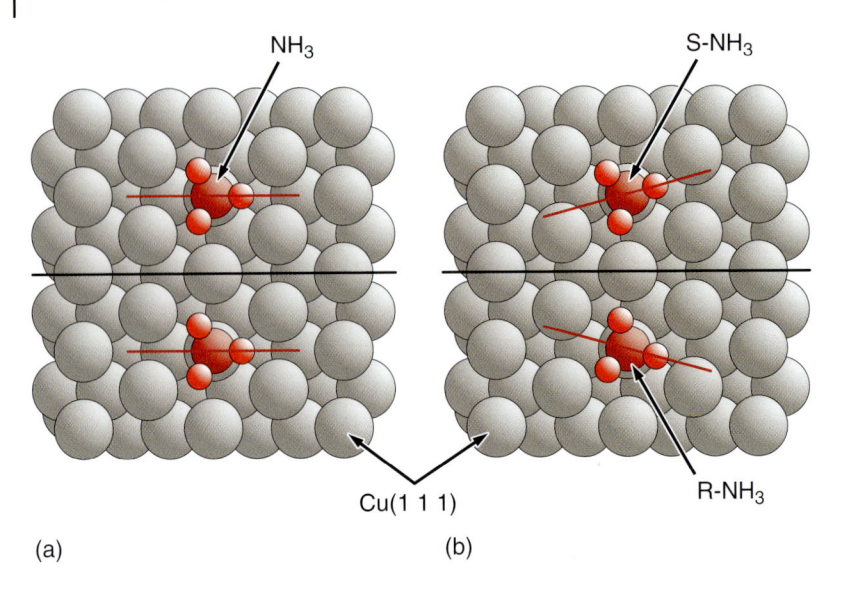

Figure 6.46 Structure of the Cu(1 1 1) + (disordered) - NH_3 adsorbate system with ammonia in different orientation, (a) achiral and (b) chiral. The upper and lower parts of the figures show the adsorbate and its mirror image. The black line indicates a mirror plane perpendicular to the substrate surface while the red lines refer to mirror planes of the separate NH_3 adsorbate.

upper red line in Figure 6.46b, denoting a molecular mirror plane, with the horizontal black line referring to the closest mirror plane of the substrate. Thus, applying a mirror operation with respect to the mirror plane of the substrate leads to an adsorbate structure shown in the lower half of Figure 6.46b with ammonia labeled R-NH_3. The two adsorbate structures are different and cannot be brought into coincidence by rotation about the surface normal and/or translation. Thus, they are chiral partners and the adsorbate system is **chiral**. The structural difference between the two enantiomeric forms may be emphasized further by relaxation of the substrate and lifting of its mirror symmetry due to the presence of the asymmetrically positioned adsorbate. While the two enantiomers are energetically equivalent there may be an energy barrier between them, which hinders easy transformation such that the two species become stable and can, at higher coverages, exist in well-separated domains. For the present Cu(1 1 1) + NH_3 adsorbate system, this is difficult to observe and one would assume an achiral structure to be preferred since the asymmetric adsorbate structure may not reflect an equilibrium. However, more complex achiral molecules are expected to yield adsorption with two enantiomeric structures that are well separated by energetic barriers and can be clearly identified.

So far, chirality was discussed as a local phenomenon that occurs near the adsorbate site (**local chirality**). Chiral structures can also be formed by atoms or achiral molecules arranged as chiral clusters or islands at achiral surfaces, which can be considered as **cooperative chirality**. This is illustrated in Figure 6.47, which

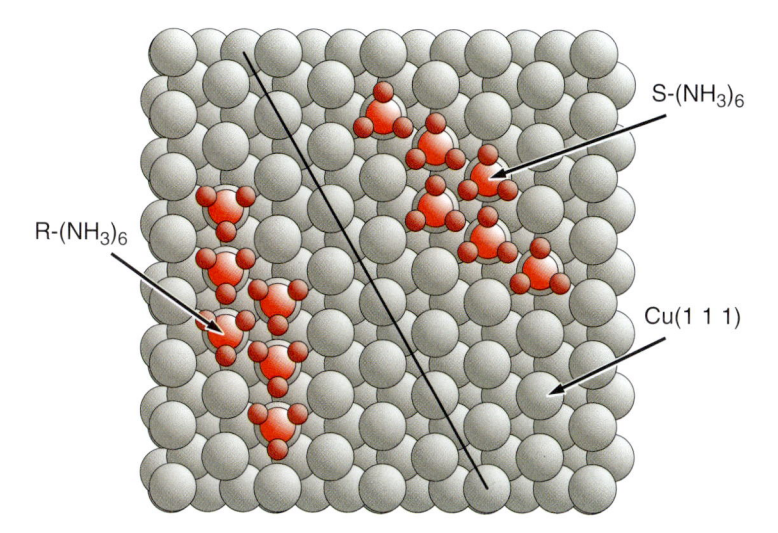

Figure 6.47 Structure of the two enantiomers of a cooperative chiral arrangement of NH_3 adsorbates on the Cu(1 1 1) surface, denoted by R-$(NH_3)_6$ and S-$(NH_3)_6$, in a view along the surface normal. The black line indicates a mirror plane perpendicular to the substrate surface.

shows a fictitious chiral arrangement of six NH_3 adsorbates on the Cu(1 1 1) surface together with its enantiomeric image where the two structures are denoted R-$(NH_3)_6$ and S-$(NH_3)_6$.

Cooperative chirality at achiral surfaces can be obtained with both chiral and achiral adsorbates and is found for large molecular networks assembling at metal surfaces. A fairly complex example is rubrene adsorbed on the Au(1 1 1) surface [176]. This aromatic molecule, $C_{42}H_{28}$, combining tetracene with four phenyl rings, is chiral and can form large symmetric flower-like supramolecular structures with up to 150 molecules on the Au(1 1 1) surface where both left- and right-handed structures have been identified [176].

The adsorption of **single chiral** molecules on an **achiral** surface will always create a chiral adsorbate system with energetically equal or different enantiomeric structures. As an example, the adsorption of tartaric acid (TA), $C_4O_6H_6$, on the Cu(1 1 0) surface has been studied experimentally in great detail [174, 177, 178]. As discussed in Section 4.4, the free TA molecule contains two chiral carbon centers and forms two enantiomers, right-handed (R,R)-TA and left-handed (S,S)-TA, see Figure 4.19. Upon adsorption, the TA molecule is deprotonated thereby losing hydrogen from its opposite COOH ends, and the resulting tartrate, $C_4O_6H_4$, is distorted, bridging between the dense copper atom rows on the (1 1 0) surface with its two COO ends binding with copper along the rows. This is illustrated in Figure 6.48 where the two tartrate adsorbates are labeled according to their free TA molecule enantiomers, (R,R)-TA and (S,S)-TA. As a consequence of the substrate symmetry, the two species are expected to be energetically equivalent if they bind in symmetrically equivalent sites.

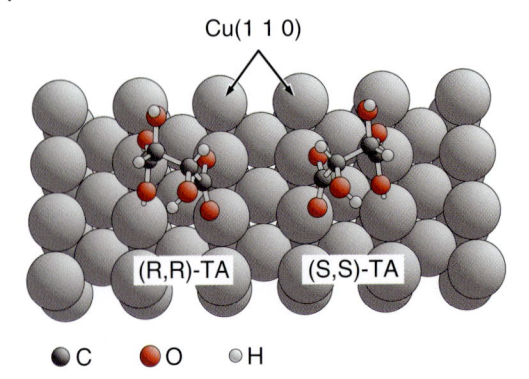

Figure 6.48 Structure of the two enantiomers of tartrate, (R,R)-TA (left) and (S,S)-TA (right), adsorbed on the Cu(1 1 0) surface. The adsorbate atoms are color coded as explained at the bottom.

At higher adsorbate coverage, chiral adsorbates can form ordered overlayer structures with different two-dimensional periodicity in larger domains. As an illustration, Figure 6.49 shows the observed structures [177, 178] of **enantiopure** overlayers of the two chiral tartrate species (R,R)-TA and (S,S)-TA, on the Cu(1 1 0) surface. They refer to an adsorbate coverage $\Theta = 16.7\%$ (1/6) and can be described as Cu(1 1 0) + (9 0 | 1 2) − 3(R,R)-TA and Cu(1 1 0) + (9 0 | −1 2) − 3(S,S)-TA in 2×2 matrix notation. When mixtures of left- and right-handed tartrate are adsorbed on the Cu(1 1 0) surface the two species are found to form well-separated enantiopure domains [178] with the geometric structures as shown in Figure 6.49. However, enantiomers may also mix at the surface to produce two-dimensionally periodic or randomly structured overlayers. The latter seems to occur for the adsorption of 1 : 1 (i.e., **racemic**) mixtures of left- and right-handed tartrate on the Cu(1 1 0) substrate where some patches suggest disorder [178].

Figure 6.48 illustrates further that the combined adsorption of the two (R,R)-TA and (S,S)-TA enantiomers on the Cu(1 1 0) surface leads to an achiral adsorbate pair and, thus, to an altogether achiral adsorbate system. This is a more general result that can be associated with **cooperative achirality** where clusters of chiral adsorbates with different chirality on an achiral substrate can result in an achiral adsorbate system.

The adsorption of single atoms or **single achiral** molecules on a **chiral** surface will always create a chiral system independent of the actual adsorption site or orientation of the adsorbate at the surface. Here, the resulting chirality is determined by that of the substrate. However, at higher coverage, achiral adsorbates may arrange as **chiral clusters** or islands reflecting cooperative chirality as discussed earlier. At chiral surfaces this leads to four different chiral conformations where the substrate/adsorbate combinations with equal handedness, $Sub^R + R\text{-Ads}$ and $Sub^S + S\text{-Ads}$, form enantiomers as do combinations with opposite handedness,

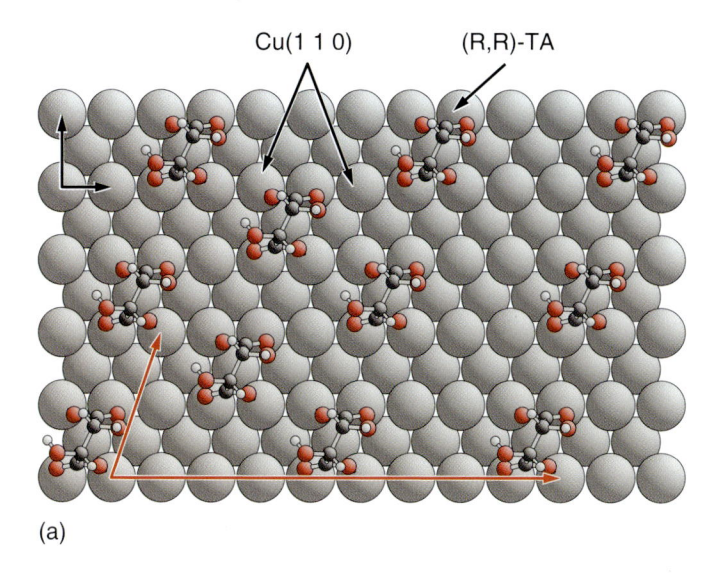

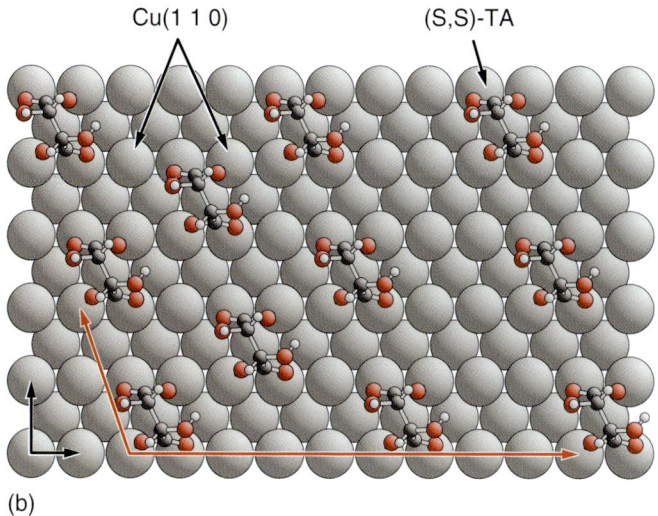

Figure 6.49 Structure of enantiopure tartrate overlayers on the Cu(1 1 0) surface, (a) Cu(1 1 0) + (R,R) - TA and (b) Cu(1 1 0) + (S,S) - TA, see text. The color coding of the adsorbate atoms is identical to that used in Figure 6.48.

Sub^R + S-Ads and Sub^S + R-Ads. As an illustration, Figure 6.50 shows kinked chiral iron surfaces described as left-handed Fe(25 20 3)^S and right-handed Fe(−25 −20 −3)^R, which form substrate enantiomers. These surfaces are both covered by two chiral islands of nine cobalt atoms each, denoted R-Co$_9$ and S-Co$_9$, which are island enantiomers. Transitions between the two island types of these fictitious systems can be achieved by surface diffusion of cobalt along the substrate terraces, oriented (1 1 0) for Fe(25 20 3)^S and (−1 −1 0) for Fe(−25 −20 −3)^R.

Fe(25 20 3)^S   **Fe(−25 −20 −3)^R**

(a)   (b)

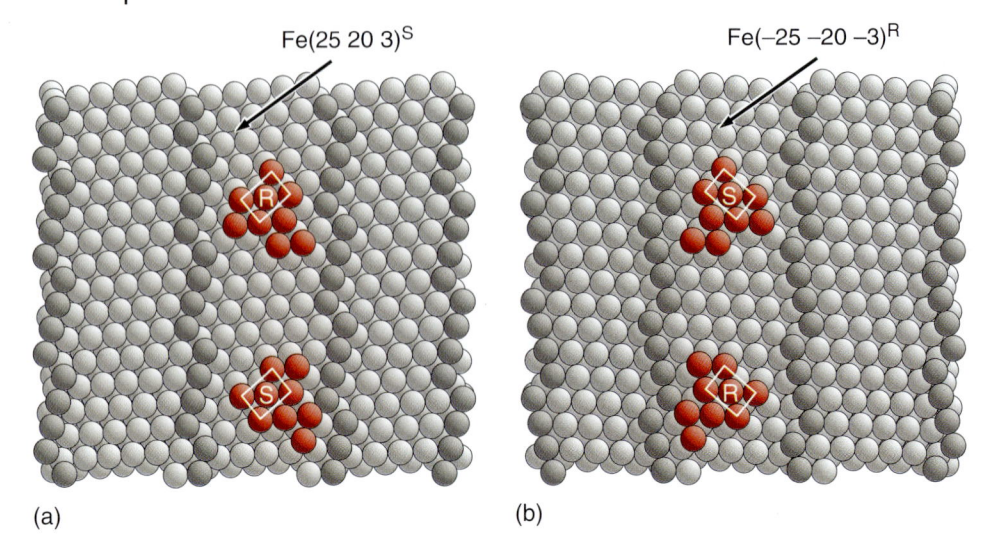

Figure 6.50 Structure of chiral cobalt islands, R-Co$_9$ and S-Co$_9$, at a kinked chiral iron surface, (a) left-handed Fe(25 20 3)^S and (b) right-handed Fe(−25 −20 −3)^R. Adsorbate (substrate) atoms are shown in red (gray) with those along the kink lines emphasized in dark gray. The islands are denoted R for right- and S for left-handed and their orientations are outlined by white rectangles.

As a result of differences in the adsorbate sites and binding, the two island enantiomers at the left-handed substrate surface, Figure 6.50a, are energetically different, which also holds for the right-handed substrate surface, Figure 6.50b. However, for symmetry reasons the S-Co$_9$ island at Fe(25 20 3)^S and the R-Co$_9$ island at Fe(−25 −20 −3)^R are energetically equivalent, which is also true for R-Co$_9$ at Fe(25 20 3)^S and S-Co$_9$ at Fe(−25 −20 −3)^R.

The adsorption of **single chiral** molecules on a **chiral** surface will always create a chiral system. Assuming adsorbates with only one chiral center leads to an overall chiral adsorbate system with four different conformers, completely analogous to the discussion for cooperative chirality of chiral clusters at chiral surfaces. In the following, we consider chiral substrates, represented by kinked vicinal surfaces, right-handed Sub$(h\,k\,l)^R$ and left-handed Sub$(h\,k\,l)^S$, and chiral adsorbates R-Ads and S-Ads. For vicinal surfaces with wide (locally achiral) terraces between the chiral kink lines we can distinguish two different scenarios. First, chiral adsorbates may stabilize at terrace sites that are far away from the kink sites. Then the handedness of the substrate may be irrelevant. Thus, the discussion of local chiral behavior is analogous to that of chiral molecules at achiral surfaces. Second, chiral adsorbates may stabilize near kink sites forming the chiral substrate centers. Here, all four conformers need to be considered and can lead to different structural behavior.

As an example, the adsorption of chiral fluoro-amino-methoxy (FAM) on the kinked copper surface Cu(8 7 4) has been examined in theoretical (DFT) studies [179] where it was found that stabilization near the chiral centers of the kink lines

S-FAM

R-FAM

○ F
● O
○ N
● C
○ H
○ Cu

R-FAM

S-FAM

(a)   (b)

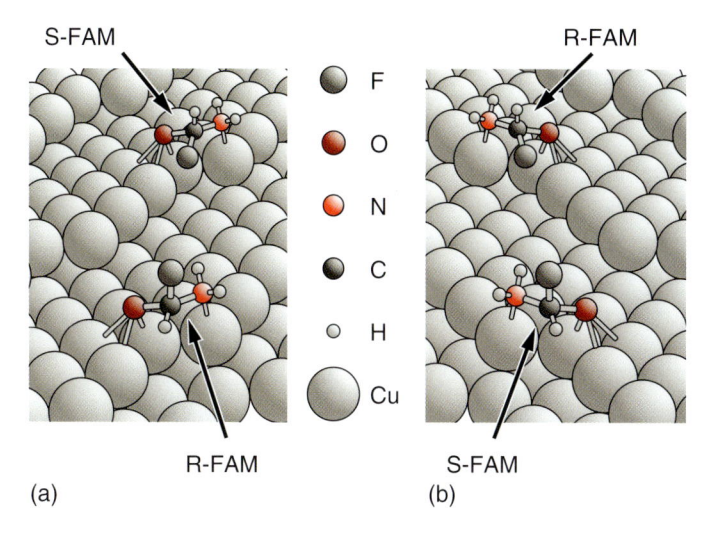

Figure 6.51 Structure of chiral fluoro-amino-methoxy (FAM) adsorbates at chiral centers of a kinked copper surface, (a) right-handed Cu(8 7 4)^R and (b) left-handed Cu(−8 −7 −4)^S. Both figures include the two adsorbate enantiomers, R-FAM and S-FAM labeled accordingly. The color coding of the atoms is explained between the figures. The figures illustrate the general adsorption behavior but do not reflect exact computed atom coordinates.

is energetically preferred. The free FAM molecule, FNH_2CHO, with a chiral carbon center, is distorted by adsorption such that its nitrogen end binds on top of a kink atom of the copper substrate while its oxygen end binds at a threefold fcc site near the kink edge, as shown on Figure 6.51. This basic binding scheme is independent of the chirality of the adsorbate. However, further structural details are quite different between the adsorbed R-FAM (right-handed-fluoro-aminomethoxy) and S-FAM (left-handed-fluoro-amino-methoxy) enantiomers. At the right-handed Cu(8 7 4)^R surface, Figure 6.51a, the left-handed S-FAM bends over the kink edge with a CH group sticking out of the surface and its C–F bond pointing toward the lower terrace. In contrast, the right-handed R-FAM, while bending over the kink edge, stabilizes with its fluorene sticking out of the surface and its CH group pointing toward the lower terrace. For both adsorbates the local surface structure is also affected by relaxation of the substrate atoms near the adsorption site. At the left-handed Cu(−8 −7 −4)^S surface, Figure 6.51b, the equilibrium structures of the adsorbates are complementary to those at the Cu(8 7 4)^R surface, with a C–F bond parallel and a C–H bond perpendicular to the substrate surface for R-FAM and vice versa for S-FAM. This creates, altogether, four adsorbate structures which can interact differently with approaching reactants.

6.8 Exercises

6.1 Consider the unrelaxed (0 0 1) surface of a tungsten single crystal (bcc lattice, lattice constant $a = 3.160$ Å) with sulfur atoms adsorbed in a c(2 × 2) overlayer. Sulfur is assumed to adsorb in hollow sites at a distance from the nearest tungsten atoms of d(W − S) = 2.456 Å. Discuss structural details of the adsorbate system. Evaluate neighbor shells of the adsorbate center.

6.2 Consider unrelaxed surfaces of a nickel single crystal (fcc lattice, lattice constant $a = 3.520$ Å) with CO adsorbed in periodic overlayers (interatomic distance d(C−O) = 1.137 Å). Discuss the adsorbate structure and symmetry at the surface for adsorbate systems

(a) Ni(1 0 0) + c(2×2) − CO; perpendicular CO in on top sites,
(b) Ni(1 1 0) − (2×1) + (2×2) − CO; Ni substrate missing row reconstructed, tilted CO in bridge sites above topmost Ni rows,
(c) Ni(1 1 1) + ($\sqrt{3}$ × $\sqrt{3}$)R30° − CO; CO in bridge sites,
(d) Ni(1 1 0) + p2mg(2×1) − 2CO; tilted CO in bridge sites above topmost Ni rows, analogous to structure shown in Figure 6.2,
(e) Ni(1 1 1) + c(4×2) − CO; perpendicular CO in hollow sites.

Assume, for all systems, CO adsorption with carbon pointing toward the substate with a distance d(Ni−C) = 1.840 Å and tilt angles of 15°.

6.3 Consider the unreconstructed (1 1 1) surface of a silicon single crystal (diamond lattice, lattice constant $a = 5.431$ Å) with hydrogen atoms in a (1×1) overlayer adsorbed at on top sites (hydrogen "terminators"). Evaluate neighbor shells of the hydrogen center up to sixth nearest neighbors assuming an adsorbate distance d(H−Si) = 1.000 Å.

6.4 Consider the (0 0 1) surface of a copper single crystal (fcc lattice, lattice constant $a = 3.610$ Å) with atomic oxygen adsorbed

(a) in an ideal c(2×2) overlayer structure with oxygen in fourfold centered hollow sites at a perpendicular distance z(O−Cu) = 0.800 Å from the topmost Cu layer, see Figure 6.52.

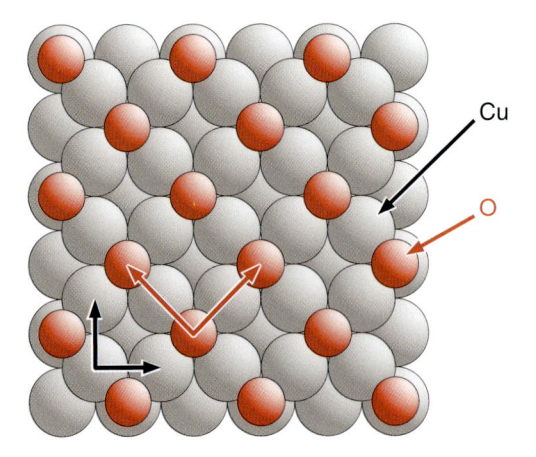

Figure 6.52 Structure of the Cu(0 0 1) + c(2×2) − O adsorbate system. Substrate and adsorbate lattice vectors are indicated in red and black.

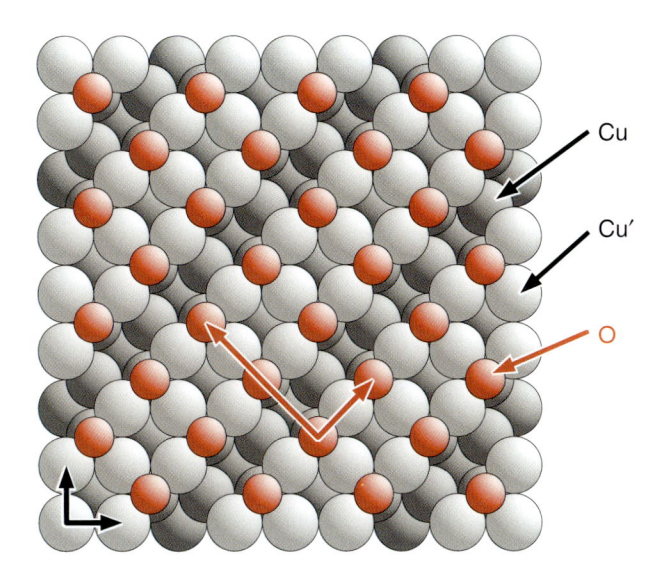

Figure 6.53 Structure of the Cu(0 0 1) – $(2\sqrt{2} \times \sqrt{2})R45° + 2O$ adsorbate system. Substrate and adsorbate lattice vectors are indicated in red and black.

    (b) in an overlayer with the topmost Cu layer reconstructed as $(2\sqrt{2} \times \sqrt{2})R45°$ missing row, see Figure 6.53. Oxygen is assumed to adsorb at threefold centered hollow sites of the reconstructed substrate layer (substrate atoms Cu′ in Figure 6.53, perpendicular distance z(O–Cu) = 0.700 Å).

Describe both systems in 2×2 matrix notation. Determine symmetry elements of the surfaces and evaluate corresponding space groups. Evaluate neighbor shells of the adsorbate centers.

6.5 Consider the (1 1 1) surface of a palladium single crystal (fcc lattice, lattice constant $a = 3.890$ Å) with atomic xenon adsorbed in a hexagonal overlayer (interatomic distance d(Xe–Xe) = 4.384 Å). One lattice vector of the Xe overlayer is assumed to be colinear with one lattice vector of the topmost Pd substrate layer. Further, one Xe atom is assumed to adsorb in on top sites. Determine minimum and maximum Xe–Pd neighbor distances for a planar Xe overlayer.

6.6 Consider the (1 1 1) surface of an aluminum single crystal (fcc lattice, lattice constant $a = 4.050$ Å) with carbon dioxide, CO_2, adsorbed at very low coverage Θ. Discuss possible structures with commensurate overlayers and determine supercells with corresponding 2×2 matrices and Wood notations. Calculate intermolecular distances at the surface. Discuss possible adsorbate sites.

6.7 Consider the stepped (1 1 3) surface of a silver single crystal (fcc lattice, lattice constant $a = 4.090$ Å) with sulfur atoms adsorbed at very low coverage

Θ. Assume adsorbate structures to be described by a model of hard spheres. Here, interatomic distances d(S–Ag) are determined by touching spheres of radii corresponding to the covalent radii r_{cov} of the atoms involved, where $r_{cov}(Ag) = 1.45$ Å, $r_{cov}(S) = 1.02$ Å. Determine perpendicular distances z of the adsorbate at different substrate sites (on top, bridge, central sites; at terraces, near steps). Here, perpendicular distances z are defined with respect to the normal vector of the (1 1 3) surface. For which structure is z smallest?

6.8 Consider sulfur adsorption with a (5 × 5) coincidence lattice on the unrelaxed Ag(1 1 1) surface in a model of hard spheres as in Exercise 6.7 using the same structure parameters. Determine possible symmetric structures with 9, 16, 25, 36, 49 sulfur atoms in the coincidence supercell. Calculate interatomic distances d(S–S) accounting for possible buckling of the sulfur layer according to the model of hard spheres.

6.9 Determine the Wood notation of the adsorbate systems defined by 2 × 2 matrix notation as

(a) Cu(1 0 0) + (1 1 | –1 1) – O
(b) Ni(1 1 1) + (1 2 | –1 1) – CO
(c) Ni(1 1 1) + (4 1 | –1 3) – NH_3
(d) Pd(1 1 0) + (1 1 | –1 2) – NO
(e) W(1 1 0) + (2 2 | –2 4) – O
(f) W(1 1 1) + (6 1 | –1 5).

Determine adsorbate coverages Θ of these systems.

6.10 Quantify Wood notation parameters p, α of commensurate surface overlayers at different crystal faces

(a) fcc(1 1 1) + ($\sqrt{3}$ × p)Rα – ···
(b) fcc(1 1 1) + (p × $\sqrt{(7/2)}$)Rα – ···
(c) fcc(1 1 1) + (p × p)R13.898° – ···
(d) fcc(1 1 1) + c(p × 4p)R7.598° – ···
(e) fcc(1 1 0) + (2$\sqrt{3}$ × $\sqrt{3}$)Rα – ···
(f) fcc(1 1 0) + (3$\sqrt{2}$ × 3$\sqrt{2}$)Rα – ···
(g) fcc(1 0 0) + ($\sqrt{5}$ × $\sqrt{5}$)Rα – ···
(h) fcc(1 0 0) + (p × 2p)R26.565° – ···
(i) bcc(1 1 0) + c(3 × 5)Rα – ···
(j) bcc(1 0 0) + (5 × 5) Rα – ···

6.11 Consider the adsorbate system Ni(1 1 1) + ($\sqrt{3}$ × $\sqrt{3}$)R30° – CO where CO adsorbs in bridge sites with its molecular axis (a) perpendicular and (b) parallel to the surface. Identify possible rotational and mirror domains. Evaluate corresponding 2 × 2 matrices connecting the lattice vectors of each domain with those of the substrate.

6.12 Consider benzene, C_6H_6, adsorption on the (1 1 1) surface of a substrate with an fcc lattice. Give examples of commensurate overlayers for different coverage and orientations of the adsorbate which allow two, three, and six domains at the surface.

6.13 Show that the (1 0 0) surface of an ideal fcc or bcc single crystal cannot have commensurate overlayers with hexagonal lattice. Hint: Use results from Appendix E.4.

6.14 Show that the (1 1 0) surface of an ideal fcc or bcc single crystal cannot have commensurate overlayers with hexagonal lattice. Hint: Use results from Appendix E.4.

6.15 Assume that a substrate surface has two overlayers with different lattice where both overlayers are commensurate with respect to the substrate. Show that the overlayers are, in general, high-order commensurate with respect to each other. When are they (first order) commensurate?

6.16 Consider an overlayer on top of a substrate layer with given Bravais lattice type (square, rectangular, hexagonal). Which lattice types are allowed for the overlayer to be commensurate?

6.17 Show that all commensurate hexagonal overlayers on top of hexagonal substrate can be described in Wood notation as $(p \times p)R\alpha$ with $p = \sqrt{(m^2 + n^2 + mn)}$ and $\tan(\alpha) = 3n/(2m+n)$ where m, n are integers. Determine corresponding 2×2 transformation matrices between overlayer and substrate lattice vectors. Discuss acute and obtuse representations of the hexagonal lattice.

6.18 Consider a commensurate overlayer (lattice vectors $\underline{R}'_{o1}$ and $\underline{R}'_{o2}$) on top of hexagonal substrate with lattice vectors $\underline{R}_{o1}$ and $\underline{R}_{o2}$ in acute representation. Show that the transformation according to

$$\begin{pmatrix} \underline{R}'_{o1} \\ \underline{R}'_{o2} \end{pmatrix} = \begin{pmatrix} m & n \\ -n & m+n \end{pmatrix} \cdot \begin{pmatrix} \underline{R}_{o1} \\ \underline{R}_{o2} \end{pmatrix}, \quad m, n \text{ integer}$$

results in a hexagonal overlayer lattice.

6.19 Consider an HOC overlayer with square lattice vectors $\underline{R}'_{o1}$ and $\underline{R}'_{o2}$ on top of substrate with square lattice vectors $\underline{R}_{o1}$ and $\underline{R}_{o2}$.

(a) Show that in the lattice vector transformation can be written as

$$\begin{pmatrix} \underline{R}'_{o1} \\ \underline{R}'_{o2} \end{pmatrix} = \begin{pmatrix} m' & n' \\ -n' & m' \end{pmatrix}^{-1} \cdot \begin{pmatrix} m & n \\ -n & m \end{pmatrix} \cdot \begin{pmatrix} \underline{R}_{o1} \\ \underline{R}_{o2} \end{pmatrix}, \quad m, n, m', n' \text{ integer}$$

(b) Assume further that the vector lengths of $\underline{R}'_{o1}$ and $\underline{R}_{o1}$ are identical and show that the matrix elements m, n, m', n' must be solutions of the Diophantine equation

$$m^2 + n^2 = m'^2 + n'^2$$

6.20 Consider an HOC overlayer with hexagonal lattice vectors $\underline{R}'_{o1}$ and $\underline{R}'_{o2}$ on top of hexagonal substrate with lattice vectors $\underline{R}_{o1}$ and $\underline{R}_{o2}$ (both lattices in acute representation). Assume further that the vector lengths of $\underline{R}'_{o1}$ and $\underline{R}_{o1}$ are identical. Show that in the corresponding transformation according to Eq. (6.27) the integer parameters m, n, m', n' must be solutions of the Diophantine equation

$$m^2 + n^2 + mn = m'^2 + n'^2 + m'n'$$

6.21 Show that a commensurate hexagonal overlayer with lattice vectors $\underline{R}'_{o1}$ and $\underline{R}'_{o2}$ on hexagonal substrate with lattice vectors $\underline{R}_{o1}$ and $\underline{R}_{o2}$ (both lattices in obtuse representation) can be described by a transformation

$$\begin{pmatrix} \underline{R}'_{o1} \\ \underline{R}'_{o2} \end{pmatrix} = \begin{pmatrix} m & n \\ -n & m-n \end{pmatrix} \cdot \begin{pmatrix} \underline{R}_{o1} \\ \underline{R}_{o2} \end{pmatrix}, \quad \text{m, n integer}$$

6.22 Consider an overlayer structure where the substrate is Wood-representable and the overlayer is stretched with respect to the substrate along direction $\underline{e}$ by a factor γ. For which directions $\underline{e}$ and Bravais lattice types is the overlayer structure Wood-representable?

6.23 Relation (6.86) can be inverted to compute the parameters γ, α of a p($\gamma \times \gamma$)Rα overlayer structure from (measured) moiré parameters λ, φ. Determine the inverted relationships and prove that

$$\gamma = \frac{\lambda}{\sqrt{\Delta'}}, \quad \cos(\alpha) = \frac{\cos(\varphi) + d\lambda}{\sqrt{\Delta'}}, \quad \Delta' = 1 + d^2\lambda^2 + 2d\lambda\cos(\varphi)$$

6.24 Well separated moiré patterns are observed for moiré factors $\lambda > 5$. Consider isotropically scaled and rotated overlayers with an integer approximant $\underline{M}_I = \underline{1}$ and evaluate the range of scaling factors γ and rotation angles α which fulfill the condition $\lambda > 5$.

7
Experimental Analysis of Real Crystal Surfaces

7.1
Experimental Methods

Truly quantitative structure determinations of single crystal surfaces and adsorbate systems by experiment are intrinsically difficult. While three-dimensionally periodic bulk crystal structures can be measured routinely with the help of X-ray diffraction methods, a complete surface structure analysis requires usually a combination of different experimental methods to yield a unique result. Methods that can contribute to a quantitative analysis of structural details of real crystal surfaces must be able to mainly probe the atoms near the surface, ignoring those of the inner substrate. This excludes standard **X-ray diffraction** methods from surface analyses. X-ray photons can penetrate deep into the bulk and, therefore, yield structural bulk information, with that from surface atoms representing only a minor perturbation. However, special geometric arrangements of the **X-ray** beam with respect to the single crystal surface, so-called **grazing incidence** geometry [34], can also yield structural information pertaining to the surface. Other diffraction methods, such as low-energy electron diffraction (**LEED**), have proven to be particularly useful in identifying surface structure. These methods rely on the interference of particles that scatter (often multiple times) from periodic arrangements of atoms at single crystal surfaces that are ordered over a relatively wide area. If the surface structure deviates strongly from periodic ordering, for example, as a result of large size imperfections or disordered adsorbate structure, local (small-area) diffraction becomes more useful. For these systems, local diffraction methods such as photoelectron diffraction (**PED**) or surface extended X-ray absorption fine structure (**SEXAFS**) can be used to obtain quantitative information about the local environments of surface atoms including coordination and binding angles.

In general, methods that can provide information about structural details of real crystal surfaces and adsorbate systems include those based on **scattering, diffraction, imaging**, as well as **spectroscopy**, and use **photons, electrons**, or **atoms** and **ions**. A detailed discussion of each method and its merits in connection with quantitative determination of surface structure is beyond the scope of this book. Corresponding methods are well documented in the surface science

Table 7.1 Experimental methods used to determine surface structure.

Method by name	Acronym
(a) Imaging methods	
Atomic Force Microscopy	AFM
Scanning Tunneling Microscopy	STM
Transmission Electron Microscopy	TEM
(b) Diffraction methods	
Atom Diffraction	AD
Grazing Incidence X-Ray Diffraction	GI-XRD
Low-Energy Electron Diffraction	LEED
Low-Energy Positron Diffraction	LEPD
Medium Energy Electron Diffraction	MEED
Photoelectron Diffraction	PED
Reflection High Energy Electron Diffraction	RHEED
(Surface) Extended X-ray Absorption Fine Structure	(S)EXAFS
Transmission Electron Diffraction	TED
X-Ray Diffraction	XRD
X-ray Standing Wavefield absorption	XSW
(c) Scattering methods	
High Energy Ion Scattering	HEIS/IS
High-Resolution Helium Atom Scattering	HRHAS
Inelastic Molecular Beam Scattering	IMBS
Low-Energy Ion Scattering	LEIS/IS
Medium Energy Ion Scattering	MEIS/IS
Thermal Energy Atom Scattering	TEAS
Time-Of-Flight Scattering And Recoiling Spectroscopy	TOF-SARS
(d) Spectroscopy methods	
(Fourier Transform) Reflection Absorption Infrared Spectroscopy	(FT)RAIRS
(High-Resolution) Electron Energy Loss Spectroscopy	(HR)EELS
Near-Edge X-ray Absorption Fine Structure or X-ray Absorption Near-Edge Spectroscopy	NEXAFS, XANES
Surface Electron Energy Loss Fine Structure	SEELFS

The methods are given by their names, where capital letters in the names are meant to explain the acronyms included in the table. Details of the different methods are given, for example, in [34].

literature, see, for example, [34, 94, 180]. In Table 7.1, we list only example methods by their names, where the list may not be exhaustive.

7.2
Surface Structure Compilations

As a result of the experimental complexity, the number of quantitatively solved surface structures is rather small compared to that of bulk crystal structures. As of 2012, more than 800 000 bulk crystal structures have been published and stored in crystal structure databases, such as CSD (Cambridge Structural Database) [181], ICSD (Inorganic Crystal Structure Database) [182], Crystmet [183], NAD (Nucleic

Acid Database) [184], or PDB (Protein Data Bank) [185]. In contrast, as of 2004 (when there were about 400 000 published bulk crystal structures) the National Institute of Science and Technology (NIST) Surface Structure Database (SSD) listed only about 1400 surface structures that are quantitatively complete in all details, with an estimated additional 150 structures published between 2004 and 2014.

A number of compilations of crystallographic information on surfaces and interfaces have been published in the literature or are available in electronic form. Early tabulations by Somorjai and Van Hove [36] give mostly **two-dimensional** structure information, that is, lateral periodicity patterns of ordered monolayers of atoms and molecules with only few results of three-dimensional parameters. Examples are adsorbate layer spacings and adsorption bond lengths. Here, the lateral periodicity was derived mainly from LEED measurements. A more detailed listing of **three-dimensional** structure parameters with many early literature references is included in the LEED book by Van Hove *et al.* [94]. The structure information in the review by Ohtani *et al.* [186] with an extensive list of references for surfaces with and without ordered adsorbates is strictly **two-dimensional** describing the surfaces only by their lateral symmetry pattern. Most of the tables of [186] can also be found in the textbook by Somorjai [12].

More recent **three-dimensional** structure parameters have been published in the **Atlas of Surface Structures** [23] based on data from the electronic **NIST** Surface Structure Database (**SSD**), Version 1, discussed below, which includes all quantitatively known surface structures until 1991. The second volume of this Atlas also gives graphical representations of the surfaces yielding a good qualitative overview of the structures examined at that time. (Most of these structures are still relevant at present.) Further, a review chapter of the **Landolt–Börnstein** Series [187] lists quantitative structure parameters from experiments on adsorbates at metal and semiconductor surfaces where the data are complete until 2002.

Another source of quantitative surface structure data is the **Surface Crystallographic Information Service** (SCIS) where the printed version [25] includes references until the end of 1987. The SCIS book was complemented by database software allowing for easy search and basic visualization of published surface structure on a personal computer. The SCIS software was updated in a second version to include references until the end of 1992. The SCIS software project was succeeded by the development of the **NIST SSD** which was published and distributed by the NIST between 1992 and 2010 with biennial updates until the end of 2003. The distribution of the latest version 5 of SSD by NIST has been discontinued but it remains available as an open source database (**open Surface Structure Database**, oSSD) and can be downloaded from the web, see Appendix H. Complete compilations of quantitative surface structure data published after 2004 have not yet appeared in the literature.

So far, the **NIST SSD** [22–24] is the only **complete critical compilation** of reliable crystallographic information available on surfaces and interfaces. The database provides access to detailed text and graphical information for 1379 experimentally determined atomic-scale structures that which been published

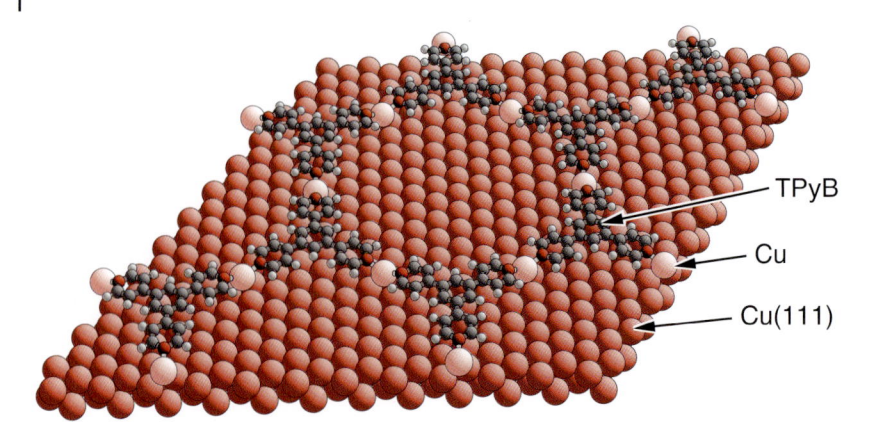

Figure 7.1 Structure of a molecular adsorbate network on copper substrate described as Cu(1 1 1) + (6√3 × 6√3)R30° − 2TPyB − 3Cu. The adsorbate atoms are shown in different size and color, small/dark gray for carbon, small/dark red for nitrogen, very small/light gray for hydrogen, and large/light red for adsorbate copper. Substrate copper is shown by large/dark red balls.

until 2004 (SSD version 5). It can be considered to cover all **classical** surface structures that have not been revised after 2004 or have only experienced very minor modifications in distances or angles since then. More recent published surface structures, which are not included in the SSD, concern mainly single crystal surfaces with rather complex reconstruction and/or large molecules or molecular networks [188, 189]. An example is shown in Figure 7.1 where scanning tunneling microscopy (STM) measurements together with theoretical density-functional theory (DFT) studies have identified large organic molecules, 1,3,5-tris(pyridyl)benzene (TPyB) forming an ordered hexagonal network on the hexagonal Cu(1 1 1) surface [190], which is described in Wood notation as Cu(1 1 1) + (6√3 × 6√3)R30° − 2TPyB − 3Cu.

SSD is a **critical** compilation of published structure data. This means, in particular, that structure information provided by experimentalists or taken from the literature was checked for **completeness** and **consistency** before being included in the database. Symmetry or qualitative geometry information only, which is available for many single crystal surfaces and adsorbate systems, did not qualify as a surface system to be entered in the SSD. This applies specifically to numerous experimental structure studies on complex overlayers where only STM is used to obtain a qualitative impression of the surface with very limited quantitative structural details.

In the following, we discuss a few results from statistical analyses of the SSD data which have been provided by M. A. Van Hove (Hong Kong Baptist University) and refer to structures solved until 2004. These results can shed some light on the types of surfaces and interfaces that are known quantitatively but also on the use of different experimental methods to obtain quantitative surface structure. First, of the 1379 entries contained in the SSD database, the majority, 1363

Table 7.2 Number of quantitatively solved surface and interface structures until 2004 taken from the Surface Structure Database (SSD, V. 5).

Element composition	Entries	Electronic property	Entries	Basic structure	Entries
Elemental	1148 (83%)	Metal	1124 (81%)	Unreconstructed	1054 (76%)
Compound	121 (9%)	Semiconductor	215 (16%)	Reconstructed	325 (24%)
Alloy	106 (8%)	Insulator	40 (3%)		
Other	4 (<1%)	Semimetal	7 (<1%)		

The data are grouped according to substrate type, element composition, electronic property, and basic structure.

(98.8%), concerns the clean substrate surface with or without adsorbate overlayers while 16 (1.2%) are interstitial structures with foreign atoms inside the substrate. Table 7.2 lists further details about the kinds of substrates that were examined. The majority of substrates appearing in the SSD are **elemental** substrates, where transition **metals**, such as Co, Ni, V, Mo, Ru, Rh, Pd, Pt, and W dominate due to the relatively easy growth and preparation of corresponding single crystals and their surfaces. In addition, substrates of elemental **semiconductors**, such as Si and Ge, and of compound semiconductors, such as GaAs, CdSe, CdTe, and InP, have been analyzed by their surface structure due to their great technological importance. They account for 16% of all entries in the SSD. Finally, the majority of entries in the SSD, 76%, refers to **unreconstructed** substrate surfaces.

Table 7.3 lists details about the kinds of adsorbates that were examined. Of all SSD entries 961 (68%) refer to adsorbate systems. Of these, the majority, 720 entries, concern **atomic** adsorbates, such as H, N, O, Cl, S, Na, K, Cs, Al, Fe, and Mn, while a smaller group includes **molecular** adsorbates, such as CO, CN, NO, N_2, PF_x, and small hydrocarbons. An even smaller group is given by **compound** adsorbate systems where the adsorbates can mix with atoms of the substrate surface forming mixed surface layers. Examples are surface alloy layers, such as those formed by Na, which penetrates into the top layer of the Al(1 1 1) surface [191]. Finally, there are a small number of entries with both atoms and molecules **coadsorbed** at the surface, such as O and C_6H_6 on the Ru(0 0 0 1) surface, as well as with metal adsorbates that form **thin overlayer films**, such as Ni films on top of a Cu(1 0 0) substrate.

As to details of the adsorbate morphology, Table 7.3 shows that the majority, 716 (52%), of all SSD entries describe adsorbates that form ordered or disordered overlayers, such as CO on the Ni(1 0 0) surface. A smaller group of entries, 101 (7%), concerns **pseudomorphic** layers of adsorbates in positions that continue the crystal structure of the substrate. Examples are metal adsorbates on a metal substrate, such as Fe films on top of a Cu(1 1 0) substrate. Another group, 74 entries (5%), is given by adsorbates that occupy positions of substrate atoms at the surface forming mixed adsorbate/substrate atom layers (**substitution**). Examples are again metal adsorbates on a metal substrate that yield surface alloy layers, such as Au adsorbates at a Cu(1 0 0) substrate. Mixed adsorbate/substrate atom layers can also occur when adsorbate atoms assume **interstitial** positions between atoms of

7 Experimental Analysis of Real Crystal Surfaces

Table 7.3 Number of quantitatively solved surface and interface structures until 2004 taken from the Surface Structure Database (SSD, V. 5).

Adsorbate type	Entries	Adsorbate structure	Entries
Atomic	720 (52%)	Overlayer	716 (52%)
Molecular	173 (13%)	Pseudomorphic	101 (7%)
Compound	38 (3%)	Substitutional	74 (5%)
Mixed atomic/molecular	18 (1%)	Epitaxial	39 (3%)
Thin films	12 (1%)	Interstitial	15 (1%)
		Other	16 (1%)

The data are grouped according to adsorbate type and adsorbate structure.

substrate layers near the surface. This group includes 15 entries (1%), where an example is given by O atoms adsorbing onto and penetrating into the Al(1 1 1) surface. Further, one group of 39 systems (3%) is described by (**epitaxial**) adsorbate overlayers of crystalline structure where the overlayer lattice does not match that of the substrate. An example is the adsorption of Xe atoms forming (slightly distorted) hexagonal overlayers on top of a Ag(1 1 1) substrate, whose substrate surface is also hexagonal but whose lattice constant differs from that of the overlayer. There are numerous examples of surface systems that do not fit uniquely into one of the different groups discussed above but the above schemes can give a sound basis for a general classification of surface structure.

Further, an inspection of the SSD data shows that of all **experimental methods** used in surface structure evaluations, Low Energy Electron Diffraction (**LEED**) is by far the **most often** applied method, which covers about 65% of all

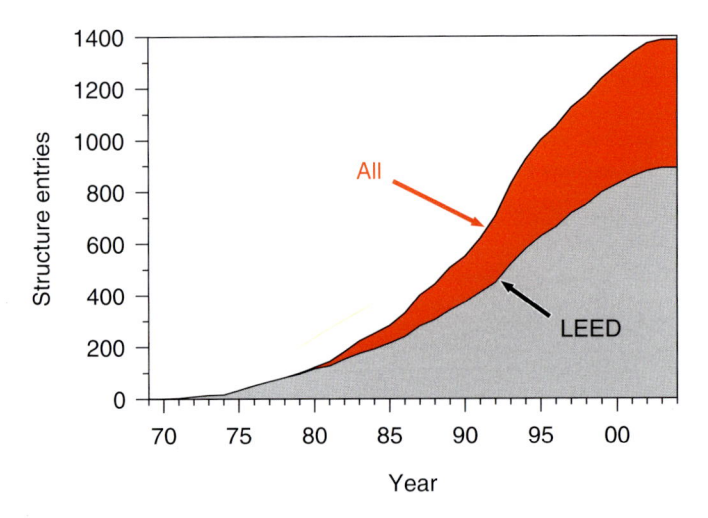

Figure 7.2 Number of quantitatively solved surface and interface structures as a function of the publication year, see text. The data are given separately for all measured structures and for those using LEED.

Table 7.4 Number of quantitatively solved surface and interface structures until 2004 as a function of the methods applied.

Method	Entries	Method	Entries	Method	Entries
LEED	891 (65%)	TOF-SARS	13 (1%)	AED	3 (<1%)
PED	132 (10%)	NEXAFS	11 (1%)	SEELFS	2 (<1%)
IS	117 (8%)	RHEED	10 (1%)	TED	1 (<1%)
SEXAFS	67 (5%)	LEPD	5 (<1%)	AD	1 (<1%)
XSW	63 (5%)	HREELS	4 (<1%)	STM	1 (<1%)
XRD	55 (4%)	MEED	3 (<1%)		

The methods are given by their acronyms explained in Table 7.1.

surface structure determinations contained in SSD. In fact, in the beginning of quantitative surface structure analysis, between 1969 and 1980, LEED was the only method available. This is illustrated in Figure 7.2, which shows the number of quantitatively solved surface and interface structures as a function of the publication year until 2004 according to the SSD data. Quantitative results from experimental methods other than LEED have started to appear in the literature only **after 1980**, and even in 2004 surface structure studies applying LEED still outnumbered those using other methods. This is also clear from a listing of all surface and interface structures solved until 2004 as a function of the methods applied as given in Table 7.4.

The table shows that **LEED, PED, IS, SEXAFS**, and **XSW** were the five topmost experimental methods applied to quantitative surface analysis covering 93% of all solved surface and interface structures.

7.3
Database Formats for Surface and Nanostructures

The collection of quantitative surface and nanostructure data is of vital importance for a scientific understanding and documentation of many properties of these systems. This is required not only for archiving purposes but also when structural data of a group of systems are analyzed by their similarities or screened in their characteristic differences. Further, quantitative structure data are needed for theoretical postprocessing when "intelligent guesses" from experiments are used as input for visualization and for numerical simulation.

In general, structure information includes quantitative numerical parameters, such as lattice constants, interatomic distances, or bond angles, as well as symmetry properties. Further, a complete characterization must contain textual descriptions, such as system preparation and experimental methods together with estimates of instrumental and methodological errors. This requires standard data formats that are commonly accepted and widely adopted. The issue has been discussed extensively in the surface science literature [192–194] of which we mention in the following only some of the most important points raised [194].

In the past, numerous data formats for documenting structural parameters derived from bulk and surface crystallography have been proposed for different purposes. Examples are three-dimensional crystal data formats, such as

- **CIF** (Crystallographic Information Framework) [195], which is adopted by the International Union of Crystallography for the documentation of bulk crystal data,
- **PDB** (Protein Data Bank) [196], which is a generally accepted documentation format for crystals of biological macromolecules,
- different formats used in simulations and theoretical studies with standard computer codes, such as **VASP** (Vienna Ab initio Simulation Package) [197],

or data formats developed specifically for surface structure, such as

- **SSD** (Surface Structure Database) [23, 27], which was proposed as a combined numerical and textual data format for submitting quantitative surface structures to the NIST database SSD [22],
- **SURVIS** (SURface VISualizer) [27], which was developed for the visualization tool Survis inside the NIST SSD project and documented in the Survis and Balsac manuals,
- **BALSAC** (Build and Analyze Lattices, Surfaces, And Clusters) [26], which is the standard format of the visualization and analysis tool Balsac,

or general data formats for documenting atom positions, such as

- **XYZ** which is widely supported by quantum chemical and solid state physics codes (although no formal specification standard has been published so far) and collects Cartesian coordinates and element specifications of all atoms in the bulk, at a surface, or in a molecule.

While the data formats mentioned above (and many others) have been used and proven valuable for information exchange within different scientific communities, no format has been generally adopted as a single data format for documenting structure and other properties of the bulk, of surfaces, or of nanoparticles. This is partly due to the fact that different scientific groups put emphasis on different aspects of structure data. Thus, formats that are complete in one community may not be flexible enough to serve the purposes of another. As a consequence, a generally acceptable structure data format has to be either developed from scratch or updated from an existing format addressing the kind of structures that are examined today and may be studied in the near future. These will be mainly nanostructures and complex surfaces including molecular networks, that is, structures that very often have lower dimensionality and order than two-dimensionally periodic surfaces.

In the following, we list the most important requirements that an ideal structure data format has to fulfill according to [194] to conveniently document experimental and theoretical results from surface structures and nanostructures. The data format should

- describe all theoretical and experimental methods in enough detail to enable reproduction of the results. This may require different accuracies depending not only on the experiment but also on the theoretical methods applied.
- allow the combination of results from multiple techniques in a single structure determination. This includes experimentally as well as theoretically obtained data.
- have a consistent scheme allowing coordinate listings of all three- and lower-dimensional structures as well as combinations thereof. Examples include carbon nanotubes or graphene layers adsorbed incommensurately on a single-crystal surface.
- allow documenting incomplete structural results that may arise in complex systems. Examples include hydrogen atom positions at surfaces, which may not be determined in a given structure while all other atoms have been evaluated. In nanostructures many more partial structure determinations are to be expected.
- be unique and machine-readable and also include human-readable tags. It should be described in the open literature and should be free of copyright protection.
- be sufficiently simple and intuitive to allow easy conversion of past published results to complement a structure collection. This could be achieved by appropriate conversion utilities.

Further, there should be an a priori agreement on the data format within the community of experimental and theoretical crystallographers, surface scientists, and nanoscientists. This can be enforced by a concensus between major journal and book publishers in publication policies.

The development of a general structure data format that obeys all or most of the above requirements is a major task that requires not only a lot of conceptual work but also the ability to convince the scientific community of its value and need. Besides, such a development project may not be "scientifically rewarding." Therefore, the authors of [194] have proposed a "hybrid universal format" in which different already existing file formats may appear in the same structure file for different parts of a complex structure; for further details see [194].

7.4
Exercises

7.1 Discuss the basic physical mechanisms that are applied in the different structure determination methods listed in Table 7.1. Explain the need for the different methods to arrive at a unique structure determination.

7.2 Consider a (fictitious) database of experimental and theoretical surface structures where each structure needs to be documented in complete form.

(a) Which structure parameters are required for each entry?
(b) Which additional information should be included for each surface structure?

7.3 Consider the entry of a surface structure in a database with the following parameters

- substrate crystal: monoatomic nickel, fcc lattice, bulk symmetry, lattice constant, lattice vectors, bulk unit cell (shape and volume), atom density.
- surface: orientation (1 0 0), square lattice, monolayer symmetry, monolayer spacing lateral, and perpendicular (bulk value), 2×2 reconstruction matrix, perpendicular relaxation of the topmost monolayers (in % of monolayer spacing), surface unit cell (shape and volume).

Which of these data entries can be generated from a minimal set of parameters? Determine a minimal set.

7.4 What are the possible uses of a surface structure database? Discuss examples.

7.5 Why are different formats for surface structure documentation needed? Discuss examples of possible advantages and disadvantages.

8
Nanotubes

Nanotubes form an important class of **nanoparticles** that may exhibit **spatial periodicity** and can, thus, be described structurally in a way analogous to crystalline material. They have become a major area of research in recent years since these systems exhibit extremely interesting and new electronic and chemical properties. Examples are nanotubes that provide new substrate material for catalytically active particles but may also act as catalysts themselves [14, 16]. The most prominent and oldest members of the nanotube family are **carbon nanotubes** observed as early as 1952 [198] but which have attracted attention only much later [199, 200]. Meanwhile, very different materials, such as Si, BN, VO_x, TiO_2, WS_2, MoS_2, and MnO have been found to form nanotubes [14–16]. While the physical and chemical properties of these systems have been widely discussed in the literature, the present section focuses on **structural aspects**. The structure of nanotubes depends very much on their **preparation** [14, 201], where the crystal structure of the corresponding bulk material may give some hints. In the following, we consider a **special class** of nanotubes derived from rolling sections of single crystal layers, which also includes carbon nanotubes. It should be emphasized that the nanotubes discussed as examples in this section, originating from NaCl crystal layers, are meant only to illustrate the basic crystallographic concepts. They have not been prepared by experiments so far and may be difficult to produce due to the highly ionic character of their constituent atoms.

8.1
Basic Definition

The construction of **nanotubes** considered in this section starts from infinitely **long strips** of finite width, cut out of a planar two-dimensional periodic layer parallel to an (*h k l*) netplane of a perfect single crystal. These strips are then rolled up by joining their parallel borders to yield long **cylindrical tubes** whose circumference equals the strip width. For the sake of simplicity, we confine ourselves first to nanotubes originating from **(*h k l*) monolayers**. As an example, Figure 8.1 shows a part of the NaCl(1 2 2) (6, 1) nanotube representing a rolled up section

Crystallography and Surface Structure: An Introduction for Surface Scientists and Nanoscientists,
Second Edition. Klaus Hermann.
© 2017 Wiley-VCH Verlag GmbH & Co. KGaA. Published 2017 by Wiley-VCH Verlag GmbH & Co. KGaA.

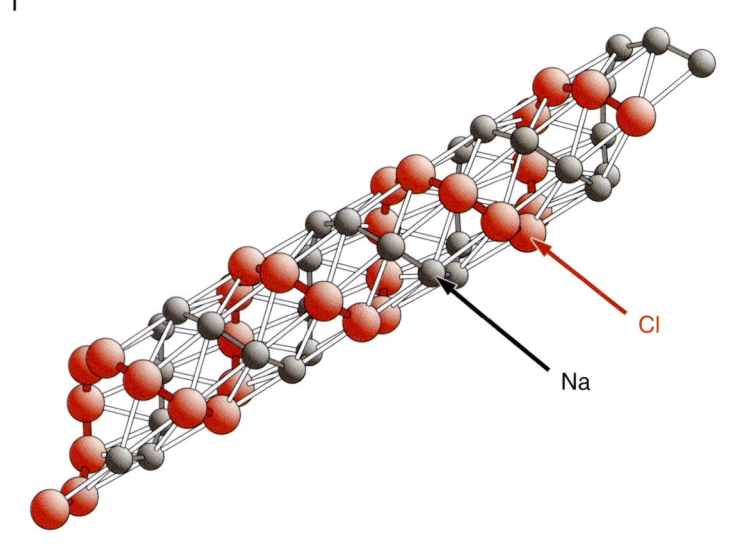

Figure 8.1 Structure of a NaCl(1 2 2) (6, 1) nanotube section.

of a (1 2 2) monolayer of the perfect cubic NaCl crystal with (6, 1) denoting the rolling direction explained in the following.

As discussed in Section 3.2, the periodicity of an $(h\,k\,l)$ monolayer is given by **netplane-adapted lattice vectors** $\underline{R}_1$ and $\underline{R}_2$, which result from linear transformations of the initial lattice vectors $\underline{R}_{o1}, \underline{R}_{o2}, \underline{R}_{o3}$ of the bulk crystal according to

$$\underline{R}_1 = t_{11}\,\underline{R}_{o1} + t_{12}\,\underline{R}_{o2} + t_{13}\,\underline{R}_{o3} \tag{8.1a}$$

$$\underline{R}_2 = t_{21}\,\underline{R}_{o1} + t_{22}\,\underline{R}_{o2} + t_{23}\,\underline{R}_{o3} \tag{8.1b}$$

where t_{ij} are (integer-valued) elements of a transformation matrix $\underline{\underline{T}}^{(h\,k\,l)}$ referring to Miller indices $(h\,k\,l)$ as given by Eqs. (3.11), (3.12), or (3.13). The definition of a nanotube requires an additional lattice vector along the monolayer, called **rolling vector** $\underline{R}_r$, which can be written as

$$\underline{R}_r = m\,\underline{R}_1 + n\,\underline{R}_2, \quad m, n \text{ integer} \tag{8.2}$$

where m, n are commonly referred to as **rolling indices**. The rolling vector, starting at any point of the $(h\,k\,l)$ monolayer, is used to construct an **infinitely** long strip extending perpendicular to vector $\underline{R}_r$ with a width equal to the length of $\underline{R}_r$. This is illustrated in Figure 8.2 for the (1 2 2) monolayer of a NaCl crystal where a strip section (emphasized by gray background) is defined in its width by a rolling vector $\underline{R}_r$ with m = 6, n = 1. **Rolling** the strip along $\underline{R}_r$ such that its two edges **coincide** creates a circular **tube** where atoms positioned exactly at one edge of the planar strip will coincide with their counterparts connected by vector $\underline{R}_r$ on the other edge. The resulting nanotube is commonly labeled by the decomposition (8.2) of the rolling vector $\underline{R}_r$ as an **(m, n) nanotube**. Thus, the strip shown in Figure 8.2 corresponds to the (6, 1) nanotube of NaCl(1 2 2) displayed in Figure 8.1.

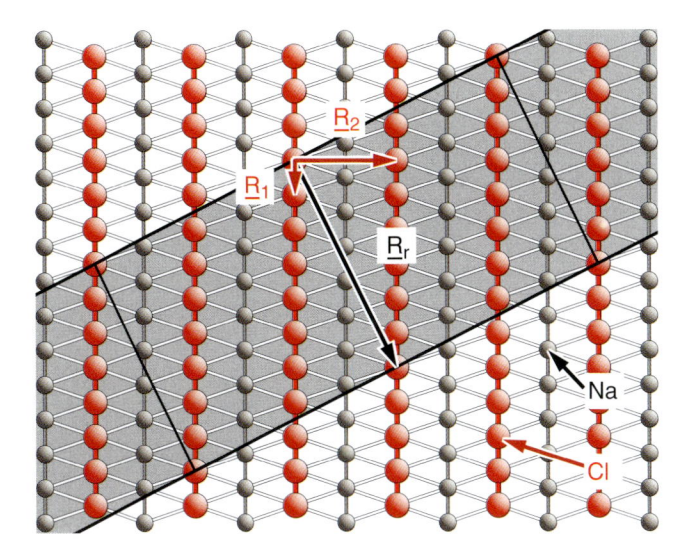

Figure 8.2 Structure of a NaCl(1 2 2) monolayer including a section of the nanotube strip defined by a rolling vector m = 6, n = 1. The strip is emphasized in gray with netplane-adapted lattice vectors $\underline{R}_1$, $\underline{R}_2$ and the rolling vector $\underline{R}_r$ labeled accordingly.

In **mathematical** terms, the rolling procedure can be achieved by a **non-linear coordinate transformation** $(x, y, z) \rightarrow (x_t, y_t, z_t)$ in Cartesian space with

$$x_t = \kappa \cos \varphi, \qquad y_t = \kappa \sin \varphi, \qquad z_t = z$$
$$\kappa = (R_r - x)/2\pi, \qquad \varphi = (y/R_r) \, 360°, \qquad x \leq R_r \tag{8.3}$$

where the Cartesian coordinate system is chosen such that the x-axis is perpendicular to the monolayer with x = 0 defining the monolayer plane. Further, the y-axis points along the rolling vector $\underline{R}_r$, and the z-axis points along the strip border in the monolayer. Inverting the rolling vector direction by going from $\underline{R}_r$ to $-\underline{R}_r$ corresponds in Eq. (8.3) to a transformation

$$(x, y, z) \rightarrow (x, -y, -z) \rightarrow (x_t, -y_t, -z_t) \tag{8.4}$$

which does not change the structure of the nanotube. Thus, rolling indices (m, n) and (−m, −n) reflect identical nanotubes.

As mentioned earlier, the oldest examples of nanotubes are **carbon nanotubes** created by rolling sections of honeycomb structured (0 0 0 1) monolayers of hexagonal graphite (so-called graphene layers), as shown in Figure 8.3. These monolayers form the basis of layer-type graphite crystals but can also exist in nature as separate graphene sheets, either in aqueous solution or adsorbed at solid surfaces. Carbon nanotubes have been **classified** according to their structure and symmetry given by the **rolling indices** (m, n) in Eq. (8.2). Here, the indices are commonly based on an acute representation of the graphene netplane lattice vectors $\underline{R}_1$ and $\underline{R}_2$ ($\angle(\underline{R}_1, \underline{R}_2) = 60°$), which is also used for the present discussion. The following three types of nanotubes can be distinguished:

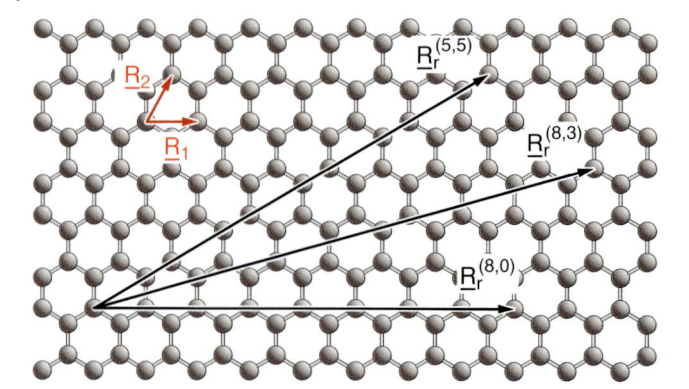

Figure 8.3 Structure of a graphite (0 0 0 1) monolayer (graphene). The monolayer includes three different nanotube rolling vectors, $\underline{R}_r^{(8,0)}$, $\underline{R}_r^{(8,3)}$, and $\underline{R}_r^{(5,5)}$. The lattice vectors $\underline{R}_1$ and $\underline{R}_2$ of the monolayer are shown in red.

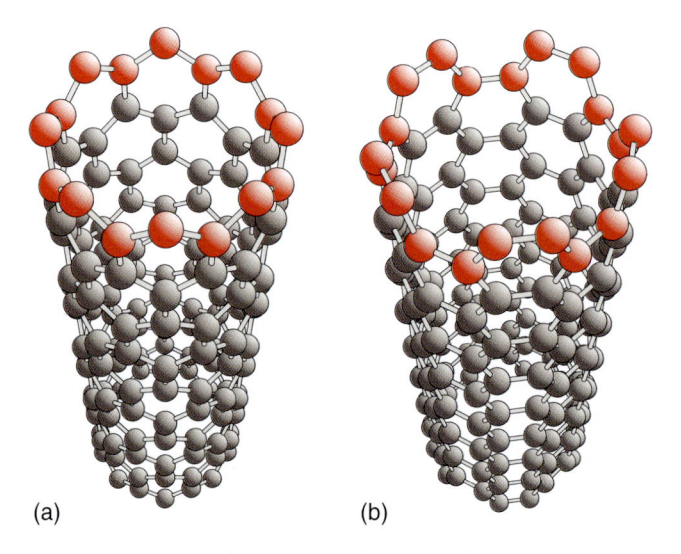

Figure 8.4 Structure of symmetric carbon nanotubes, (a) zigzag (8, 0) and (b) armchair (5, 5) tube. The zigzag and armchair cuts are emphasized by red atom balls at the top.

a) For $\mathbf{m \neq 0, n = 0}$, nanotubes are described as **zigzag** tubes, which is evident from the direction of the corresponding rolling vectors $\underline{R}_r$ pointing along zigzag carbon rows of the graphene sheet. As an illustration, Figure 8.4a shows a section of the (8, 0) carbon nanotube defined by the rolling vector $\underline{R}_r^{(8,0)}$ sketched in Figure 8.3. Due to the sixfold rotational symmetry of the graphene layer there are 3 equiv. zigzag nanotubes described by $\pm(m, 0)$, $\pm(0, m)$, and $\pm(m, -m)$.

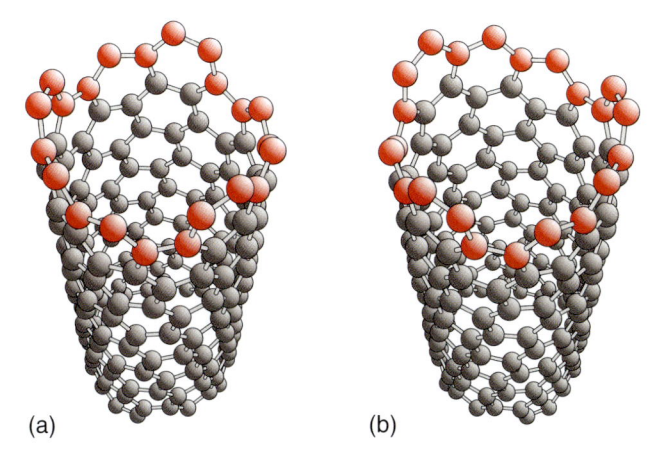

Figure 8.5 Chiral pair of carbon nanotubes, (a) (8, 3) and (b) (3, 8) tube. The spiral structures are emphasized by red atom balls at the top.

b) For $m = n \neq 0$, nanotubes are described as **armchair** tubes, which is also clear from the direction of the corresponding rolling vectors $\underline{R}_r$ pointing along meandering carbon rows of the graphene sheet. As an illustration, Figure 8.4b shows a section of the (5, 5) carbon nanotube defined by the rolling vector $\underline{R}_r^{(5,5)}$ sketched in Figure 8.3. Due to the sixfold rotational symmetry of the graphene layer there are 6 equiv. armchair nanotubes described by $\pm(m, m)$, $\pm(-m, 2m)$, and $\pm(2m, -m)$.

c) For $m, n \neq 0$, $m \neq n$, nanotubes are described by spiral networks of carbon honeycombs without mirror symmetry along the tube axis. This implies that there are always **chiral pairs** of nanotubes, where one arises from the other by mirroring with respect to a plane along the nanotube axis. As an illustration, Figure 8.5 shows a sections of a chiral pair of carbon nanotubes, denoted (8, 3) and (3, 8); the rolling vector $R_r^{(8,3)}$ is sketched in Figure 8.3. Here, the spiral networks of the two nanotubes proceed in different directions. Due to the mirror symmetry of the graphene layer there are three equivalent chiral pairs of nanotubes for each (m, n) described by (m, n) with $\pm(n, m)$, $\pm(m + n, -n)$, and with $\pm(-m, m + n)$.

8.2
Nanotubes and Symmetry

Ideal linear nanotubes can also be characterized by their **symmetry properties**, which derive from translational and point symmetry of their defining **monolayers**. As a result of translational symmetry of these layers, defined by lattice vectors $\underline{R}_1$ and $\underline{R}_2$, nanotubes can exhibit combined translational and rotational symmetry.

Considering a general lattice vector $\underline{R}$ of the netplane with

$$\underline{R} = k_1 \underline{R}_1 + k_2 \underline{R}_2, \quad k_1, k_2 \text{ integer} \tag{8.5}$$

the rolling transformation (8.3) shows immediately that changes of its components **along** the rolling vector $\underline{R}_r$, pointing along coordinate y, and affecting only angle φ in Eq. (8.3), are transformed to **rotational increments** on the tube. In contrast, changes of its components **perpendicular** to $\underline{R}_r$, pointing along coordinate z in Eq. (8.3), are transformed to linear shifts **parallel to the axis** of the tube. This means, in particular, that rows of atoms at equal distances along vector $\underline{R}$, with components both parallel and perpendicular to $\underline{R}_r$, on the monolayer, yield **spiral** arrangements on the corresponding nanotube. This is illustrated by the NaCl(1 2 2) (6, 1) nanotube in Figure 8.1 where the Na and Cl spirals, relating to dense atom rows along $\underline{R}_1$ on the monolayer, see Figure 8.2, are evident. In contrast, if vector $\underline{R}$ points **along** the rolling vector $\underline{R}_r$ then atom rows along $\underline{R}$ on the monolayer lead to **rings** on the nanotube. This is shown in Figure 8.6a for the NaCl(1 2 2) (8, 0) nanotube with its the Na and Cl rings. On the other hand, if vector $\underline{R}$ is **perpendicular** to $\underline{R}_r$ then atom rows along $\underline{R}$ on the monolayer result in rows **parallel** to the axis of the nanotube, as demonstrated in Figure 8.6b for the NaCl(1 2 2) (0, 6) nanotube showing Na and Cl rows.

Nanotubes originating from **general monolayers** appear usually as **chiral pairs** of tubes, where one derives from the other by **mirroring** with respect to a plane along the nanotube axis. The two tubes can be thought of as arising from rolling monolayer strips **above** and **below** the layer, which corresponds to the coordinate transformation (8.3) as well as to the complementary transformation

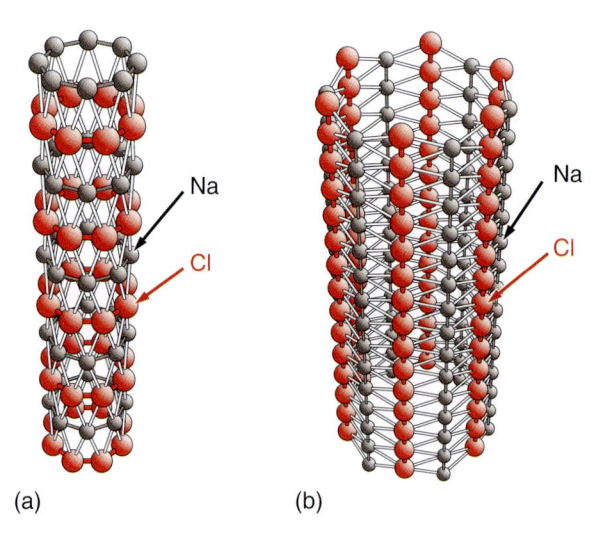

Figure 8.6 Structures of symmetric NaCl(1 2 2) nanotubes, (a) (8, 0) and (b) (0, 6) tubes.

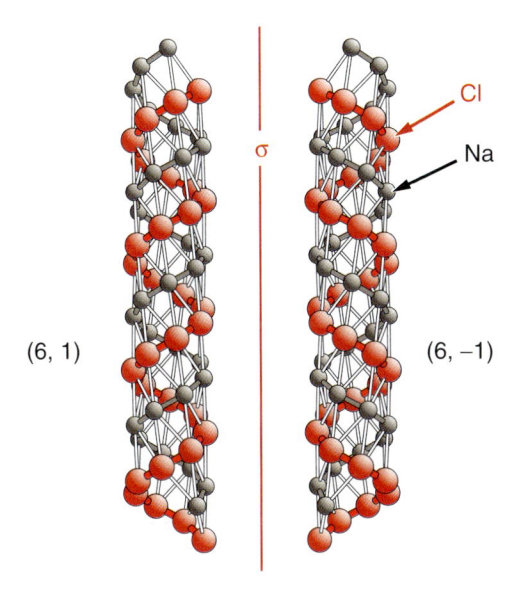

Figure 8.7 Chiral pair of NaCl(1 2 2) nanotubes, (6, 1) to the left and (6, −1) to the right. The red line between the tubes denoted σ is meant to indicate the mirror symmetry between the two tubes.

$(x, y, z) \rightarrow (x'_t, y'_t, z'_t)$ in Cartesian space with

$$x'_t = \kappa \cos \varphi, \qquad y'_t = \kappa \sin \varphi, \qquad z'_t = z$$
$$\kappa = (R_r + x)/2\pi, \qquad \varphi = (-y/R_r)\,360°, \qquad x \geq -R_r \qquad (8.6)$$

For monolayers with **mirror symmetry** (mirror lines in the corresponding netplane) chiral pairs of nanotubes can also be obtained by two different rolling vectors $\underline{R}_r$ and $\underline{R}'_r$ (rolling at the same side of the layer) where $\underline{R}'_r$ is a mirror image of $\underline{R}_r$. As an illustration, Figure 8.7 shows the chiral pairs of NaCl(1 2 2) (6, 1) and (6, −1) nanotubes.

If a **mirror line** σ exists **perpendicular** to the **rolling vector** $\underline{R}_r$ in the monolayer then there is always a **second** mirror line σ′ parallel to the first at a distance 1/2 R_r which can be proven analogous to the discussion of mirror lines in Section 3.6.4. Then the rolling transformation converts the two mirror lines, σ and σ′, on the monolayer into a mirror plane that goes through the axis of the corresponding nanotube. As a result, the nanotube exhibits **mirror symmetry** along its tube axis and is **achiral**. This is illustrated in Figure 8.8 for a (1 1 1) monolayer of an fcc crystal and a rolling vector $\underline{R}_r^{(6,0)}$. The mirror lines, σ and σ′ of the monolayer in Figure 8.8a lie on a mirror plane of the nanotube along its axis, see Figure 8.8b, illustrating its mirror symmetry and achirality.

Mirror lines σ **parallel** to the **rolling vector** $\underline{R}_r$ in the monolayer are converted by the rolling transformation to mirror planes pointing **perpendicular** to the nanotube **axis**. Thus, corresponding nanotubes exhibit **mirror symmetry** again and are **achiral**. Figure 8.8 can also be used to illustrate this behavior. The fcc(1 1 1)

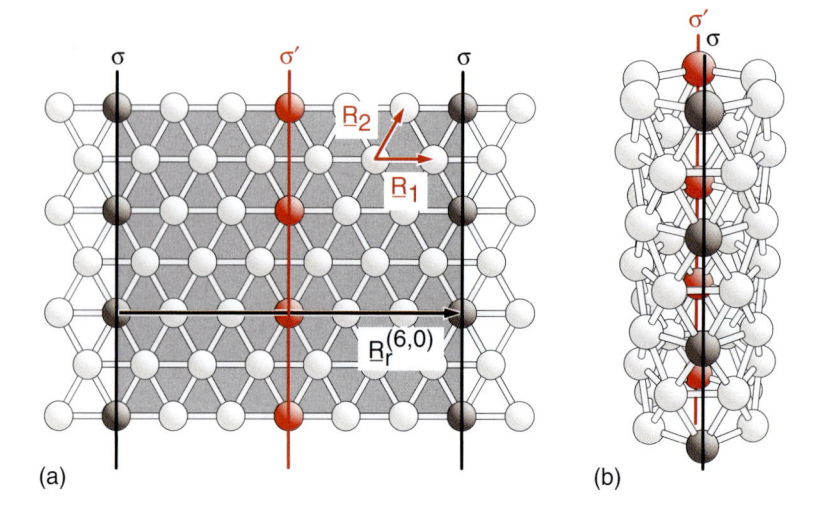

Figure 8.8 (a) Structure of an fcc(1 1 1) monolayer including a nanotube strip (emphasized in gray) with rolling vector $\underline{R}_r^{(6,0)}$. The netplane-adapted lattice vectors $\underline{R}_1$ and $\underline{R}_2$ are shown in red and mirror lines σ, σ' as thick lines. (b) fcc(1 1 1) (6, 0) nanotube corresponding to the nanotube strip in (a).

monolayer contains, in addition to its mirror lines perpendicular to $\underline{R}_r^{(6,0)}$ indicated in Figure 8.8a, mirror lines parallel to $\underline{R}_r^{(6,0)}$. The resulting mirror symmetry of the nanotube perpendicular to its axis is quite clear from Figure 8.8b.

According to Eqs. (8.3) and (8.6), the netplane coordinate z **perpendicular** to the **rolling vector** $\underline{R}_r$ in the monolayer is transformed by the tube rolling procedure to yield coordinate z_t or z'_t **along** the **nanotube axis**. Thus, **translational periodicity** of the nanotube along its axis is connected with periodicity of the corresponding monolayer along the direction perpendicular to the rolling vector. This means, in particular, that the monolayer must contain **lattice vectors** $\underline{R}_s$ **perpendicular** to the **rolling vector** $\underline{R}_r$ to yield a nanotube with translational symmetry. If vector $\underline{R}_s$, represented as

$$\underline{R}_s = p\,\underline{R}_1 + q\,\underline{R}_2, \quad p,\,q \text{ integer} \tag{8.7}$$

denotes the smallest of these perpendicular vectors then according to Eq. (8.2)

$$(\underline{R}_s\,\underline{R}_r) = (p\,\underline{R}_1 + q\,\underline{R}_2)\,(m\,\underline{R}_1 + n\,\underline{R}_2)$$
$$= [m\,R_1^2 + n\,(\underline{R}_1\,\underline{R}_2)]\,p + [n\,R_2^2 + m\,(\underline{R}_1\,\underline{R}_2)]\,q = 0 \tag{8.8}$$

Together with representation (8.1) of the lattice vectors $\underline{R}_1$ and $\underline{R}_2$ this yields equations

$$F_1\,p + F_2\,q = 0 \quad \text{or} \quad F_1/F_2 = -q/p \tag{8.9}$$

where

$$F_1 = \sum_{i=1}^{3}\sum_{j=1}^{3} (m\,t_{1i}t_{1j} + n\,t_{1i}t_{2j})\,(\underline{R}_{oi}\,\underline{R}_{oj}) \tag{8.10a}$$

$$F_2 = \sum_{i=1}^{3}\sum_{j=1}^{3} (n\,t_{2i}t_{2j} + m\,t_{1i}t_{2j})(\underline{R}_{oi}\,\underline{R}_{oj}) \qquad (8.10b)$$

Thus, **translational periodicity** of a nanotube along its axis requires that the **ratio** F_1/F_2 assumes a **rational** value. This requirement can be satisfied for **all** $(h\,k\,l)$ **monolayers** of crystals whose lattices are

a) **cubic** (sc, fcc, bcc), since

$\quad R_{oi}^2 = a^2, \qquad (\underline{R}_{oi}\,\underline{R}_{oj}) = 0, \qquad i \neq j, \quad$ for sc lattices

$\quad R_{oi}^2 = 1/2\,a^2, \quad (\underline{R}_{oi}\,\underline{R}_{oj}) = 1/4\,a^2, \quad i \neq j, \quad$ for fcc lattices

$\quad R_{oi}^2 = 3/4\,a^2, \quad (\underline{R}_{oi}\,\underline{R}_{oj}) = -1/4\,a^2, \; i \neq j, \quad$ for bcc lattices

b) **hexagonal close-packed** (hcp) (i.e., hexagonal with a ratio $c/a = \sqrt{(8/3)}$, since

$\quad R_{o1}^2 = R_{o2}^2 = a^2, \quad R_{o3}^2 = 8/3\,a^2, \quad (\underline{R}_{o1}\,\underline{R}_{o2}) = 1/2\,a^2, \quad (\underline{R}_{oi}\,\underline{R}_{o3}) = 0, \quad i \neq 3$

c) **primitive orthorhombic** or **tetragonal** with lattice constants a, b, c, resulting in rational values of $(b/a)^2$ and $(c/a)^2$, since

$\quad R_{o1}^2 = a^2, \quad R_{o2}^2 = (b/a)^2 a^2, \quad R_{o3}^2 = (c/a)^2 a^2, \quad (\underline{R}_{oi}\,\underline{R}_{oj}) = 0, \quad i \neq j$

Further, **translational periodicity** of nanotubes is also guaranteed for **selected** $(h\,k\,l)$ monolayers of crystals with other Bravais lattices. As examples, we mention

(d) (0 0 0 1) monolayers of **hexagonal** lattices,
(e) (0 0 1) monolayers of **tetragonal** lattices,
(f) (1 1 1) monolayers of **trigonal** lattices.

In general, a nanotube can be **achiral** only if it exhibits **mirror symmetry** where possible mirror planes are either perpendicular to the tube axis or contain the axis. Thus, unrolling an achiral nanotube will always create a strip inside a $(h\,k\,l)$ monolayer that contains a mirror line σ either perpendicular or parallel to the rolling vector $\underline{R}_r$. This means, in particular, that symmetry properties of the corresponding $(h\,k\,l)$ monolayer must be described by a two-dimensional space group that contains at least one mirror line. This excludes, according to Table 3.13 and Figure 3.49, monolayers with symmetry described by space groups 1, 2 (oblique), 4, 8 (p-rectangular), 10 (square), and 13, 16 (hexagonal).

8.3 Complex Nanotubes

Nanotubes can assume much more complex geometric arrangements compared with those defined by rolling single crystal monolayers, which have been discussed so far. In particular, structural details of **thicker nanotubes** do not need

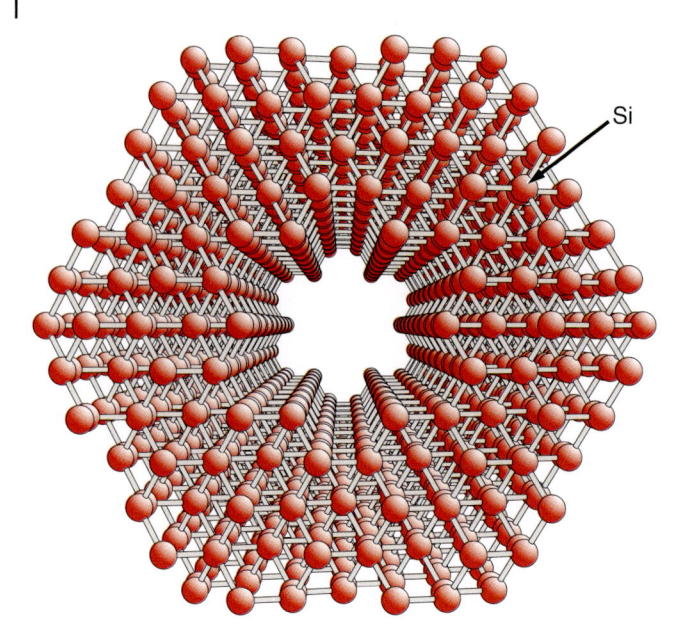

Figure 8.9 Structure of a thick silicon nanotube of hexagonal cross section simulated by a balls-and-sticks model based on the crystal structure of bulk silicon.

to be immediately connected with rolling single crystal layers. As an example, silicon nanotubes with thicker walls have been proposed to possess polygonal rather than circular cross-sections consisting of distorted crystalline material and are described as **hollow nanowires** [201]. This is illustrated by Figure 8.9, which shows a model of a silicon nanotube with thick walls made of single crystal bulk silicon and a hexagonal cross section.

On the other hand, rolled nanotubes can also exhibit rather complex structures. As an example, we mention **multi-walled** nanotubes of carbon, which have been observed [202]. Here, foreign atoms may be inserted between the walls to yield **intercalation nanotubes**. As an illustration, Figure 8.10 shows a fictitious double-walled carbon nanotube composed of a (12, 3) tube (outer wall) and a (7, 3) tube (inner wall).

Rolled nanotubes can be composed of even **thicker crystal layers**. This is of particular interest for crystals with **layer-type lattices**, where strong chemical binding exists inside physical layers combining several monolayers. These **physical layers** can then be **rolled** to form **complex nanotubes**. As an example, vanadium pentoxide, V_2O_5, discussed in Section 4.1, is described by a layer-type orthorhombic lattice with 14 atoms (4 vanadium, 10 oxygen atoms) in the unit cell. Here, (0 1 0) oriented physical layers (of eight monolayers each, see Figure 4.4) are loosely coupled to form the layer-type crystal. These physical layers may serve as building units for V_2O_5 nanotubes, which has also been confirmed experimentally [203]. As an example, Figure 8.11 shows the model of a section of a V_2O_5(0 1 0) (0, 5) nanotube (referring to netplane-adapted lattice vectors $\underline{R}_1$ and $\underline{R}_2$ given in Figure 4.4),

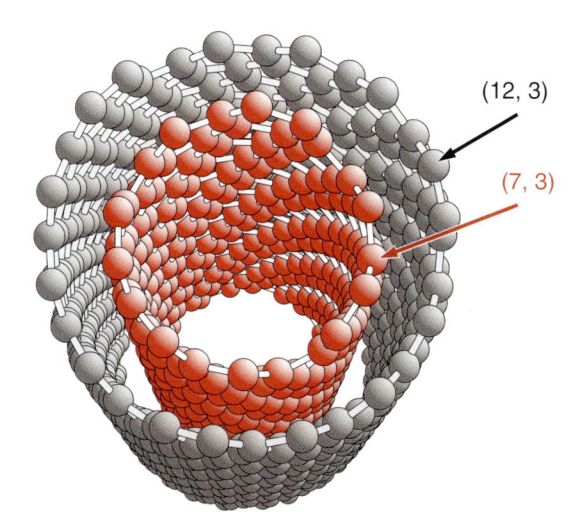

Figure 8.10 Structure of a fictitious double-walled carbon nanotube combining a (12, 3) tube (outer wall, gray) with a (7, 3) tube (inner wall, red).

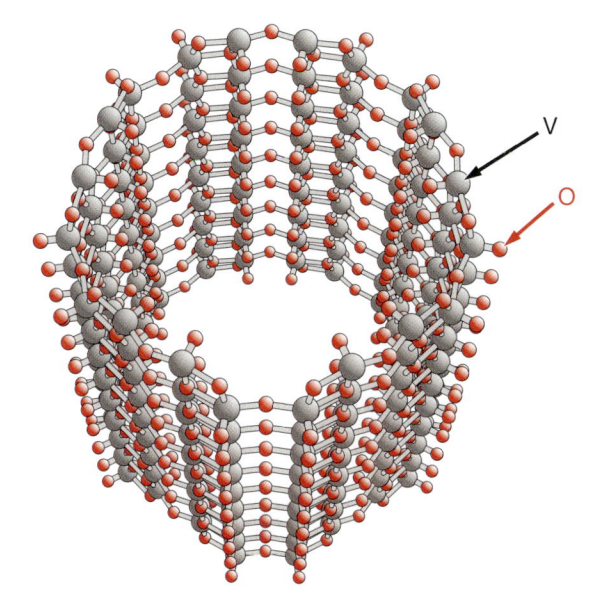

Figure 8.11 Structure of a (0, 5) nanotube of a (0 1 0) oriented physical layer of the V_2O_5 crystal.

which arises from one physical layer. The singly coordinated vanadyl oxygen atoms sticking out of this nanotube surface are assumed to be catalytically active.

Even more complex shapes of nanotubes have been observed by experiment [202], in particular, for carbon. Examples are flexible ("**spaghetti**" type) carbon nanotubes, nanotube **junctions** (reminding of junctions of tree branches), or

8.4
Exercises

8.1 Consider a (1 1 0) oriented monolayer of an fcc crystal (lattice constant a) and nanotubes originating from the monolayer with rolling vectors $\underline{R}_r^{(m,n)}$ according to Eq. (8.2).

(a) Determine radii $R_{tube}(m, n)$ of the nanotubes as a function of a, m, n. Hint: the required netplane-adapted lattice vectors $\underline{R}_1$ and $\underline{R}_2$ can be represented by

$$\underline{R}_1 = a/\sqrt{2}\,(1,0,0), \qquad \underline{R}_2 = a\,(0,\,1,\,0)$$

(b) Which of the (m, n) nanotubes are translationally periodic? Along which direction?

(c) Determine the indices (m, n) and (m', n') of chiral pairs of nanotubes. Which (m, n) nanotubes are achiral?

8.2 Show that (m, n) nanotubes of (0 0 0 1) monolayers of graphite are always translationally periodic along the tube axis. Calculate periodicity lengths L(m, n). Calculate tube radii $R_{tube}(m, n)$.

8.3 Discuss symmetry elements of (m, n) nanotubes of fcc(1 1 1) and bcc(1 1 0) monolayers.

8.4 Consider a (monoatomic) monolayer with Minkowski-reduced lattice vectors $\underline{R}_1$ and $\underline{R}_2$ describing a Bravais lattice with symmetry and a rolling vector $\underline{R}_r^{(m,n)}$ according to Eq. (8.2).

(a) Calculate radii $R_{tube}(m, n)$ and periodicity lengths L(m, n) (if applicable) of translationally periodic (m, n) nanotubes referring to different Bravais lattices.

(b) Find constraints for the lattice vectors $\underline{R}_1$ and $\underline{R}_2$ such that there are no translationally periodic (m, n) nanotubes. Discuss examples for different Bravais lattices.

8.5 Consider a nanotube constructed from a p-rectangular monolayer with symmetry properties according to space group no. 6, see Table 3.13, and a rolling vector $\underline{R}_r^{(m,n)}$ according to Eq. (8.2). Show that all chiral partners of the (m, n) nanotube are given by $(m', n') = \pm(m, -n)$. For which (m, n) values are the nanotubes achiral?

8.6 Consider a nanotube constructed from a c-rectangular monolayer with symmetry properties according to space group no. 9, see Table 3.13, and a rolling vector $\underline{R}_r^{(m,n)}$ according to Eq. (8.2). Show that with orthogonal lattice vectors all chiral partners of the (m, n) nanotube are given by (m′, n′) = ±(m, −n). For which (m, n) values are the nanotubes achiral?

8.7 Consider a nanotube constructed from a square monolayer with symmetry properties according to space group no. 11, see Table 3.13, and a rolling vector $\underline{R}_r^{(m,n)}$ according to Eq. (8.2). Show that all chiral partners of the (m, n) nanotube are given by (m′, n′) = ±(m, −n), ±(n, −m). For which (m, n) values are the nanotubes achiral?

8.8 Consider a nanotube constructed from a hexagonal monolayer with symmetry properties according to space group no. 17 of ITC (International Tables of Crystallography) [33], see Table 3.13, and a rolling vector $\underline{R}_r^{(m,n)}$ according to Eq. (8.2). Show that all chiral partners of the (m, n) nanotube are given by
acute lattice vectors: (m′, n′) = ±(n, m), ±(m + n, −n), ±(−m, m + n)
obtuse lattice vectors: (m′, n′) = ±(n, m), ±(m − n, −n), ±(m, m − n)
For which (m, n) values are the nanotubes achiral?

8.9 Discuss chiral pairs of (m, n) nanotubes of (1 0 0)-, (1 1 0)-, and (1 1 1)-oriented monolayers of gold crystals. For which values of m, n are the corresponding nanotubes achiral?

8.10 Compare (m, n) nanotubes of (1 1 0)-oriented monolayers of nickel crystals with those of vanadium. Which nanotubes are achiral for both nickel and vanadium?

8.11 Consider fictitious (m, n) nanotubes of dense (h k l) monolayers of NaCl and CsCl crystals. Which nanotubes include both elements? Discuss examples.

8.12 Hexagonal boron nitride, BN, is described by a layer-type crystal structure, which is analogous to that of graphite. Thus, ignoring layer buckling, (0 0 0 1) monolayers of BN have a honeycomb appearance and can be described approximately by a two-dimensional lattice with lattice and lattice basis vectors

$$\underline{R}_1 = a\,(1,\,0), \quad \underline{R}_2 = a\,(1/2,\,\sqrt{3}/2), \quad \underline{r}_1^B = (0,\,0), \quad \underline{r}_2^N = (1/3, 1/3)$$

Determine achiral (m, n) nanotubes of (0 0 0 1) BN and show that there are only two different types.

8.13 Consider a monolayer with lattice vectors of a square netplane and two different elements A, B in the unit cell, described by

$$\underline{R}_1 = a\,(1,0), \quad \underline{R}_2 = a\,(0,1), \quad \underline{r}_1^A = (0,0), \quad \underline{r}_2^B = (x, y), \quad 0 < x, y < 1$$

Determine values x, y which allow achiral (m, n) nanotubes.

Appendix A: Sketches of High-Symmetry Adsorbate Sites

This appendix gives an overview of the most common sites that are considered for adsorbates stabilizing at single crystal surfaces of high symmetry. Each site is sketched in (a) a view along the surface normal and (b) a view parallel to the surface. The sites are grouped according to bulk crystal structure, referring to Sections A.1 (fcc), A.2 (bcc), A.3 (hcp), A.4 (diamond), and A.5 (zincblende), and according to surface orientation $(h\ k\ l)$. Here, the substrate is shown as a collection of gray and light red shaded atom balls near the adsorption site where the gray shading distinguishes the different atom layers becoming darker for layers that are deeper below the surface. Neighboring atoms of each layer are labeled Nx according to their layer index ($N = 1$ (topmost), 2, 3, ...) and distance from the adsorption site ($x = $ a (nearest), b, c, ...). The adsorption site is shown by a small dark red ball representing the adsorbate center and larger light red balls referring to the substrate atoms nearest to the adsorbate. The dashed line in each view (a) denotes the cutting plane used for the view in (b).

A.1 Face-Centered Cubic (fcc) Surface Sites

Details of the sketches are explained in the beginning of this appendix (Figures A.1–A.15).

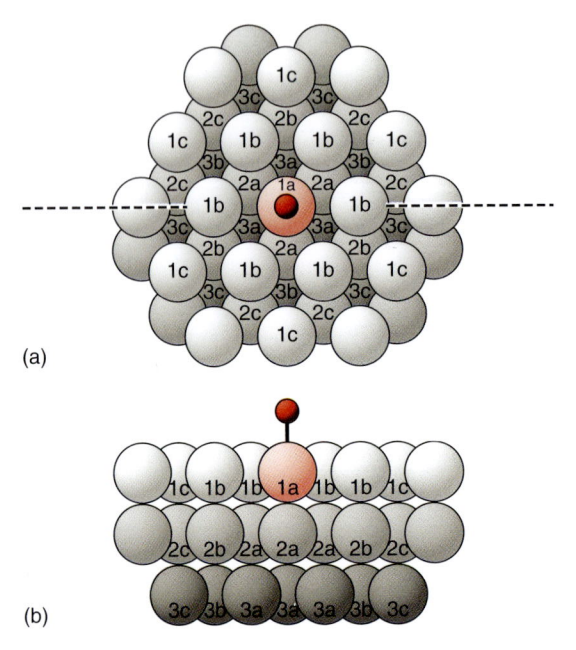

Figure A.1 (a, b) fcc(1 1 1) Top.

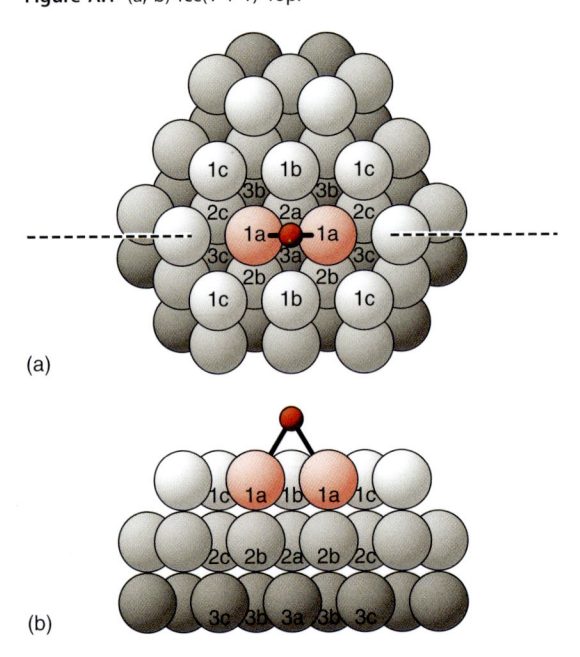

Figure A.2 (a, b) fcc(1 1 1) Bridge.

A.1 Face-Centered Cubic (fcc) Surface Sites | 331

(a)

(b)

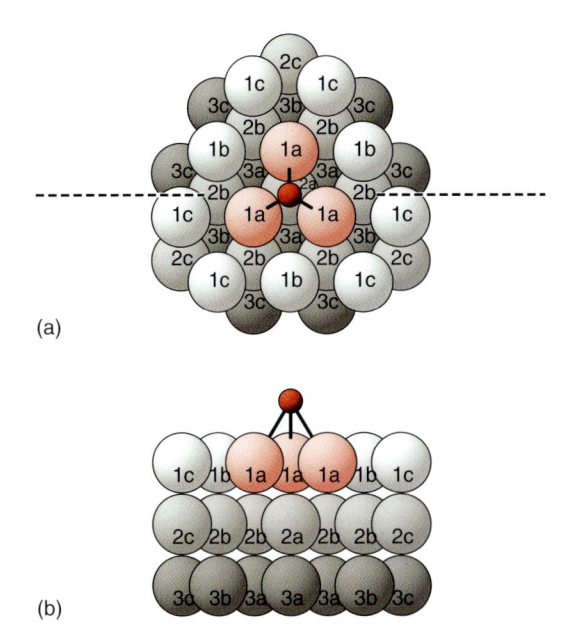

Figure A.3 (a, b) fcc(1 1 1) hcp Hollow.

(a)

(b)

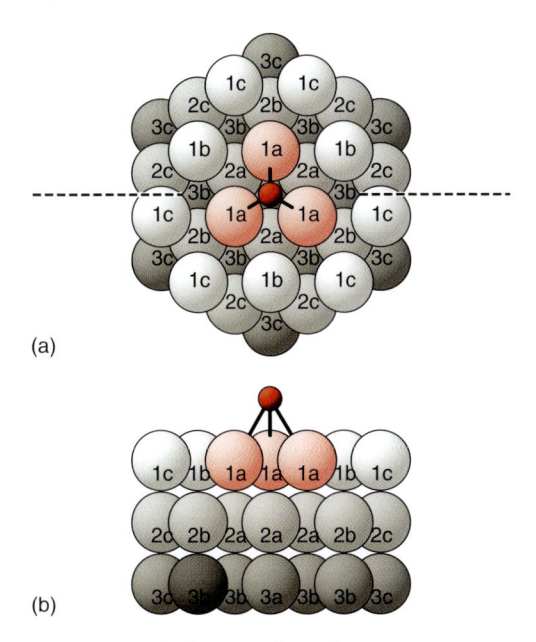

Figure A.4 (a, b) fcc(1 1 1) fcc Hollow.

332 | *Appendix A: Sketches of High-Symmetry Adsorbate Sites*

(a)

(b)

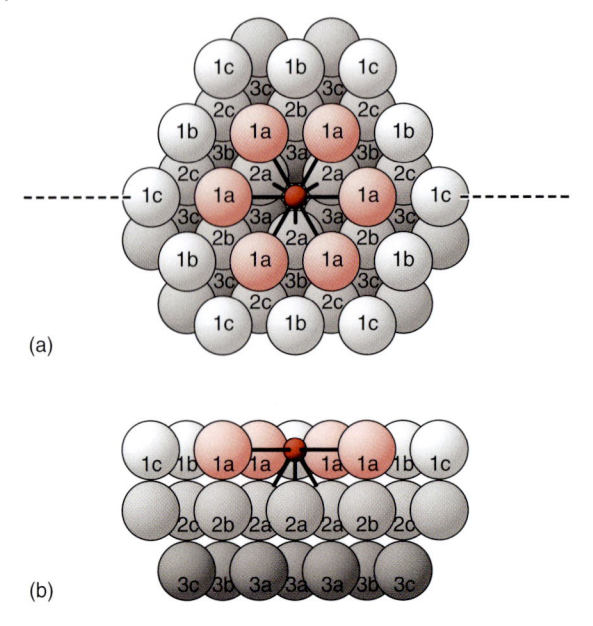

Figure A.5 (a, b) fcc(1 1 1) Substitutional.

(a)

(b)

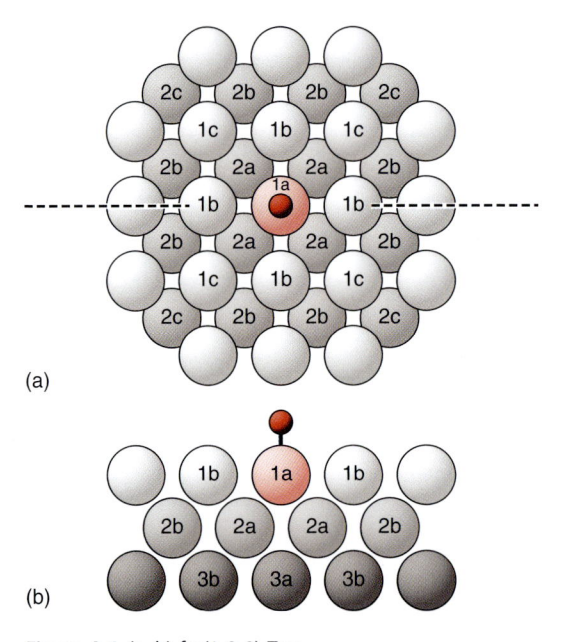

Figure A.6 (a, b) fcc(1 0 0) Top.

A.1 Face-Centered Cubic (fcc) Surface Sites | 333

(a)

(b)

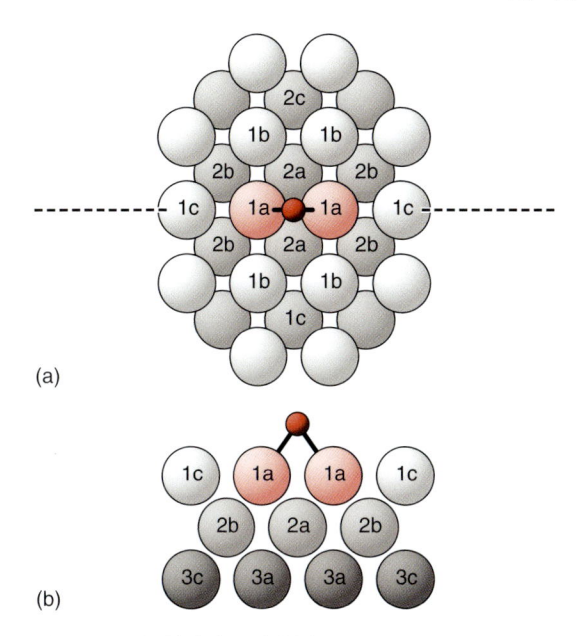

Figure A.7 (a, b) fcc(1 0 0) Bridge.

(a)

(b)

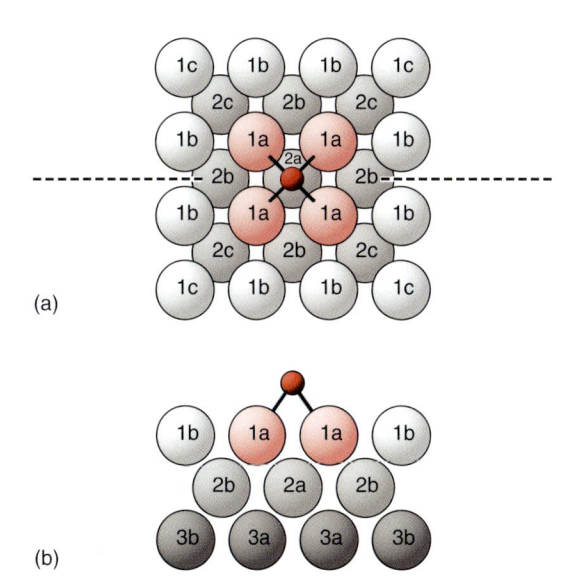

Figure A.8 (a, b) fcc(1 0 0) Fourfold hollow.

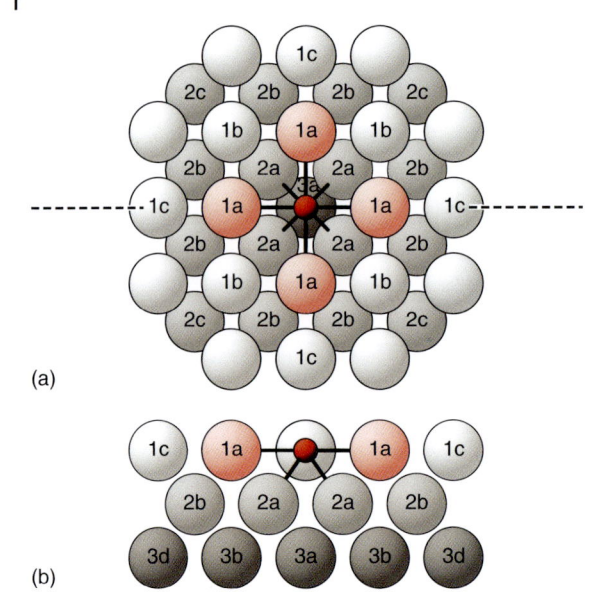

Figure A.9 (a, b) fcc(1 0 0) Substitutional.

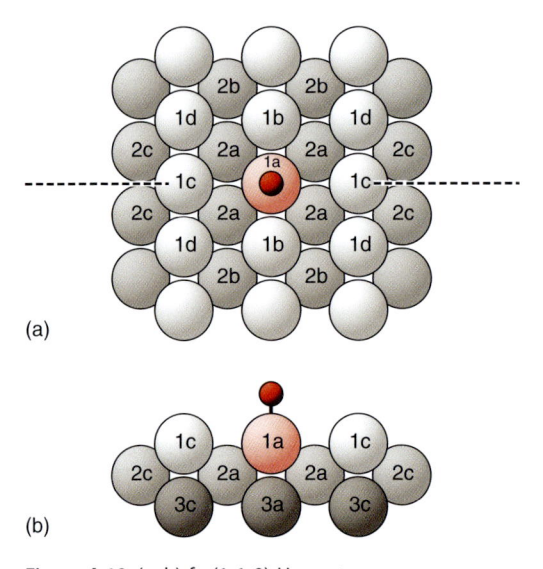

Figure A.10 (a, b) fcc(1 1 0) Upper top.

A.1 Face-Centered Cubic (fcc) Surface Sites | 335

(a)

(b)

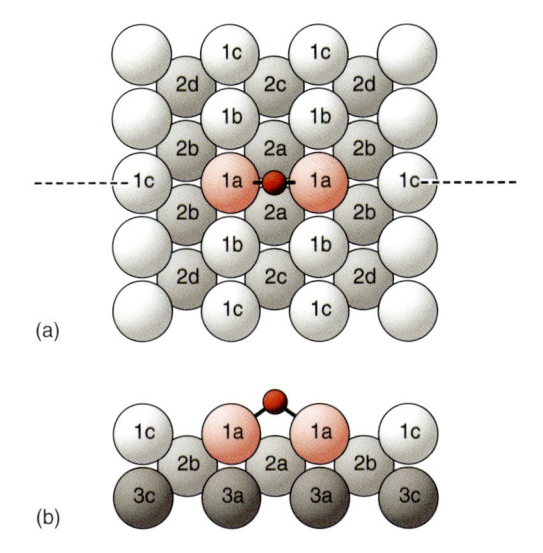

Figure A.11 (a, b) fcc(1 1 0) Long bridge.

(a)

(b)

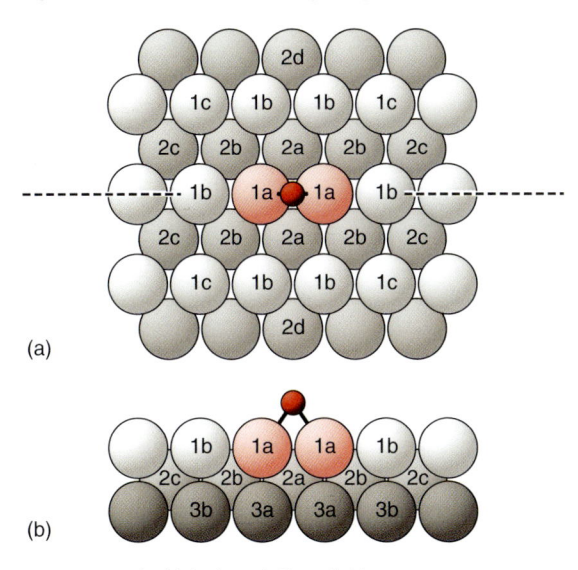

Figure A.12 (a, b) fcc(1 1 0) Short bridge.

336 | *Appendix A: Sketches of High-Symmetry Adsorbate Sites*

(a)

(b)

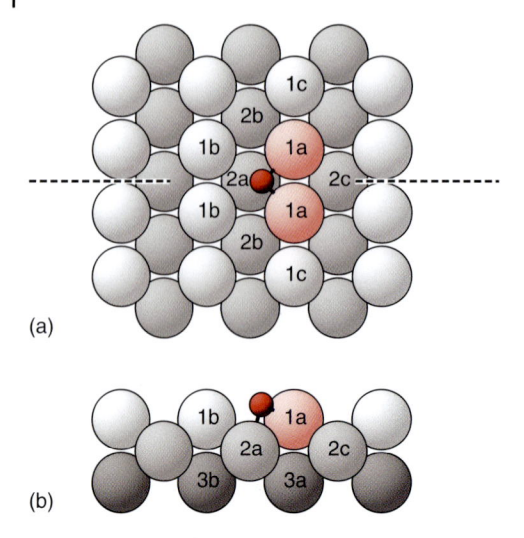

Figure A.13 (a, b) fcc(1 1 0) Threefold hollow.

(a)

(b)

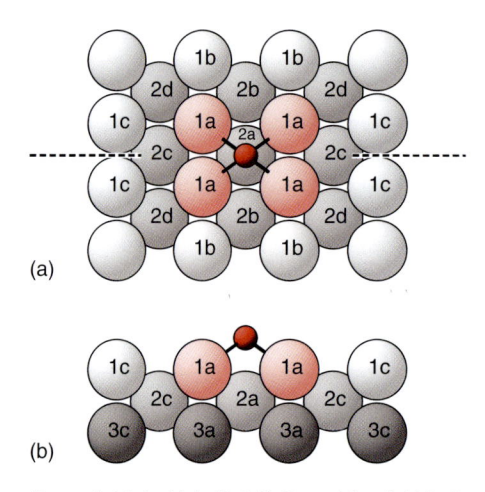

Figure A.14 (a, b) fcc(1 1 0) Central fourfold hollow.

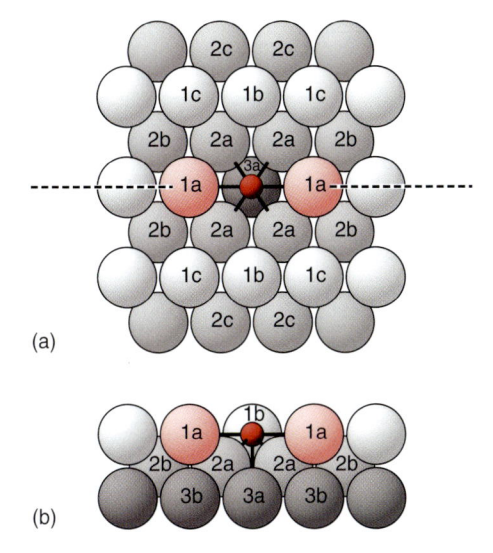

Figure A.15 (a, b) fcc(1 1 0) Upper substitutional.

A.2 Body-Centered Cubic (bcc) Surface Sites

Details of the sketches are explained in the beginning of this appendix (Figures A.16–A.22).

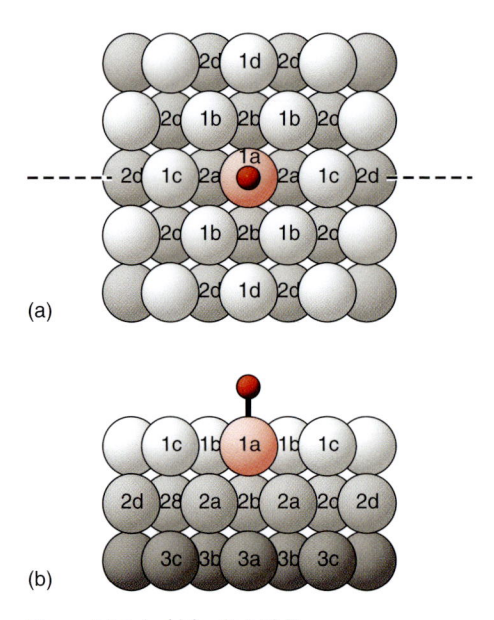

Figure A.16 (a, b) bcc(1 1 0) Top.

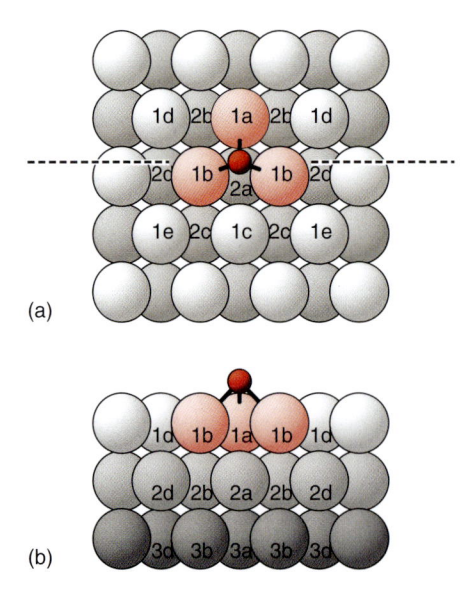

Figure A.17 (a, b) bcc(1 1 0) Threefold hollow.

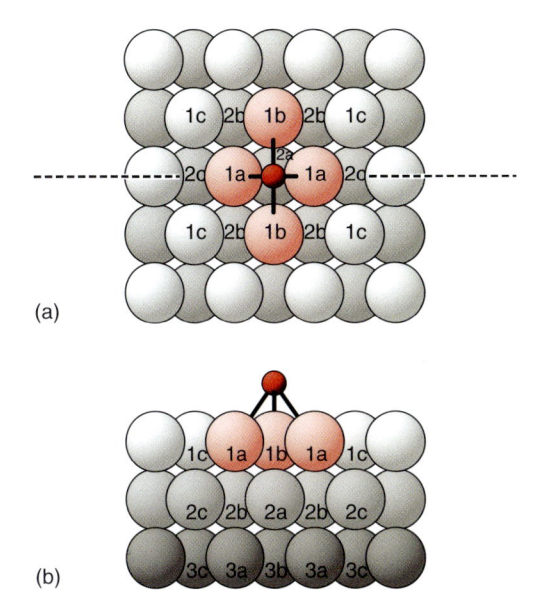

Figure A.18 (a, b) bcc(1 1 0) Central hollow.

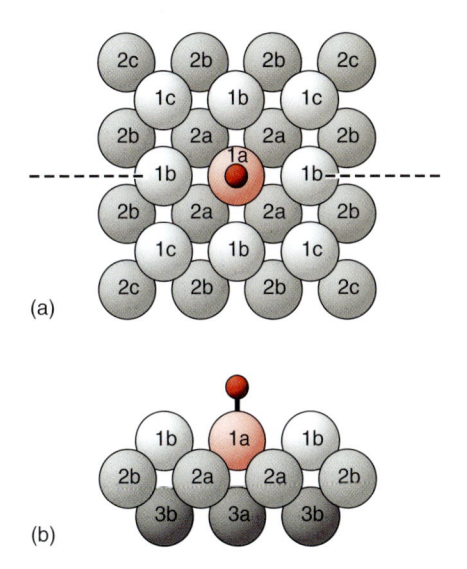

Figure A.19 (a, b) bcc(1 0 0) Top.

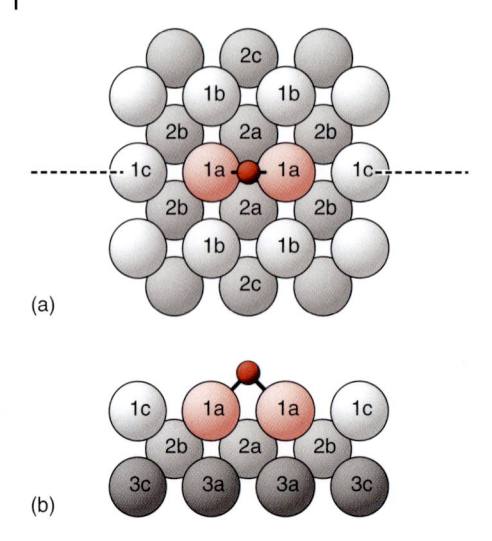

Figure A.20 (a, b) bcc(1 0 0) Bridge.

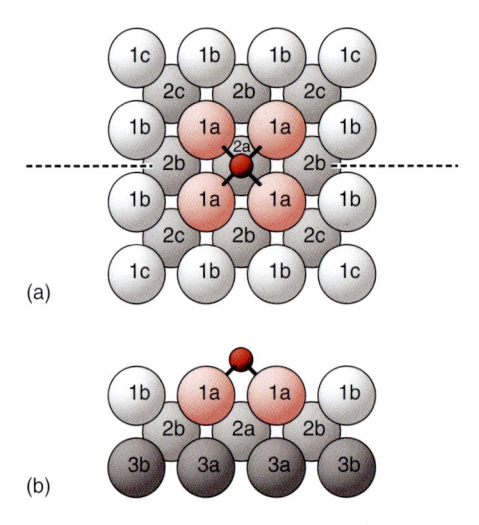

Figure A.21 (a, b) bcc(1 0 0) Fourfold hollow.

(a)

(b)

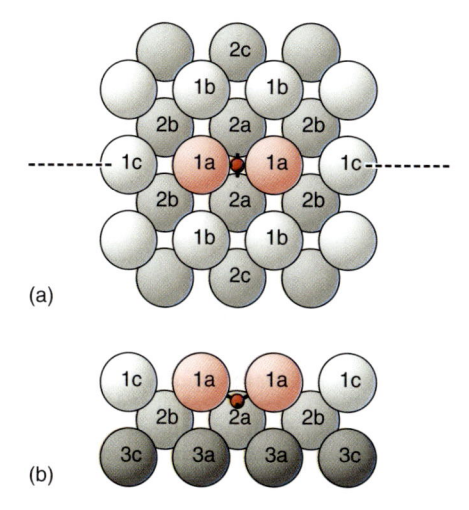

Figure A.22 (a, b) bcc(1 0 0) Fourfold interstitial.

Appendix A: Sketches of High-Symmetry Adsorbate Sites

A.3 Hexagonal Close-Packed (hcp) Surface Sites

Details of the sketches are explained in the beginning of this appendix (Figures A.23–A.30).

(a)

(b)

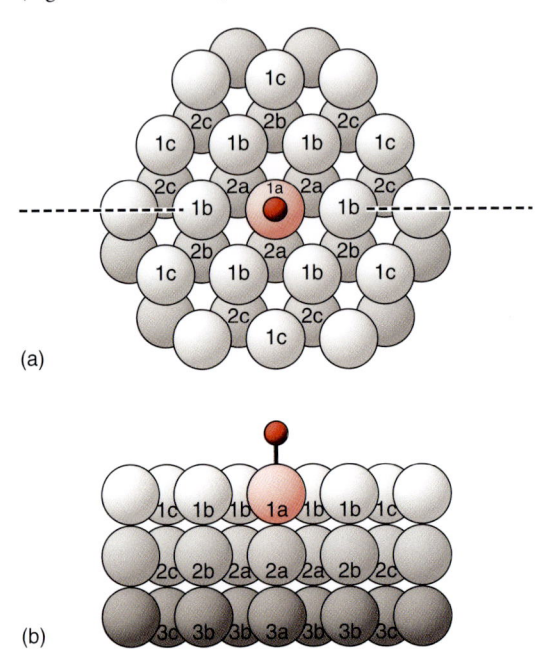

Figure A.23 (a, b) hcp(0 0 0 1) Top.

(a)

(b)

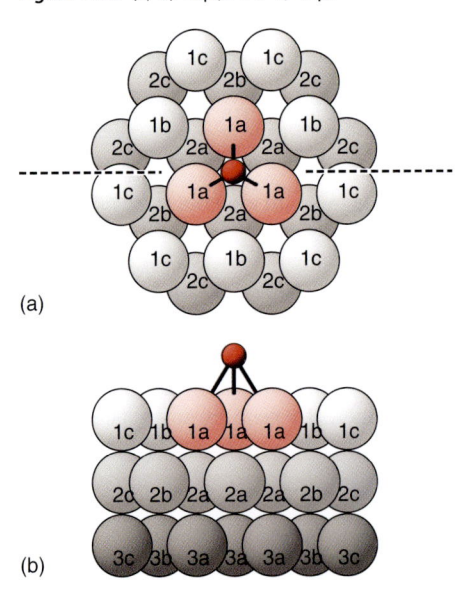

Figure A.24 (a, b) hcp(0 0 0 1) fcc Hollow.

A.3 Hexagonal Close-Packed (hcp) Surface Sites | **343**

(a)

(b)

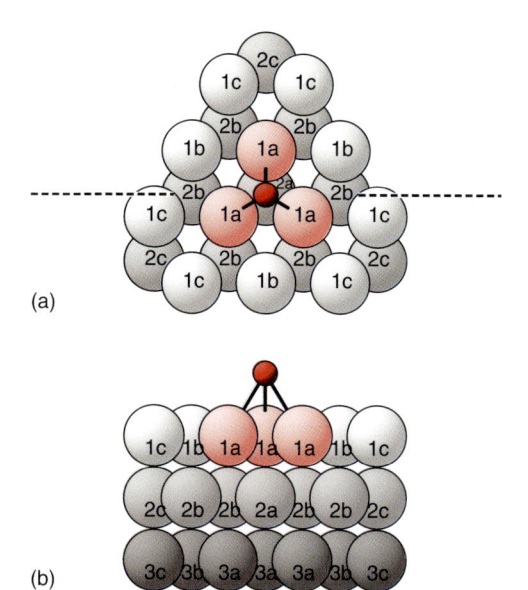

Figure A.25 (a, b) hcp(0 0 0 1) hcp Hollow.

(a)

(b)

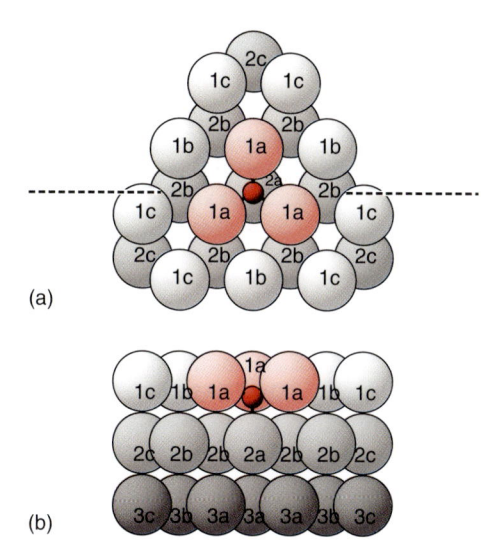

Figure A.26 (a, b) hcp(0 0 0 1) Tetrahedral interstitial.

344 | *Appendix A: Sketches of High-Symmetry Adsorbate Sites*

(a)

(b)

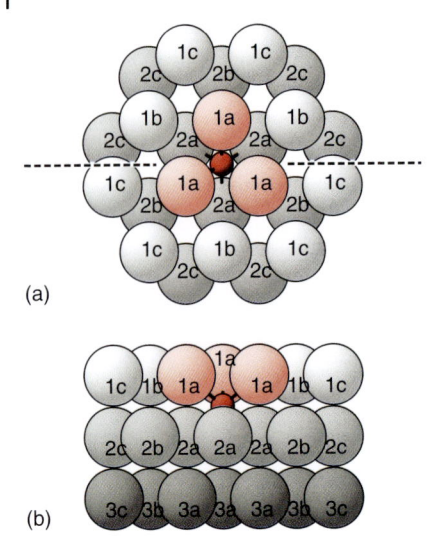

Figure A.27 (a, b) hcp(0 0 0 1) Octahedral interstitial.

(a)

(b)

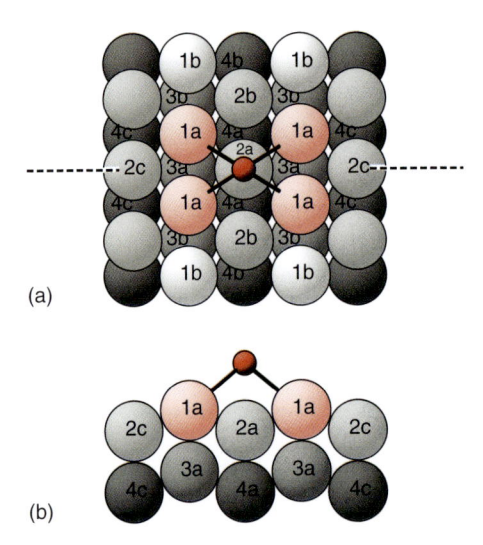

Figure A.28 (a, b) hcp(1 0 −1 0) Fourfold center.

A.3 Hexagonal Close-Packed (hcp) Surface Sites | 345

(a)

(b)

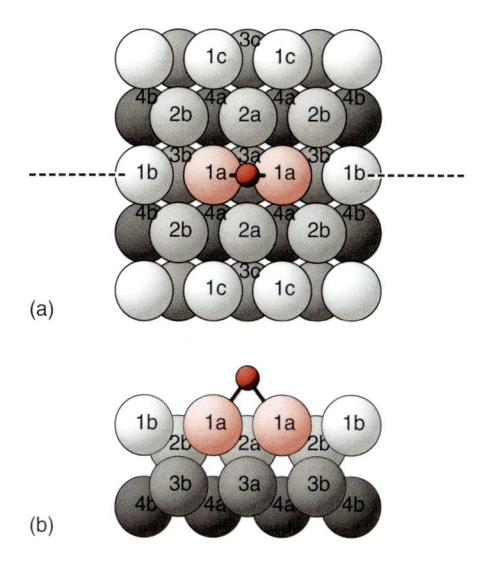

Figure A.29 (a, b) hcp(1 0 −1 0) Short bridge.

(a)

(b)

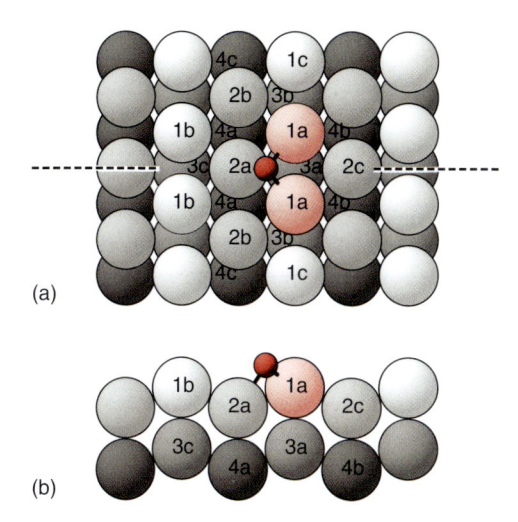

Figure A.30 (a, b) hcp(1 0 −1 0) Threefold hollow.

A.4 Diamond Surface Sites

Details of the sketches are explained in the beginning of this appendix (Figures A.31–A.36).

(a)

(b)

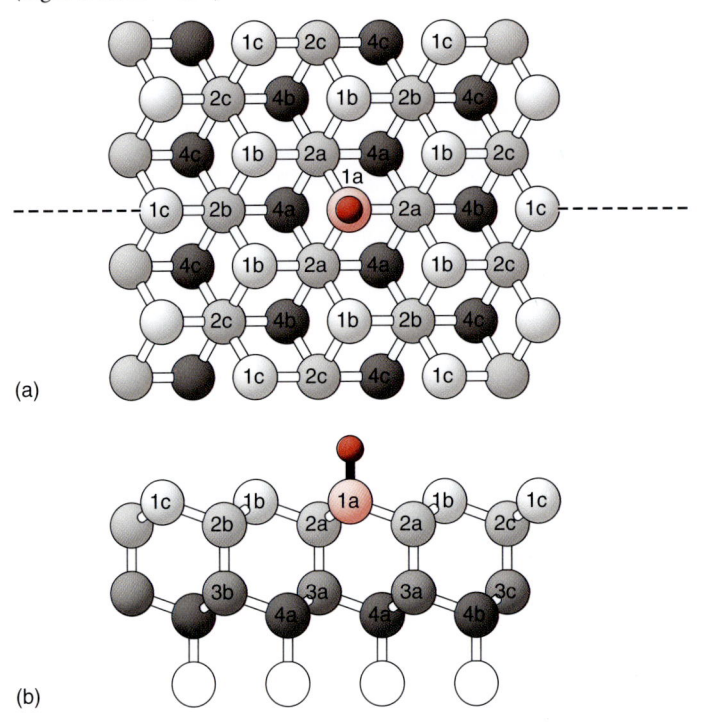

Figure A.31 (a, b) Diamond(1 1 1) top.

(a)

(b)

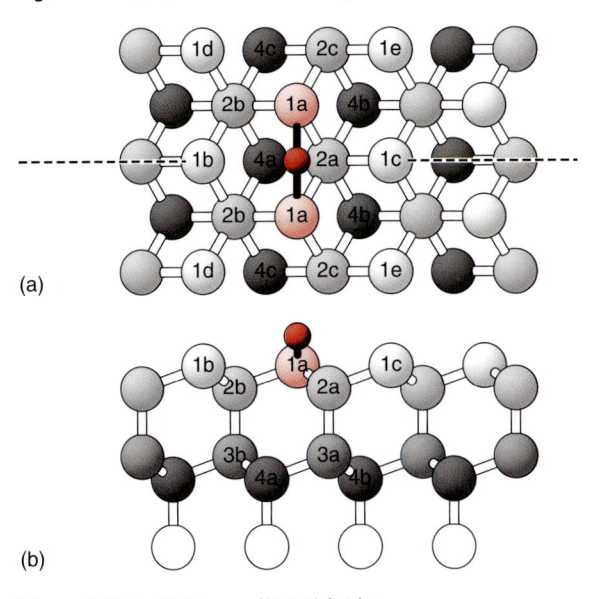

Figure A.32 (a, b) Diamond(1 1 1) bridge.

A.4 Diamond Surface Sites | 347

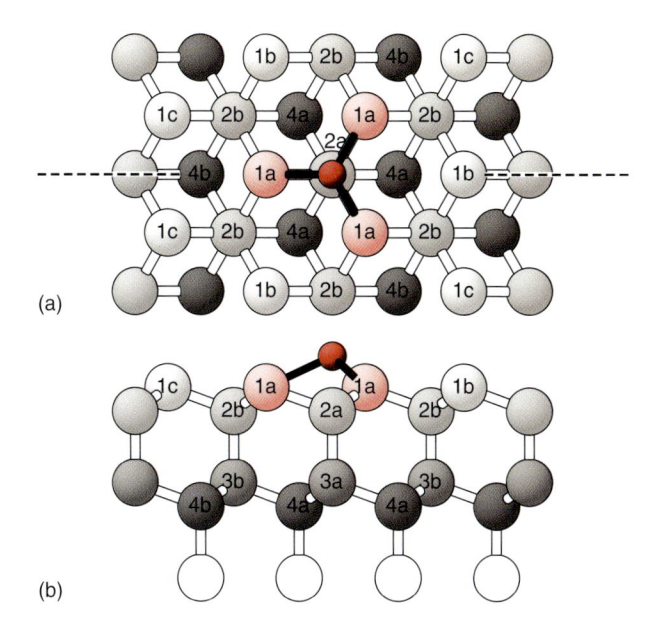

Figure A.33 (a, b) Diamond(1 1 1) T4.

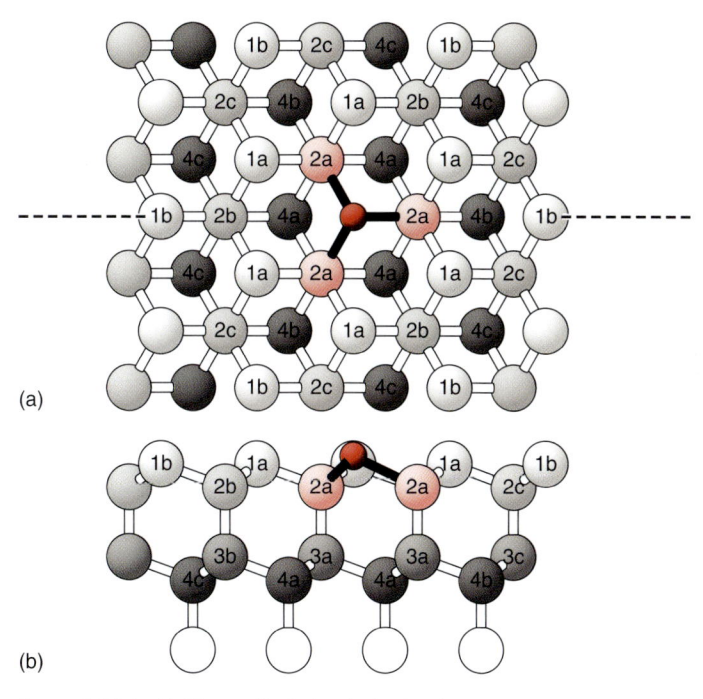

Figure A.34 (a, b) Diamond(1 1 1) substitutional top.

348 *Appendix A: Sketches of High-Symmetry Adsorbate Sites*

(a)

(b)

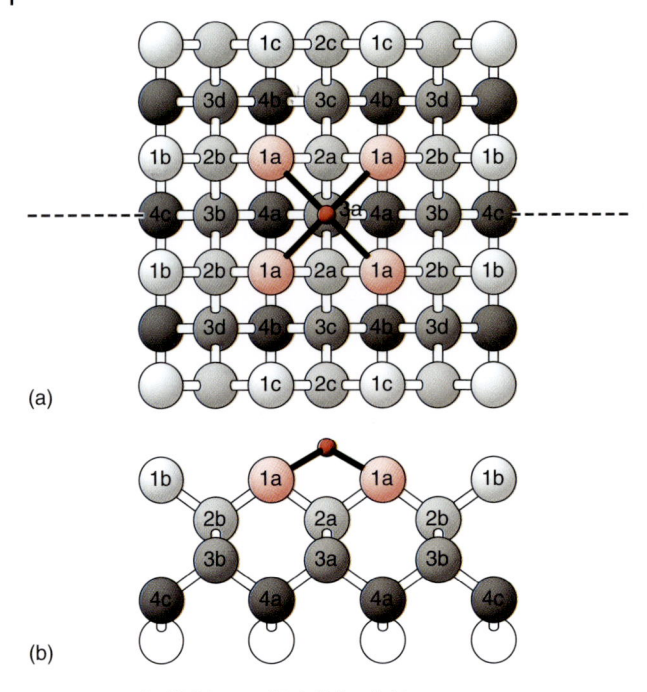

Figure A.35 (a, b) Diamond(1 0 0) fourfold center.

(a)

(b)

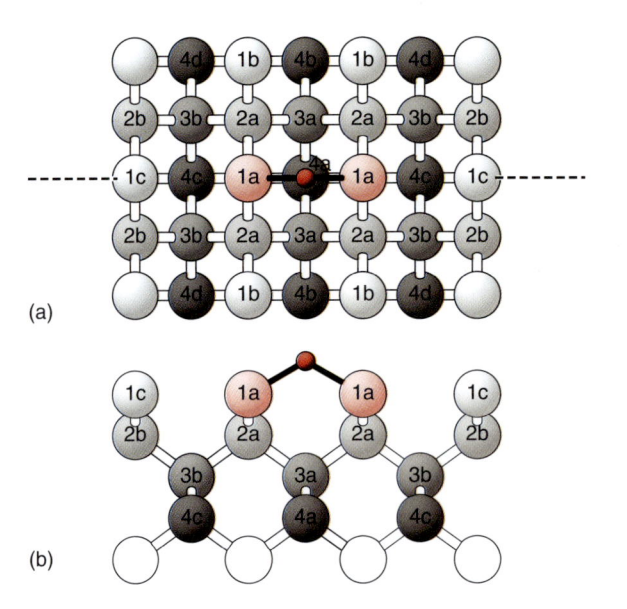

Figure A.36 (a, b) Diamond(1 0 0) continuation bridge.

A.5 Zincblende Surface Sites

Details of the sketches are explained in the beginning of this appendix (Figures A.37 and A.38).

(a)

(b)

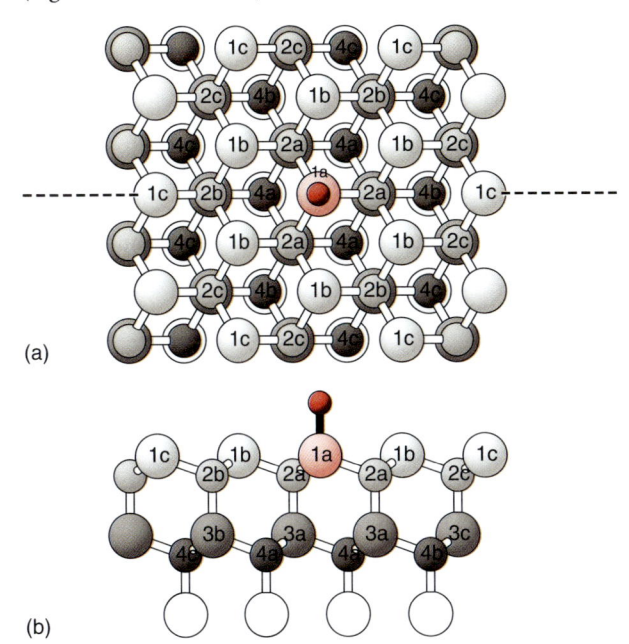

Figure A.37 (a, b) Zincblende(1 1 1) top.

(a)

(b)

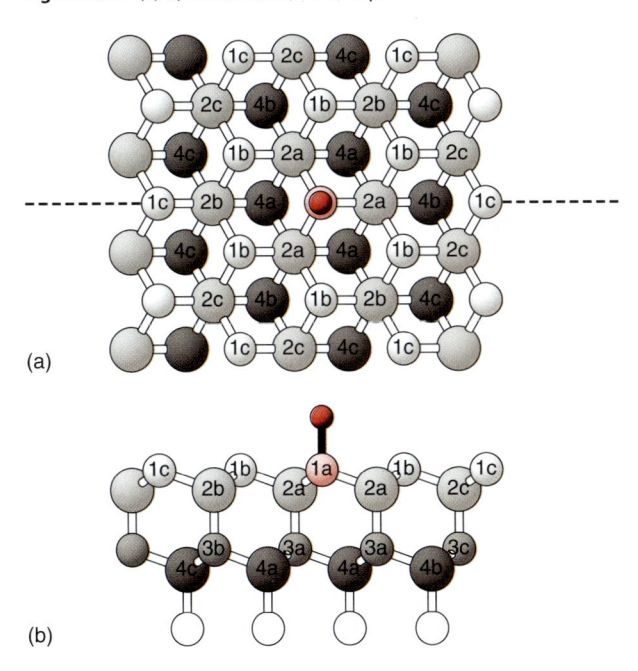

Figure A.38 (a, b) Zincblende(-1 -1 -1) top.

Appendix B: Parameter Tables of Crystals

This appendix lists the most common lattice types and corresponding lattice parameters of elemental crystals at standard temperature and pressure (Tables B.1–B.3).

Table B.1 Lattices of elemental single crystals.

Element	Lattice	Element	Lattice	Element	Lattice	Element	Lattice
1 H	hex	26 Fe	bcc	51 Sb	rhl	76 Os	hex
2 He	hex	27 Co	hex	52 Te	hex	77 Ir	fcc
3 Li	bcc	28 Ni	fcc	53 I	ort	78 Pt	fcc
4 Be	hex	29 Cu	fcc	54 Xe	fcc	79 Au	fcc
5 B	tet	30 Zn	hex	55 Cs	bcc	80 Hg	rhl
6 C	dia,hex	31 Ga	ort	56 Ba	bcc	81 Tl	hex
7 N	hex	32 Ge	dia	57 La	hex	82 Pb	fcc
8 O	mcl	33 As	rhl	58 Ce	fcc	83 Bi	rhl
9 F	mcl	34 Se	hex	59 Pr	hex	84 Po	cub
10 Ne	fcc	35 Br	ort	60 Nd	hex	85 At	—
11 Na	bcc	36 Kr	fcc	61 Pm	—	86 Rn	(fcc)
12 Mg	hex	37 Rb	bcc	62 Sm	rhl	87 Fr	(bcc)
13 Al	fcc	38 Sr	fcc	63 Eu	bcc	88 Ra	—
14 Si	dia	39 Y	hex	64 Gd	hex	89 Ac	fcc
15 P	tcl	40 Zr	hex	65 Tb	hex	90 Th	fcc
16 S	ort	41 Nb	bcc	66 Dy	hex	91 Pa	tet
17 Cl	ort	42 Mo	bcc	67 Ho	hex	92 U	ort
18 Ar	fcc	43 Tc	hex	68 Er	hex	93 Np	ort
19 K	bcc	44 Ru	hex	69 Tm	hex	94 Pu	mcl
20 Ca	fcc	45 Rh	fcc	70 Yb	fcc	95 Am	—
21 Sc	hex	46 Pd	fcc	71 Lu	hex	96 Cm	—
22 Ti	hex	47 Ag	fcc	72 Hf	hex	97 Bk	—
23 V	bcc	48 Cd	hex	73 Ta	bcc	98 Cf	—
24 Cr	bcc	49 In	tet	74 W	bcc	99 Es	—
25 Mn	cub	50 Sn	tet	75 Re	hex	100 Fm	—

The Bravais lattice types are abbreviates as follows: bcc for body-centered cubic; cub for cubic; dia, for diamond; fcc for face-centered cubic; hex for hexagonal; mcl for monoclinic; ort for orthorhombic; rhl for rhombohedral; tcl for triclinic; and tet for tetragonal.

Crystallography and Surface Structure: An Introduction for Surface Scientists and Nanoscientists,
Second Edition. Klaus Hermann.
© 2017 Wiley-VCH Verlag GmbH & Co. KGaA. Published 2017 by Wiley-VCH Verlag GmbH & Co. KGaA.

Table B.2 Lattice constants of face- and body-centered cubic single crystals.

Element	a (Å)	Element	a (Å)
(a) Face-centered cubic (fcc)			
10 Ne	4.43	54 Xe	6.20
13 Al	4.05	58 Ce	5.16
18 Ar	5.26	70 Yb	5.49
20 Ca	5.58	77 Ir	3.84
28 Ni	3.52	78 Pt	3.92
29 Cu	3.61	79 Au	4.08
36 Kr	5.72	82 Pb	4.95
38 Sr	6.08	86 Rn	—
45 Rh	3.80	89 Ac	5.31
46 Pd	3.89	90 Th	5.08
47 Ag	4.09		
(b) Body-centered cubic (bcc)			
3 Li	3.49	42 Mo	3.15
11 Na	4.23	55 Cs	6.05
19 K	5.23	56 Ba	5.02
23 V	3.02	63 Eu	4.61
24 Cr	2.88	73 Ta	3.31
26 Fe	2.87	74 W	3.16
37 Rb	5.59	87 Fr	—
41 Nb	3.30		

The fcc lattice definitions in Cartesian coordinates are
$\underline{R}_1^{fcc} = a/2\,(0, 1, 1);\ \underline{R}_2^{fcc} = a/2\,(1, 0, 1);\ \underline{R}_3^{fcc} = a/2\,(1, 1, 0)$.
The bcc lattice definitions in Cartesian coordinates are
$\underline{R}_1^{bcc} = a/2\,(-1, 1, 1);\ \underline{R}_2^{bcc} = a/2\,(1, -1, 1);\ \underline{R}_3^{bcc} = a/2\,(1, 1, -1)$.

Table B.3 Lattice constants of hexagonal single crystals.

Element	a (Å)	c/a	Element	a (Å)	c/a
1 H	3.75	1.731	52 Te	4.45	1.330
2 He	3.57	1.633	57 La	3.75	1.619
4 Be	2.29	1.567	59 Pr	3.67	1.614
7 N	4.039	1.651	60 Nd	3.66	1.614
12 Mg	3.21	1.624	64 Gd	3.64	1.588
21 Sc	3.31	1.594	65 Tb	3.60	1.581
22 Ti	2.95	1.588	66 Dy	3.59	1.573
27 Co	2.51	1.622	67 Ho	3.58	1.570
30 Zn	2.66	1.856	68 Er	3.56	1.570
34 Se	4.36	1.136	69 Tm	3.54	1.570
39 Y	3.65	1.571	71 Lu	3.51	1.585
40 Zr	3.23	1.593	72 Hf	3.20	1.582
43 Tc	2.74(?)	1.604(?)	75 Re	2.76	1.615
44 Ru	2.70	1.584	76 Os	2.74	1.579
48 Cd	2.98	1.886	81 Tl	3.46	1.599

Lattice constant ratios c/a may be compared with the ideal value $(c/a)^{hcp} = \sqrt{(8/3)} = 1.63299$ for a hexagonal close-packed (hcp) crystal. Note that some the crystals may contain several atoms in the primitive unit cell, two atoms for hcp crystals with lattice basis vectors given in the following. The lattice definitions in Cartesian coordinates are

$\underline{R}_1^{hex} = a\,(1, 0, 0)$, $\underline{R}_2^{hex} = a\,(-1/2, \sqrt{3}/2, 0)$, $\underline{R}_3^{hex} = a\,(0, 0, c/a)$.
$\underline{r}_1^{hcp} = a\,(0, 0, 0)$, $\underline{r}_2^{hcp} = a\,(1/2, 1/\sqrt{12}, \sqrt{(2/3)})$.

Appendix C: Mathematics of the Wood Notation

This appendix gives further mathematical details of the Wood notation [150] introduced in Section 6.3 to denote the structure of reconstructed single crystal surfaces as well as of adsorbate layers [94]. As explained in Section 6.3 the formal definition of the Wood notation reads as

$$\text{Sub}(h\,k\,l) - \kappa(\gamma_1 \times \gamma_2)\text{R}\alpha \quad \text{for reconstructed surfaces} \tag{C1a}$$

and

$$\text{Sub}(h\,k\,l) - \kappa(\gamma_1 \times \gamma_2)\text{R}\alpha - \eta\text{Ovl} \quad \text{for adsorbate surfaces} \tag{C1b}$$

where, for the sake of simplicity and because of rare occurrence, we ignore possible substrate reconstruction in the presence of an adsorbate treated by notation (6.4).

C.1 Basic Formalism and Examples

The periodicity information of the Wood notation given in Eq. (C1) may be expressed alternatively by a more general 2×2 matrix transformation according to Eq. (5.3). Surface-adapted lattice vectors $\underline{R}_1$ and $\underline{R}_2$ of the substrate, forming an angle ω, can be represented by orthonormal unit vectors $\underline{e}_1$ and $\underline{e}_2$ and written in matrix notation as

$$\begin{pmatrix} \underline{R}_1 \\ \underline{R}_2 \end{pmatrix} = \begin{pmatrix} R_1 & 0 \\ R_2 \cos(\omega) & R_2 \sin(\omega) \end{pmatrix} \cdot \begin{pmatrix} \underline{e}_1 \\ \underline{e}_2 \end{pmatrix} \tag{C2}$$

This relation can be inverted to yield

$$\begin{pmatrix} \underline{e}_1 \\ \underline{e}_2 \end{pmatrix} = \frac{1}{\sin(\omega)} \begin{pmatrix} R_1^{-1} \sin(\omega) & 0 \\ -R_1^{-1} \cos(\omega) & R_2^{-1} \end{pmatrix} \cdot \begin{pmatrix} \underline{R}_1 \\ \underline{R}_2 \end{pmatrix} \tag{C3}$$

Crystallography and Surface Structure: An Introduction for Surface Scientists and Nanoscientists,
Second Edition. Klaus Hermann.
© 2017 Wiley-VCH Verlag GmbH & Co. KGaA. Published 2017 by Wiley-VCH Verlag GmbH & Co. KGaA.

Rotating the orthonormal vector set $\underline{e}_1, \underline{e}_2$ anticlockwise by an angle α corresponds to a transformation

$$\begin{pmatrix}\underline{e}_1\\ \underline{e}_2\end{pmatrix} \rightarrow \begin{pmatrix}\underline{e}'_1\\ \underline{e}'_2\end{pmatrix} = \begin{pmatrix}\cos(\alpha) & \sin(\alpha)\\ -\sin(\alpha) & \cos(\alpha)\end{pmatrix} \cdot \begin{pmatrix}\underline{e}_1\\ \underline{e}_2\end{pmatrix} \quad (C4)$$

Therefore, rotating the two lattice vectors $\underline{R}_1$ and $\underline{R}_2$ anticlockwise by an angle α and scaling each by a factor γ_1 and γ_2, respectively, leads to transformed overlayer lattice vectors $\underline{R}'_1$ and $\underline{R}'_2$, given by

$$\begin{pmatrix}\underline{R}'_1\\ \underline{R}'_2\end{pmatrix} = \begin{pmatrix}\gamma_1 R_1 & 0\\ \gamma_2 R_2 \cos(\omega) & \gamma_2 R_2 \sin(\omega)\end{pmatrix} \cdot \begin{pmatrix}\underline{e}'_1\\ \underline{e}'_2\end{pmatrix}$$

$$= \begin{pmatrix}\gamma_1 R_1 & 0\\ \gamma_2 R_2 \cos(\omega) & \gamma_2 R_2 \sin(\omega)\end{pmatrix} \cdot \begin{pmatrix}\cos(\alpha) & \sin(\alpha)\\ -\sin(\alpha) & \cos(\alpha)\end{pmatrix} \cdot \begin{pmatrix}\underline{e}_1\\ \underline{e}_2\end{pmatrix}$$

$$= \begin{pmatrix}\gamma_1 R_1 & 0\\ \gamma_2 R_2 \cos(\omega) & \gamma_2 R_2 \sin(\omega)\end{pmatrix} \cdot \begin{pmatrix}\cos(\alpha) & \sin(\alpha)\\ -\sin(\alpha) & \cos(\alpha)\end{pmatrix} \cdot$$

$$\frac{1}{\sin(\omega)}\begin{pmatrix}R_1^{-1}\sin(\omega) & 0\\ -R_1^{-1}\cos(\omega) & R_2^{-1}\end{pmatrix} \cdot \begin{pmatrix}\underline{R}_1\\ \underline{R}_2\end{pmatrix}$$

$$= \frac{1}{\sin(\omega)}\begin{pmatrix}\gamma_1 \sin(\omega-\alpha) & \gamma_1 q^{-1}\sin(\alpha)\\ -\gamma_2 q \sin(\alpha) & \gamma_2 \sin(\omega+\alpha)\end{pmatrix} \cdot \begin{pmatrix}\underline{R}_1\\ \underline{R}_2\end{pmatrix}$$

$$= \underline{\underline{M}}_p \cdot \begin{pmatrix}\underline{R}_1\\ \underline{R}_2\end{pmatrix}, \quad q = \frac{R_2}{R_1} \quad (C5)$$

This proves representation (6.6a) for the transformation matrix $\underline{\underline{M}}_p$, which connects the substrate lattice vectors $\underline{R}_1$ and $\underline{R}_2$ with those, $\underline{R}'_1$ and $\underline{R}'_2$, of the overlayer in the case of **primitive** overlayers denoted by "… – $p(\gamma_1 \times \gamma_2)R\alpha$ – …." It should be noted that matrix $\underline{\underline{M}}_p$ appearing in Eq. (C5) is identical to the transformation matrix (5.16), describing rotational overlayers for scaling factors $\gamma_1 = \gamma_2 = 1$.

Corresponding lattice vectors $\underline{R}''_1$ and $\underline{R}''_2$ of centered overlayers with respect to $\underline{R}'_1$ and $\underline{R}'_2$ of the primitive overlayers are described by

$$\underline{R}''_1 = (\underline{R}'_1 + \underline{R}'_2)/2, \quad \underline{R}''_2 = (-\underline{R}'_1 + \underline{R}'_2)/2 \quad (C6)$$

which yields a transformation

$$\begin{pmatrix}\underline{R}''_1\\ \underline{R}''_2\end{pmatrix} = \frac{1}{2}\begin{pmatrix}1 & 1\\ -1 & 1\end{pmatrix} \cdot \begin{pmatrix}\underline{R}'_1\\ \underline{R}'_2\end{pmatrix} = \frac{1}{2}\begin{pmatrix}1 & 1\\ -1 & 1\end{pmatrix} \cdot \underline{\underline{M}}_p \cdot \begin{pmatrix}\underline{R}_1\\ \underline{R}_2\end{pmatrix} = \underline{\underline{M}}_c \cdot \begin{pmatrix}\underline{R}_1\\ \underline{R}_2\end{pmatrix} \quad (C7)$$

and, using Eq. (C5), results in

$$\underline{\underline{M}}_c = \frac{1}{2\sin(\omega)}\begin{pmatrix}\gamma_1\sin(\omega-\alpha) - \gamma_2 q\sin(\alpha) & \gamma_1 q^{-1}\sin(\alpha) + \gamma_2\sin(\omega+\alpha)\\ -\gamma_1\sin(\omega-\alpha) - \gamma_2 q\sin(\alpha) & -\gamma_1 q^{-1}\sin(\alpha) + \gamma_2\sin(\omega+\alpha)\end{pmatrix}$$

$$(C8)$$

This proves relation (6.6b) for the transformation matrix $\underline{\underline{M}}_c$, connecting the substrate lattice vectors $\underline{R}_1$ and $\underline{R}_2$ with those, $\underline{R}_1''$ and $\underline{R}_2''$, of the overlayer in the case of **centered** overlayers denoted by " ... $- c(\gamma_1 \times \gamma_2)R\alpha - $ ...."

In the following, we discuss special cases of primitive overlayers, described by transformation matrices $\underline{\underline{M}}_p$ of Eq. (C5), where q is the ratio of the lengths of the periodicity vectors $\underline{R}_1$ and $\underline{R}_2$, that is,

$$q = R_2/R_1 \tag{C9}$$

1) For **primitive rectangular** substrate lattices the periodicity vectors $\underline{R}_1$ and $\underline{R}_2$ are orthogonal ($\omega = 90°$), which yields for $p(\gamma_1 \times \gamma_2)R\alpha$

$$\underline{\underline{M}}_p = \begin{pmatrix} m_{11} & m_{12} \\ m_{21} & m_{22} \end{pmatrix} = \begin{pmatrix} \gamma_1 \cos(\alpha) & \gamma_1 q^{-1} \sin(\alpha) \\ -\gamma_2 q \sin(\alpha) & \gamma_2 \cos(\alpha) \end{pmatrix} \tag{C10}$$

Here **commensurate** overlayers are subject to **integer-valued** elements m_{ij} where

$$\gamma_1 \cos(\alpha) = m_{11} \tag{C11a}$$

$$\gamma_1 q^{-1} \sin(\alpha) = m_{12} \tag{C11b}$$

$$\gamma_2 q \sin(\alpha) = -m_{21} \tag{C11c}$$

$$\gamma_2 \cos(\alpha) = m_{22} \tag{C11d}$$

Simple examples are
a. Rotation angle $\alpha = 0°$ and integer $\gamma_i = n_i$ leading to $p(n_1 \times n_2)R0°$ or $(n_1 \times n_2)$ with

$$\underline{\underline{M}}_p = \begin{pmatrix} n_1 & 0 \\ 0 & n_2 \end{pmatrix} \tag{C12}$$

b. Rotation angle $\alpha = 90°$ and $\gamma_1 = n_1 q$, $\gamma_2 = n_2/q$, n_1, n_2 integer, leading to $p(\gamma_1 \times \gamma_2)R90°$ with

$$\underline{\underline{M}}_p = \begin{pmatrix} 0 & n_1 \\ -n_2 & 0 \end{pmatrix} \tag{C13}$$

Other combinations of γ_1, γ_2, and α lead in many cases to **incommensurate** overlayers.

2) For **centered rectangular** substrate lattices the periodicity vectors $\underline{R}_1$ and $\underline{R}_2$ are of equal length (q = 1) and angle ω differs from 60°, 90°, and 120° which yields

$$\underline{\underline{M}}_p = \begin{pmatrix} m_{11} & m_{12} \\ m_{21} & m_{22} \end{pmatrix} = \frac{1}{\sin(\omega)} \begin{pmatrix} \gamma_1 \sin(\omega - \alpha) & \gamma_1 \sin(\alpha) \\ -\gamma_2 \sin(\alpha) & \gamma_2 \sin(\omega + \alpha) \end{pmatrix} \tag{C14}$$

Here **commensurate** overlayers are subject to **integer-valued** elements m_{ij} where

$$\gamma_1 \sin(\omega - \alpha) = m_{11} \sin(\omega) \tag{C15a}$$

$$\gamma_1 \sin(\alpha) = m_{12} \sin(\omega) \tag{C15b}$$

$$\gamma_2 \sin(\alpha) = -m_{21} \sin(\omega) \tag{C15c}$$

$$\gamma_2 \sin(\omega + \alpha) = m_{22} \sin(\omega) \tag{C15d}$$

This means in particular that

$$\gamma_1 \cos(\alpha) = m_{11} + m_{12} \cos(\omega) \quad \text{and} \quad \gamma_2 \cos(\alpha) = m_{22} + m_{21} \cos(\omega) \tag{C16}$$

A simple example is
a. Rotation angle $\alpha = 0°$ and integer $\gamma_i = n_i$ leading to $p(n_1 \times n_2)$ or $(n_1 \times n_2)$ with

$$\underline{\underline{M}}_p = \begin{pmatrix} n_1 & 0 \\ 0 & n_2 \end{pmatrix} \tag{C17}$$

Other combinations of γ_1, γ_2, and α lead in many cases to **incommensurate** overlayers.

3) For **square** substrate lattices the periodicity vectors $\underline{R}_1$ and $\underline{R}_2$ are orthogonal ($\omega = 90°$) and of equal length ($q = 1$) which yields

$$\underline{\underline{M}}_p = \begin{pmatrix} m_{11} & m_{12} \\ m_{21} & m_{22} \end{pmatrix} = \begin{pmatrix} \gamma_1 \cos(\alpha) & \gamma_1 \sin(\alpha) \\ -\gamma_2 \sin(\alpha) & \gamma_2 \cos(\alpha) \end{pmatrix} \tag{C18}$$

Here **commensurate** overlayers are subject to **integer-valued** elements m_{ij} where

$$\gamma_1 \cos(\alpha) = m_{11} \tag{C19a}$$

$$\gamma_1 \sin(\alpha) = m_{12} \tag{C19b}$$

$$\gamma_2 \sin(\alpha) = -m_{21} \tag{C19c}$$

$$\gamma_2 \cos(\alpha) = m_{22} \tag{C19d}$$

Simple examples are
a. Rotation angle $\alpha = 0°$ and integer $\gamma_i = n_i$ leading to $p(n_1 \times n_2)$ or $(n_1 \times n_2)$ with

$$\underline{\underline{M}}_p = \begin{pmatrix} n_1 & 0 \\ 0 & n_2 \end{pmatrix} \tag{C20}$$

b. Rotation angle $\alpha = 90°$ and integer $\gamma_i = n_i$, leading to $p(n_1 \times n)R90°$ with

$$\underline{\underline{M}}_p = \begin{pmatrix} 0 & n_1 \\ -n_2 & 0 \end{pmatrix} \tag{C21}$$

c. Rotation angles $-90° < \alpha < 90°$, $\alpha \neq 0°$, with

$$\cos(\alpha) = m/g, \quad \sin(\alpha) = n/g, \quad g = (m^2 + n^2)^{1/2}$$

$$\gamma_1 = a\,g, \quad \gamma_2 = b\,g, \quad a, b > 0 \text{ integer}$$

$$m, n > 0 \text{ integer, gcd}(m, n) = 1$$

leading to $p(\gamma_1 \times \gamma_2)R\alpha$ with

$$\underline{\underline{M}}_p = \begin{pmatrix} am & an \\ -bn & bm \end{pmatrix} \tag{C22}$$

Other combinations of γ_1, γ_2, and α lead in many cases to **incommensurate** overlayers.

4) For **hexagonal** substrate lattices in **acute** representation ($\omega = 60°$) the periodicity vectors $\underline{R}_1$ and $\underline{R}_2$ are of equal length ($q = 1$), which yields

$$\underline{\underline{M}}_p = \begin{pmatrix} m_{11} & m_{12} \\ m_{21} & m_{22} \end{pmatrix} = \frac{2}{\sqrt{3}} \begin{pmatrix} \gamma_1 \sin(60° - \alpha) & \gamma_1 \sin(\alpha) \\ -\gamma_2 \sin(\alpha) & \gamma_2 \sin(60° + \alpha) \end{pmatrix} \tag{C23}$$

Here, **commensurate** overlayers are subject to **integer-valued** elements m_{ij} where

$$\gamma_1 \sin(60° - \alpha) = c\,m_{11}, \quad c = \sqrt{3/2} \tag{C24a}$$

$$\gamma_1 \sin(\alpha) = c\,m_{12} \tag{C24b}$$

$$\gamma_2 \sin(\alpha) = -c\,m_{21} \tag{C24c}$$

$$\gamma_2 \sin(60° + \alpha) = c\,m_{22} \tag{C24d}$$

This means in particular that

$$\gamma_1 \cos(\alpha) = m_{11} + m_{12}/2 \quad \text{and} \quad \gamma_2 \cos(\alpha) = m_{22} + m_{21}/2 \tag{C25}$$

Simple examples are

a. Rotation angle $\alpha = 0°$ and integer $\gamma_i = n_i$ leading to $p(n_1 \times n_2)$ or $(n_1 \times n_2)$ with

$$\underline{\underline{M}}_p = \begin{pmatrix} n_1 & 0 \\ 0 & n_2 \end{pmatrix} \tag{C26}$$

b. Rotation angle $\alpha = 60°$ and integer $\gamma_i = n_i$ leading to $p(n_1 \times n_2)R60°$ with

$$\underline{\underline{M}}_p = \begin{pmatrix} 0 & n_1 \\ -n_2 & n_2 \end{pmatrix} \tag{C27}$$

c. Rotation angle $\alpha = 120°$ and integer $\gamma_i = n_i$ leading to $p(n_1 \times n_2)R120°$ with

$$\underline{\underline{M}}_p = \begin{pmatrix} -n_1 & n_1 \\ -n_2 & 0 \end{pmatrix} \tag{C28}$$

d. Assuming $\gamma_i = n_i g$, $\cos(\alpha) = (a+b/2)/g$, $g = (a^2 + ab + b^2)^{1/2}$ with a, b, and n_i integer, leading to $\mathbf{p}(\gamma_1 \times \gamma_2)\mathbf{R}\alpha$ with

$$\underline{\underline{M}}_p = \begin{pmatrix} n_1 a & n_1 b \\ -n_2 b & n_2(a+b) \end{pmatrix} \qquad (C29)$$

This includes cases of
- $a = b = 1$, $n_1 = n_2 = 1$, $\underline{\underline{M}}_p = \begin{pmatrix} 1 & 1 \\ -1 & 2 \end{pmatrix}$: $p(\sqrt{3} \times \sqrt{3})R30°$
- $a = 2$, $b = 1$, $n_1 = n_2 = 1$, $\underline{\underline{M}}_p = \begin{pmatrix} 2 & 1 \\ -1 & 3 \end{pmatrix}$: $p(\sqrt{7} \times \sqrt{7})R19.1°$
- $a = 3$, $b = 1$, $n_1 = n_2 = 1$, $\underline{\underline{M}}_p = \begin{pmatrix} 3 & 1 \\ -1 & 4 \end{pmatrix}$: $p(\sqrt{13} \times \sqrt{13})R13.1°$

Other combinations of γ_1, γ_2, and α lead in many cases to **incommensurate** overlayers.

5) For **hexagonal** substrate lattices in **obtuse** representation ($\omega = 120°$) the periodicity vectors $\underline{R}_1$ and $\underline{R}_2$ are of equal length ($q = 1$), which results in

$$\underline{\underline{M}}_p = \begin{pmatrix} m_{11} & m_{12} \\ m_{21} & m_{22} \end{pmatrix} = \frac{2}{\sqrt{3}} \begin{pmatrix} \gamma_1 \sin(60° + \alpha) & \gamma_1 \sin(\alpha) \\ -\gamma_2 \sin(\alpha) & \gamma_2 \sin(60° - \alpha) \end{pmatrix} \qquad (C30)$$

Here **commensurate** overlayers are subject to **integer-valued** elements m_{ij} where

$$\gamma_1 \sin(60° + \alpha) = c\, m_{11}, \qquad c = \sqrt{3/2} \qquad (C31a)$$

$$\gamma_1 \sin(\alpha) = c\, m_{12} \qquad (C31b)$$

$$\gamma_2 \sin(\alpha) = -c\, m_{21} \qquad (C31c)$$

$$\gamma_2 \sin(60° - \alpha) = c\, m_{22} \qquad (C31d)$$

This means in particular that

$$\gamma_1 \cos(\alpha) = m_{11} - m_{12}/2 \text{ and } \gamma_2 \cos(\alpha) = m_{22} - m_{21}/2 \qquad (C32)$$

Simple examples are
a. Rotation angle $\alpha = 0°$ and integer $\gamma_i = n_i$ leading to $\mathbf{p}(n_1 \times n_2)$ or $(n_1 \times n_2)$ with

$$\underline{\underline{M}}_p = \begin{pmatrix} n_1 & 0 \\ 0 & n_2 \end{pmatrix} \qquad (C33)$$

b. Rotation angle $\alpha = 120°$ and integer $\gamma_i = n_i$ leading to $\mathbf{p}(n_1 \times n_2)\mathbf{R}120°$ with

$$\underline{\underline{M}}_p = \begin{pmatrix} 0 & n_1 \\ -n_2 & -n_2 \end{pmatrix} \qquad (C34)$$

c. Rotation angle $\alpha = 240°$ and integer $\gamma_i = n_i$ leading to $\mathbf{p}(n_1 \times n_2)\mathbf{R}240°$ with

$$\underline{\underline{M}}_p = \begin{pmatrix} -n_1 & -n_1 \\ n_2 & 0 \end{pmatrix} \qquad (C35)$$

d. Assuming $\gamma_i = n_i\, g$, $\cos(\alpha) = (a + b/2)/g$, $g = (a^2 - ab + b^2)^{1/2}$ with a, b, and n_i integer, leading to $p(\gamma_1 \times \gamma_2)R\alpha$ with

$$\underline{\underline{M}}_p = \begin{pmatrix} n_1 a & n_1 b \\ -n_2 b & n_2(a-b) \end{pmatrix} \tag{C36}$$

This includes cases of

- $a = 2,\ b = 1,\ n_1 = n_2 = 1,\ \underline{\underline{M}}_p = \begin{pmatrix} 2 & 1 \\ -1 & 1 \end{pmatrix}$: $p(\sqrt{3} \times \sqrt{3})R30°$
- $a = 3,\ b = 1,\ n_1 = n_2 = 1,\ \underline{\underline{M}}_p = \begin{pmatrix} 3 & 1 \\ -1 & 2 \end{pmatrix}$: $p(\sqrt{7} \times \sqrt{7})R19.1°$
- $a = 4,\ b = 1,\ n_1 = n_2 = 1,\ \underline{\underline{M}}_p = \begin{pmatrix} 4 & 1 \\ -1 & 3 \end{pmatrix}$: $p(\sqrt{13} \times \sqrt{13})R13.1°$

Other combinations of γ_1, γ_2, and α lead in many cases to incommensurate overlayers.

C.2 Wood-Representability

Lattice structures given by Wood notation (6.2) and (C1) make it always possible to construct the corresponding 2×2 transformations matrices $\underline{\underline{M}}_{p,c}$ according to Eqs. (C5) and (C8). However, not all 2×2 matrices refer to transformations, which can be characterized by a Wood notation. The mathematical reason behind it is that 2×2 matrices are determined by four parameters, the matrix elements m_{ij}, allowing four degrees of freedom whereas the Wood notation allows only three, given by the two stretch parameters, γ_1 and γ_2, and the rotation angle α. The fourth degree of freedom does not appear because the Wood notation assumes that the angle ω between the lattice vectors $\underline{R}'_1$ and $\underline{R}'_2$ of the overlayer agrees with that between the lattice vectors $\underline{R}_1$ and $\underline{R}_2$ of the substrate surface. This constraint can be used to select all 2×2 matrices that correspond to transformations allowing Wood notation and will be called **Wood representable** in the following.

In a general lattice transformation according to Eq. (5.3),

$$\begin{pmatrix} \underline{R}'_1 \\ \underline{R}'_2 \end{pmatrix} = \begin{pmatrix} m_{11} & m_{12} \\ m_{21} & m_{22} \end{pmatrix} \cdot \begin{pmatrix} \underline{R}_1 \\ \underline{R}_2 \end{pmatrix} = \underline{\underline{M}} \cdot \begin{pmatrix} \underline{R}_1 \\ \underline{R}_2 \end{pmatrix} \tag{C37}$$

the area of the periodicity cell of the overlayer, given by $F' = |\underline{R}'_1 \times \underline{R}'_2|$, can be written as

$$F' = |\underline{R}'_1 \times \underline{R}'_2| = R'_1 R'_2 \sin(\omega') = \det(\underline{\underline{M}})\, |\underline{R}_1 \times \underline{R}_2| = \det(\underline{\underline{M}})\, R_1 R_2 \sin(\omega) \tag{C38}$$

where ω and ω' are the angles formed by vectors $\underline{R}_1$ and $\underline{R}_2$ and by $\underline{R}'_1$ and $\underline{R}'_2$, respectively. If transformation (C37) allows a **primitive Wood notation** then angles ω and ω' must be equal. Thus, Eq. (C38) leads to

$$\det(\underline{\underline{M}})^2 = \left(\frac{R'_1 R'_2}{R_1 R_2}\right)^2 = (m_{11} m_{22} - m_{12} m_{21})^2 \tag{C39}$$

and hence with

$$\left(\frac{R'_1}{R_1}\right)^2 = m_{11}^2 + m_{12}^2 q^2 + 2 m_{11} m_{12} q \cos(\omega), \quad q = \frac{R_2}{R_1} \tag{C40}$$

$$\left(\frac{R'_2}{R_2}\right)^2 = m_{21}^2 q^{-2} + m_{22}^2 + 2 m_{21} m_{22} q^{-1} \cos(\omega) \tag{C41}$$

relation (C39) yields

$$(m_{11} m_{22} - m_{12} m_{21})'^2 = \{m_{11}^2 + m_{12}^2 q^2 + 2 m_{11} m_{12} q \cos(\omega)\}$$
$$\cdot \{m_{21}^2 q^{-2} + m_{22}^2 + 2 m_{21} m_{22} q^{-1} \cos(\omega)\} \tag{C42}$$

Relation (C42) is the **mathematical constraint** for matrices $\underline{M}$ to be Wood-representable yielding transformations according to the **primitive** Wood notation, denoted as $p(\gamma_1 \times \gamma_2) R\alpha$ where

$$\gamma_1 = \frac{R'_1}{R_1} = \sqrt{m_{11}^2 + m_{12}^2 q^2 + 2 m_{11} m_{12} q \cos(\omega)}, \quad q = \frac{R_2}{R_1} \tag{C43a}$$

$$\gamma_2 = \frac{R'_2}{R_2} = \sqrt{m_{21}^2 q^{-2} + m_{22}^2 + 2 m_{21} m_{22} q^{-1} \cos(\omega)} \tag{C43b}$$

$$\cos(\alpha) = \frac{(R'_1 R_1)}{R'_1 R_1} = [m_{11} + m_{12} q \cos(\omega)]/\gamma_1 = [m_{22} + m_{21} q^{-1} \cos(\omega)]/\gamma_2 \tag{C43c}$$

These relations depend, apart from all matrix elements m_{ij}, on the geometry specifications of the substrate lattice given by q and ω, which distinguishes between Bravais lattice types.

1) For **primitive rectangular** substrate lattices, $q \neq 1$ and $\omega = 90°$, Wood-representable matrices $\underline{M}$ according to Eq. (C42) are subject to

$$m_{11} m_{21} + q^2 m_{12} m_{22} = 0 \tag{C44}$$

yielding

$$\gamma_1 = \sqrt{m_{11}^2 + m_{12}^2 q^2}, \quad \gamma_2 = \sqrt{m_{21}^2 q^{-2} + m_{22}^2}$$
$$\cos(\alpha) = m_{11}/\gamma_1 = m_{22}/\gamma_2 \tag{C45}$$

2) For **centered rectangular** substrate lattices, $q = 1$ and $\omega \neq 60°, 90°, 120°$, Wood-representable matrices $\underline{M}$ according to Eq. (C42) are subject to

$$(m_{11} m_{22} - m_{12} m_{21})^2 = \{m_{11}^2 + m_{12}^2 + 2 m_{11} m_{12} \cos(\omega)\}$$
$$\cdot \{m_{21}^2 + m_{22}^2 + 2 m_{21} m_{22} \cos(\omega)\} \tag{C46}$$

yielding

$$\gamma_i = \sqrt{m_{i1}^2 + m_{i2}^2 + 2 m_{i1} m_{i2} \cos(\omega)}, \quad i = 1, 2$$
$$\cos(\alpha) = [m_{11} + m_{12} \cos(\omega)]/\gamma_1 = [m_{22} + m_{21} \cos(\omega)]/\gamma_2 \tag{C47}$$

3) For **square** substrate lattices, $q = 1$ and $\omega = 90°$, Wood-representable matrices $\underline{M}$ according to Eq. (C42) are subject to

$$m_{11} m_{21} + m_{12} m_{22} = 0 \tag{C48}$$

yielding

$$\gamma_i = \sqrt{m_{i1}^2 + m_{i2}^2}, \quad i = 1, 2, \quad \cos(\alpha) = m_{11}/\gamma_1 = m_{22}/\gamma_2 \tag{C49}$$

4) For **hexagonal** substrate lattices in **acute** representation, $q = 1$ and $\omega = 60$, Wood-representable matrices $\underline{M}$ according to Eq. (C42) are subject to

$$(m_{11} m_{22} - m_{12} m_{21})^2 = \{m_{11}^2 + m_{12}^2 + m_{11} m_{12}\} \{m_{21}^2 + m_{22}^2 + m_{21} m_{22}\} \tag{C50}$$

yielding

$$\gamma_i = \sqrt{m_{i1}^2 + m_{i2}^2 + m_{i1} m_{i2}}, \quad i = 1, 2$$

$$\cos(\alpha) = \left(m_{11} + m_{12}/2\right)/\gamma_1 = \left(m_{22} + m_{21}/2\right)/\gamma_2 \tag{C51}$$

5) For **hexagonal** substrate lattices in **obtuse** representation, $q = 1$ and $\omega = 120$, Wood-representable matrices $\underline{M}$ according to Eq. (C42) are subject to

$$(m_{11} m_{22} - m_{12} m_{21})^2 = \{m_{11}^2 + m_{12}^2 - m_{11} m_{12}\} \{m_{21}^2 + m_{22}^2 - m_{21} m_{22}\} \tag{C52}$$

yielding

$$\gamma_i = \sqrt{m_{i1}^2 + m_{i2}^2 - m_{i1} m_{i2}}, \quad i = 1, 2$$

$$\cos(\alpha) = \left(m_{11} - m_{12}/2\right)/\gamma_1 = \left(m_{22} - m_{21}/2\right)/\gamma_2 \tag{C53}$$

If transformation (C37) allows a **centered Wood notation**, described as $c(\gamma_1 \times \gamma_2)R\alpha$, then its transformation matrix $\underline{M}_c$ can be written as the product of a simple transformation matrix $\underline{T}$ and matrix $\underline{M}_p$, which provides the corresponding transformation according to a primitive Wood notation, $p(\gamma_1 \times \gamma_2)R\alpha$, that is,

$$\underline{M}_c = \underline{T} \cdot \underline{M}_p, \quad \underline{M}_p = \underline{T}^{-1} \cdot \underline{M}_c \tag{C54}$$

where

$$\underline{T} = \frac{1}{2}\begin{pmatrix} 1 & 1 \\ -1 & 1 \end{pmatrix}, \; \frac{1}{2}\begin{pmatrix} -1 & 1 \\ -1 & -1 \end{pmatrix}, \; \frac{1}{2}\begin{pmatrix} -1 & -1 \\ 1 & -1 \end{pmatrix}, \; \frac{1}{2}\begin{pmatrix} 1 & -1 \\ 1 & 1 \end{pmatrix} \tag{C55}$$

$$\underline{T}^{-1} = \begin{pmatrix} 1 & -1 \\ 1 & 1 \end{pmatrix}, \; \begin{pmatrix} -1 & -1 \\ 1 & -1 \end{pmatrix}, \; \begin{pmatrix} -1 & 1 \\ -1 & -1 \end{pmatrix}, \; \begin{pmatrix} 1 & 1 \\ -1 & 1 \end{pmatrix} \tag{C56}$$

Therefore, testing a matrix $\underline{\underline{M}}_c$ for Wood-representability according to the centered Wood notation can be achieved by first determining the corresponding matrix $\underline{\underline{M}}_p$ using Eqs. (C54–C56) followed by testing $\underline{\underline{M}}_p$ applying the criteria given in Eqs. (C42) and (C44)–(C53).

Combining Wood-representable transformations. Lattice transformation defined in Wood notation (6.2) or (6.4) and their corresponding 2×2 matrices $\underline{\underline{M}}_{p,c}$ according to Eqs. (C5) and (C8) can be combined to yield other lattice transformations. Here, we restrict the discussion to the **primitive Wood notation** $p(\gamma_1 \times \gamma_2)R\alpha$ and 2×2 matrices $\underline{\underline{M}}_p$ given by Eq. (C5).

Applying two Wood-representable transformations, $p(\gamma_1' \times \gamma_2')R\alpha'$ and $p(\gamma_1 \times \gamma_2)R\alpha$ in sequence corresponds to transformations

$$\begin{pmatrix} R_1' \\ R_2' \end{pmatrix} = \frac{1}{\sin(\omega)} \begin{pmatrix} \gamma_1 \sin(\omega - \alpha) & \gamma_1 q^{-1} \sin(\alpha) \\ -\gamma_2 q \sin(\alpha) & \gamma_2 \sin(\omega + \alpha) \end{pmatrix} \cdot \begin{pmatrix} R_1 \\ R_2 \end{pmatrix}, \quad q = \frac{R_2}{R_1} \quad (C57)$$

followed by

$$\begin{pmatrix} R_1'' \\ R_2'' \end{pmatrix} = \frac{1}{\sin(\omega)} \begin{pmatrix} \gamma_1' \sin(\omega - \alpha') & \gamma_1' q'^{-1} \sin(\alpha') \\ -\gamma_2' q' \sin(\alpha') & \gamma_2' \sin(\omega + \alpha') \end{pmatrix} \cdot \begin{pmatrix} R_1' \\ R_2' \end{pmatrix}, \quad q' = \frac{R_2'}{R_1'} = \frac{\gamma_2}{\gamma_1} q \quad (C58)$$

such that after some basic calculus

$$\begin{pmatrix} R_1'' \\ R_2'' \end{pmatrix} = \frac{1}{\sin(\omega)} \begin{pmatrix} \gamma_1' \gamma_1 \sin(\omega - \alpha - \alpha') & \gamma_1' \gamma_1 q'^{-1} \sin(\alpha + \alpha') \\ -\gamma_2' \gamma_2 q' \sin(\alpha + \alpha') & \gamma_2' \gamma_2 \sin(\omega + \alpha + \alpha') \end{pmatrix} \cdot \begin{pmatrix} R_1 \\ R_2 \end{pmatrix} \quad (C59)$$

which is a Wood-representable transformation given by

$$p(\gamma_1' \times \gamma_2')R\alpha' * p(\gamma_1 \times \gamma_2)R\alpha = p(\gamma_1'\gamma_1 \times \gamma_2'\gamma_2)R(\alpha' + \alpha) \quad (C60)$$

Thus, the application of two Wood-representable transformations yields again a Wood-representable transformation. In addition, swapping the sequence of application of the two transformations yields the same final transformation. Therefore, the order in which the two transformations are applied does not affect the final result.

Next, for every Wood-representable transformation one can construct an inverse transformation. Considering $\underline{\underline{M}}_p$ defined by Eq. (C5) we find

$$\det(\underline{\underline{M}}_p) = \gamma_1 \gamma_2 \quad (C61)$$

and hence

$$\underline{\underline{M}}_p^{-1} = \frac{1}{\gamma_1 \gamma_2 \sin(\omega)} \begin{pmatrix} \gamma_2 \sin(\omega + \alpha) & -\gamma_1 q^{-1} \sin(\alpha) \\ \gamma_2 q \sin(\alpha) & \gamma_1 \sin(\omega - \alpha) \end{pmatrix}$$

$$= \frac{1}{\sin(\omega)} \begin{pmatrix} \gamma_1^{-1} \sin(\omega - (-\alpha)) & \gamma_2^{-1} q^{-1} \sin(-\alpha) \\ -\gamma_1^{-1} q \sin(-\alpha) & \gamma_2^{-1} \sin(\omega + (-\alpha)) \end{pmatrix} \quad (C62)$$

which is again a Wood-representable matrix of a transformation referring to $p(\gamma_1^{-1} \times \gamma_2^{-1})R(-\alpha)$ such that together with Eq. (C60)

$$p(\gamma_1 \times \gamma_2)R\alpha * p(\gamma_1^{-1} \times \gamma_2^{-1})R(-\alpha) = p(1 \times 1) \quad (C63)$$

where p(1×1) with $\underline{\underline{M}}_p = \underline{\underline{1}}$ can be considered the Wood-representable unit transformation. Altogether, the set of all Wood-representable transformations described as p($\gamma_1 \times \gamma_2$)Rα and applied to a given Bravais lattice form an (infinite) Abelian group.

Appendix D: Mathematics of the Minkowski Reduction

This appendix discusses the mathematical details of the Minkowski reduction used to obtain **symmetrically appropriate** lattice vectors $\underline{R}_{o1}$ and $\underline{R}_{o2}$ from initial lattice vectors $\underline{R}_1$ and $\underline{R}_2$, see Section 3.3.

Let us assume that $\underline{R}_1$ and $\underline{R}_2$ are lattice vectors of a two-dimensional lattice with lengths R_1 and R_2. Then vector $\underline{r}$ inside the plane spanned by $\underline{R}_1$ and $\underline{R}_2$ and defined by

$$\underline{r} = \underline{R}_2 - x\,\underline{R}_1 \tag{D1}$$

has a squared length

$$r^2 = |\underline{r}|^2 = R_2^2 + x^2 R_1^2 - 2x(\underline{R}_1\underline{R}_2) \tag{D2}$$

which, by varying x, reaches its minimum for

$$x = x_{min} = (\underline{R}_1\underline{R}_2)/R_1^2 \tag{D3}$$

The resulting smallest vector $\underline{r} = \underline{r}_{min}$ with

$$\underline{r}_{min} = \underline{R}_2 - x_{min}\underline{R}_1 = \underline{R}_2 - ((\underline{R}_1\underline{R}_2)/R_1^2)\underline{R}_1 \tag{D4}$$

is perpendicular to $\underline{R}_1$, since

$$(\underline{R}_1\underline{r}_{min}) = (\underline{R}_1\underline{R}_2) - ((\underline{R}_1\underline{R}_2)/R_1^2)(\underline{R}_1\underline{R}_1) = 0 \tag{D5}$$

However, in general, $\underline{r}_{min}$ will not be a lattice vector itself. The lattice vector $\underline{R}_2'$ represented by Eq. (D1), nearest to $\underline{r}_{min}$ in length, is given by

$$\underline{R}_2' = \underline{R}_2 - \{x_{min}\}\underline{R}_1 \tag{D6}$$

where function {x} denotes the integer nearest to a real number x as introduced in Appendix E.1. Then for the length square $|\underline{R}_2'|^2$ we obtain

$$|\underline{R}_2'|^2 = R_2^2 + \{x_{min}\}^2 R_1^2 - 2\{x_{min}\}(\underline{R}_1\underline{R}_2)$$
$$= R_2^2 - \{x_{min}\}(2x_{min} - \{x_{min}\})R_1^2 = R_2^2 - p(x_{min})R_1^2 \tag{D7}$$

where the mixing factor

$$p(x) = \{x\}(2x - \{x\}) \tag{D8}$$

Crystallography and Surface Structure: An Introduction for Surface Scientists and Nanoscientists,
Second Edition. Klaus Hermann.
© 2017 Wiley-VCH Verlag GmbH & Co. KGaA. Published 2017 by Wiley-VCH Verlag GmbH & Co. KGaA.

is symmetric in x and assumes only positive values as discussed in Appendix E.1, that is,

$$p(-x) = p(x) \geq 0 \tag{D9}$$

Thus, the reduction $\underline{R}_2 \rightarrow \underline{R}'_2$ yields according to Eq. (D7)

$$|\underline{R}'_2|^2 \leq |\underline{R}_2|^2 \tag{D10}$$

which proves that it can only decrease the length of vector $\underline{R}_2$ or leave its length unchanged. The latter means that

$$\{x_{min}\} = 0 \quad \text{or} \quad -1/2 \leq x_{min} < 1/2 \tag{D11}$$

and, together with Eq. (D3), leads to

$$-R_1^2 \leq 2\,(\underline{R}_1\,\underline{R}_2) < R_1^2 \tag{D12}$$

In addition, transformation (D6) will always lead to a vector $\underline{R}'_2$ of finite length, $|\underline{R}'_2| > 0$, since $\underline{R}_1$ and $\underline{R}_2$ are assumed to be linearly independent. A reduction analogous to Eq. (D1) can be applied to reduce the length of $\underline{R}_1$, yielding

$$\underline{R}'_1 = \underline{R}_1 - \{x'_{min}\}\,\underline{R}_2, \quad \text{with} \quad x'_{min} = (\underline{R}_1\,\underline{R}_2)/R_2^2 \tag{D13}$$

where the reduction can only decrease the length of vector $\underline{R}_1$ or leave its length unchanged. The latter results in

$$-R_2^2 \leq 2\,(\underline{R}_1\,\underline{R}_2) < R_2^2 \tag{D14}$$

Relations (D6) and (D12–D14) form the basis of an **iterative** algorithm, the **Minkowski reduction**, which can be used to determine a lattice vector set $\underline{R}_{o1}$, $\underline{R}_{o2}$ with vectors that are the smallest in length. We consider an iterative transformation starting with

$$(k = 0) \quad \underline{R}_1^{(k)} = \underline{R}_1, \quad \underline{R}_2^{(k)} = \underline{R}_2 \tag{D15}$$

where each iteration step contains two reductions. First, vector $\underline{R}_2^{(k)}$ is reduced according to Eq. (D6), which can be written as a linear transformation $(\underline{R}_1^{(k)}, \underline{R}_2^{(k)}) \rightarrow (\underline{R}_1^{(k)}, \underline{R}_2^{(k+1)})$ with

$$\begin{pmatrix} \underline{R}_1^{(k)} \\ \underline{R}_2^{(k+1)} \end{pmatrix} = \begin{pmatrix} 1 & 0 \\ -\{x_k\} & 1 \end{pmatrix} \cdot \begin{pmatrix} \underline{R}_1^{(k)} \\ \underline{R}_2^{(k)} \end{pmatrix}, \quad x_k = (\underline{R}_1^{(k)}\,\underline{R}_2^{(k)})/(R_1^{(k)})^2 \tag{D16}$$

Then vector $\underline{R}_1^{(k)}$ is reduced according to Eq. (D13), which can be written as a linear transformation $(\underline{R}_1^{(k)}, \underline{R}_2^{(k+1)}) \rightarrow (\underline{R}_1^{(k+1)}, \underline{R}_2^{(k+1)})$ with

$$\begin{pmatrix} \underline{R}_1^{(k+1)} \\ \underline{R}_2^{(k+1)} \end{pmatrix} = \begin{pmatrix} 1 & -\{x'_k\} \\ 0 & 1 \end{pmatrix} \cdot \begin{pmatrix} \underline{R}_1^{(k)} \\ \underline{R}_2^{(k+1)} \end{pmatrix}, \quad x'_k = (\underline{R}_1^{(k)}\,\underline{R}_2^{(k+1)})/(R_2^{(k+1)})^2 \tag{D17}$$

Since the transformation matrices in Eqs. (D16) and (D17) are integer-valued and their inverse matrices

$$\begin{pmatrix} 1 & 0 \\ -\{x_k\} & 1 \end{pmatrix}^{-1} = \begin{pmatrix} 1 & 0 \\ \{x_k\} & 1 \end{pmatrix}, \quad \begin{pmatrix} 1 & -\{x'_k\} \\ 0 & 1 \end{pmatrix}^{-1} = \begin{pmatrix} 1 & \{x'_k\} \\ 0 & 1 \end{pmatrix} \tag{D18}$$

exist and are also integer-valued, the reduced lattice vectors $\underline{R}_1^{(k+1)}$ and $\underline{R}_2^{(k+1)}$ of Eq. (D17) provide the same lattice description as $\underline{R}_1^{(k)}$ and $\underline{R}_2^{(k)}$. However, the vector lengths of $\underline{R}_1^{(k+1)}$ and $\underline{R}_2^{(k+1)}$ are equal or smaller compared with those of $\underline{R}_1^{(k)}$ and $\underline{R}_2^{(k)}$.

Since the two vector lengths $R_1^{(k+1)}$ and $R_2^{(k+1)}$ have both a finite lower bound, the continued reductions (D16) and (D17) will, after N iteration steps, converge to a final vector set

$$\underline{R}_{o1} = \underline{R}_1^{(N)}, \qquad \underline{R}_{o2} = \underline{R}_2^{(N)} \tag{D19}$$

which cannot be reduced further. This means in particular that the transformation matrices in Eqs. (D16) and (D17) will become unit matrices resulting in

$$\{x_N\} = \{x_N'\} = 0 \tag{D20}$$

which, according to Eqs. (D11), (D16), and (D17), leads to

$$-R_{o1}^2 \leq 2\,(\underline{R}_{o1}\,\underline{R}_{o2}) < R_{o1}^2 \tag{D21a}$$

$$-R_{o2}^2 \leq 2\,(\underline{R}_{o1}\,\underline{R}_{o2}) < R_{o2}^2 \tag{D21b}$$

or finally to

$$-\min(R_{o1}^2,\,R_{o2}^2) \leq 2\,(\underline{R}_{o1}\,\underline{R}_{o2}) < \min(R_{o1}^2,\,R_{o2}^2) \tag{D22}$$

Relation (D22) forms the **basic condition** for **Minkowski-reduced** (MR) lattice vectors, see also Eq. (3.16).

Assuming that $\underline{R}_{o1}$ and $\underline{R}_{o2}$ form MR lattice vectors according to Eq. (D22), the squared distance d^2 of any lattice vector in this lattice can be written as

$$d^2 = (n_1\,\underline{R}_{o1} + n_2\,\underline{R}_{o2})^2 = n_1^2\,R_{o1}^2 + n_2^2\,R_{o2}^2 + 2\,n_1\,n_2\,(\underline{R}_{o1}\,\underline{R}_{o2}) \tag{D23}$$

If $n_1 n_2 \geq 0$ and $(\underline{R}_{o1}\,\underline{R}_{o2}) \geq 0$ or if $n_1 n_2 < 0$ and $(\underline{R}_{o1}\,\underline{R}_{o2}) \leq 0$ this leads together with Eq. (D22) to

$$d^2 \geq (n_1^2 + n_2^2)\,\min(R_{o1}^2, R_{o2}^2) \geq \min(R_{o1}^2, R_{o2}^2) \tag{D24}$$

If, on the other hand, $n_1 n_2 < 0$ and $(\underline{R}_{o1}\,\underline{R}_{o2}) \geq 0$ or if $n_1 n_2 > 0$ and $(\underline{R}_{o1}\,\underline{R}_{o2}) \leq 0$ then we obtain

$$d^2 = n_1^2\,R_{o1}^2 + n_2^2\,R_{o2}^2 - 2\,|n_1|\,|n_2|\,|(\underline{R}_{o1}\,\underline{R}_{o2})|$$

$$\geq (n_1^2 + n_2^2 - |n_1|\,|n_2|)\,\min(R_{o1}^2,\,R_{o2}^2)$$

$$\geq 1/2\,[(|n_1| - |n_2|)^2 + n_1^2 + n_2^2]\,\min(R_{o1}^2,\,R_{o2}^2)$$

$$\geq \min(R_{o1}^2,\,R_{o2}^2) \tag{D25}$$

This proves that at least one of the two lattice vectors $\underline{R}_{o1}$ and $\underline{R}_{o2}$ connects lattice points of smallest distance in the lattice.

If vectors $\underline{R}_{o1}$ and $\underline{R}_{o2}$ span an angle γ then Eq. (D22) can be written as

$$-\min(q,\,1/q) \leq 2\cos(\gamma) < \min(q,\,1/q), \qquad q = R_{o2}/R_{o1} \tag{D26}$$

which leads to absolute limits

$$-1/2 \leq \cos(\gamma) < 1/2 \qquad 60° < \gamma \leq 120° \qquad (D27)$$

Thus, MR lattice vectors can span only angles γ between 60° and 120°.

Analogs of the Minkowski reduction to bulk lattices have been proposed [42, 43, 49] where for acute lattice representations the reduced lattice vectors $\underline{R}_{o1}, \underline{R}_{o2}, \underline{R}_{o3}$ are found to obey

$$-\min(R_{oi}^2, R_{oj}^2) \leq 2(\underline{R}_{oi}\,\underline{R}_{oj}) < \min(R_{oi}^2, R_{oj}^2), \quad i, j = 1, 2, 3, \quad i \neq j \qquad (D28)$$

Then the squared distance d^2 of any lattice vector in this lattice can be written as

$$\begin{aligned} d^2 &= (n_1 \underline{R}_{o1} + n_2 \underline{R}_{o2} + n_3 \underline{R}_{o3})^2 \\ &= n_1^2 R_{o1}^2 + n_2^2 R_{o2}^2 + n_3^2 R_{o3}^2 + 2 n_1 n_2 s_{12} + 2 n_1 n_3 s_{13} + 2 n_2 n_3 s_{23} \end{aligned} \qquad (D29)$$

with

$$s_{ij} = (\underline{R}_{oi}\,\underline{R}_{oj}) > 0, \quad i, j = 1, 2, 3, \quad i \neq j \text{ (acute representation)} \qquad (D30)$$

Relation (D29) together with (D28) yields for $n_1 \geq 0$, $n_2 \geq 0$, and $n_3 \geq 0$ or for $n_1 < 0$, $n_2 < 0$, and $n_3 < 0$

$$d^2 \geq n_1^2 R_{o1}^2 + n_2^2 R_{o2}^2 + n_3^2 R_{o3}^2 \geq (n_1^2 + n_1^2 + n_1^2) s^2 \geq s^2 \qquad (D31)$$

with

$$s^2 = \min(R_{o1}^2, R_{o2}^2, R_{o3}^2) \qquad (D32)$$

For $n_1 \geq 0$, $n_2 \geq 0$, and $n_3 < 0$ or for $n_1 < 0$, $n_2 < 0$, and $n_3 \geq 0$ relation (D29) together with Eq. (D28) yields

$$\begin{aligned} d^2 &= n_1^2 R_{o1}^2 + n_2^2 R_{o2}^2 + n_3^2 R_{o3}^2 + 2 |n_1| |n_2| s_{12} \\ &\quad - 2 |n_1| |n_3| s_{13} - 2 |n_2| |n_3| s_{23} \\ &\geq 1/2 \left[(|n_1| - |n_3|)^2 + n_1^2 + n_3^2 \right] \min(R_{o1}^2, R_{o3}^2) \\ &\quad + 1/2 \left[(|n_2| - |n_3|)^2 + n_2^2 + n_3^2 \right] \min(R_{o2}^2, R_{o3}^2) \\ &\geq 1/2 \left[(|n_1| - |n_3|)^2 + n_1^2 + n_3^2 + (|n_2| - |n_3|)^2 + n_2^2 + n_3^2 \right] s^2 \\ &\geq s^2 \end{aligned} \qquad (D33)$$

For all other combinations of positive and negative mixing factors n_i, $i = 1-3$, proofs analogous to Eq. (D33) yield the same result such that, altogether, we obtain

$$d^2 \geq \min(R_{o1}^2, R_{o2}^2, R_{o3}^2) \qquad (D34)$$

which shows that for acute lattice representations, relation (D28) guarantees that the reduced lattice vector set contains the lattice vector of smallest length.

Appendix E: Details of Number Theory

In different sections of this book number theoretical methods are applied and functions used. This appendix discusses only some basic details. Further information and additional proofs may be obtained from the mathematical literature [204–206].

E.1 Basic Definitions and Functions

The present definitions deal with the most basic functions and their properties, which are necessary for the number theoretical algorithms discussed in Appendices E.2–E.5.

The **integer truncation function** $f(x) = [x]$ (written with square brackets) is defined for real numbers x as the largest integer n with $n \leq x$. Thus, if

$$n = [x] \quad \text{then} \quad n \leq x < (n+1) \tag{E1}$$

Examples are

$$[1.98] = 1, [5] = 5, [4.5] = 4, [-0.6] = -1, [-3.5] = -4, [-7.3] = -8$$

Note that in some textbooks the integer truncation function for negative x is defined as

$$x < 0 : [x] = -[-x] \tag{E2}$$

which would yield $[-0.6] = 0$, $[-3.5] = -3$, $[-7.3] = -7$, leading to a negative integer value increased by 1 compared to the present definition.

The **nearest integer function** $g(x) = \{x\}$ (written with curly brackets) is defined for real numbers x as the integer n nearest to x, where values $x = m + 1/2$, m integer, are defined as $\{x\} = m + 1$. Thus, the nearest integer function can be expressed by the integer truncation function given by Eq. (E1) as

$$g(x) = \{x\} = [x + 1/2] \tag{E3}$$

Examples are

$$\{1.98\} = 2, \{5\} = 5, \{2.5\} = 3, \{-0.6\} = -1, \{-3.5\} = -3, \{-7.3\} = -7$$

The **modulo function** $\mod(n, m) = n \mid m$ is defined for positive integers m, n as the (integer) remainder of an integer division of n by m, that is, by

$$\mod(n, m) = n \mid m = n - [n/m]\, m \tag{E4}$$

This definition includes cases where $0 < n < m$ yielding

$$n \mid m = n \quad \text{for } 0 \leq n < m \tag{E5}$$

Examples are

$$8 \mid 3 = 2,\ 19 \mid 7 = 5,\ 5 \mid 6 = 5,\ 64 \mid 8 = 0,\ 181 \mid 9 = 1,\ 9 \mid 12 = 9$$

The **greatest common divisor** $\gcd(m, n)$ of two integers m, n is defined as the largest (positive) integer factor (divisor) that is common to both numbers, that is, if

$$m = c\,p, \quad n = c\,q \quad \text{with integer } c > 0,\ p,\ q \tag{E6}$$

and c is the largest of all factors, then

$$\gcd(m, n) = \gcd(n, m) = c \tag{E7}$$

If either m or n (or both) are **negative** integers, then we set

$$\gcd(m, n) = \gcd(|m|, |n|) = c \tag{E8}$$

which yields always **positive** values for $\gcd(m, n)$. Further, if

- the two integers m, n are **equal** then

$$\gcd(m, n) = m \tag{E9}$$

- two different integers m, n do not contain a common factor >1 then

$$\gcd(m, n) = 1 \tag{E10}$$

and the integers are called **coprime**.
- one of two integers m, n equals **zero,** for example, $n = 0$, then we define

$$\gcd(m, 0) = \gcd(0, m) = m \tag{E11}$$

- m, n are integers and m is a **multiple** of n then

$$\gcd(m, n) = \gcd(n, m) = n \tag{E12}$$

- m, n are integers then

$$\gcd(m^2, n^2) = \gcd(m, n)^2 \tag{E13}$$

- m, n are integers with $a = \gcd(m, n)$, then

$$\gcd(m, m + n) = a\, \gcd(m/a,\ (m + n)/a) \tag{E14}$$

The greatest common divisor function can be **generalized** to more than two integers by a recursive procedure, where

$$\begin{aligned} \gcd(n_1, n_2, \ldots, n_N) &= \gcd(n_1, \gcd(n_2, \ldots, n_N)) \\ &= \gcd(n_1, \gcd(n_2, \gcd(n_3, \ldots, n_N))) \\ &= \gcd(n_1, \gcd(n_2, \gcd(n_3, \gcd(\ldots, \gcd(n_{N-1}, n_N))\ldots)) \end{aligned} \tag{E15}$$

Examples are

$$\gcd(87, 9) = 3, \quad \gcd(147, 49) = 49, \quad \gcd(122, 11) = 1, \quad \gcd(18, 192) = 6$$
$$\gcd(18, 192, 333) = 3$$

In the discussion of the Minkowski reduction of Appendix D, the nearest integer function appears in a **composite** function $p(x)$ for real-valued x, see Eqs. (D7) and (D8), which is defined by

$$p(x) = \{x\}(2x - \{x\}) \tag{E16}$$

The definition (E3) of the nearest integer function $\{x\}$ yields

$$\{x\} = n, \; n \text{ integer} \quad \text{for } n - 1/2 \le x < n + 1/2 \tag{E17}$$

Thus we can write function $p(x)$ as a linear function

$$p(x) = n(2x - n) \quad \text{for } n - 1/2 \le x < n + 1/2 \tag{E18}$$

At the boundary $x_b = n + 1/2$ between the intervals $[n - 1/2, n + 1/2)$ and $[n + 1/2, n + 3/2)$ function $p(x)$ yields values

from the left: $\quad p(x_{b-}) = n(2x - n) = n(n + 1)$
from the right: $\quad p(x_{b+}) = (n + 1)(2x - n - 1) = n(n + 1) \tag{E19}$

which shows that $p(x)$ is a **continuous** function. Further, if $x \ne n + 1/2$ then

$$\{x\} = n \quad \text{implies} \quad \{-x\} = -n \tag{E20}$$

resulting in

$$p(-x) = p(x) \tag{E21}$$

For $x = n + 1/2$ we obtain

$$\{x\} = n + 1 \quad \text{and, hence,} \quad \{-x\} = -n - 1 + 1 = -n \tag{E22}$$

and thus

$$p(-x) = (-n - 1)(-n) = n(n + 1) = p(x) \tag{E23}$$

This shows that function $p(x)$ is **symmetric** with respect to x. Finally, relation (E18) shows that $p(x)$ varies linearly with x inside the interval $[n - 1/2, n + 1/2)$ where

$$p(n - 1/2) = n(n - 1) \le p(x) \le p(n + 1/2) = n(n + 1) \tag{E24}$$

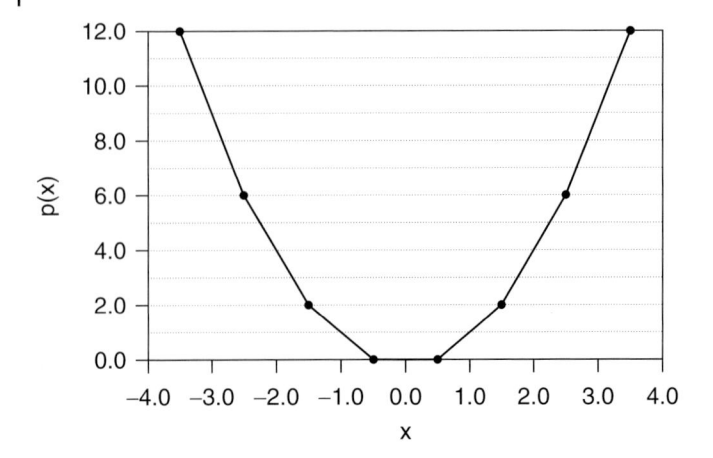

Figure E.1 Graph of function p(x) inside the interval [−3.5, +3.5].

Altogether, p(x) has been proven to assume always **positive** values, see also Figure E.1. Further, p(x) is close to a parabolic shape, which is clear from the fact that the parabolic function $f(x) = x^2 - 1/4$ coincides with p(x) at all points $x = n + 1/2$.

The **least common multiple** lcm(m, n) of two integers m, n is defined as the smallest integer that contains both numbers as factors, that is, if

$$m = a\, p, \; n = b\, p, \quad \text{where } p = \gcd(m, n) \tag{E25}$$

then

$$\text{lcm}(m, n) = a\, b\, p = m\, n\, /\, \gcd(m, n) \tag{E26}$$

Further, if

- two integers m, n are equal, then

$$\text{lcm}(m, n) = m \tag{E27}$$

- two different integers m, n are coprime, then

$$\text{lcm}(m, n) = m\, n \tag{E28}$$

- one of two integers m, n equals zero then lcm(m, n) is undefined
- one of two integers m, n equals 1, then

$$\text{lcm}(m, 1) = \gcd(1, m) = m \tag{E29}$$

Analogous to the greatest common divisor, the least common multiple function can be generalized to more than two integers by a recursive procedure, where

$$\begin{aligned}
\text{lcm}(n_1, n_2, \ldots, n_N) &= \text{lcm}(n_1, \text{lcm}(n_2, \ldots, n_N)) \\
&= \text{lcm}(n_1, \text{lcm}(n_2, \text{lcm}(n_3, \ldots, n_N))) \\
&= \text{lcm}(n_1, \text{lcm}(n_2, \text{lcm}(n_3, \text{lcm}(\ldots, \text{lcm}(n_{N-1}, n_N))\ldots)
\end{aligned} \tag{E30}$$

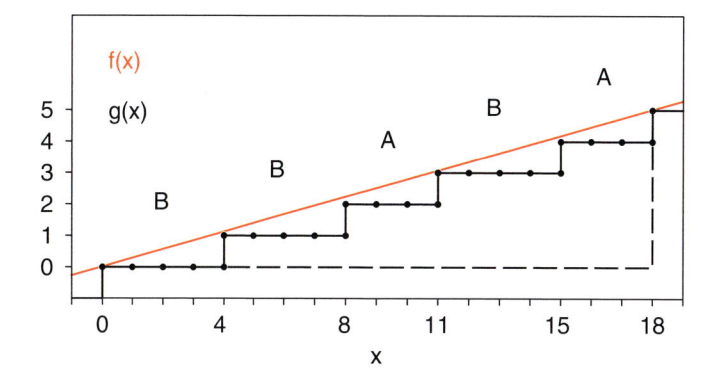

Figure E.2 Graph of linear function f(x) (red) and step function g(x) (black) for x inside [0, 18], see text. The short and long steps are labeled A and B, respectively.

Examples are

$$\text{lcm}(87, 9) = 261, \quad \text{lcm}(147, 49) = 147, \quad \text{lcm}(122, 11) = 1342$$
$$\text{lcm}(18, 81, 6) = 162.$$

The **partitioning** of multiple-atom-height step regions into subterraces A and B separated by single-height steps, discussed in Section 4.3, can be phrased mathematically as the problem to subdivide the positive integer w into h integers $w_i < w$ where w_i are of smallest variation. (In Eq. (4.13) n_s corresponds to h and n_t to w.) This is equivalent to approximating a linear function $f(x) = (h/w) x$ with integer h and w by a step function $g(x) = [(h/w) x]$ where w_i, i = 1, h refers to the intervals of constant g(x) as sketched in Figure E.2 for w = 18, h = 5.

First, we note that

$$w = p\,h + r, \quad p = [w/h] \quad \text{with } 0 \leq r < h \tag{E31}$$

which suggests a partitioning of number w into two kinds, (h − r) numbers A of size $w_i = p$ and r numbers B of size $w_i = (p + 1)$. This guarantees that the sum of all w_i equals w and, in addition, that the variation between the different w_i is smallest. If w is an integer multiple of h, then r = 0 in Eq. (E31) and there are only type A numbers. Otherwise, there are type A and type B numbers which differ by 1.

While the number of different w_i values is clear the sequence of type A and B numbers is not regular and can be obtained from the separations between the step positions x_i where

$$x_i = [(w/h) i] + 1 = p\,i + 1 + [(r/h)\,i], \quad i = 1, \ldots (h-1)$$
$$x_i = 0, \quad x_h = w \tag{E32}$$

assuming w and h to be coprime and hence

$$w_i = x_i - x_{i-1} = p + [(r/h)\,i] - [(r/h)\,(i-1)], \quad i = 2, \ldots (h-1)$$
$$w_1 = [w/h] + 1 = p + 1, \quad w_h = w - [(w/h)(h-1)] - 1 = [w/h] = p \tag{E33}$$

Therefore, the first number, w_1, is always of type **B** and the last, w_h, is of type **A**. However, the intermediate numbers w_i, $1 < i < h$, alternate between p and $p+1$ in an irregular way. As an example, the combination $w = 18$, $h = 5$, $p = 3$, $r = 3$ yields the sequence BBABA shown in Figure E.2.

E.2 Euclid's Algorithm

There is a simple number theoretical method to find the **greatest common divisor** gcd(a, b) of two integers a, b, see Appendix E.1, usually referred to as Euclid's algorithm and discussed in this appendix.

First, we note that we can restrict ourselves to **positive** integers a, b using definition (E8) for negative integers. Second, for $a = b$ the greatest common divisor gcd(a, b) is equal to the arguments, that is,

$$\gcd(a, a) = a \tag{E34}$$

which does not require further evaluation. Further, $\gcd(a, 0) = a$ and $\gcd(a, 1) = 1$ do not merit any consideration. In the more general case of $b > a > 0$ the greatest common divisor $c = \gcd(a, b)$ implies that

$$a = c\,p, \quad b = c\,q \quad \text{with integer p, q, } c > 0 \tag{E35}$$

Thus, we obtain for the auxiliary parameter b'

$$b' = b \mid a = b - [b/a]\,a = (q - [q/p]\,p)\,c = c\,q' \tag{E36}$$

(Definitions of the modulo and the integer truncation function are given in Section E.1.) Hence, parameter b' contains $c = \gcd(a, b)$ as a factor and, on the other hand, must be smaller then a, according to the definition of the modulo function. As a result,

$$\gcd(a, b) = \gcd(a, b') \quad \text{where } b > a > b' \geq 0 \tag{E37}$$

If $b' = 0$ then b is a multiple of a, and according to Eq. (E12), $\gcd(a, b) = a$. Otherwise, the reduction can be continued by reducing parameter a to yield a' analogous to Eq. (E36), which yields an even smaller pair of numbers a', b' with the greatest common divisor being equal to that of the initial pair. Further reductions will eventually lead to one of the two numbers assuming the value zero with the other to yield gcd(a, b). This **finishes** the algorithm usually attributed to the Greek mathematician **Euclid**. Its computational procedure can be formally described by the **iteration**

$$
\begin{aligned}
&a_0 = a > 0, \quad b_0 = b > 0, \quad k = 0 \\
&\text{if } (a_0 = b_0) \text{ then } \gcd(a, b) = a_0 \qquad\qquad\qquad \text{finish} \\
&(*) \quad a_{k+1} = a_k - [a_k/b_k]\,b_k, \quad b_{k+1} = b_k - [b_k/a_k]\,a_k \\
&\qquad \text{if } (a_{k+1} = 0) \text{ then } \gcd(a, b) = b_{k+1} \qquad \text{finish} \\
&\qquad \text{if } (b_{k+1} = 0) \text{ then } \gcd(a, b) = a_{k+1} \qquad \text{finish} \\
&\qquad k = k + 1 \qquad\qquad\qquad\qquad\qquad\qquad\qquad \text{goto } (*)
\end{aligned}
\tag{E38}
$$

For example, finding gcd(333, 90) with this algorithm reads

$$(a, b) = (333, 90) \rightarrow (63, 90) \rightarrow (63, 27) \rightarrow (9, 27) \rightarrow (9, 0)$$

and hence gcd(333, 90) = 9.

Euclid's algorithm can be used recursively to find the greatest common divisor of a set of integers $n_1, n_2, \ldots, n_N$ according to definition (E15).

E.3 Linear Diophantine Equations

In Section 3.2, it was shown that netplane-adapted lattice vectors can be obtained by matrix transformations, see Eqs. (3.11–3.13), where corresponding matrix elements result from solutions of **inhomogeneous linear** Diophantine equations in **two** variables, that is, equations of the type

$$a x + b y = c \tag{E39}$$

with given integer constants a, b, c, and unknown integer variables x and y to be determined. This appendix discusses an iterative method to find solutions for Eq. (E39), where the method is closely connected with **Euclid's algorithm** of Appendix E.2.

First, we can restrict ourselves to positive constants a, b, and c, since Eq. (E39) can always be written with positive constants by changing the corresponding signs of the solutions x, y, for example,

$$a < 0, \; b > 0, \; \text{and} \; c \geq 0 \; \text{leads to} \quad |a| \, (-x) + b \, y = c \tag{E40a}$$

$$a < 0, \; b < 0, \; \text{and} \; c \geq 0 \; \text{leads to} \quad |a| \, (-x) + |b| \, (-y) = c \tag{E40b}$$

$$a > 0, \; b > 0, \; \text{and} \; c < 0 \; \text{leads to} \quad |a| \, (-x) + |b| \, (-y) = |c| \tag{E40c}$$

For c = 0 and finite a, b Eq. (E39) becomes a **homogeneous** linear Diophantine equation, which possesses an infinite number of integer solutions x_o, y_o given by

$$x_o = p \, (b / g), \quad y_o = -p \, (a / g), \quad p \; \text{integer}, \quad g = \gcd(a, b) \tag{E41}$$

Thus, any particular integer solution x_p, y_p of the corresponding **inhomogeneous** equation (E39) with $c \neq 0$ can be used to construct an **infinite** set of solutions, $x_p + x_o, y_p + y_o$ by adding those of the homogeneous equation, given by Eq. (E41), since

$$a \, (x_p + x_o) + b \, (y_p + y_o) = a \, x_p + b \, y_p = c \tag{E42}$$

In the following, we will discuss an algorithm to find a **particular** integer solution x, y of Eq. (E39) where we can restrict ourselves to solutions for c = 1. A solution of Eq. (E39) for $c \neq 1$ can be easily obtained from that for c = 1 by scaling, since equation

$$a x + b y = 1 \tag{E43}$$

can be transformed to

$$a x' + b y' = c \quad \text{with } x' = c x, \; y' = c y \tag{E44}$$

Constants a and b of Eq. (E43) must be coprime, that is, constrained to

$$\gcd(a, b) = 1 \tag{E45}$$

Otherwise, Eq. (E43) has no solution. As an example,

$$117 x + 18 y = 9 (13 x + 2 y) = 1 \tag{E46}$$

has no integer solution x, y since the right hand side of Eq. (E46) is not a multiple of 9.

In order to determine a particular solution of Eq. (E43) we consider a transformation

$$\begin{pmatrix} x' \\ y' \end{pmatrix} = \begin{pmatrix} 1 & [b/a] \\ [a/b] & 1 \end{pmatrix} \cdot \begin{pmatrix} x \\ y \end{pmatrix} = \underline{T} \cdot \begin{pmatrix} x \\ y \end{pmatrix}, \quad \underline{T} = \begin{pmatrix} 1 & [b/a] \\ [a/b] & 1 \end{pmatrix} \tag{E47}$$

and, thus,

$$\begin{pmatrix} x \\ y \end{pmatrix} = \underline{T}^{-1} \cdot \begin{pmatrix} x' \\ y' \end{pmatrix} = \begin{pmatrix} 1 & -[b/a] \\ -[a/b] & 1 \end{pmatrix} \cdot \begin{pmatrix} x' \\ y' \end{pmatrix} \tag{E48}$$

where for $a \neq b$ either [b/a] or [a/b] must equal zero. This transforms Eq. (E43) into

$$a x + b y = a (x' - [b/a] y') + b (y' - [a/b] x')$$
$$= (a - [a/b] b) x' + (b - [b/a] a) y' = a'x' + b'y' = 1 \tag{E49}$$

yielding a modified Diophantine equation of the same structure as Eq. (E43) but with changed constants a', b' where

$$\begin{pmatrix} a' \\ b' \end{pmatrix} = \begin{pmatrix} 1 & -[a/b] \\ -[b/a] & 1 \end{pmatrix} \cdot \begin{pmatrix} a \\ b \end{pmatrix} \tag{E50}$$

These relations are analogous to the reduction step (E36) of Euclid's algorithm for finding the greatest common divisor of a and b, which suggests to apply the same iterative procedure to finding solutions for Eq. (E43).

Thus, in analogy to transformations (E47) and (E48) we consider an iterative sequence of transformations starting with

$$(k = 0) \quad a_k = a, \quad b_k = b, \quad x_k = x, \quad y_k = y \tag{E51}$$

and given by

$$\begin{pmatrix} x_{k+1} \\ y_{k+1} \end{pmatrix} = \begin{pmatrix} 1 & [b_k/a_k] \\ [a_k/b_k] & 1 \end{pmatrix} \cdot \begin{pmatrix} x_k \\ y_k \end{pmatrix} = \underline{T}_k \cdot \begin{pmatrix} x_k \\ y_k \end{pmatrix} \tag{E52}$$

This transforms equation

$$a_k x_k + b_k y_k = 1 \tag{E53}$$

into

$$a_{k+1} x_{k+1} + b_{k+1} y_{k+1} = 1 \tag{E54}$$

where the coefficients a_k and b_k transform to a_{k+1} and b_{k+1} according to

$$\begin{pmatrix} a_{k+1} \\ b_{k+1} \end{pmatrix} = \begin{pmatrix} 1 & -[a_k/b_k] \\ -[b_k/a_k] & 1 \end{pmatrix} \cdot \begin{pmatrix} a_k \\ b_k \end{pmatrix} = \underline{\underline{T}}_k^{-1} \cdot \begin{pmatrix} a_k \\ b_k \end{pmatrix} \quad (E55)$$

This transformation can be continued iteratively, where coefficients a_k and b_k become successively smaller as a result of reduction (E55), that is,

$$0 \leq a_{k+1} \leq a_k \quad \text{and} \quad 0 \leq b_{k+1} \leq b_k \quad (E56a)$$

while according to Euclid's algorithm

$$\gcd(a_{k+1}, b_{k+1}) = \gcd(a_k, b_k) = 1 \quad (E56b)$$

The iteration finishes when one of the coefficients a_{k+1} or b_{k+1} becomes zero, which happens after a finite number of steps since the iteration deals with integers. Assuming that for $k+1=N$ coefficient a_{k+1} vanishes we obtain from Eq. (E53) together with Eq. (E45)

$$a_N x_N + b_N y_N = b_N y_N = \gcd(a, b) y_N = y_N = 1 \quad (E57)$$

Setting $x_N = y_N = 1$ we can iterate backward using relation

$$\begin{pmatrix} x_k \\ y_k \end{pmatrix} = \begin{pmatrix} 1 & -[b_k/a_k] \\ -[a_k/b_k] & 1 \end{pmatrix} \cdot \begin{pmatrix} x_{k+1} \\ y_{k+1} \end{pmatrix} = \underline{\underline{T}}_k^{-1} \cdot \begin{pmatrix} x_{k+1} \\ y_{k+1} \end{pmatrix} \quad (E58)$$

to find the solution for the initial equation (E43) according to

$$\begin{pmatrix} x \\ y \end{pmatrix} = \begin{pmatrix} x_0 \\ y_0 \end{pmatrix} = \underline{\underline{T}}_0^{-1} \cdot \begin{pmatrix} x_1 \\ y_1 \end{pmatrix} = \underline{\underline{T}}_0^{-1} \cdot \underline{\underline{T}}_1^{-1} \cdot \begin{pmatrix} x_2 \\ y_2 \end{pmatrix} = \cdots$$

$$= (\underline{\underline{T}}_0^{-1} \cdot \underline{\underline{T}}_1^{-1} \cdots \underline{\underline{T}}_{N-1}^{-1}) \cdot \begin{pmatrix} x_N \\ y_N \end{pmatrix} = (\underline{\underline{T}}_0^{-1} \cdot \underline{\underline{T}}_1^{-1} \cdots \underline{\underline{T}}_{N-1}^{-1}) \cdot \begin{pmatrix} 1 \\ 1 \end{pmatrix} \quad (E59)$$

If for $k+1=N$ coefficient b_{k+1} vanishes we can apply the same reasoning as before and obtain the same iterative solution (E59). Altogether, the iteration defined by Eqs. (E55), (E57), and (E58) with starting values (E51) allows the iterative calculation of a particular solution of the linear Diophantine equation (E43), where the iteration requires only a finite number of steps.

Solutions of linear Diophantine equations with **n > 2 variables**

$$a_1 x_1 + a_2 x_2 + \cdots + a_n x_n = 1 \quad (E60)$$

can be formally reduced to the $n=2$ problem by rewriting Eq. (E60) as

$$a_1 x_1 + a_2 x_2 = 1 - a_3 x_3 - \cdots - a_n x_n = C \quad (E61)$$

where setting $x_3, \ldots, x_n$ equal to an appropriate combination of integers defines an auxiliary constant C. Then, the above described procedure yields a particular solution x_1, x_2 of Eq. (E61) which, together with the predefined values $x_3, \ldots, x_n$, results in a particular solution of Eq. (E60). Analogous to the constraint (E45), integer solutions of Eq. (E60) impose a constraint on the greatest common divisor of all coefficients $a_1, a_2, \ldots, a_n$, where

$$\gcd(a_1, a_2, \ldots, a_n) = 1 \quad (E62)$$

E.4 Quadratic Diophantine Equations

There is extensive literature on quadratic Diophantine equations [204, 205], which will not be reviewed in this appendix. Here, we focus only on solutions of **specific** equations that arise in connection with **neighbor shells** in crystals of high symmetry and with **overlayer reconstruction** at single crystal surfaces.

1) **Diophantine equations** of the type

$$n_1^2 + n_2^2 + n_3^2 = N, \quad n_i, N \text{ integer}, \quad N \geq 0 \quad \text{(E63)}$$

appear in connection with neighbor shells discussed in Section 2.6. These equations do not have solutions for every value of parameter N, which can be proven using an octal representation of integers. Here, a positive integer a is written as

$$a = (8p + r), \quad p, r \text{ integer with } 0 \leq r < 8 \quad \text{(E64)}$$

and, hence, the square of an integer can be represented as

$$a^2 = 8(8p^2 + 2pr) + r^2 \quad \text{(E65)}$$

Thus, using the definition of the modulo function $a|b$ of Appendix E.1, this yields

$$a^2 | 8 = r^2 | 8 \quad \text{(E66)}$$

where the following table lists all possible values of $r^2 | 8$.

(a | 8)	(r^2 | 8)
0	0
1	1
2	4
3	1
4	0
5	1
6	4
7	1

(E67)

As a result, ($r^2 | 8$) allows only three values, 0, 1, and 4. Further, the sum of three squares can be written in octal representation ($n_i = 8p_i + r_i$) as

$$n_1^2 + n_2^2 + n_3^2 = 64(p_1^2 + p_2^2 + p_3^2) + 16(p_1 r_1 + p_2 r_2 + p_3 r_3) + (r_1^2 + r_2^2 + r_3^2)$$

and hence

$$(n_1^2 + n_2^2 + n_3^2) | 8 = (r_1^2 + r_2^2 + r_3^2) | 8 \quad \text{(E68)}$$

with possible values according to Eq. (E67) in canonical order

r_i^2, r_j^2, r_k^2	$(r_1^2 + r_2^2 + r_3^2)\,|\,8$
0, 0, 0	0
0, 0, 1	1
0, 0, 4	4
0, 1, 1	2
0, 1, 4	5
0, 4, 4	0
1, 1, 1	3
1, 1, 4	6
1, 4, 4	1
4, 4, 4	4

(E69)

This shows that

$$(n_1^2 + n_2^2 + n_3^2)\,|\,8 \neq 7 \quad \text{or} \quad n_1^2 + n_2^2 + n_3^2 \neq 8\,q + 7, \quad q \text{ integer} \quad \text{(E70)}$$

Next we note that if all numbers n_i in Eq. (E63) are even numbers, then their squares are multiples of 4 and, as a consequence, N in Eq. (E63) must be a multiple of 4. Thus, we can write

$$n_1^2 + n_2^2 + n_3^2 = (2\,m_1)^2 + (2\,m_2)^2 + (2\,m_3)^2 = N = 4\,M \quad \text{(E71)}$$

which can be reduced to

$$m_1^2 + m_2^2 + m_3^2 = M \quad \text{(E72)}$$

Therefore, if Eq. (E72) has no solutions for $M = (8\,q + 7)$ then Eq. (E71) has no solutions equal to $4\,(8\,q + 7)$. This can be continued to yield the final result that Eq. (E63) does not have solutions for

$$N = 4^p\,(8\,q + 7), \quad p, q \text{ integer} \quad \text{(E73)}$$

For N values different from those given by Eq. (E73), solutions for Eq. (E63) can be determined by trial-and-error since values n_i, $i = 1-3$, are restricted to a finite number of integers $|n_i| \leq [\sqrt{N}]$. For any given solution (n_1, n_2, n_3) the sets of triplets $(\pm n_1, \pm n_2, \pm n_3)$ and $(\pm n_1', \pm n_2', \pm n_3')$ with n_i' denoting permutations of n_i offer other solutions counted by the symmetry degeneracy m_{sym} where $m_{sym} = 6, 8, 12, 24, 48$ depending on the actual values of n_i. Thus, Eq. (E63) has always multiple solutions for $N \geq 1$. In addition, this equation can have solutions (n_1, n_2, n_3) and (n_1', n_2', n_3') where the absolute values of the numbers n_i and n_i' differ beyond simple permutations. As an example we mention (1, 3, 4) and (1, 0, 5) yielding $N = 26$. These solutions are counted by the accidental degeneracy m_{acc}. While m_{sym} for each solution (n_1, n_2, n_3) is limited to a maximum of $m_{sym} = 48$, values of m_{acc} have no upper bound. As an illustration, Eq. (E63) for $N = 972\,221$ yields $m_{acc} = 521$ where for 520 different solutions $m_{sym} = 48$ and for one solution $m_{sym} = 12$ amounting, altogether, to 24 972 different solutions.

It may be mentioned in passing that **Diophantine equations** of the type

$$n_1^2 + n_2^2 + n_3^2 + n_4^2 = N, \quad n_i, \ N \text{ integer} \tag{E74}$$

have (multiple) solutions for any value of $N \geq 0$, which is known as Lagrange's theorem [206].

2) **Diophantine equations** of the type

$$n_1^2 + n_2^2 = N^2, \quad n_1, n_2, \ N \text{ integer} \tag{E75}$$

appear in connection with rotationally reconstructed overlayers at single crystal surfaces with **square** geometry discussed in detail in Section 5.2. Equation (E75) is the integer version of the well-known Pythagorean equation defining so-called **Pythagorean triplets** (N, n_1, n_2). These numbers can be constructed by the generalized **Euclid formula.** Setting

$$n_1 = k\,(2\,m\,n), \quad n_2 = k\,(m^2 - n^2), \quad k, m, n \text{ integer with } m > n \tag{E76}$$

and letting m, n, k assume all possible values yields

$$N = \sqrt{(n_1^2 + n_2^2)} = k\,(m^2 + n^2) \tag{E77}$$

It can be shown that Eq. (E76) generates all possible Pythagorean triplets. Setting $k = 1$ in Eq. (E76) yields so-called **primitive triplets,** which are defined by $\gcd(m, n) = 1$ with either odd m and even n or vice versa. In this case the representation of (N, n_1, n_2) by m, n is **unique,** that is, there are no other values m, n, which yield the same triplet (N, n_1, n_2). Further, Eq. (E76) shows that there are infinitely many Pythagorean triplets. In addition, the numbers n_1, n_2 must always be different (if the trivial cases $n_1 = 0$ and/or $n_2 = 0$ are excluded), since no square number can be represented by $(2\,N^2)$.

The following table lists the 10 smallest primitive Pythagorean triplets (N, n_1, n_2) with $n_1 > n_2 > 0$.

	(N, n_1, n_2)
1	(5, 4, 3)
2	(13, 12, 5)
3	(17, 15, 8)
4	(25, 24, 7)
5	(29, 21, 20)
6	(37, 35, 12)
7	(39, 36, 15)
8	(41, 40, 9)
9	(51, 45, 24)
10	(53, 45, 28)

3) **Diophantine equations** of the type

$$m^2 + mn + n^2 = N^2, \quad m, n, N \text{ integer} \tag{E78}$$

appear in connection with rotationally reconstructed overlayers at single crystal surfaces with **hexagonal** geometry, see Section 6.4. If the hexagonal surface is described (in Cartesian coordinates) by lattice vectors in acute representation, that is, by

$$\underline{R}_1 = a(1, 0), \quad \underline{R}_2 = a(1/2, \sqrt{3}/2) \tag{E79}$$

then rotated overlayer lattice vectors $\underline{R}'_1, \underline{R}'_2$ must be of equal length compared with those of $\underline{R}_1, \underline{R}_2$. Assuming a coincidence lattice overlayer, see Section 5.2, where

$$\begin{pmatrix} \underline{R}'_1 \\ \underline{R}'_2 \end{pmatrix} = \frac{1}{N} \begin{pmatrix} m_{11} & m_{12} \\ m_{21} & m_{22} \end{pmatrix} \cdot \begin{pmatrix} \underline{R}_1 \\ \underline{R}_2 \end{pmatrix}, \quad m_{ij}, N \text{ integer} \tag{E80}$$

we obtain for the vector lengths with Eq. (E79)

$$N^2 |\underline{R}'_i|^2 = (m_{i1} \underline{R}_1 + m_{i2} \underline{R}_2)^2 = a^2 (m_{i1}^2 + m_{i1} m_{i2} + m_{i2}^2)$$
$$= N^2 |\underline{R}_i|^2 = a^2 N^2 \tag{E81}$$

where setting

$$m_{i1} = m, \quad m_{i2} = n \tag{E82}$$

leads to Diophantine equations of the type (E78). These equations are quadratic and of the elliptic type with multiple solutions for all values N. We note that

- for integer $N \neq 0$ there are always six trivial solutions,

$$(m, n) = \pm(N, 0), \quad \pm(0, N), \quad \pm(N, -N) \tag{E83}$$

- if $m_o \neq 0, n_o \neq 0$ are solutions of Eq. (E78) then
  - m_o, n_o must be different in value, since, assuming $m_o = n_o$, Eq. (E78) would read $3 n_o^2 = N^2$, which cannot be solved for integers n_o, N.
  - there are altogether 12 different solutions

$$(m, n) = \pm(m_o, n_o), \quad \pm(m_o, -m_o - n_o),$$
$$\pm(m_o + n_o, -m_o), \quad \pm(n_o, m_o),$$
$$\pm(n_o, -m_o - n_o), \quad \pm(m_o + n_o, -n_o) \tag{E84}$$

- Equation (E78) can also be written as

$$m = -\frac{n}{2} \pm \sqrt{N^2 - \frac{3}{4} n^2} \quad \text{or} \quad n = -\frac{m}{2} \pm \sqrt{N^2 - \frac{3}{4} m^2} \tag{E85}$$

This restricts the range of integer solutions (m, n) to

$$\sqrt{N^2 - \frac{3}{4} m^2} \geq 0, \quad \sqrt{N^2 - \frac{3}{4} n^2} \geq 0 \quad \rightarrow \quad |m|, |n| \leq \left[\frac{2}{\sqrt{3}} N\right] \tag{E86}$$

and offers a trial-and-error method to determine solutions (m, n) for given N > 0. Letting integer n run from $-[(2/\sqrt{3})\,N]$ to $[(2/\sqrt{3})\,N]$ and checking the validity of Eq. (E85) for integers m yields the corresponding solutions.

The following table lists the 10 smallest triplets (N, m, n) with $m > n > 0$ (gcd(m, n) = 1)

	(N, m, n)
1	(7, 5, 3)
2	(13, 8, 7)
3	(19, 16, 5)
4	(31, 24, 11)
5	(37, 33, 7)
6	(43, 35, 13)
7	(49, 39, 16)
8	(61, 56, 9)
9	(67, 45, 32)
10	(73, 63, 17)

4) **Diophantine equations** of the type

$$m^2 + 2n^2 = N^2, \quad m,\ n,\ N \text{ integer} \tag{E87}$$

appear in connection with rotationally reconstructed overlayers at single crystal surfaces with **rectangular** geometry, see Section 6.4. If the rectangular surface is described (in Cartesian coordinates) by lattice vectors

$$\underline{R}_1 = a\,(1,\ 0), \quad \underline{R}_2 = a\,(0,\ \sqrt{2}) \tag{E88}$$

then each of the two rotated overlayer lattice vectors $\underline{R}'_i$ must be of equal length compared with that of $\underline{R}_i$. Assuming a coincidence lattice overlayer, see Section 5.2, with lattice vectors defined by Eq. (E80), we obtain for the vector lengths together with Eq. (E88)

$$N^2 |\underline{R}'_1|^2 = (m_{11}\underline{R}_1 + m_{12}\underline{R}_2)^2 = a^2\,(m_{11}^2 + 2m_{12}^2) = N^2\,|\underline{R}_1|^2$$
$$= a^2\,N^2 \tag{E89}$$

The orthogonality of $\underline{R}'_1,\ \underline{R}'_2$ and the vector ratio $R'_2/R'_1 = \sqrt{2}$ of the rotated lattice is guaranteed by setting $m_{21} = -2\,m_{12}$ and $m_{22} = m_{11}$. Further, setting

$$m_{11} = m, \quad m_{12} = n \tag{E90}$$

leads to Diophantine equations of the type (E87) that are quadratic and of the elliptic type. Equation (E87) can also be written as

$$m = \pm\sqrt{N^2 - 2n^2} \tag{E91}$$

which restricts the range of integer solutions (m, n) to

$$\sqrt{N^2 - 2n^2} \geq 0 \quad \rightarrow \quad |n| \leq \left[\frac{1}{\sqrt{2}}N\right] \tag{E92}$$

Thus, solutions can be obtained by trial-and-error methods analogous to those used for hexagonal surfaces described earlier. Letting integer n run from $-[N/\sqrt{2}]$ to $[N/\sqrt{2}]$ and checking the validity Eq. (E91) for integers m yields corresponding solutions.

The following table lists solution triplets (N, m, n) for the 10 smallest N values (gcd(m, n) = 1)

	(N, m, n)
1	(3, 1, 2)
2	(9, 7, 4)
3	(11, 7, 6)
4	(17, 1, 12)
5	(19, 17, 6)
6	(27, 23, 10)
7	(33, 17, 20)
8	(33, 31, 8)
9	(41, 23, 24)
10	(43, 7, 30)

5) **Diophantine equations** of the type

$$a\,m^2 = b\,n^2, \quad a, b, m, n \text{ integer} \tag{E93}$$

appear in connection with symmetry constraints of nanotubes, see Section 8.2. This equation can also be written as

$$m = p\,n \text{ with } p = \sqrt{(b/a)}, \quad a, b, m, n \text{ integer} \tag{E94}$$

Thus, for Eq. (E94) to have integer solutions, parameter p must be a rational number which is only possible if (b/a) equals the square of a rational number, that is, if

$$b/a = (w/v)^2 = w^2/v^2, \quad w, v \text{ integer} \tag{E95}$$

Applying Eq. (E95) to the initial equation (E93) yields

$$v\,m = \pm w\,n \quad \text{with integer solutions} \quad m = q\,w, \; n = q\,v, \; q \text{ integer} \tag{E96}$$

Constraint Eq. (E95) means, in particular, that the Diophantine equation

$$3\,m^2 = n^2, \quad m, n \text{ integer} \tag{E97}$$

has no solutions. This is the mathematical basis of proving that ideal substrate surfaces with a square lattice, like fcc(1 0 0) or bcc(1 0 0), cannot have commensurate overlayers with a hexagonal lattice. Likewise, the Diophantine equation

$$3\,m^2 = 2\,n^2, \quad m, n \text{ integer} \tag{E98}$$

has no solutions, which is essential for proving that primitive rectangular substrate surfaces described by fcc(1 1 0) cannot have commensurate overlayers with a hexagonal lattice.

E.5 Number Theory and 2 × 2 Matrices

This appendix discusses a few number theoretical details connected with two-dimensional matrices whose elements are integers or fractional numbers.

The **transposed** matrix $\underline{\underline{M}}^+$ and the **inverted** matrix $\underline{\underline{M}}^{-1}$ of an integer-valued 2×2 matrix $\underline{\underline{M}}$ are given by

$$\underline{\underline{M}} = \begin{pmatrix} m_{11} & m_{12} \\ m_{21} & m_{22} \end{pmatrix}, \quad \underline{\underline{M}}^+ = \begin{pmatrix} m_{11} & m_{21} \\ m_{12} & m_{22} \end{pmatrix}, \quad \underline{\underline{M}}^{-1} = \frac{1}{\Delta} \begin{pmatrix} m_{22} & -m_{12} \\ -m_{21} & m_{11} \end{pmatrix} \quad (E99)$$

where

$$\Delta = \det(\underline{\underline{M}}) = m_{11}m_{22} - m_{12}m_{21} \quad (E100)$$

is the **determinant** of matrix $\underline{\underline{M}}$. The determinant Δ and matrix $\underline{\underline{M}}^+$ are integer-valued while matrix $\underline{\underline{M}}^{-1}$ is fractional except for $\Delta = 1$. However, the product ($\Delta \, \underline{\underline{M}}^{-1}$) is always integer-valued.

There is an infinite set of integer-valued 2×2 matrices $\underline{\underline{M}}$ whose inverse matrix is an integer, that is, where

$$\Delta = \det(\underline{\underline{M}}) = m_{11}m_{22} - m_{12}m_{21} = 1 \quad (E101)$$

This means in particular that m_{i1} and m_{i2} as well as m_{1i} and m_{2i}, $i = 1, 2$ must be coprime to yield a solution of Eq. (E101), that is,

$$\gcd(m_{i1}, m_{i2}) = \gcd(m_{1i}, m_{2i}) = 1, \quad i = 1, 2 \quad (E102)$$

Matrices $\underline{\underline{M}}$ can be obtained by selecting one of the four pairs of coprime integer elements (m_{i1}, m_{i2}) or (m_{1i}, m_{2i}). Then, Eq. (E101) can be understood as a linear Diophantine equation for the complementing matrix elements (m_{j1}, m_{j2}) or (m_{1j}, m_{2j}) for which an infinite set of solutions can be evaluated following the procedure based on Euclid's algorithm as discussed in Appendix E.3. As an example, the infinite set of matrices with predefined $m_{11} = 1$, $m_{12} = 2$ is given by

$$\underline{\underline{M}} = \begin{pmatrix} 1 & 2 \\ p & 2p+1 \end{pmatrix}, \quad p \text{ integer} \quad (E103)$$

If $\underline{\underline{A}}$ is a real-valued 2×2 matrix, then its **integer approximant** $\{\underline{\underline{A}}\}$ is defined by an integer-valued 2×2 matrix

$$\underline{\underline{A}} = \begin{pmatrix} a_{11} & a_{12} \\ a_{21} & a_{22} \end{pmatrix} \Rightarrow \{\underline{\underline{A}}\} = \begin{pmatrix} \{a_{11}\} & \{a_{21}\} \\ \{a_{12}\} & \{a_{22}\} \end{pmatrix} \quad (E104)$$

where $\{a_{ij}\}$ denotes the integer nearest to a_{ij}, see Appendix E.1. As an example, the rotation matrix

$$\underline{\underline{A}} = \begin{pmatrix} \cos\alpha & \sin\alpha \\ -\sin\alpha & \cos\alpha \end{pmatrix} \stackrel{\text{def.}}{=} (\cos\alpha, \sin\alpha \mid -\sin\alpha, \cos\alpha) \quad (E105)$$

with $\det(\underline{A}) = 1$ has as integer approximants $\{\underline{A}\}$ with their determinants $\det(\{\underline{A}\})$ for ranges of angles α given in the following table.

α	$\{\underline{A}\}$, $\det(\{\underline{A}\})$	α	$\{\underline{A}\}$, $\det(\{\underline{A}\})$
[0°, 30°)	(1, 0 | 0, 1), 1	[210°, 240°]	(−1, −1 | 1, −1), 2
[30°, 60°]	(1, 1 | −1, 1), 2	(240°, 300°)	(0, −1 | 1, 0), 1
(60°, 120°)	(0, 1 | −1, 0), 1	[300°, 330°]	(1, −1 | 1, 1), 2
[120°, 150°]	(−1, 1 | −1, −1), 2	(330°, 360°]	(1, 0 | 0, 1), 1
(150°, 210°)	(−1, 0 | 0, −1), 1		

(E106)

In connection with reducing cell sizes of commensurate lattices of high-order commensurate (HOC) overlayer structures discussed in Section 6.4 the **reduction** of two 2 × 2 **integer matrices** to yield a coprime pair is of interest.

The starting point are two integer matrices $\underline{A}$ and $\underline{B}$ where $\underline{B}$ is diagonal, that is,

$$\underline{A} = \begin{pmatrix} a_{11} & a_{12} \\ a_{21} & a_{22} \end{pmatrix}, \quad \underline{B} = \begin{pmatrix} b_1 & 0 \\ 0 & b_2 \end{pmatrix}, \quad a_{ij}, b_i \text{ integer} \quad (E107)$$

These matrices form a **coprime pair** if their determinants are coprime, that is, if

$$g = \gcd(|\det(\underline{A})|, |\det(\underline{B})|) = 1 \quad (E108)$$

For $g > 1$, we consider a pair of reduced integer matrices $\underline{A}'$ and $\underline{B}'$ resulting from a joint integer-valued reduction matrix $\underline{T}$ with

$$\underline{A}' = \underline{T}^{-1} \cdot \underline{A}, \quad \underline{B}' = \underline{T}^{-1} \cdot \underline{B} \quad (E109)$$

This yields for the determinants

$$\det(\underline{A}) = \det(\underline{T}) \det(\underline{A}'), \quad \det(\underline{B}) = \det(\underline{T}) \det(\underline{B}') \quad (E110)$$

Thus, the integer $\det(\underline{T})$ must be a common divisor of both integers $\det(\underline{A})$ and $\det(\underline{B})$ with an upper limit given by

$$|\det(\underline{T})| \leq g \quad (E111)$$

according to Eq. (E108). For $\det(\underline{T}) = g$ we obtain

$$|\det(\underline{A})| = g |\det(\underline{A}')|, \quad |\det(\underline{B})| = g |\det(\underline{B}')| \quad (E112)$$

and hence with Eq. (E108)

$$\gcd(|\det(\underline{A}')|, |\det(\underline{B}')|) = \gcd(|\det(\underline{A})|, |\det(\underline{B})|)/g = 1 \quad (E113)$$

Thus, the reduced matrices $\underline{\underline{A}}'$ and $\underline{\underline{B}}'$ form a coprime pair that can be evaluated by a simple **trial-and-error** procedure. From

$$\underline{\underline{A}}' = \underline{\underline{T}}^{-1} \cdot \underline{\underline{A}} = \underline{\underline{B}}' \cdot \underline{\underline{B}}^{-1} \cdot \underline{\underline{A}} \tag{E114}$$

together with Eq. (E107) we obtain for the matrix elements

$$\begin{pmatrix} a'_{11} & a'_{12} \\ a'_{21} & a'_{22} \end{pmatrix} = \begin{pmatrix} \frac{a_{11}}{b_1}b'_{11} + \frac{a_{21}}{b_2}b'_{12} & \frac{a_{12}}{b_1}b'_{11} + \frac{a_{22}}{b_2}b'_{12} \\ \frac{a_{11}}{b_1}b'_{21} + \frac{a_{21}}{b_2}b'_{22} & \frac{a_{12}}{b_1}b'_{21} + \frac{a_{22}}{b_2}b'_{22} \end{pmatrix}, \tag{E115}$$

which yields four Diophantine equations

$$a'_{ij} = \frac{a_{1j}}{b_1}b'_{i1} + \frac{a_{2j}}{b_2}b'_{i2}, \quad i, j = 1, 2 \tag{E116}$$

transforming between integers a'_{ij} and b'_{ij} with integers b_i and a_{ij} being determined by the initial matrices $\underline{\underline{A}}$ and $\underline{\underline{B}}$. The search for solutions for Eq. (E116), and, thus for matrices $\underline{\underline{A}}'$ and $\underline{\underline{B}}'$, can be restricted to element values

$$0 \leq b'_{ij} < b_j, \quad j = 1, 2 \tag{E117}$$

since for larger or smaller values $b''_{ij} = b'_{ij} + n_j b_j$, Eq. (E116) read

$$\begin{aligned} a''_{ij} &= \frac{a_{1j}}{b_1}b''_{i1} + \frac{a_{2j}}{b_2}b''_{i2} = \frac{a_{1j}}{b_1}b'_{i1} + \frac{a_{2j}}{b_2}b'_{i2} + n_1 a_{1j} + n_2 a_{2j} \\ &= a'_{ij} + n_1 a_{1j} + n_2 a_{2j} \end{aligned} \tag{E118}$$

such that according to Eq. (E118) the complete set of solutions, yielding matrix elements of $\underline{\underline{A}}'$, can be determined from those inside the region Eq. (E117) by additive corrections involving only elements a_{ij} of the initial matrix $\underline{\underline{A}}$. Thus, the search procedure can be performed by inserting the $b_1 b_2$ different integers combinations (b'_{i1}, b'_{i2}) into Eq. (E116) and checking for integer-valued a'_{ij}. The resulting matrices $\underline{\underline{A}}'$ and $\underline{\underline{B}}'$ can then be tested for

$$\gcd(|\det(\underline{\underline{A}}')|, |\det(\underline{\underline{B}}')|) = 1 \tag{E119}$$

where the validity yields a coprime pair of reduced matrices $\underline{\underline{A}}'$ and $\underline{\underline{B}}'$.

As an example, we consider the transformation (6.17) discussed in Section 6.4 where

$$\underline{\underline{M}} = \frac{1}{13}\begin{pmatrix} 14 & -5 \\ 5 & 14 \end{pmatrix} = \underline{\underline{B}}^{-1} \cdot \underline{\underline{A}}, \quad \underline{\underline{A}} = \begin{pmatrix} 14 & -5 \\ 5 & 14 \end{pmatrix}, \quad \underline{\underline{B}} = \begin{pmatrix} 13 & 0 \\ 0 & 13 \end{pmatrix}, \quad g = 13$$

which leads to Diophantine equations according to Eq. (E116)

$$a'_{i1} = \frac{a_{11}}{b_1}b'_{i1} + \frac{a_{21}}{b_2}b'_{i2} = \frac{14}{13}b'_{i1} + \frac{5}{13}b'_{i2}$$

$$a'_{i2} = \frac{a_{12}}{b_1}b'_{i1} + \frac{a_{22}}{b_2}b'_{i2} = \frac{-5}{13}b'_{i1} + \frac{14}{13}b'_{i2}$$

with solutions

$$\underline{A}' = \begin{pmatrix} 4 & 1 \\ -1 & 4 \end{pmatrix}, \quad \underline{B}' = \begin{pmatrix} 3 & 2 \\ -2 & 3 \end{pmatrix}$$

and

$$\gcd(|\det(\underline{A}')|, |\det(\underline{B}')|) = \gcd(17, 13) = 1$$

Appendix F: Details of Vector Calculus and Linear Algebra

This appendix discusses a few mathematical details connected with vector calculus, which are needed in several places in this book.

In three dimensions, a **vector** $\underline{R}$ is defined by its Cartesian components x_i along the x-, y-, and z-axes as

$$\underline{R} = (x_1, x_2, x_3) \quad \text{with its } \textbf{length} \ R = |\underline{R}| = \sqrt{(x_1^2 + x_2^2 + x_3^2)} \tag{F1}$$

The **scalar product** between two vectors $\underline{R}$, $\underline{R}'$ is a real number given by

$$(\underline{R}\,\underline{R}') = x_1 x_1' + x_2 x_2' + x_3 x_3' = (\underline{R}'\,\underline{R}) = R\,R' \cos(\gamma) \tag{F2}$$

with γ defining the angle between the two vectors.

The **vector product**, $\underline{R} \times \underline{R}'$ is a vector given in Cartesian coordinates by

$$\underline{R} \times \underline{R}' = (x_2 x_3' - x_3 x_2',\ x_3 x_1' - x_1 x_3',\ x_1 x_2' - x_2 x_1') = -(\underline{R}' \times \underline{R}) \tag{F3}$$

with its length

$$|\underline{R} \times \underline{R}'| = \sqrt{\{(x_2 x_3' - x_3 x_2')^2 + (x_3 x_1' - x_1 x_3')^2 + (x_1 x_2' - x_2 x_1')^2\}} =$$
$$= R\,R' \sin(\gamma) \tag{F4}$$

where γ defines the angle between the two vectors.

The **volume product** $(\underline{R} \times \underline{R}')\,\underline{R}''$ is a real number given by

$$(\underline{R} \times \underline{R}')\,\underline{R}'' = \det \begin{pmatrix} x_1 & x_2 & x_3 \\ x_1' & x_2' & x_3' \\ x_1'' & x_2'' & x_3'' \end{pmatrix} = (\underline{R}' \times \underline{R}'')\,\underline{R} = (\underline{R}'' \times \underline{R})\,\underline{R}'$$

$$= (x_2 x_3' - x_3 x_2')\,x_1'' + (x_3 x_1' - x_1 x_3')\,x_2'' + (x_1 x_2' - x_2 x_1')\,x_3'' \tag{F5}$$

Consider a **linear transformation** between lattice vectors $\underline{R}_1$, $\underline{R}_2$, $\underline{R}_3$ and $\underline{R}_1'$, $\underline{R}_2'$, $\underline{R}_3'$, where

$$\begin{pmatrix} \underline{R}_1' \\ \underline{R}_2' \\ \underline{R}_3' \end{pmatrix} = \begin{pmatrix} t_{11} & t_{12} & t_{13} \\ t_{21} & t_{22} & t_{23} \\ t_{31} & t_{32} & t_{33} \end{pmatrix} \cdot \begin{pmatrix} \underline{R}_1 \\ \underline{R}_2 \\ \underline{R}_3 \end{pmatrix} = \underline{\underline{T}} \cdot \begin{pmatrix} \underline{R}_1 \\ \underline{R}_2 \\ \underline{R}_3 \end{pmatrix} \tag{F6}$$

Crystallography and Surface Structure: An Introduction for Surface Scientists and Nanoscientists,
Second Edition. Klaus Hermann.
© 2017 Wiley-VCH Verlag GmbH & Co. KGaA. Published 2017 by Wiley-VCH Verlag GmbH & Co. KGaA.

Then, products of transformed vectors $\underline{R}'_i$, $\underline{R}'_j$; can be expressed by those of the initial vectors $\underline{R}_i$, $\underline{R}_j$ where

- **Scalar products**, $(\underline{R}'_i \underline{R}'_j)$, yield

$$(\underline{R}'_i \underline{R}'_j) = (t_{i1}\underline{R}_1 + t_{i2}\underline{R}_2 + t_{i3}\underline{R}_3)(t_{j1}\underline{R}_1 + t_{j2}\underline{R}_2 + t_{j3}\underline{R}_3)$$

$$= t_{i1}t_{j1}|\underline{R}_1|^2 + t_{i2}t_{j2}|\underline{R}_2|^2 + t_{i3}t_{j3}|\underline{R}_3|^2 + (t_{i1}t_{j2} + t_{i2}t_{j1})(\underline{R}_1\underline{R}_2)$$

$$+ (t_{i1}t_{j3} + t_{i3}t_{j1})(\underline{R}_1\underline{R}_3) + (t_{i2}t_{j3} + t_{i3}t_{j2})(\underline{R}_2\underline{R}_3) \tag{F7}$$

- **Vector products**, $\underline{R}'_i \times \underline{R}'_j$, yield

$$\underline{R}'_i \times \underline{R}'_j = (t_{i1}\underline{R}_1 + t_{i2}\underline{R}_2 + t_{i3}\underline{R}_3) \times (t_{j1}\underline{R}_1 + t_{j2}\underline{R}_2 + t_{j3}\underline{R}_3)$$

$$= (t_{i1}t_{j2} - t_{i2}t_{j1})(\underline{R}_1 \times \underline{R}_2) + (t_{i1}t_{j3} - t_{i3}t_{j1})(\underline{R}_1 \times \underline{R}_3)$$

$$+ (t_{i2}t_{j3} - t_{i3}t_{j2})(\underline{R}_2 \times \underline{R}_3) \tag{F8}$$

- **Volume products** $(\underline{R}'_1 \times \underline{R}'_2)\underline{R}'_3$ yield

$$(\underline{R}'_1 \times \underline{R}'_2)\underline{R}'_3 =$$

$$= [(t_{11}\underline{R}_1 + t_{12}\underline{R}_2 + t_{13}\underline{R}_3) \times (t_{21}\underline{R}_1 + t_{22}\underline{R}_2 + t_{23}\underline{R}_3)]\underline{R}'_3$$

$$= (t_{11}t_{22} - t_{12}t_{21})(\underline{R}_1 \times \underline{R}_2)\underline{R}'_3$$

$$+ (t_{11}t_{23} - t_{13}t_{21})(\underline{R}_1 \times \underline{R}_3)\underline{R}'_3$$

$$+ (t_{12}t_{23} - t_{13}t_{22})(\underline{R}_2 \times \underline{R}_3)\underline{R}'_3$$

$$= t_{33}(t_{11}t_{22} - t_{12}t_{21})(\underline{R}_1 \times \underline{R}_2)\underline{R}_3$$

$$+ t_{32}(t_{11}t_{23} - t_{13}t_{21})(\underline{R}_1 \times \underline{R}_3)\underline{R}_2$$

$$+ t_{31}(t_{12}t_{23} - t_{13}t_{22})(\underline{R}_2 \times \underline{R}_3)\underline{R}_1$$

$$= \det(\underline{T})(\underline{R}_1 \times \underline{R}_2)\underline{R}_3 \tag{F9}$$

Consider **reciprocal lattice vectors**, defined in Section 2.5 and given by

$$\underline{G}_1 = \beta(\underline{R}_2 \times \underline{R}_3), \quad \underline{G}_2 = \beta(\underline{R}_3 \times \underline{R}_1), \quad \underline{G}_3 = \beta(\underline{R}_1 \times \underline{R}_2)$$

$$\beta = 2\pi/[(\underline{R}_1 \times \underline{R}_2)\underline{R}_3] \tag{F10}$$

and **reciprocal of the reciprocal** lattice vectors given by

$$\underline{H}_1 = \gamma(\underline{G}_2 \times \underline{G}_3), \quad \underline{H}_2 = \gamma(\underline{G}_3 \times \underline{G}_1), \quad \underline{H}_3 = \gamma(\underline{G}_1 \times \underline{G}_2)$$

$$\gamma = 2\pi/[(\underline{G}_1 \times \underline{G}_2)\underline{G}_3] \tag{F11}$$

Then, using relation

$$(\underline{a} \times \underline{b}) \times \underline{c} = (\underline{a}\,\underline{c})\underline{b} - (\underline{b}\,\underline{c})\underline{a} \tag{F12}$$

from basic vector calculus, we obtain

$$\underline{H}_1 = \gamma(\underline{G}_2 \times \underline{G}_3) = \gamma\beta^2\,[(\underline{R}_3 \times \underline{R}_1) \times (\underline{R}_1 \times \underline{R}_2)]$$
$$= \gamma\beta^2\,(2\pi/\beta)\underline{R}_1 = \underline{R}_1 \qquad (F13)$$

since according to Eq. (2.99)

$$2\pi\gamma\beta = (2\pi)^3/(\underline{G}_1 \times \underline{G}_2)\underline{G}_3 \,/\, (\underline{R}_1 \times \underline{R}_2)\underline{R}_3) = 1 \qquad (F14)$$

Likewise, we obtain

$$\underline{H}_2 = \gamma(\underline{G}_3 \times \underline{G}_1) = \gamma\beta^2\,[(\underline{R}_1 \times \underline{R}_2) \times (\underline{R}_2 \times \underline{R}_3)]$$
$$= \gamma\beta^2\,(2\pi/\beta)\underline{R}_2 = \underline{R}_2 \qquad (F15)$$

$$\underline{H}_3 = \gamma(\underline{G}_1 \times \underline{G}_2) = \gamma\beta^2\,[(\underline{R}_2 \times \underline{R}_3) \times (\underline{R}_3 \times \underline{R}_1)]$$
$$= \gamma\beta^2\,(2\pi/\beta)\underline{R}_3 = \underline{R}_3 \qquad (F16)$$

As a result, the reciprocal of the reciprocal lattice **agrees** with the initial lattice.

Further, the **vector relation**

$$[(\underline{a} \times \underline{b})\underline{c}]\,[(\underline{a}' \times \underline{b}')\underline{c}'] = \det\begin{pmatrix} \underline{a}\underline{a}' & \underline{a}\underline{b}' & \underline{a}\underline{c}' \\ \underline{b}\underline{a}' & \underline{b}\underline{b}' & \underline{b}\underline{c}' \\ \underline{c}\underline{a}' & \underline{c}\underline{b}' & \underline{c}\underline{c}' \end{pmatrix} \qquad (F17)$$

together with the orthogonality relation (2.96) for lattice and reciprocal lattice vectors, that is,

$$(\underline{G}_i\,\underline{R}_j) = 2\pi\delta_{ij}; \quad i,j = 1, 2, 3 \qquad (F18)$$

yields

$$[(\underline{G}_1 \times \underline{G}_2)\underline{G}_3]\,[(\underline{R}_1 \times \underline{R}_2)\underline{R}_3] = \det\begin{pmatrix} \underline{G}_1\underline{R}_1 & \underline{G}_1\underline{R}_2 & \underline{G}_1\underline{R}_3 \\ \underline{G}_2\underline{R}_1 & \underline{G}_2\underline{R}_2 & \underline{G}_2\underline{R}_3 \\ \underline{G}_3\underline{R}_1 & \underline{G}_3\underline{R}_2 & \underline{G}_3\underline{R}_3 \end{pmatrix}$$

$$= \det\begin{pmatrix} 2\pi & 0 & 0 \\ 0 & 2\pi & 0 \\ 0 & 0 & 2\pi \end{pmatrix} = (2\pi)^3 \qquad (F19)$$

This gives another proof of relation (2.99).

The **vector relation**

$$(\underline{a}\,\underline{m})((\underline{b} \times \underline{c})\underline{m}) + (\underline{b}\,\underline{m})((\underline{c} \times \underline{a})\underline{m}) + (\underline{c}\,\underline{m})((\underline{a} \times \underline{b})\underline{m})$$
$$= m^2((\underline{a} \times \underline{b})\underline{c}) \qquad (F20)$$

can be used in connection with mirror symmetry operations. If the lattice vectors $\underline{R}_1, \underline{R}_2, \underline{R}_3$ are subject to a mirror plane operation

$$\underline{R}_i \rightarrow \underline{R}'_i = \underline{R}_i - 2(\underline{R}_i\,\underline{m})\underline{m}, \quad i = 1, 2, 3, \quad m^2 = 1 \qquad (F21)$$

then

$$(\underline{R}_1' \times \underline{R}_2')\underline{R}_3' = ([\underline{R}_1 - 2(\underline{R}_1\underline{m})\underline{m}] \times [\underline{R}_2 - 2(\underline{R}_2\underline{m})\underline{m}])[\underline{R}_3 - 2(\underline{R}_3\underline{m})\underline{m}]$$

$$= ((\underline{R}_1 \times \underline{R}_2) - 2(\underline{R}_1\underline{m})(\underline{m} \times \underline{R}_2)$$

$$- 2(\underline{R}_2\underline{m})(\underline{R}_1 \times \underline{m}))[\underline{R}_3 - 2(\underline{R}_3\underline{m})\underline{m}]$$

$$= (\underline{R}_1 \times \underline{R}_2)\underline{R}_3 - 2(\underline{R}_3\underline{m})(\underline{R}_1 \times \underline{R}_2)\underline{m}$$

$$- 2(\underline{R}_1\underline{m})(\underline{m} \times \underline{R}_2)\underline{R}_3 - 2(\underline{R}_2\underline{m})(\underline{R}_1 \times \underline{m})\underline{R}_3$$

$$= (\underline{R}_1 \times \underline{R}_2)\underline{R}_3 - 2[(\underline{R}_1\underline{m})(\underline{R}_2 \times \underline{R}_3)\underline{m}$$

$$+ (\underline{R}_2\underline{m})(\underline{R}_3 \times \underline{R}_1)\underline{m} + (\underline{R}_3\underline{m})(\underline{R}_1 \times \underline{R}_2)\underline{m}] \qquad (F22)$$

and applying Eq. (F20) to the expression in square brackets yields

$$(\underline{R}_1' \times \underline{R}_2')\underline{R}_3' = (\underline{R}_1 \times \underline{R}_2)\underline{R}_3 - 2\underline{m}^2[(\underline{R}_1 \times \underline{R}_2)\underline{R}_3] = -(\underline{R}_1 \times \underline{R}_2)\underline{R}_3 \qquad (F23)$$

The changed sign in Eq. (F23) shows that mirror operations change the handedness of the coordinate system. However, the corresponding cell volume remains unchanged since

$$V_{el}' = |(\underline{R}_1' \times \underline{R}_2')\underline{R}_3'| = |-(\underline{R}_1 \times \underline{R}_2)\underline{R}_3| = V_{el} \qquad (F24)$$

Appendix G: Details of Fourier Theory

Fourier theory plays a major role in the approximation of functions in many fields of science and engineering. As a result, there is a vast amount of literature available at all mathematical levels, see, for example, [207, 208]. In this appendix, we will discuss only a few simple issues dealing with spatially periodic functions, which are relevant for some topics treated in this book. For further details, we refer to the existing literature.

In the following text we consider continuous functions $f(\underline{r})$ where $\underline{r}$ is a three-dimensional coordinate in real space. In the harmonic analysis (**Fourier analysis**) functions $f(\underline{r})$ are approximated by weighted superpositions of harmonic functions $h_{\underline{G}}(\underline{r})$. These **harmonic functions** are of given periodicity defined by a wave vector $\underline{G}$ in three-dimensional reciprocal space (whose dimensions are [length^{-1}]) and can be written as

$$h_{\underline{G}}^c(\underline{r}) = \cos(\underline{G}\,\underline{r}) \quad \text{and} \quad h_{\underline{G}}^s(\underline{r}) = \sin(\underline{G}\,\underline{r}) \tag{G1}$$

where the direction of periodicity is given by that of wave vector $\underline{G}$ and the periodicity length L is determined by

$$L = 2\pi/|\underline{G}| \tag{G2}$$

The combined set of sine and cosine functions in Eq. (G1) can be written more elegantly using complex variables and the definition of a complex valued exponential $\exp(i\,x)$, where, according to Euler's formula, complex harmonic functions are given by

$$h_{\underline{G}}(\underline{r}) = h_{\underline{G}}^c(\underline{r}) + i\,h_{\underline{G}}^s(\underline{r}) = \cos(\underline{G}\,\underline{r}) + i\sin(\underline{G}\,\underline{r}) = \exp(i\,\underline{G}\,\underline{r}) \tag{G3}$$

(i being the imaginary unit number) which will be used in the following.

The approximation of a continuous function $f(\underline{r})$ by a weighted superposition of harmonic functions $h_{\underline{G}}(\underline{r})$ must include, in general, an infinite set of superposition functions for all wave vectors $\underline{G}$ where the approximation function $f_{app}(\underline{r})$ is written as an integral

$$f_{app}(\underline{r}) = \iiint c_{\underline{G}} \exp(i\,\underline{G}\,\underline{r})\,d^3 G \tag{G4}$$

Crystallography and Surface Structure: An Introduction for Surface Scientists and Nanoscientists,
Second Edition. Klaus Hermann.
© 2017 Wiley-VCH Verlag GmbH & Co. KGaA. Published 2017 by Wiley-VCH Verlag GmbH & Co. KGaA.

with the integration extending over the full three-dimensional reciprocal space. It can be shown mathematically that the approximation function $f_{app}(\underline{r})$ reproduces the exact function $f(\underline{r})$ if the expansion coefficients $c_{\underline{G}}$ are chosen as

$$c_{\underline{G}} = \iiint f(\underline{r}) \exp(-i\underline{G}\,\underline{r})\, d^3r \tag{G5}$$

Thus, expansion (G4) together with Eq. (G5) can be used to represent any continuous function $f(\underline{r})$.

If the continuous function $f(\underline{r})$ is periodic in three different directions described by a three-dimensional lattice with lattice vectors $\underline{R}_{o1}, \underline{R}_{o2}, \underline{R}_{o3}$, that is,

$$f(\underline{r} + \underline{R}) = f(\underline{r}) \quad \text{with} \quad \underline{R} = n_1 \underline{R}_{o1} + n_2 \underline{R}_{o2} + n_3 \underline{R}_{o3}, \quad n_i \text{ integer} \tag{G6}$$

then the Fourier expansion given by integrals (G4) and (G5) reduces to a discrete sum of harmonic functions where the converged expansion is

$$f(\underline{r}) = \sum_{k_1, k_2, k_3} c_{\underline{G}} \exp(i\underline{G}\,\underline{r}) \quad \text{with} \quad \underline{G} = k_1 \underline{G}_{o1} + k_2 \underline{G}_{o2} + k_3 \underline{G}_{o3}, \quad k_i \text{ integer} \tag{G7}$$

Here, the (infinite) summation extends over all vectors $\underline{G}$ of the reciprocal lattice $\underline{G}_{o1}, \underline{G}_{o2}, \underline{G}_{o3}$ defined in Section 2.5. The expansion coefficients $c_{\underline{G}}$ of the exact function $f(\underline{r})$ are then given by

$$c_{\underline{G}} = \frac{1}{V_{el}} \iiint_{V_{el}} f(\underline{r}) \exp(-i\underline{G}\,\underline{r})\, d^3r \tag{G8}$$

where the three-dimensional integration is carried out over the elementary cell V_{el} spanned by $\underline{R}_{o1}, \underline{R}_{o2}, \underline{R}_{o3}$ of the real space lattice. Relation (G8) can be derived from Eq. (G7) by integration

$$\iiint_{V_{el}} f(\underline{r}) \exp(-i\underline{G}'\,\underline{r})\, d^3r = \sum_{k_1, k_2, k_3} c_{\underline{G}} \iiint_{V_{el}} \exp(i[\underline{G} - \underline{G}']\underline{r})\, d^3r$$

$$= c'_{\underline{G}} V_{el} \tag{G9}$$

where

$$\iiint_{V_{el}} \exp(i[\underline{G} - \underline{G}']\underline{r})\, d^3r = \begin{cases} V_{el} & \text{if } \underline{G} = \underline{G}' \\ 0 & \text{if } \underline{G} \neq \underline{G}' \end{cases} \tag{G10}$$

has been applied.

According to the orthogonality theorem (2.96) of real and reciprocal space lattice vectors we obtain

$$(\underline{G}\,\underline{R}) = (k_1 \underline{G}_{o1} + k_2 \underline{G}_{o2} + k_3 \underline{G}_{o3})(n_1 \underline{R}_{o1} + n_2 \underline{R}_{o2} + n_3 \underline{R}_{o3})$$
$$= 2\pi(k_1 n_1 + k_2 n_2 + k_3 n_3) = 2\pi N, \quad N \text{ integer} \tag{G11}$$

and thus

$$\exp(i\underline{G}\,\underline{R}) = \exp(i 2\pi N) = 1 \tag{G12}$$

which guarantees the periodicity of the expansion since

$$f(\underline{r}+\underline{R}) = \sum_{\underline{G}} c_{\underline{G}} \exp(i\underline{G}(\underline{r}+\underline{R})) = \sum_{\underline{G}} c_{\underline{G}} \exp(i\underline{G}\,\underline{r})\exp(i\underline{G}\,\underline{R}) = f(\underline{r}) \quad (G13)$$

Next we consider the case where function $f(\underline{r})$ in three-dimensional space is periodic in two different directions, described by a two-dimensional lattice with lattice vectors $\underline{R}_{o1}$ and $\underline{R}_{o2}$, and depends on a third coordinate $\underline{r}_3$ perpendicular to $\underline{R}_{o1}$ and $\underline{R}_{o2}$ in a nonperiodic fashion. Here, we write the three-dimensional coordinate vector $\underline{r}$ as a sum of a projected vector $\underline{r}_{12}$ parallel to $\underline{R}_{o1}$ and $\underline{R}_{o2}$ and the third vector $\underline{r}_3$, that is, $\underline{r} = \underline{r}_{12} + \underline{r}_3$. Then the periodicity constraint reads

$$f(\underline{r}_{12} + \underline{R} + \underline{r}_3) = f(\underline{r}_{12} + \underline{r}_3) \quad \text{with} \quad \underline{R} = n_1 \underline{R}_{o1} + n_2 \underline{R}_{o2}, \quad n_i \text{ integer} \quad (G14)$$

and the Fourier expansion given by integrals (G4) and (G5) reduces to a discrete sum of harmonic functions where the converged expansion is

$$f(\underline{r}) = \sum_{k_1, k_2} c_{\underline{G}}(\underline{r}_3) \exp(i\underline{G}\,\underline{r}_{12}) \quad \text{with} \quad \underline{G} = k_1 \underline{G}_{o1} + k_2 \underline{G}_{o2}, \quad k_i \text{ integer.} \quad (G15)$$

Here, the (infinite) summation extends over all vectors $\underline{G}$ of the two-dimensional reciprocal lattice defined by vectors $\underline{G}_{o1}$ and $\underline{G}_{o2}$ where the latter can be derived from the orthogonality theorem

$$(\underline{G}_{oi}\underline{R}_{oj}) = \begin{cases} 2\pi & \text{if } i = j \\ 0 & \text{if } i \neq j \end{cases}, \quad i, j = 1, 2 \quad (G16)$$

yielding

$$\underline{G}_{o1} = 2\pi/\Delta \left(R_{o2}^2 \underline{R}_{o1} - (\underline{R}_{o1}\underline{R}_{o2})\underline{R}_{o2} \right) \quad (G17a)$$

$$\underline{G}_{o2} = 2\pi/\Delta \left(R_{o1}^2 \underline{R}_{o2} - (\underline{R}_{o1}\underline{R}_{o2})\underline{R}_{o1} \right), \quad \Delta = R_{o1}^2 R_{o2}^2 - (\underline{R}_{o1}\underline{R}_{o2})^2 \quad (G17b)$$

The dependence of $f(\underline{r})$ on the third coordinate $\underline{r}_3$ in expansion (G15) is expressed by all expansion coefficients $c_{\underline{G}}$ being functions of $\underline{r}_3$ rather than scalar constants where

$$c_{\underline{G}}(\underline{r}_3) = \frac{1}{A_{el}} \iint_{A_{el}} f(\underline{r}_{12}, \underline{r}_3) \exp(-i\underline{G}\,\underline{r}_{12}) d^2 r_{12} \quad (G18)$$

and the two-dimensional integration is carried out over the area A_{el} of the elementary cell spanned by the lattice vectors $\underline{R}_{o1}$ and $\underline{R}_{o2}$ of the real space lattice such that the integration does not affect the third coordinate $\underline{r}_3$.

Appendix H: List of Surface Web Sites

This appendix lists a few web sites that are relevant for surface crystallography.

Surface Structure Information, different program codes
http://www.icts.hkbu.edu.hk/surfstructinfo/
The ICSOS Web Site
http://www3.lut.fi/projectsites/icsos/
The Nanotube Site
http://www.pa.msu.edu/cmp/csc/nanotube.html
NIST SSD, Surface Structure Database V. 5 (Windows XP/7+)
NIST has discontinued the distribution of SSD, see oSSD (database, V. 5)
    http://www.fhi-berlin.mpg.de/KHsoftware/ssdin5/index.html
    (SSD structure input, SURVIS visualizer, V. 5).
open SSD (oSSD), based on NIST SSD V. 5 (Windows XP/7+)
http://www.fhi-berlin.mpg.de/KHsoftware/oSSD/index.html
oSSD is identical to NIST SSD V. 5 in its content and handling and is available as an open source database.
SURFACE EXPLORER, surface visualization (WWW)
http://surfexp.fhi-berlin.mpg.de
LEEDpat4, LEED symmetry pattern simulator (Windows XP/7+)
http://www.fhi-berlin.mpg.de/KHsoftware/LEEDpat/index.html
SARCH, LATUSE, PLOT3D, surface visualization and analysis (DOS, outdated)
http://www.fhi-berlin.mpg.de/KHsoftware/SLP/index.html
BALSAC, surface visualization, and analysis (Windows XP/7+, Linux)
http://www.fhi-berlin.mpg.de/KHsoftware/Balsac/balpam.html (pamphlet)
http://www.fhi-berlin.mpg.de/KHsoftware/Balsac/index.html
    (program download)
http://www.fhi-berlin.mpg.de/KHsoftware/Balsac/Balsac4.pdf
    (Balsac manual, V. 4.00)
http://www.fhi-berlin.mpg.de/KHsoftware/Balsac/pictures.html
    (Balsac picture gallery)
The Linux version of Balsac (latest version 2.16) is obsolete and will not be developed further.

ANA-ROD, analysis by surface X-ray diffraction
http://www.esrf.eu/computing/scientific/joint_projects/ANA-ROD/index.html

Appendix I: List of Surface Structures

This appendix lists all examples of measured clean and adsorbate covered surfaces that have been used to illustrate structural details in this book. Each example, given in Wood notation, is quoted by its figure number, by its literature reference, and by its SSD database reference "n.m". Here, n.m refers to the entry number of the corresponding surface structure in the Surface Structure Database (NIST Version 5 or oSSD), see Section 7.2. For more extended structure compilations consult the references in Section 7.2.

Wood notation	Figure	Reference	SSD entry
Ag(1 1 0) + (2×1) − O	6.45	[172]	—
Ag(1 1 0) + (2×1) − O	6.13	[140]	47.8.4
Ag(1 1 1) + Xe(incommensurate)	6.6	[128]	47.54.1
Al(1 1 1) + (1×1) − O	6.39	[168]	13.8.19
Au(1 0 0) − hex	5.8	[112]	79.80
Au(1 1 1) − ($\sqrt{3}$×22)rect	5.6	[111]	—
Cu(1 0 0) + c(2×2) − Cl	6.19	[131]	29.17.7
Cu(1 1 0) + c(2×2) − Mn	6.17	[146]	29.25.8
Cu(1 1 0) + (2×3) − 4N	6.18	[148]	29.7.10
Cu(1 1 1) + (4×4) − C_{60}	6.3	[126]	—
Cu(1 1 1) + ($\sqrt{3}$×$\sqrt{3}$)R30° − In	5.13	[121]	—
Cu(1 1 1) + (disordered) − C_2H_2	6.11	[138]	29.6.1.6
Cu(1 1 1) + (disordered) − NH_3	6.1, 6.46	[124]	29.7.1.3
Ni(1 0 0) + c(2×2) − CO	6.8	[130]	28.6.8.8
Ni(1 1 0) + c(2×2) − CN	6.42	[170]	28.6.7.2
Ni(1 1 0) + p2mg(2×1) − 2CO	6.2	[125]	28.6.8.45
Ni(1 1 1) + c(4×2) − 2NO	6.21	[137]	28.7.8.8
Ni(1 1 1) + (1.155×1.155)R2.2° − Ag	6.26	[154]	—
Pb(1 1 1) + ($\sqrt{403}/7$ × $\sqrt{403}/7$)R22.8° − C_{60}	6.25	[151]	—
Pd(1 1 1) + ($\sqrt{3}$×$\sqrt{3}$)R30° − CO	6.20	[136]	46.6.8.13
Pt(1 1 0) − (1×2)	5.2	[106]	78.77
Pt(1 1 0) + c(2×2) − Br	6.14	[106]	78.35.1
Pt(1 1 1) + c(4×2) − 2CO	6.40	[144]	78.6.8.4
Rh(1 1 0) + (1×3) − H	6.44	[171]	45.1.5

Crystallography and Surface Structure: An Introduction for Surface Scientists and Nanoscientists,
Second Edition. Klaus Hermann.
© 2017 Wiley-VCH Verlag GmbH & Co. KGaA. Published 2017 by Wiley-VCH Verlag GmbH & Co. KGaA.

(continued)

Wood notation	Figure	References	SSD entry
Rh(1 1 0) + p2mg(2 × 1) − 2O	6.37	[167]	45.8.7
Ru(0 0 0 1) + (0.92 × 0.92) − Gra(phene)	6.5	[127]	—
Si(1 0 0) − (2 × 1)	5.4	[107]	14.203
Si(1 0 0) − c(4 × 2)	5.4	[108]	14.182b
Si(1 1 1) − (7 × 7)	5.5	[110]	14.132
W(1 0 0) − c(2 × 2)	5.3	[22]	74.14
W(1 1 0) + (2 × 1) − O	6.16	[143]	74.8.1

Glossary and Abbreviations

In this section different keywords and abbreviations that have been commonly used in the book are briefly explained. The keywords are arranged in three groups.
- Bulk crystals and three-dimensional
- Monolayers, surfaces, and two-dimensional
- Miscellaneous

Within each group the order is alphabetic. Bold-faced words in the explanatory text are usually also keywords of the glossary.

Bulk Crystals and Three-dimensional

Achiral	A three-dimensional object is achiral if it can be superimposed onto its mirror image by simple rotation and shifting. Otherwise, it is **chiral**.
Acute representation	Lattice vectors $\underline{R}_1, \underline{R}_2, \underline{R}_3$, which all form mutual angles $0° < \alpha, \beta, \gamma \leq 90°$.
Bain path	Geometry variation of a continuous phase transition between face- and body-centered cubic crystals. The intermediate lattice type is described as centered tetragonal (**ct**).
Basis	Collection of atom positions inside the three-dimensional (morphological) unit cell of a **crystal** or **monolayer**.
bcc	Body-centered cubic, also called cubic-I, a lattice type of cubic crystals.
Bravais lattice	Lattice type defined by specific translational and point symmetry, given by lattice vectors $\underline{R}_1, \underline{R}_2, \underline{R}_3$. Overall, there are 14 three-dimensional Bravais lattices. Additional (point) symmetry properties

Brillouin zone	are described by the corresponding 230 three-dimensional **space groups**. Compact polyhedral unit cell of the reciprocal lattice corresponding to the **Wigner–Seitz cell** of the real space lattice.
Buerger cell	Primitive morphological unit cell spanned by lattice vectors $\underline{R}_1, \underline{R}_2, \underline{R}_3$ of a three-dimensional lattice where $\underline{R}_i$ are smallest in length and $|\underline{R}_1| \leq |\underline{R}_2| \leq |\underline{R}_3|$.
BZ	See **Brillouin zone**.
Centering	Augmenting a lattice by additional lattice points (in the centers of morphological unit cells or at their faces). This may lead to a different Bravais lattice type, for example, centered tetragonal versus primitive tetragonal.
Chirality	Also referred to as **handedness**. Symmetry property of a three-dimensional object. An object is called **chiral** if it cannot be superimposed onto its mirror image by simple rotation and shifting. Otherwise, it is **achiral**. Crystals are chiral if their symmetry does not include a mirror plane.
Coordination number	Number of nearest neighbor atoms (including all atoms of the **coordination shell**) with respect to a given atom center in a crystal.
Coordination shell	Set of atoms of (about) the smallest distance from a center (usually an atom position, **nearest neighbor shell**) in a crystal.
Crystal	Strictly defined (IUCr) as a material with a discrete sharp diffraction pattern. This includes three-dimensionally periodic arrangements of atoms. Their periodicity is defined by a **lattice**, corresponding symmetry properties by a three-dimensional **space group**, and atom positions inside the morphological unit cell are given by the **basis**. The **two-dimensional** analog is a **monolayer**.

Crystal system	Three-dimensional lattice classification resulting in the 14 primitive and centered **Bravais lattices**.
ct	Centered tetragonal, also called tetragonal-I, a three-dimensional lattice type of crystals.
Cubic Miller indices	See **Miller indices** and **sc notation**.
fcc	Face-centered cubic, also called cubic-F, a three-dimensional lattice type of cubic crystals.
Four-index notation	See **Miller–Bravais indices**.
General lattice vector	Integer-valued linear combination of the three lattice vectors $\underline{R}_1, \underline{R}_2, \underline{R}_3$ of a bulk lattice.
Generic Miller indices	See **Miller indices**.
Gibbs–Wulff theorem	Basis of a quasi-continuum model to determine the shape of polyhedral crystalline particles confined by planar sections (**facets**) of $(h\,k\,l)$-oriented monolayers.
Gra	Short-hand writing for **graphene**, a $(0\,0\,0\,1)$-oriented monolayer of graphite exhibiting a honeycomb structure.
Graphene	Monolayer of graphite with $(0\,0\,0\,1)$ orientation exhibiting a honeycomb structure. Graphene sheets also exist in nature as flakes or as adsorbed films. This carbon structure is sometimes abbreviated as **Gra**.
Handedness	See **Chirality**.
hcp	Hexagonal close-packed, a three-dimensional crystal structure with hexagonal lattice, a lattice constant ratio $c/a = \sqrt{(8/3)}$, and two atoms in the morphological unit cell. In practice, lattices of hexagonal crystals with ratios c/a near $\sqrt{(8/3)}$ are also called hcp type.
Hermann–Mauguin	Notation used to define symmetry operations and symmetry groups describing crystals and monolayers. This notation is preferred by crystallographers, see also **Schönflies**.
hex (hcp)	A hexagonal lattice with a lattice constant ratio c/a of $\sqrt{(8/3)} = 1.63299$. Lattice of the **hcp** crystal structure.

Incommensurate composite crystal	One of three classes of aperiodic bulk systems with specific long-range order and local symmetry. The other types are **modulated structures** and **quasicrystals**.
ITC	International Tables for Crystallography, general reference for two- and three-dimensional space groups.
Lattice	Definition of periodicity (translational symmetry) in three dimensions by lattice vectors $\underline{R}_1, \underline{R}_2, \underline{R}_3$. Additional point symmetry is given by a corresponding three-dimensional **space group**. The two-dimensional equivalent of a lattice is a **netplane**, while of a space group it is a **plane group**. The term **lattice** is also used sometimes to define periodicity in one and two dimensions.
Lattice basis vectors	Position vectors $\underline{r}_1, \underline{r}_2, \ldots, \underline{r}_p$ of all p atoms in the primitive unit cell of a crystal.
Lattice constants	Scaling parameters a, b, c of the three lattice vectors $\underline{R}_1, \underline{R}_2, \underline{R}_3$ of a crystal where $a = |\underline{R}_1|, b = |\underline{R}_2|, c = |\underline{R}_3|$.
Lattice vectors	Periodicity vectors $\underline{R}_1, \underline{R}_2, \underline{R}_3$ of a lattice.
Miller–Bravais indices	Referring to the **four-index notation**. Integer quadruplets $(l\ m\ n\ q)$ characterizing orientations of netplanes in hexagonal lattices. The definition is based on the reciprocal (hexagonal) lattice with symmetry considerations. The four-index notation is equivalent to the common three-index notation $(h\ k\ l)$ of generic **Miller indices**.
Miller indices	Integer triplets h, k, l, such as in $(h\ k\ l)$ and $[h\ k\ l]$ characterizing orientations of netplanes and directions in a lattice, also called **generic Miller indices**. The definition is based on lattice vectors of the **reciprocal lattice**. For centered cubic lattices (fcc, bcc), Miller indices of the simple cubic lattice, also called **simple cubic (sc) Miller indices**, are often used. For hexagonal lattices an alternative

	four-index notation ($l\ m\ n\ q$), also called **Miller–Bravais indices**, is often used.
Modulated structure	One of three classes of aperiodic bulk systems with specific long-range order and local symmetry. The other types are **incommensurate composite crystals** and **quasicrystals**.
Monoatomic crystal	Crystal with one atom in its primitive morphological unit cell.
Morphological unit cell	Also called **unit cell**. Six-faced polyhedron (parallelepiped) spanned by lattice vectors $\underline{R}_1, \underline{R}_2, \underline{R}_3$ of a three-dimensional lattice. The cell is **primitive** if it is of smallest volume.
Motif	A cluster of atoms recurring in a **crystal**.
Multiplicity	Number of atom members in **neighbor shells**.
Nanotube	Hollow cylindrical cluster of atoms or molecules. A nanotube may be constructed as a rolled up strip of a planar sheet of atomic or molecular components in a periodic arrangement.
Neighbor shell	Set of atoms of (about) the same distance from a center (usually an atom position) in a crystal.
Netplane-adapted lattice	Lattice description by lattice vectors $\underline{R}_1, \underline{R}_2, \underline{R}_3$ where $\underline{R}_1, \underline{R}_2$ point parallel to a given netplane of the lattice and $\underline{R}_3$ connects adjacent netplanes.
n-fold rotation	Rotation by an angle $\alpha = 360°/n$ about an axis. For lattices with translational symmetry only values $n = 1, 2, 3, 4, 6$ are allowed.
Non-symmorphic space group	Space group whose generating symmetry elements also include those combining point and translational symmetry, such as glide reflection or rototranslation.
Niggli cell	Primitive morphological unit cell spanned by lattice vectors $\underline{R}_1, \underline{R}_2, \underline{R}_3$ of a three-dimensional lattice where $\underline{R}_i$ are smallest in length and $|\underline{R}_1| \leq |\underline{R}_2| \leq |\underline{R}_3|$. Additional constraints are applied to make the lattice definition unique.

Obtuse representation	Lattice vectors $\underline{R}_1, \underline{R}_2, \underline{R}_3$, which form mutual angles $\alpha, \beta, \gamma \geq 90°$ with at least one angle $> 90°$.
Point symmetry group	Collection of three-dimensional point symmetry operations, such as inversion, rotation, mirroring, forming a mathematical group. Crystals are invariant with respect to all operations of a corresponding point symmetry group.
Polyatomic crystal	Crystal with several atoms in its primitive morphological unit cell defined by lattice vectors $\underline{R}_1, \underline{R}_2, \underline{R}_3$. The atoms may be of different element type and/or are placed at different positions $\underline{r}$ inside the cell with $\underline{r} = x_1 \underline{R}_1 + x_2 \underline{R}_2 + x_3 \underline{R}_3$ and $0 \leq x_i < 1$.
Primitive lattice	Crystal lattice with lattice vectors $\underline{R}_1, \underline{R}_2, \underline{R}_3$ forming the primitive morphological unit cell containing the smallest number of non-equivalent atoms.
Quasicrystal	One of three classes of aperiodic bulk systems with specific long-range order and local symmetry. The latter may not be compatible with allowed symmetries appearing in periodic crystals. The other types are **modulated structures** and **incommensurate composite crystals**.
Reciprocal lattice	Lattice defined by lattice vectors $\underline{G}_1, \underline{G}_2, \underline{G}_3$ of dimension [inverse length], which are connected with vectors $\underline{R}_1, \underline{R}_2, \underline{R}_3$ of the real space lattice by orthogonality relations $(\underline{G}_i \underline{R}_j) = 2\pi \delta_{ij}$.
Relative coordinates	Representation of lattice basis vectors by linear combinations of lattice vectors rather than by absolute Cartesian coordinates.
sc	Simple cubic, also called primitive cubic or cubic-P, the basic lattice type of cubic crystals.
sc notation	See **Simple cubic notation**.
Schönflies	Notation used to define symmetry operations and symmetry groups describing crystals and monolayers. This notation is preferred by physicists, see also **Hermann–Mauguin**.
Shell multiplicity	Number of atoms in **neighbor shells**.

Simple cubic notation	Notation of Miller indices for face- and body-centered cubic lattices referring to the simple cubic lattice.
Single crystal	Perfect crystal with exact three-dimensional periodicity and symmetry.
Space group	Collection of all symmetry properties (translational and point symmetry elements) available for a given crystal with its periodicity described by a Bravais lattice. There are 230 different space groups for (three-dimensional) crystals. The two-dimensional equivalent of a space group is a **plane group**.
Superlattice	Description of a crystal lattice by lattice vectors $\underline{R}_1, \underline{R}_2, \underline{R}_3$ (and corresponding unit cells), which are larger than (often integer multiples of) those suggested by the basic periodicity of the crystal.
Symmorphic (space) group	Space group whose generating symmetry elements include only true point symmetry operations and true translations, that is, no combinations, such as glide planes or screw axes.
Unit cell	See **Morphological unit cell**.
Voronoi cell	See **Wigner–Seitz cell, WSC**.
Wigner–Seitz cell	Compact polyhedral unit cell of a real lattice, sometimes called **Voronoi cell**.
WSC	See **Wigner–Seitz cell**.
Wulff construction	See **Gibbs–Wulff theorem**.

Monolayers, Surfaces, and Two-dimensional

2 × 2 Matrix notation	Notation of the periodicity and orientation of reconstructed surfaces and adsorbate systems.
Achiral surface	Single crystal surface which is symmetric with respect to at least one mirror plane perpendicular to it. Otherwise, the surface is called **chiral**.
Acute representation	Lattice vectors $\underline{R}_1$ and $\underline{R}_2$, which form an angle $0° < \gamma \leq 90°$ ($= 60°$ for hexagonal lattices).

Additivity theorem	Mathematical theorem connecting **Miller indices** of a **stepped** or **kinked surface** with those of its terraces steps, and kinks.
Basis	Collection of atom positions inside the two-dimensional (morphological) unit cell.
Bravais lattice	Lattice type defined by specific translational and point symmetry, given by lattice vectors $\underline{R}_1$ and $\underline{R}_2$. Overall, there are 5 two-dimensional Bravais lattices. Additional (point) symmetry properties are described by the corresponding 17 two-dimensional **space groups**.
Centering	Augmenting a netplane by additional lattice points (in the center of the unit cell). This leads to a different Bravais netplane type only for rectangular netplanes.
Chiral surface	Single crystal surface, which does not possess symmetry with respect to a mirror plane perpendicular to it. Otherwise, the surface is called **achiral**.
Coincidence (super)lattice	Commensurate reconstruction type where a two-dimensional periodicity cell of the overlayer is shared with the substrate layers. However, this cell is larger than the smallest possible unit cell of each of the two separate subsystems. The reconstruction type is also referred to as high-order commensurate (**HOC**) or **scaled commensurate** reconstruction.
Commensurate reconstruction	Surface reconstruction where netplanes of the topmost monolayers are commensurate with those of the substrate layers. Corresponding netplane transformations are described by integer-valued 2×2 matrices.
c-rectangular	Centered rectangular, a netplane type of rectangular monolayers.
Crystallographic plane	See **Netplane**, **Monolayer**.
Crystal system	Netplane classification resulting in the five primitive and centered **Bravais lattices**.

Cut-and-project	A method to create aperiodic linear atom arrangements (**Fibonacci chains**) by projecting a two-dimensional square lattice.
DAS model	Dimer-adatom-stacking-fault model of the reconstructed $Si(1\,1\,1)-(7\times 7)$ surface.
Domain formation	A substrate surface with symmetry may allow differently oriented overlayers that are energetically equivalent. This can lead to large but finite patches of overlayers— so-called domains – corresponding to the different orientations.
Facet	Finite flat region at a single crystal surface described by Miller indices $(h_f\,k_f\,l_f)$. The facet orientation does not need to coincide with that of the long-range surface characterized by $(h\,k\,l)$.
Fibonacci chain	Aperiodic linear atom arrangement of alternating short and long interatomic distances characterized by self-similarity. Used to simulate one-dimensional quasicrystals.
FM growth mode	See **Frank–Van-der-Merwe**.
Frank–Van-der-Merwe	Growth mode at surfaces referring to **layer-by-layer** growth.
General lattice vector	Integer-valued linear combination of the two lattice vectors $\underline{R}_1$ and $\underline{R}_2$ of a netplane.
High-Miller-index surface	See **Vicinal surface**.
High-order commensurate	See **Coincidence (super)lattice**.
HOC reconstruction	See **Coincidence (super)lattice**.
Incommensurate reconstruction	Surface reconstruction where netplanes of the topmost monolayers are not commensurate with those of the corresponding substrate layers. Corresponding netplane transformations are described by 2×2 matrices containing irrational elements.
Interference lattice	Lattice structure at adsorbate covered surfaces with quasi-periodic long-range order expressed by one- and two-dimensional **moiré patterns**.

Kinked surface	Crystal surface composed of terrasses separated by steps analogous to a **stepped surface**. However, the step lines are broken in a periodic or nonperiodic fashion ("stepped steps" or "kinked steps"). If the kink and step distribution is regular, that is, periodic at the surface, the kinked surface can be described in its orientation by large Miller indices, see also **Vicinal surface**.
Lattice basis vectors	Position vectors $\underline{r}_1, \underline{r}_2, \ldots, \underline{r}_p$ of all p atoms in the primitive unit cell of a monolayer.
Lattice gas	Amorphous monolayer of atoms derived from an ideal (periodic) monolayer by occupying lattice sites in a random fashion.
Lattice vectors	Periodicity vectors $\underline{R}_1$ and $\underline{R}_2$ of a netplane.
Microfacet notation	Formal notation of the structure and orientation of **stepped** and **kinked surfaces**.
Microfacetted surface	A surface structure built of different **facets** that are finite, usually small, in one or two dimensions, and combine to form the global surface.
Minkowski reduction	Iterative method to determine symmetrically appropriate lattice vectors $\underline{R}_{o1}$ and $\underline{R}_{o2}$ of a netplane from an initial set $\underline{R}_1, \underline{R}_2$.
Moiré pattern	Spatial interference pattern originating from superimposing two-dimensionally periodic objects whose periodicity differs only slightly. Simple examples are identical parallel monolayers that are scaled or rotated by a small angle with respect to each other.
Monolayer	**Two-dimensionally** periodic arrangement of atoms. Its periodicity is defined by a **netplane**, corresponding symmetry properties by a two-dimensional **plane group**, and atom positions inside the unit cell are given by the **basis**. The **three-dimensional** analog is a **crystal**. The present definition of a

	monolayer **deviates** from definitions used elsewhere which refer to adsorbate overlayers of a given atom density, one adsorbate per substrate atom or unit cell.
Morphological unit cell	Also called **unit cell**. Four-sided polygon (parallelogram) spanned by lattice vectors $\underline{R}_1$ and $\underline{R}_2$ of a two-dimensional netplane. The cell is **primitive** if it is the unit cell of smallest area.
Motif	A cluster of atoms recurring in a **monolayer**.
Netplane	Definition of periodicity (translational symmetry) in two dimensions by vectors $\underline{R}_1$ and $\underline{R}_2$. Additional point symmetry is given by a corresponding two-dimensional **plane group**. The three-dimensional equivalent of a netplane is a **lattice**, of a plane group it is a three-dimensional **space group**. The present definition of a netplane is more **strict** than definitions used elsewhere which refer sometimes to both periodicity and atom basis defining a **monolayer** in this book.
n-fold rotation	Rotation by an angle $\alpha = 360°/n$ about an axis perpendicular to the netplane. For netplanes with translational symmetry, only values n = 1, 2, 3, 4, 6 are allowed.
Obtuse representation	Lattice vectors $\underline{R}_1$ and $\underline{R}_2$, which form an angle $\gamma > 90°$ (=120° for hexagonal lattices).
Penrose tiling	Procedure of covering a plane completely without holes or overlaps using tiles of a finite set of different polygons, often rhombi, where the tiling yields an aperiodic pattern.
Plane group	Also referred to as two-dimensional **space group**. Collection of all symmetry properties (translational and point symmetry elements) available for a given monolayer with its periodicity described by a Bravais lattice. There are 17 different plane groups for (two-dimensional) monolayers. The three-dimensional

Point symmetry group	equivalent of a plane group it is a three-dimensional **space group**. Collection of two-dimensional point symmetry operations, such as inversion, rotation, and mirroring, forming a mathematical group. Netplanes are invariant with respect to all operations of the corresponding point symmetry group.
Rolling vector	General lattice vector inside a $(h\,k\,l)$ monolayer used to define a nanotube by rolling a monolayer strip along the vector.
p-rectangular	Primitive rectangular, also called rectangular, rect. A two-dimensional lattice type of rectangular monolayers.
Scaled commensurate	See **Coincidence (super)lattice**.
SK growth mode	See **Stranski–Krastanov**.
Space group	See **Plane group**.
Step notation	Formal notation of the structure and orientation of **stepped surfaces**.
Stepped surface	Crystal surface composed of terraces of monolayers (with orientations defined by Miller indices $(h_t\,k_t\,l_t)$), separated by steps (with orientations of step sides defined by Miller indices $(h_s\,k_s\,l_s)$). If the step distribution is regular, that is, periodic at the surface, the stepped surface can be described in its orientation by large Miller indices, see also **Vicinal surface**.
Stranski–Krastanov	Growth mode at surfaces referring to mixed **layer-by-layer** and **three-dimensional cluster** growth.
Superlattice	Description of a netplane by lattice vectors $\underline{R}_1$ and $\underline{R}_2$ (and corresponding unit cells), which are larger than (often integer multiples of) those suggested by the basic periodicity of the netplane. At surfaces, superlattices apply to netplanes of surface-adapted lattice vectors $\underline{R}_1$ and $\underline{R}_2$ (and corresponding unit cells).
Surface reconstruction	Structural modification of a single crystal surface where monolayers near the surface are structurally changed. This can yield changed two-dimensional periodicity compared with that of the

	bulk termination and/or different atom composition and placement.
Surface relaxation	Structural modification of a single crystal surface where whole monolayers near the surface are shifted. Shifts can occur both perpendicular and parallel to the surface.
Surface termination	Structure of the topmost atom layers of the surface of a **single crystal**. In particular, for polyatomic crystals, this specifies which atoms terminate the bulk structure at the surface.
Unit cell	See **Morphological unit cell**.
Vicinal surface	Surface of a single crystal finishing with monolayers whose orientation in the crystal is close to but not identical with those of densest monolayers. Vicinal surfaces are often stepped or kinked. Their orientations are usually described by large values of **Miller indices** ($h\,k\,l$). Thus, vicinal surfaces are often called **High-Miller-index surfaces**.
Volmer–Weber	Growth mode at surfaces referring to **three-dimensional cluster** growth.
Voronoi cell	See **Wigner–Seitz cell**.
VW growth mode	See **Volmer–Weber**.
Wigner–Seitz cell	Compact polygonal ((distorted) hexagonal, rectangular, or square) unit cell of a netplane, sometimes called **Voronoi cell**.
Wood notation	Formal notation of the structure and orientation of reconstructed surfaces or adsorbate systems.
WSC	See **Wigner–Seitz cell**.

Miscellaneous

CIF	Crystallographic Information Framework.
CSD	Cambridge Structural Database.
DFT	Density-functional theory, a quantum mechanical method to examine properties of many-electron systems, such as atoms, molecules, and solids with/without surfaces.
Diophantine equations	Equations using only integer valued constants and variables.

Enantiomer	A chiral molecule and its mirror image are enantiomers (enantiomer pairs).
Enantiopure	Gas or liquid of chiral molecules, which contain only one type, left- or right-handed.
Euclid's algorithm	Algorithm to determine the greatest common divisor gcd(a, b) of two integers a, b.
Fibonacci numbers	Infinite series of integers 0, 1, 1, 2, 3, 5, 8, 13, … defined by element a_k being the sum of the two previous elements, a_{k-1}, a_{k-2} with $a_0 = 0$, $a_1 = 1$.
gcd	Greatest common divisor of two or more integers.
Golden mean	See **Golden ratio**.
Golden ratio	Also referred to as **Golden mean**. Mathematical constant $\varphi = (1 + \sqrt{5})/2 = 1.618034$, appearing in **quasicrystal** structures.
ICSD	Inorganic Crystal Structure Database.
lcm	Least common multiple of two or more integers.
LEED	Low-energy electron diffraction, an experimental method for surface structure determination.
Magic numbers	Total number of atoms in a compact atom cluster with closed polygonal shells.
Moiré	French for "of wavy watery appearance."
NAD	Nucleic Acid Database.
Number theoretical methods	Mathematical methods dealing with integer numbers.
oSSD	Open access version of SSD, see **Surface Structure Database**.
PDB	Protein Data Bank.
Racemic mixture	Gas or liquid of chiral molecules, which contains both types, left- and right-handed, in equal amounts.
SSD	See **Surface Structure Database**.
Surface Structure Database	Also referred to as **SSD**. Database of experimentally known surface structures. The latest version 5 of SSD has been made public as open access SSD (**oSSD**).

References

1. Slater, J.C. (1972) *Symmetry and Energy Bands in Crystals*, Dover Publications, New York.
2. Ashcroft, N.W. and Mermin, N.D. (1976) *Solid State Physics*, Holt-Saunders International, New York.
3. Garcia-Moliner, F. and Flores, F. (1976) *J. Phys. C: Solid State Phys.*, **9**, 1609.
4. Wang, X., Zhang, R.Q., Niehaus, T.A., and Frauenheim, T. (2007) *J. Phys. Chem. C*, **111**, 2394.
5. Arai, T. and Tomitori, M. (2004) *Phys. Rev. Lett.*, **93**, 256101.
6. Rurali, R. and Lorente, N. (2005) *Phys. Rev. Lett.*, **94**, 026805.
7. Hermann, K. and Witko, M. (2001) in *The Chemical Physics of Solid Surfaces*, Oxide Surfaces, vol. **9** (ed. D.P. Woodruff), Elsevier, Amsterdam, p. 136.
8. Rau, C., Liu, C., Schmalzbauer, A., and Xing, G. (1986) *Phys. Rev. Lett.*, **57**, 2311.
9. Biedermann, A., Tscheließnig, R., Schmid, M., and Varga, P. (2001) *Phys. Rev. Lett.*, **87**, 086103.
10. Takahashi, Y., Gotoh, Y., Akimoto, J., Mizuta, S., Tokiwa, K., and Watanabe, T. (2002) *J. Solid State Chem.*, **164**, 1.
11. Ertl, G., Knötzinger, H., Schüth, F., and Weitkamp, J. (2008) *Handbook of Heterogeneous Catalysis*, Wiley-VCH Verlag GmbH, Weinheim.
12. Somorjai, G.A. (1994) *Introduction to Surface Chemistry and Catalysis*, Wiley-Interscience, New York.
13. Marković, N.M. and Ross Jr., P.N. (2002) *Surf. Sci. Rep.*, **45**, 117.
14. Cao, G. (2004) *Nanostructures and Nanomaterials: Synthesis, Properties and Applications*, Imperial College Press, London.
15. Tang, Z. and Sheng, P. (2003) *Nano Science and Technology*, Taylor & Francis, London.
16. Harris, P.J.F. (1999) *Carbon Nanotubes and Related Structures*, Cambridge University Press, Cambridge.
17. Kawazoe, Y., Kondow, T., and Ohno, K. (eds) (2002) *Clusters and Nanomaterials, Theory and Experiment*, Springer, Berlin.
18. Langa, F. and Nierengarten, J.-F. (eds) (2007) *Fullerenes: Principles and Applications*, RSC Publishing, Cambridge.
19. Rogach, A.L. (ed) (2008) *Semiconductor Nanocrystal Quantum Dots: Synthesis, Assembly, Spectroscopy and Applications*, Springer, New York.
20. Gong, K., Du, F., Xia, Z., Durstock, M., and Dai, L. (2009) *Science*, **323**, 760.
21. Ma, D.D.D., Lee, C.S., Au, F.C.K., Tong, S.Y., and Lee, S.T. (2003) *Science*, **299**, 1874.
22. Watson, P.R., Van Hove, M.A., and Hermann, K. (2003) NIST Surface Structure Database (SSD), Standard Reference Database 42, Software Version 5, available as open SSD (oSSD), see Appendix H for web reference.
23. Watson, P.R., Van Hove, M.A., and Hermann, K. (1994) *Atlas of Surface Crystallography based on the NIST Surface Structure Database (SSD)*, J. Phys. Chem. Ref. Data, Monograph Series, vol. **5a, b**, American Chemical Society, Washington.

24. Van Hove, M.A., Hermann, K., and Watson, P.R. (2002) *Acta Crystallogr., Sect. B: Struct. Sci.*, **58**, 338.
25. Maclaren, J.M., Pendry, J.B., Rous, P.J., Saldin, D.K., Somorjai, G.A., Van Hove, M.A., and Vvedenski, D.D. (1987) *Surface Crystallographic Information Service*, Reidel, Dordrecht (Update 1992).
26. Hermann, K. (2016) Balsac software, Fritz-Haber-Institut, Berlin, see Appendix H for web reference.
27. Hermann, K. (2004) SSDIN and Survis software, Fritz-Haber-Institut, Berlin, see Appendix H for web reference.
28. Giacovazzo, C. *et al.* (1998) *Fundamentals of Crystallography*, IUCr Texts on Crystallography, vol. **2**, Oxford Science Publishing.
29. Megaw, H.D. (1973) *Crystal Structures: A Working Approach*, W.B. Saunders Co., Philadelphia, PA.
30. Tilley, R.J.D. (2006) *Crystals and Crystal Structures*, John Wiley & Sons, Ltd., Chichester.
31. Wyckoff, R.W.G. (1963) *Crystal Structures*, vol. **1-6**, Interscience Pub., New York.
32. Burns, G. and Glazer, A.M. (1990) *Space Groups for Solid State Scientists*, 2nd edn, Academic Press, New York.
33. Hahn, T. (ed) (1965, 1983, 1987) *International Tables for Crystallography*, vol. **A**, Reidel Publishing, Boston, MA.
34. Woodruff, D.P. and Delchar, T.A. (1994) *Modern Techniques of Surface Science*, 2nd edn, Cambridge University Press, Cambridge.
35. Ibach, H. (2006) *Physics of Surfaces and Interfaces*, Springer, Berlin.
36. Somorjai, G.A. and Van Hove, M.A. (1979) *Adsorbed Monolayers on Solid Surfaces*, Structure and Bonding, vol. **38**, Springer, New York.
37. Woodruff, D.P. (ed) (2015) *The Chemical Physics of Solid Surfaces*, vol. **1–12**, Elsevier Science, Amsterdam.
38. Zangwill, A. (1990) *Physics at Surfaces*, Cambridge University Press, Cambridge.
39. Henrich, V.E. and Cox, P.A. (1994) *The Surface Science of Metal Oxides*, Cambridge University Press, Cambridge.
40. Dresselhaus, M.S., Dresselhaus, G., and Jorio, A. (2008) *Group Theory, Application to Physics of Condensed Matter*, Springer, Heidelberg.
41. Niggli, P. (1928) *Handbuch der Experimentalphysik*, vol. **7**, Part 1, Akademische Verlagsgesellschaft, Leipzig.
42. Santoro, A. and Mighell, A.D. (1970) *Acta Crystallogr., Sect. A*, **26**, 124.
43. Gruber, B. (1973) *Acta Crystallogr., Sect. A*, **29**, 433.
44. Merz, L. and Ernst, K.-H. (2010) *Surf. Sci.*, **604**, 1049.
45. Huang, J., Zhang, G., Huang, Y., Fang, D., and Zhang, D. (2006) *J. Magn. Magn. Mater.*, **299**, 480.
46. Kellou, A. and Aourag, H. (2003) *Phys. Status Solidi B*, **236**, 166.
47. Puska, M.J., Pöykkö, S., Pesola, M., and Nieminen, R.M. (1998) *Phys. Rev. B*, **58**, 1318.
48. see e.g. Srivastava, G.P. (1999) *Theoretical Modelling of Semiconductor Surfaces: Microscopic Studies of Electrons and Phonons*, World Scientific, Singapore.
49. Buerger, M.J. (1957) *Z. Kristallogr.*, **109**, 42; *Z. Kristallogr.* **113** (1960) 52.
50. Delaunay, B.N. (1983) *Z. Kristallogr.*, **84**, 109.
51. Bain, E.C. (1924) *Trans. Am. Inst. Min. Metall. Eng.*, **70**, 25.
52. Turner, R.D. and Inkson, J.C. (1978) *J. Phys. C: Solid State Phys.*, **11**, 3961.
53. Hass, K.C. (1992) *Phys. Rev. B*, **46**, 139.
54. Quintanar, C., Caballero, R., and Köster, A.M. (2003) *Int. J. Quantum Chem.*, **96**, 483.
55. Fournier, R. (2001) *J. Chem. Phys.*, **115**, 2165.
56. Johnson, R.L. (2002) *Atomic and Molecular Clusters*, Taylor & Francis, London.
57. Teo, B.K. and Sloane, N.J.A. (1985) *Inorg. Chem.*, **24**, 4545.
58. Wulff, G. (1901) *Z. Kristallogr.*, **34**, 449.
59. Gibbs, J.W. (1928) *The Collected Works*, Longmans-Green Company, New York.
60. Herring, C. (1951) *Phys. Rev.*, **82**, 87.
61. Dobrushin, R.L., Kotecky, R., and Shlosman, S.B. (1993) *Wulff Construction: A Global Shape from Local Interaction*, http://www.cpt.univ-mrs

.fr/dobrushin/DKS-book.pdf (accessed 8 January 2016).
62. International Union of Crystallography Report (1991) *Acta Crystallogr., Sect. A*, **48** (1992), 922.
63. Van Smaalen, S. (2009) *Incommensurate Crystallography*, Oxford University Press, New York.
64. Van Aalst, W., Den Hollander, J., Peterse, W.J.A.M., and De Wolff, P.M. (1976) *Acta Crystallogr., Sect. B*, **32**, 47.
65. Arakcheeva, A., Pattison, P., Chapuis, G., Rossell, M., Filaretov, A., Morozov, V., and Van Tendeloo, G. (2008) *Acta Crystallogr., Sect. B*, **64**, 160.
66. Jobst, A. and Van Smaalen, S. (2002) *Acta Crystallogr., Sect. B*, **58**, 179.
67. Nuss, J., Pfeiffer, S., Van Smaalen, S., and Jansen, M. (2010) *Acta Crystallogr., Sect. B*, **66**, 27.
68. Shechtman, D., Blech, I., Gratias, D., and Cahn, J.W. (1984) *Phys. Rev. Lett.*, **53**, 1951.
69. Steinhardt, P.J. and Ostlund, S. (eds) (1987) *The Physics of Quasicrystals*, World Scientific, Singapore.
70. DiVincenzo, D.P. and Steinhardt, P.J. (eds) (1999) *Quasicrystals: The State of the Art*, Directions in Condensed Matter Physics, 2nd edn, vol. **16**, World Scientific, Singapore.
71. Trebin, H.-R. (ed) (2003) *Quasicrystals, Structure and Physical Properties*, Wiley-VCH Verlag GmbH, Weinheim.
72. Janot, C. (1992) *Quasicrystals: A Primer*, Clarendon Press, Oxford.
73. Stadnik, Z.M. (1999) *Physical Properties of Quasicrystals*, Springer, Berlin.
74. Suck, B., Schreiber, M., and Haussler, P. (eds) (2002) *Quasicrystals: An Introduction to Structure, Physical Properties and Applications*, Springer, Berlin.
75. Goldman, A.I., Sordelet, D.J., Thiel, P.A., and Dubois, J.M. (eds) (1997) *New Horizons in Quasicrystals*, World Scientific Press, Singapore.
76. Thiel, P.A. (2008) *Annu. Rev. Phys. Chem.*, **59**, 129.
77. Dubois, J.M. (2005) *Useful Quasicrystals*, World Scientific Press, Singapore.
78. Senechal, M. (1995) *Quasicrystals and Geometry*, Cambridge University Press, Cambridge.
79. Steurer, W. and Deloudi, S. (2009) *Crystallography of Quasicrystals: Concepts, Methods and Structures*, Springer, Berlin.
80. Penrose, R. (1974) *Bull. Inst. Math. Appl.*, **10**, 266.
81. Yamamoto, A. (2008) *Sci. Technol. Adv. Mater.*, **9**, 013001.
82. Calvayrac, Y., Quivy, A., Bessière, M., Lefebvre, S., Cornier-Quiquandon, M., and Gratias, D. (1990) *J. Phys. Fr.*, **51**, 417.
83. Cai, T., Shi, F., Shen, Z., Gierer, M., Goldman, A.I., Kramer, M.J., Jenks, C.J., Lograsso, T.A., Delaney, D.W., Thiel, P.A., and Van Hove, M.A. (2001) *Surf. Sci.*, **495**, 19.
84. Hermann, K. (1997) *Surf. Rev. Lett.*, **4**, 1063.
85. Minkowski, H. (1911) *Gesammelte Abhandlungen*, vol. **1**, Teubner Verlag, Leipzig, p. 145, 153, 217; vol. **2**, p. 78.
86. Bradshaw, A.M. and Richardson, N.V. (1996) *Pure Appl. Chem.*, **68**, 457.
87. Cracknell, A.P. (1974) *Thin Solid Films*, **21**, 107.
88. Czekaj, I., Hermann, K., and Witko, M. (2003) *Surf. Sci.*, **525**, 33.
89. Nicholas, J.F. (1965) *An Atlas of Models of Crystal Surfaces*, Gordon and Breach, New York. (out of print).
90. Nicholas, J.F. (1993) in *Physics of Solid Surfaces: Structure*, Landolt-Börnstein, New Series, vol. **III/24a**, Springer, Heidelberg.
91. Somorjai, G.A. and Van Hove, M.A. (1989) *Prog. Surf. Sci.*, **30**, 201.
92. Eisner, D.R. and Einstein, T.L. (1993) *Surf. Sci. Lett.*, **286**, L559.
93. Lang, B., Joyner, R.W., and Somorjai, G.A. (1972) *Surf. Sci.*, **30**, 454.
94. Van Hove, M.A., Weinberg, W.H., and Chan, C.M. (1986) *Low Energy Electron Diffraction*, Springer Series in Surface Science, vol. **6**, Springer, New York.
95. Van Hove, M.A. and Somorjai, G.A. (1980) *Surf. Sci.*, **92**, 489.
96. Jenkins, S.J. and Pratt, S.J. (2007) *Surf. Sci. Rep.*, **62**, 373.

97. Wagnière, G.H. (2007) *On Chirality and the Universal Asymmetry*, Wiley-VCH Verlag GmbH, Weinheim.
98. Potočka, J. and Dvořak, A. (2004) *J. Appl. Biomed.*, **2**, 95.
99. Horvath, J.D. and Gellman, A.J. (2003) *Top. Catal.*, **25**, 9.
100. Barlow, S. and Raval, R. (2003) *Surf. Sci. Rep.*, **50**, 201.
101. IUPAC Rules for the Nomenclature of Organic Chemistry, Section E: Stereochemistry (1976) *Pure Appl. Chem.*, **45**, 11.
102. Mucha, D., Stadnicka, K., Kaminsky, W., and Glazer, A.M. (1997) *J. Phys.: Condens. Matter*, **9**, 10829.
103. McFadden, C.F., Cremer, P.S., and Gellman, A.J. (1996) *Langmuir*, **12**, 2483.
104. Sholl, D.S., Asthagiri, A., and Power, T.D. (2001) *J. Chem. Phys. B*, **105**, 4771.
105. Jiang, P., Jona, F., and Marcus, P.M. (1987) *Phys. Rev. B*, **35**, 7952.
106. Blum, V., Hammer, L., Heinz, K., Franchini, C., Redinger, J., Swamy, K., Deisl, C., and Bertel, E. (2002) *Phys. Rev. B*, **65**, 165408.
107. Wang, Y., Shi, M., and Rabalais, J.W. (1993) *Phys. Rev. B*, **48**, 1678.
108. Zhao, R.G., Jia, J.F., Li, Y.F., and Yang, W.S. (1991) *Springer Ser. Surf. Sci.*, **24**, 517.
109. Takayanagi, K., Tanishiro, Y., Takahashi, M., and Takahashi, S. (1985) *J. Vac. Sci. Technol. A*, **3**, 1502.
110. Tong, S.Y., Huang, H., Wei, C.M., Packard, W.E., Men, F.K., Glander, G., and Webb, M.B. (1988) *J. Vac. Sci. Technol. A*, **6**, 615.
111. Van Hove, M.A., Koestner, R.J., Stair, P.C., Bibérian, J.P., Kesmodel, L.L., Bartoš, I., and Somorjai, G.A. (1981) *Surf. Sci.*, **103**, 189.
112. Ocko, B.M., Gibbs, D., Huang, K.G., Zehner, D.M., and Mochrie, S.G.J. (1991) *Phys. Rev. B*, **44**, 6429.
113. Bauer, E. (1958) *Z. Kristallogr.*, **110**, 372; *Z. Kristallogr.*, **110**, 395.
114. Schmailzl, P., Schmidt, K., Bayer, P., Döll, R., and Heinz, K. (1994) *Surf. Sci.*, **312**, 73.
115. Schmitz, D., Charton, C., Scholl, A., Carbone, C., and Eberhardt, W. (1999) *Phys. Rev. B*, **59**, 4327.
116. Koch, R. (1994) *J. Phys.: Condens. Matter*, **6**, 9519.
117. Venables, J.A., Spiller, G.D.T., and Hanbücken, M. (1984) *Rep. Prog. Phys.*, **47**, 399.
118. Bauer, E., Poppa, H., Todd, G., and Davis, P.R. (1977) *J. Appl. Phys.*, **48**, 3773.
119. Hanbücken, M., Neddermeyer, H., and Rupieper, P. (1982) *Thin Solid Films*, **90**, 37.
120. see e.g. Woodruff, D.P. (ed) (2002) *The Chemical Physics of Solid Surfaces*, Surface Alloys and Alloy Surfaces, vol. 10, Elsevier Science, Amsterdam.
121. Wider, H., Gimple, V., Evenson, W., Schatz, G., Jaworski, J., Prokop, J., and Marszałek, M. (2003) *J. Phys.: Condens. Matter*, **15**, 1909.
122. Henrich, V.E. (1976) *Surf. Sci.*, **57**, 385.
123. Zaremba, E. and Kohn, W. (1977) *Phys. Rev. B*, **15**, 1769.
124. Baumgärtel, P., Lindsay, R., Giessel, T., Schaff, O., Bradshaw, A.M., and Woodruff, D.P. (2000) *J. Phys. Chem. B*, **104**, 3044.
125. Peters, K.F., Walker, C.J., Steadman, P., Robach, O., Isern, H., and Ferrer, S. (2001) *Phys. Rev. Lett.*, **86**, 5325.
126. Pai, W.W., Jeng, H.T., Cheng, C.-M., Lin, C.-H., Xiao, X., Zhao, A., Zhang, X., Xu, G., Shi, X.Q., Van Hove, M.A., Hsue, C.-S., and Tsuei, K.-D. (2010) *Phys. Rev. Lett.*, **104**, 036103.
127. Moritz, W., Wang, B., Bocquet, M.-L., Brugger, T., Greber, T., Wintterlin, J., and Günther, S. (2010) *Phys. Rev. Lett.*, **104**, 136102.
128. Cohen, P.I., Unguris, J., and Webb, M.B. (1976) *Surf. Sci.*, **58**, 429.
129. Charrier, E. and Buzer, L. (2009) *Discrete Appl. Math.*, **157**, 3473.
130. Tong, S.Y., Maldonado, A., Li, C.H., and Van Hove, M.A. (1980) *Surf. Sci.*, **94**, 73.
131. Citrin, P.H., Hamann, D.R., Mattheiss, L.F., and Rowe, J.E. (1982) *Phys. Rev. Lett.*, **49**, 1712.
132. Johnson, K., Ge, Q., Titmuss, S., and King, D.A. (2000) *J. Chem. Phys.*, **112**, 10460.

133. Lindgren, S.A., Wallden, L., Rundgren, J., Westrin, P., and Neve, J. (1983) *Phys. Rev. B*, **28**, 6707.
134. Gierer, M., Barbieri, A., Van Hove, M.A., and Somorjai, G.A. (1997) *Surf. Sci.*, **391**, 176.
135. Vu Grimsby, D.T., Wu, Y.K., and Mitchell, K.A.R. (1990) *Surf. Sci.*, **232**, 51.
136. Zasada, I. and Van Hove, M.A. (2000) *Surf. Sci.*, **457**, L421.
137. Materer, N., Barbieri, A., Gardin, D., Starke, U., Batteas, J.D., Van Hove, M.A., and Somorjai, G.A. (1994) *Surf. Sci.*, **303**, 319.
138. Bao, S., Schindler, K.-M., Hofmann, P., Fritzsche, V., Bradshaw, A.M., and Woodruff, D.P. (1993) *Surf. Sci.*, **291**, 295.
139. Hofmann, P., Schindler, K.-M., Bao, S., Fritzsche, V., Bradshaw, A.M., and Woodruff, D.P. (1995) *Surf. Sci.*, **337**, 169.
140. Puschmann, A. and Haase, J. (1984) *Surf. Sci.*, **144**, 559.
141. Yokoyama, T., Hamamatsu, H., Kitajima, Y., Takata, Y., Yagi, S., and Ohta, T. (1994) *Surf. Sci.*, **313**, 197.
142. Shih, H.D., Jona, F., Jepsen, D.W., and Marcus, O.M. (1981) *Phys. Rev. Lett.*, **46**, 731.
143. Van Hove, M.A. and Tong, S.Y. (1975) *Phys. Rev. Lett.*, **35**, 1092.
144. Ogletree, D.F., Van Hove, M.A., and Somorjai, G.A. (1986) *Surf. Sci.*, **173**, 351.
145. Wu, Z.Q., Lu, S.H., Wang, Z.Q., Lok, C.K.C., Quinn, J., Li, Y.S., Tian, D., Jona, F., and Marcus, P.M. (1988) *Phys. Rev. B*, **38**, 5363.
146. Ross, C., Schirmer, B., Wuttig, M., Gauthier, Y., Bihlmayer, G., and Blügel, S. (1998) *Phys. Rev. B*, **57**, 2607.
147. Helgesen, G., Gibbs, D., Baddorf, A.P., Zehner, D.M., and Mochrie, S.G.J. (1993) *Phys. Rev. B*, **48**, 15320.
148. Baddorf, A.P., Zehner, D.M., Helgesen, G., Gibbs, D., Sandy, A.R., and Mochrie, S.G.J. (1993) *Phys. Rev. B*, **48**, 9013.
149. Nishiyama, A., terHost, G., Lohmeier, M., Molenbroek, A.M., and Frenken, J.W.M. (1994) *Surf. Sci.*, **321**, 261.
150. Wood, E.A. (1964) *J. Appl. Phys.*, **35**, 1306.
151. Li, H.I., Franke, K.J., Pascual, J.I., Bruch, L.W., and Diehl, R.D. (2009) *Phys. Rev. B*, **80**, 085415.
152. Tkatchenko, A. (2006) *Phys. Rev. B*, **74**, 035428; *Phys. Rev. B*, **75** (2007) 235411.
153. Brugger, T., Ma, H., Iannuzzi, M., Berner, S., Winkler, A., Hutter, J., Osterwalder, J., and Greber, T. (2010) *Angew. Chem. Int. Ed.*, **49**, 6120; *Angew. Chem. Int. Ed.*, **49**, 6256.
154. Chambon, C., Creuze, J., Coati, A., Sauvage-Simkin, M., and Garreau, Y. (2009) *Phys. Rev. B*, **79**, 125412.
155. Hermann, K. (2012) *J. Phys.: Condens. Matter*, **24**, 314210.
156. Hermann, K. (2016) *J. Phys.: Condens. Matter*, submitted.
157. Preobrajenski, A.B., Ng, M.-L., Vinogradov, A.S., and Mårtensson, N. (2008) *Phys. Rev. B*, **78**, 073401.
158. N'Diaye, A.T., Coraux, J., Plasa, T.N., Busse, C., and Michely, T. (2008) *New J. Phys.*, **10**, 043033.
159. Loginova, E., Nie, S., Thürmer, K., Bartelt, N.C., and McCarty, K.F. (2009) *Phys. Rev. B*, **80**, 085430.
160. Wang, B. and Bocquet, M.-L. (2011) *J. Phys. Chem. Lett.*, **2**, 2341.
161. Gao, L., Guest, J.R., and Guisinger, N.P. (2010) *Nano Lett.*, **10**, 3512.
162. Zhao, W., Kozlov, S.M., Höfert, O., Gotterbarm, K., Lorenz, M.P.A., Viñes, F., Papp, C., Görling, A., and Steinrück, H.-P. (2011) *J. Phys. Chem. Lett.*, **2**, 759.
163. Zeller, P. and Günther, S. (2014) *New J. Phys.*, **16**, 083028.
164. Bernhardt, T.M., Kaiser, B., and Rademann, K. (1998) *Surf. Sci.*, **408**, 86.
165. Xue, J., Sanchez-Yamagishi, J., Bulmash, D., Jacquod, P., Deshpande, A., Watanabe, K., Taniguchi, T., Jarillo-Herrero, P., and LeRoy, B.J. (2011) *Nat. Mater.*, **10**, 282.
166. Coraux, J., N'Diaye, A.T., Busse, C., and Michely, T. (2008) *Nano Lett.*, **8**, 565.
167. Batteas, J.D., Barbieri, A., Starkey, E.K., Van Hove, M.A., and Somorjai, G.A. (1995) *Surf. Sci.*, **339**, 142.

168. Kerkar, M., Fisher, D., Woodruff, D.P., and Cowie, B. (1992) *Surf. Sci.*, **271**, 45.
169. Hermann, K. (2016) LEEDpat4 software, see Appendix H for web reference.
170. Booth, N.A., Davis, R., Woodruff, D.P., Chrysostomou, D., McCabe, T., Lloyd, D.R., Schaff, O., Fernandez, V., Bau, S., Schindler, K.-M., Lindsay, R., Hoeft, J.T., Terborg, R., Baumgärtel, P., and Bradshaw, A.M. (1998) *Surf. Sci.*, **416**, 448.
171. Lehnberger, K., Nichtl-Pecher, W., Oed, W., Heinz, K., and Müller, K. (1989) *Surf. Sci.*, **217**, 511.
172. Pascal, M., Lamont, C.L.A., Baumgärtel, P., Terborg, R., Hoeft, J.T., Schaff, O., Polčik, M., Bradshaw, A.M., Toomes, R.L., and Woodruff, D.P. (2000) *Surf. Sci.*, **464**, 83.
173. Baddeley, C.J. and Richardson, N.V. (2010) in *Scanning Tunneling Microscopy in Surface Science, Nanoscience, and Catalysis* (eds M. Bowker and P.R. Davis), Wiley-VCH Verlag GmbH, Weinheim, p. 1.
174. Raval, R. (2009) *Chem. Soc. Rev.*, **38**, 707.
175. Gellman, A.J. (2010) in *Model Systems in Catalysis* (ed R. Rioux), Springer, New York, p. 75.
176. Blüm, M.-C., Cavar, E., Pivetta, M., Patthey, F., and Schneider, W.-D. (2005) *Angew. Chem. Int. Ed.*, **44**, 5334.
177. Ortega Lorenzo, M., Haq, S., Bertrams, T., Murray, P., Raval, R., and Baddeley, C.J. (1999) *J. Phys. Chem. B*, **103**, 10661.
178. Mark, A.G., Forster, M., and Raval, R. (2011) *ChemPhysChem*, **12**, 1474.
179. Bhatia, B. and Sholl, D.S. (2005) *Angew. Chem. Int. Ed.*, **44**, 7761.
180. Stöhr, J. (1992–2003) *NEXAFS Spectroscopy*, Springer Series in Surface Science, vol. **25**, Springer, Heidelberg.
181. Allen, F.H. (2002) *Acta Crystallogr., Sect. B*, **58**, 380.
182. Belsky, A., Hellenbrandt, M., Lynn, V., Luksch, K., and Luksch, P. (2002) *Acta Crystallogr., Sect. B*, **58**, 364.
183. White, P.S., Rodgers, J.R., and Le Page, Y. (2002) *Acta Crystallogr., Sect. B*, **58**, 343.
184. Berman, H.M., Westbrook, J., Feng, Z., Iype, L., Schneider, B., and Zardecki, C. (2002) *Acta Crystallogr., Sect. D*, **58**, 889.
185. Berman, H.M., Battistuz, T., Bhat, T.N., Bluhm, W.F., Bourne, P.E., Burkhardt, K., Feng, Z., Gilliland, G.L., Iype, L., Jain, S., Fagan, P., Marvin, J., Padilla, D., Ravichandran, V., Schneider, B., Thanki, N., Weissig, H., Westbrook, J.D., and Zardecki, C. (2002) *Acta Crystallogr., Sect. D*, **58**, 899.
186. Ohtani, H., Kao, C.-T., Van Hove, M.A., and Somorjai, G.A. (1986) *Prog. Surf. Sci.*, **23**, 155.
187. Van Hove, M.A., Hermann, K., and Watson, P.R. (2002) *Landolt-Börnstein, Physics of Covered Solid Surfaces*, vol. **III/42**, New Series, Springer, Heidelberg.
188. Barth, J.V. (2007) *Annu. Rev. Phys. Chem.*, **58**, 375.
189. Elemans, J.A.A., Lei, S., and De Feyter, S. (2009) *Angew. Chem. Int. Ed.*, **48**, 7298.
190. Wang, W., Shi, X., Wang, S., Liu, J., Van Hove, M.A., Liu, P.N., Zhang, R.Q., and Lin, N. (2013) *Phys. Rev. Lett.*, **110**, 046802.
191. Burchardt, J., Nielsen, M.M., Adams, D.L., Lundgren, E., and Andersen, J.N. (1994) *Phys. Rev. B*, **50**, 4718.
192. Campbell, C.T. (2010) *Surf. Sci.*, **604**, 877.
193. Marks, L.D. (2010) *Surf. Sci.*, **604**, 878.
194. Van Hove, M.A., Hermann, K., Watson, P.R., Woodruff, D.P., Tong, S.Y., Diehl, R.D., Heinz, K., Minot, C., and Tochihara, H. (2010) *Surf. Sci.*, **604**, 1544.
195. Hall, S.R., Allen, F.H., and Brown, I.D. (1991) *Acta Crystallogr., Sect. A*, **47**, 655, see also http://www.iucr.org/resources/cif (accessed 12 January 2015).
196. Berman, H.M., Henrick, K., and Nakamura, H. (2003) *Nat. Struct. Biol.*, **10**, 908, see also http://www.wwpdb.org/documentation/file-format.php (accessed 12 January 2016).
197. Kresse, G. and Furthmüller, J. (1996) *Phys. Rev. B*, **54**, 11169, see also http://www.vasp.at/index.php/documentation (accessed 12 January 2016).

198. Radushkevich, L.V. and Lukyanovich, V.M. (1952) *Zurn. Fisic. Chim.*, **111**, 24.
199. Oberlin, A., Endo, M., and Koyama, T. (1976) *J. Cryst. Growth*, **32**, 335.
200. Ijima, S. and Ichihashi, T. (1993) *Nature*, **363**, 603.
201. Wang, N., Cai, Y., and Zhang, R.Q. (2008) *Mater. Sci. Eng., R*, **60**, 1.
202. The Nanotube Site (2016) http://www.pa.msu.edu/cmp/csc/nanotube.html (accessed 12 January 2016).
203. Chen, W., Peng, J., Mai, L., Zhu, Q., and Xu, Q. (2004) *Mater. Lett.*, **58**, 2275.
204. Mordell, L.J. (1969) *Diophantine Equations*, Academic Press, New York.
205. Andreescu, T. and Andrica, D. (2009) *Quadratic Diophantine Equations*, Springer Monographs in Mathematics, Springer, Heidelberg.
206. Hary, G.H. and Wright, E.M. (2008) *An Introduction to the Theory of Numbers*, 6th edn, Oxford University Press, Oxford.
207. Pinski, M. (2002) *Introduction to Fourier Analysis and Wavelets*, Brooks/Cole, Pacific Grove, CA.
208. Grafakos, L. (2008) *Classical Fourier Analysis*, Graduate Texts in Mathematics, 2nd edn, Springer, New York.

Index

a
achiral 403
– definition 192, 194
– surface 197, 409
acute representation
– two-dim. 409
– three-dim. 403
– hexagonal 21, 45, 109
additivity theorem 410
adsorbate layer 235
– clusters 236
– coverage 235
– fully disordered 235
– interference lattice 263
– – anisotropic scaling 279
– – basic formalism 264
– – chiral pairs 279
– – first order 267
– – Fourier expansion 265
– – high- order 268
– – HOC 271
– – isotropic scaling 267, 275
– – isotropic scaling + rotation 275, 277
– – modulation factor 267, 269
– – non-periodic 266
– – order matrix 269
– – periodic 266
– – rotation 275
– – shifting 281
– – stretching 280
– – warping 264
– – Wood notation 272
– islands 236
– lattice gas 236
– partially disordered 236
– periodic overlayer 236
– – coincidence lattice 238
– – commensurate 237
– – high-order commensurate 258, 261, 262
– – – hexagonal 261
– – – square 262
– – high-order commensurate (HOC) 238
– – incommensurate 74, 240
– – modulation 239
– – warping 239
adsorbate sites 241
– centered rectangular lattice 247
– – threefold hollow 248
– – fourfold hollow 247
– – top 247
– hexagonal lattice 243
– – twofold bridge 245
– – threefold hollow 243
– – top 243
– high-symmetry sites 241
– mixed 249
– primitive rectangular lattice 245
– – twofold long bridge 245
– – twofold short bridge 246
– – fourfold hollow 247
– – top 245
– sketches 329
– – bcc surface 338
– – diamond surface 346
– – fcc surface 330
– – hcp surface 342
– – zincblende surface 349
– square lattice 242
– – twofold bridge 243
– – fourfold hollow 242
– – top 242
– substitutional 249
adsorbates 235
adsorption
– chiral 293
– – cooperative 294, 296

Crystallography and Surface Structure: An Introduction for Surface Scientists and Nanoscientists,
Second Edition. Klaus Hermann.
© 2017 Wiley-VCH Verlag GmbH & Co. KGaA. Published 2017 by Wiley-VCH Verlag GmbH & Co. KGaA.

Index

– – enantiopure 296
– – enantioselective 293
– – local 293
– – racemic 296
anti-ferromagnetic ordering 26

b

Bain path 37, 403
Balsac XII, 4, 399
basis 403, 410
bcc 16, 403
Bravais lattice (two-dim.) 410
– centered rectangular 125, 127, 129, 136, 138, 145
– hexagonal 129, 130, 144
– oblique 144
– overview 146
– primitive rectangular 123, 125, 126, 136, 138, 144
– square 128, 144
Bravais lattice (three-dim.) 40, 41, 403
– cubic-F 44, 51
– cubic-I 44, 51
– cubic-P 44, 51
– hexagonal-P 45, 51
– monoclinic-A 42
– monoclinic-B 42, 51
– monoclinic-C 42
– monoclinic-P 42, 51
– orthorhombic-C 43, 51
– orthorhombic-F 43, 51
– orthorhombic-I 43, 51
– orthorhombic-P 43, 51
– overview 46
– tetragonal-I 43, 51
– tetragonal-P 43, 51
– triclinic-P 41, 51
– trigonal-R (rhombohedral) 45, 51
BrClFCH 192
Brillouin zone (BZ) 404
Buerger cell 35, 404
bulk crystals 7
– representation 11

c

$C_4O_6H_6$ 194, 295
Cahn-Ingold-Prelog rules 195
carbon nanotube 317
– armchair 319
– chiral pairs 319
– zigzag 318
Cartesian coordinates 15
catalysis 2
centering 31, 404, 410

cesium chloride 12, 58
chemical binding 1
chiral
– centers 194
– definition 192, 194
– molecule 192
– partner 195
– surface 192, 410
chirality 192, 404
CIF 415
classification scheme
– crystal lattices 40, 46
– netplanes 144, 149, 163
clusters 63
coincidence (super)lattice 410
commensurate reconstruction 410
compatibility constraints
– two-dim. 110, 115, 123, 127, 133
– three-dim. 40
coordinate transformation (two-dim.)
– glide reflection 132
– inversion 112
– mirroring (reflection) 120
– rotation 114
coordination
– number 53, 404
– shell 52, 404
c-rectangular 410
crystal 404
– aperiodic 71, 72
– decomposition 12
– definition 9, 11
– growth 222
– growth mode 222
– – three-dim cluster 224
– – alloying 225
– – Frank-Van-der-Merwe (FM) 223, 224
– – layer-by-layer 223
– – mixed layer, cluster 224
– – Stranski–Krastanov (SK) 224
– – Volmer–Weber (VW) 224, 225
– hexagonal close-packed (hcp) 21, 165
– incommensurate 71
– incommesurate composite 73, 75
– modulated structures 71
– periodic 72
– quasicrystal 71, 76, 78
– – $Al_{65}Cu_{20}Fe_{15}$ 82
– – dihedral 80
– – i-AlCuFe 82
– – icosahedral 80
crystal parameters 351
crystal systems (two-dim.) 144, 410

Index | 427

– hexagonal   144
– oblique   144
– overview   146
– rectangular   144
– square   144
crystal systems (three-dim.)   40, 405
– cubic   44
– hexagonal   44
– monoclinic   42
– orthorhombic   43
– rhombohedral   45
– tetragonal   43
– triclinic   41
– trigonal   45
crystallites   63, 69
– quasi-continuum models   69
crystallographic plane   92, 410
CsCl   58, 201
CSD   415
ct   405
cubic Miller Indices   405
cuboctahedral clusters   65
cut-and-project method   80, 411

d

DAS model   214, 411
density-functional theory   415
DFT   263, 415
diamond   13
Diophantine equations   415
– linear   31, 97, 177, 189, 377
– – (in)homogeneous   377
– – iterative solution   377
– quadratic   54, 57, 59, 380
– – $a\,m^2 = b\,n^2$ 385
– – $m^2 + 2\,n^2 = N^2$ 384
– – $m^2 + m\,n + n^2 = N^2$ 383
– – $n_1^2 + n_2^2 = N$ 382
– – $n_1^2 + n_2^2 + n_3^2 = N$ 380
– – $n_1^2 + n_2^2 + n_3^2 + n_4^2 = N$ 382
dodecahedral clusters   68
domain formation   411

e

electronic properties   1
enantiomer   192, 416
enantiopure   296, 416
Euclid formula   382
Euclid's algorithm   376, 377, 416
experimental methods   305
– IS   310
– LEED   305, 310
– PED   305, 310
– SEXAFS   305, 310

– table   306
– X-ray grazing incidence   305
– XSW   310

f

facet   69, 411
faceting   226
fcc   18, 405
ferromagnetic ordering   26
Fibonacci
– chain   81, 411
– numbers   79, 416
five-fold symmetry   78, 82
Fourier theory   51, 213, 395
– harmonic functions   395
four-index notation   108, 405, 406
Frank-Van-der-Merwe   411
FW growth mode   411

g

gallium arsenite, GaAs   86, 172
gcd   416
generic Miller Indices   405
Gibbs-Wulff theorem   69, 405
glossary
– bulk and three-dim.   403
– miscellaneous   415
– surface and two-dim.   409
glossary, abbreviations   403
golden mean   416
golden ratio   77, 416
Gra   405
grains   69
graphene   317, 405

h

handedness   30, 192, 195, 405
– left-handed   196
– right-handed   14, 196
hcp   21, 405
Hermann-Mauguin notation   40, 115, 151, 405
hex(hcp)   405
hexagonal close-packed (hcp) crystal   61
hexagonal graphite   83
high-Miller-index surface   411
high-order commensurate (HOC)   411
HOC reconstruction   214, 411

i

icosahedral
– clusters   64
– symmetry   64

ICSD 416
incommensurate composite crystal 406
incommensurate reconstruction 411
intercept factors 106
interference lattice 411
international notation 40
international tables of crystallography, ITC 40, 151
inversion symmetry 172
ITC 40, 151, 406
IUCr crystal definition 71
IUPAC 222

k
kinked surface 412

l
label 10
Lagrange's theorem 382
lattice 406
− alternative description 12, 14, 30
− basis 11
− centered 31
− − A-centered 33
− − B-centered 34
− − body centered 16, 33, 44, 50, 57, 101
− − − cubic (bcc) 16, 44, 50, 57, 101
− − C-centered 34
− − face centered 18, 34, 44, 50, 58, 100
− − − cubic (fcc) 18, 44, 50, 58, 100
− − F-centered 34
− − I-centered 33
− classification 38
− definition 9, 11
− geometric constraints 29
− hexagonal 21, 104
− hexagonal sublattice (trigonal) 23
− linear transformations 29
− perovskite 187
− reciprocal 49
− − Bravais lattices 51
− − double reciprocal lattice 392
− − lattice vectors 49
− − orthogonality relations 50, 265
− − unit cell volume 50
− rhombohedral 22
− simple cubic (sc) 16, 44, 50, 54, 100
− symmetrically appropriate vectors 98
− symmetry 38
− trigonal 22, 104
lattice basis vectors 10, 92, 93, 406, 412
lattice constant 9, 406
lattice gas 412
lattice points 9

lattice vectors 9, 406, 412
− acute 14, 100
− general 91, 228, 405, 411
− netplane-adapted 96, 316
− obtuse 14, 100
− reduced 15
lcm 416
LEED 82, 416
LEEDpat 286, 399
linear algebra 391
− linear transformation 391

m
magic numbers 64, 65, 416
magnetism 2, 26
matrix notation (2×2) 221, 409
MgO 3, 93, 170, 200, 226
microfacet notation 185, 412
microfaceted surface 414
Miller indices 91, 406
− additivity theorem
− − kinked surface 182, 183
− − stepped surface 178, 179
− alternative definition 106, 107
− cubic lattices 100
− decomposition 178, 187
− direction 96
− direction family 96
− family 96
− four-index notation 108
− generic 94
− large values 175, 178, 182
− negative values 94, 169
− simple cubic notation 101, 102, 169
− trigonal (hexagonal notation) 105
Miller-Bravais indices 106, 108, 406
Minkowski reduction 98, 99, 368, 412
− basic condition 369
− mathematics 367
modulated structure 407
moiré 416
− angle 274
− factor 268, 274
− lattice vectors 267, 269
− matrix 267, 270, 274
− moirons 264
− pattern 263, 412
− stripes 280
molecular adsorbates 249
monoatomic crystal 407
monolayer 91, 412
− atom density 94
− definition 92, 93
− distance 94

– NaCl(1 2 2)   316
morphological unit cell   10, 35, 93, 110
– two-dim.   413
– three-dim.   407
motif   151
– two-dim.   413
– three-dim.   407

n
NaCl   60, 170, 200, 226
NAD   416
nanoparticle   2, 63, 315
nanotube   315, 407
– achiral   323
– basic definition   315
– carbon   315
– chiral pairs   320
– complex nanotubes   323
– coordinate transformation   317
– crystallography   315
– rolling indices (m, n)   316
– rolling vector   316
– symmetry   319
– translational periodicity   322
nanowire   324
neighbor shells   52, 407
– complete set   53
– Diophantine equations   380
– evaluation   61
– polyhedral shells   64
– shell center   53
– shell multiplicity   53, 407, 408
– – accidental   55
– – symmetry related   55, 58, 60
– – total   55, 58, 60
– shell radius   52, 53
– shell range   53
– shell thickness   53
netplane   91, 413
– centered   110, 145
– definition   92
– distance   94
– normal direction   93, 104
– symmetry   109
netplane symmetry
– centered rectangular   155
– classification   146
– hexagonal   158
– oblique   151
– primitive rectangular   151
– square   157
netplane-adapted
– lattice   407
– lattice vectors   16, 96

– matrix   93
– transformation   93
n-fold rotation   40, 407, 413
Niggli cell   35, 407
non-primitive
– lattice vectors   10, 32
– unit cell   10
non-symmorphic space group   407
number theory   97, 371, 416
– basic definitions   371
– composite function p(x)   373
– coprime   372
– greatest common divisor, gcd   98, 372, 376
– least common multiple, lcm   374
– matrices   386
– – coprime pair   387
– – integer approximant   266, 386
– – reduction   387
– modulo function   372
– nearest integer function   371
– truncation function   371

o
obtuse representation
– two-dim.   413
– three-dim.   408
– hexagonal   21, 45, 107

p
PDB   416
Penrose tiling   76, 413
periodicity cells   35
plane group   143, 413
point symmetry group   142
– two-dim.   414
– three-dim.   408
– associativity   142
– highest   151
– inverse element   142
– list of groups   142
– product   142
– subgroup   143
– unit element   142
point symmetry operations
– two-dim.   110
– three-dim.   39
polyatomic crystal   408
p-rectangular   414
primitive
– lattice   10, 408
– lattice vectors   10
– unit cell   10
prototiles   76

Index

Pythagorean
- equation 220
- triplets 382

q

quantitative structure determination 305
quasicrystal 76, 408
quasiperiodic crystal 76

r

racemic mixture 296, 416
reciprocal lattice 408
relative coordinates 11, 408
repeated slab geometry 25, 27
rhombohedral graphite 83
rolling vector 414
rotation
- angles 115
- anti-clockwise 356
- two-, three-, four-, and sixfold 116
rutile 86

s

sc 16, 408
- notation 101, 102, 408
scaled commensurate 414
Schönflies notation 40, 41, 115, 140, 151, 408
self-similarity 79
shell models 52
simple cubic notation 409
single crystal 409
SK growth mode 414
sodium chloride 60, 170
space group (two-dim.) 150, 414
- non-symmorphic 143, 150
- overview 163
- simple 143
- symmorphic 143, 150
space group (three-dim.) 49, 409
$SrTiO_3$ 187
SSD 418
- experimental methods 310
- NIST SSD 242, 251, 307, 312, 399, 401
- oSSD 242, 307, 312, 399, 401, 416
- SSDIN 4, 399
- statistical analysis 308
step notation 180, 414
stepped surface 414
Stranski-Krastanov 414
strontium titanate 187
structure 10
structure database
- critical 308
- Crystmet 306

- CSD 306
- format 311
- – Balsac 312
- – CIF 312
- – PDB 312
- – requirements 312
- – SSD 312
- – Survis 312
- – XYZ 312
- ICSD 306
- NAD 306
- PDB 307
- SCIS 307
supercell 25
superlattice 25, 409, 414
- methods 16
surface
- achiral 197
- – bcc crystal 202
- – fcc crystal 201
- – hexagonal crystal 203
- bulk truncation 169, 210
- chiral 192, 195
- ideal 169
- kinked 175
- – multiple-atom-height 185
- microfacet notation 185
- microfaceted 231
- moiré pattern 216, 218
- morphology 175
- orientation 169
- polyatomic crystal 187
- real 209
- step notation 180
- stepped 175
- – multiple-atom-height 176, 181, 375
- – – partitioning 375
- – step edges 176
- – subterraces 181
- surface atoms 173
- symmetry 283
- termination 170
- terrace 175
- vicinal 175, 179, 183, 415
surface domain 283, 289
- anti-phase 293
- glide line 290
- mirrored 289
- rotational 289
- translational 290
surface facet
- angle 228
- edge vector 227
- negative 227

Index | 431

– – positive   227
surface free energy   69
surface reconstruction   210, 414
– buckling   213, 215
– coincidence lattice   214, 218
– commensurate   211
– disordered   210
– displacive   213
– high-order commensurate   214
– incommensurate   217, 221
– matrix   211, 214, 217
– modulation function   213
– reconstruction matrix   211
– rotational superlattice   218
– scaled commensurate   214
– superlattice   211
surface relaxation   209, 415
– inwards/outwards   209
surface structure
– compilations   306
– examples
– – Ag(1 1 0) + (2 × 1) − O   245, 291
– – Ag(1 1 1) + Xe hex disordered   240
– – Al(1 1 1) + (1 × 1) − O   286
– – Au(1 0 0) hex disordered   217
– – Au(1 1 1) − ($\sqrt{3} \times 22$)rect   215, 280
– – Au(1 1 1) + rubrene   295
– – Co(1 0 -1 5)   203
– – Cu(0 0 1) + ($2\sqrt{2} \times \sqrt{2}$)R45° − 2O   301
– – Cu(1 0 0) + ($\sqrt{2} \times \sqrt{2}$)R45° − Cl   251
– – Cu(1 0 0) + c(2 × 2) − Cl   243
– – Cu(1 0 0) + c(2 × 2) − Pd   249
– – Cu(1 1 0) − (1 × 1)   250
– – Cu(1 1 0) + (9 0 | 1 2) − 3(R,R)-TA   296
– – Cu(1 1 0) + (9 0 | −1 2) − 3(S,S)-TA   296
– – Cu(1 1 0) + (2 × 1) − CO   245
– – Cu(1 1 0) + (2 × 3) − 4N   250
– – Cu(1 1 0) + c(2 × 2) − Mn   249
– – Cu(1 1 1) + ($6\sqrt{3} \times 6\sqrt{3}$)R30° − 2TPyB − 3Cu   308
– – Cu(1 1 1) + (disordered) − NH_3 236, 293
– – Cu(1 1 1) + (2 × 2) − Cs   243
– – Cu(1 1 1) + (4 × 4) − C_{60} 238
– – Cu(1 1 1) + (disordered) − C_2H_2 245
– – Cu(1 1 1) + (R,R)-TA   295
– – Cu(1 1 1) + (S,S)-TA   295
– – Cu(1 1 1) + $C_4O_6H_6$ 295
– – Cu(1 1 1) + R-$(NH_3)_6$ 295
– – Cu(1 1 1) + S-$(NH_3)_6$ 295
– – Cu(8 7 4)^R + R/S-FAM   299
– – Cu(−8 −7 −4)^S + R/S-FAM   299
– – fcc(15 15 23)   181
– – fcc(3 3 1)   198
– – fcc(3 3 5)   178

– – fcc(37 25 17)   186
– – fcc(5 6 8)   175
– – fcc(6 1 1)   176
– – fcc(7 1 1) facets   227
– – fcc(7 7 9)   175
– – fcc($h\,k\,l$) crystal sphere   231
– – Fe(110) + (2 × 2) − S   247
– – Fe(1 2 3)   202
– – Fe(−25 −20 −3)^R 297
– – Fe(−25 −20 −3)^R + R-Co_9 298
– – Fe(−25 −20 −3)^R + S-Co_9 298
– – Fe(25 20 3)^S 297
– – Fe(25 20 3)^S + R-Co_9 298
– – Fe(25 20 3)^S + S-Co_9 298
– – GaAs(1 1 1)   173
– – GaAs(−1 −1 −1)   173
– – Gra + (1 × 1)R3.5° − Gra   275
– – h-BN + (0.97 × 0.97)R3° − Gra   275
– – Ir(1 0 0) + (1 × 2) − O   243
– – Ir(1 1 1) − $\left(\sqrt{124/151} \times \sqrt{124/151}\right)$ R29.5° − Gra   270, 277
– – Ir(1 1 1) − (0.9 × 0.9)R29.5° − Gra   270
– – Ir(1 1 1) + (0.906 × 0.906) − Gra   283
– – Ir(1 1 1) + (0.906 × 0.906)R5° − Gra   283
– – Me + (γ × γ) − Gra   267
– – MgO(1 1 1) facets   227
– – MgO(15 11 9)   187
– – NaCl($h\,k\,l$)   170
– – Ni(1 0 0) + c(2 × 2) − CO   242, 250
– – Ni(1 1 0) + c(2 × 2) − S   247
– – Ni(1 1 0) + c(2 × 2) − CN   290
– – Ni(1 1 0) + p2mg(2 × 1) − 2CO   237, 247, 250, 253
– – Ni(1 1 1) − (1 × 1) − Gra   268
– – Ni(1 1 1) + (1.155 × 1.155)R2.2° − Ag   264
– – Ni(1 1 1) + (1.167 × 1.167)R2.4° − Ag   264
– – Ni(1 1 1) + (2 × 2) − O   243
– – Ni(1 1 1) + c(4 × 2) − 2NO   244, 249
– – Ni($h\,k\,l$)   170
– – Pb(1 1 1) + ($\sqrt{403}/7 \times \sqrt{403}/7$) R22.85° − C_{60} 260
– – Pd(1 1 1) + ($\sqrt{3} \times \sqrt{3}$)R30° − CO   243, 251
– – Pt(1 1 0) − (1 × 2) missing row   211, 251
– – Pt(1 1 0) + c(2 × 2) − Br   246
– – Pt(1 1 1) + (0.89 × 0.89) − Gra   268, 275
– – Pt(1 1 1) + c(4 × 2) − 2CO   249, 289
– – Rh(1 1 0) + (1 × 3) − H   290
– – Rh(1 1 0) + p2mg(2 × 1) − 2O   284
– – Rh(1 1 1) + ($\sqrt{3} \times \sqrt{3}$)R30° − CO   243

– – Rh(1 1 1) + (12/13 × 12/13) – B_3N_3 264
– – Ru(0 0 0 1) + (12/13 × 12/13) – Gra   239, 263
– – Ru(0 0 0 1) + (23/25 × 23/25) – Gra   263, 272
– – sc(0 0 1) – (1 × 1)R36.87°   219
– – Si(1 0 0) – (2 × 1) symmetric dimer   213
– – Si(1 0 0) – c(4 × 2) buckled dimer   213
– – Si(1 1 1) – (7 × 7) DAS model   214, 250
– – Si(1 1 1) + (1 × 1) – H   250
– – $SrTiO_3$(0 1 8)   187
– – V_2O_3(0 0 0 1)   171
– – V_2O_5(0 1 0), (0 0 1)   172
– – W(1 0 0) – c(2 × 2)   212
– – W(1 1 0) + (2 × 1) – O   248
– – W(2 1 1) microfaceted   231
– listing   401
– web sites   399
Surface Structure Database (SSD)   4, 416
surface symmetry
– allowed space groups   286
surface termination   415
symmetry group
– two-dim.   110, 139, 142
– three-dim.   49
symmetry operation (two-dim.)
– glide reflection   110, 131, 138
– identity   141
– inversion   110, 111, 114
– mirroring (reflection)   110, 119, 130, 140
– rotation   110, 114, 119, 140
symmetry operation (three-dim.)
– glide reflection   40
– inversion   39, 193
– mirroring (reflection)   39, 193
– rotation   39, 194
– rotoinversion   39
– rotoreflection   39
– rototranslation (screw operation)   40
symmorphic space group   409

t

tartaric acid (TA)   295
thin films   222
TiO_2 86
titanium dioxide   86
transformation matrix
– integer approximant   264
– supercell   25
translation
– group   143
– operation   39
– symmetry   39

u

unit cell   10
– two-dim.   415
– three-dim.   409
– origin   11
– primitive   10, 93

v

V_2O_3 171
V_2O_5 172
vanadium pentoxide   172
vanadium sesquioxide   171
vector calculus   391
– scalar product   391, 392
– vector product   392
– volume product   393, 392
vicinal partner   179, 183
Volmer-Weber   415
Voronoi cell   36
– two-dim.   415
– three-dim.   409
VW growth mode   415

w

web sites   399
Wigner-Seitz cell (WSC)
– two-dim.   147, 415
– three-dim.   36, 409
Wood notation   212, 222, 251, 415
– centered   363
– combined transformation   364
– definition   253
– examples   254, 357
– mathematics   355
– matrix transformation   254, 272
– primitive   361
– simplified   253
– symmetry information   253
– Wood-representable   361
Wulff
– construction   69, 71, 409
– polyhedron   69

y

$YBa_2Cu_3O_7$ 7, 12

z

zincblende   86